The Theory and Applications of Iteration Methods

I0036340

The Theory and Applications of Iteration Methods

Second Edition

Ioannis K. Argyros

CRC Press
Taylor & Francis Group
Boca Raton London New York

CRC Press is an imprint of the
Taylor & Francis Group, an **informa** business

Second edition published 2022
by CRC Press
6000 Broken Sound Parkway NW, Suite 300, Boca Raton, FL 33487-2742

and by CRC Press
2 Park Square, Milton Park, Abingdon, Oxon, OX14 4RN

© 2022 Ioannis K. Argyros

First edition published by CRC Press 1993

CRC Press is an imprint of Taylor & Francis Group, LLC

Reasonable efforts have been made to publish reliable data and information, but the author and publisher cannot assume responsibility for the validity of all materials or the consequences of their use. The authors and publishers have attempted to trace the copyright holders of all material reproduced in this publication and apologize to copyright holders if permission to publish in this form has not been obtained. If any copyright material has not been acknowledged please write and let us know so we may rectify in any future reprint.

Except as permitted under U.S. Copyright Law, no part of this book may be reprinted, reproduced, transmitted, or utilized in any form by any electronic, mechanical, or other means, now known or hereafter invented, including photocopying, microfilming, and recording, or in any information storage or retrieval system, without written permission from the publishers.

For permission to photocopy or use material electronically from this work, access www.copyright.com or contact the Copyright Clearance Center, Inc. (CCC), 222 Rosewood Drive, Danvers, MA 01923, 978-750-8400. For works that are not available on CCC please contact mpkbookspermissions@tandf.co.uk

Trademark notice: Product or corporate names may be trademarks or registered trademarks and are used only for identification and explanation without intent to infringe.

ISBN: 978-0-367-65101-5 (hbk)
ISBN: 978-0-367-65330-9 (pbk)
ISBN: 978-1-003-12891-5 (ebk)

DOI: 10.1201/9781003128915

Publisher's note: This book has been prepared from camera-ready copy provided by the authors.

Dedication

---

*The author dedicates this book to his beloved parents, **Anastasia** and **Konstantinos**, wife **Diana**, and children **Christopher, Gus, Michael**, and **Stacey***

Contents

Preface

This text book and its first edition are written for students in engineering, the physical sciences, mathematics, and economics at an upper division undergraduate or graduate level. Prerequisites for using the text are calculus, linear algebra, elements of functional analysis, and the fundamentals of differential equations. Student with some knowledge of the principles of numerical analysis and optimization will have an advantage, since the general schemes and concepts can be easily followed if particular methods, special cases, are already known. However, such knowledge is not essential in understanding the material of this book.

A plethora of problems in mathematics and also in engineering are solved by finding the solutions of certain equations. For example, dynamics systems are mathematical model by difference or differential equations, and their solutions usually represent the states of the systems. Assume for the sake of simplicity that a time-invariant system is driven by equation $\dot{x} = f(x)$, where x is the state. Then the equilibrium states are determined by solving the equation $f(x) = 0$. Similar equations are used in the case of discrete systems. The unknowns of engineering equations can be functions (difference, differential, and integral equations), vectors (systems of linear or non-linear algebraic equations), or real or complex numbers (single algebraic equations with single unknowns). Except in the special cases, the most commonly used solution methods are iterative – when starting from one or several initial approximations a sequence is constructed that converges to a solution of the equation. Iteration methods are also applied for solving optimization problems. In such cases, the iteration sequences converge to an optimal solution of the problem at hand. Since all of these methods have the same recursive structure, they can be introduced and discussed in a general framework.

In recent years, the study of general iteration schemes has included a substantial effort to identity properties of iteration schemes that will guarantee their convergence is some sense. A number of these results have used an abstract iteration scheme that consists of the recursive application of a point-to-set mapping. In this book, we are concerned with these types of results.

Each chapter contains several new theoretical results and important applications in engineering, in dynamic economic systems, in input-output systems, in the solution of non-linear and linear differential equations, and in optimization problems.

Chapter 1 gives an outline of general iteration schemes in which the convergence of such schemes is examined. We also show that our conditions are very general: most classical results can be obtained as special cases and, if the conditions are weakened slightly, then our results may not hold. In Chapter 2 the discrete time-scale Liapunov theory is extended to time dependent, higher order, non-linear difference equations. In addition, the speed of convergence is estimated in most cases. The monotone convergence of the solution is examined in Chapter 3. It is also shown that our results generalize well-known classical theorem such as contraction mapping principle, the lemma of Kantovich, the famous Gronwall lemma, and the well-known stability theorem of Uzawa. Application in: Neural networks, Reliability engineering and Economic models are contained in Chapter 4. The local as well as the semi-local convergence of specialized iterative methods is developed in the rest of the chapters. The rest of the chapters have not appeared in the first edition of the book.

Acknowledgement

The author would like to express his gratitude to Dr. Ference Szidarovsky for his contributions and guidance in the first four chapters and Researcher Samundra Regmi, Mr. Christopher I. Argyros, Mr. Gus I. Argyros, and Mr. Michael I. Argyros for their contributions and technical assistance.

Author

Ioannis K. Argyros is a professor of Mathematics in the Department of Mathematical Sciences, Cameron University, Lawton, Oklahoma, USA. He received his B.Sc. degree in 1979 from the University of Athens, Greece. In March 1982, he started his graduate studies and received his M.Sc. and Ph.D. in 1983 and 1984, respectively from the University of Georgia, Athens, GA, U.S.A.

Professor Argyros has published 31 books, over 1600 papers in Computational Sciences, and is an editor in a plethora of journals and an active reviewer of papers send by AMS and other peer reviewed journals. Meanwhile, his primary interests are in numerical functional analysis, numerical analysis, approximation theory, optimization theory, and applied mathematics.

1 The Convergence of Algorithmic Models

This chapter introduces the fundamentals of the theory of general iteration algorithms. The convergence of such algorithmic models is examined under very general conditions.

1.1 ALGORITHMIC MODELS

Let X denote an abstract set. The elements of X may be real or complex numbers, vectors, functions, etc. Introduce the following notation:

$$X^1 = X, X^2 = X \times X, \ldots, X^k = X^{k-1} \times X, \quad k \geq 2.$$

Assume that l is a positive integer, and for all $k \geq l-1$, the point-to-set mappings f_k are defined on X^{k+1}. Furthermore, for all $(x^{(1)}, \ldots, x^{(k+1)}) \in X^{k+1}$ and $x \in f_k(x^{(1)}, \ldots, x^{(k+1)}), x \in X$; that is, $f_k(x^{(1)}, \ldots, x^{(k+1)})$ is a subset of X. For the sake of brevity, we use the notation $f_k : X^{k+1} \to 2^X$, where 2^X denotes the set of all subsets of X.

Definition 1.1. Select $x_0, x, \ldots, x_{l-1} \in X$ arbitrarily, and construct the sequence

$$x_{k+1} \in f_k(x_0, x_1, \ldots, x_k) \quad (k \geq l-1), \tag{1.1}$$

where arbitrary point from the set $f_k(x_0, x_1, \ldots, x_k)$ can be accepted as the successor of x_k. Recursion (1.1) is called the general algorithmic model.

Remark 1.1. Because the domain of f_k is X^{k+1} and $f_k(x_0, x_1, \ldots, x_k) \subseteq X$, the recursion is well defined for all $k \geq l-1$. Points $x_0, \ldots, x_{l-1}$ are called as initial approximations, and the maps f_k are called as iteration mappings.

Definition 1.2. The algorithmic model (1.1) is called an l-step process if for all $k \geq l-1$, mapping f_k does not depend explicitly on $x_0, x_1, \ldots, x_{k-1}$, i.e., if algorithm (1.1) has the special form

$$x_{k+1} \in f_k(x_{k-l+1}, \ldots, x_{k-1}, x_k). \tag{1.2}$$

It is easy to show that any l-step process is equivalent to a certain single-step process defined on X^l. For $k \geq 0$, introduce vectors $z_k = (z_k^{(1)}, z_k^{(2)}, \ldots, z_k^{(l)})$. Starting from the initial approximation

$$z_0 = (x_0, x_1, \ldots, x_{l-1}),$$

DOI: 10.1201/9781003128915-1

consider the single-step algorithmic model:

$$z_{k+1}^{(1)} = z_k^{(2)}$$
$$z_{k+1}^{(2)} = z_k^{(3)}$$
$$\dots$$
$$z_{k+1}^{(l-1)} = z_k^{(l)}$$
$$z_{k+1}^{(l)} \in f_k(z_k^{(1)}, \dots, z_k^{(l)}). \qquad (1.3)$$

This iteration algorithm is a single-step process, and obviously it is equivalent to the algorithmic model (1.2), because for all $k \geq 0$,

$$z_k^{(1)} = x_k, \quad z_k^{(2)} = x_{k+1}, \dots, z_k^{(1)} = x_{k+l-1}$$

This equivalence is the main reason that only single-step iteration methods are discussed in most publications, because any result on single-step processes automatically applies to multistep processes.

Reduction (1.3) is illustrated in the following example.

Example 1.1. Consider the second-order difference equation

$$x_{k+1} = x_k + x_{k-1}, \quad x_0 = x_1 = 1.$$

The solution $x_k (k \geq 0)$ of this equation is called the kth Fibonacci number. In this case, equation (1.3) can be written as

$$z_{k+1}^{(1)} = z_k^{(2)}$$
$$z_{k+1}^{(2)} = z_k^{(1)} + z_k^{(2)}$$

with the initial condition $z_0^{(1)} = z_0^{(2)} = 1$.

Definition 1.3. An l-step process is called stationary if mappings f_k do not depend on k; otherwise the process is called non-stationary. (Note that the process of Example 1.1 is stationary.)

Iteration models in the most general form (1.1) are of great importance in certain optimization methods. For example, in using cutting plane algorithms very early cuts can still remain in the latter stages of the process by assuming that they are not dominated by later cuts (Higle and Sen [17]). Hence, the optimization problem of each step may depend on the solutions of very early steps. Multistep processes are also used in many other fields of applied mathematics. As an example, the secant method for solving non-linear equations (Szidarovszky and Yakowitz [27]; Argyros [1–11]) is a special two-step method. We briefly return to these applications at the end of this chapter. Non-stationary methods have great practical importance in analyzing the global asymptotic stability of dynamic economic systems, when the state transition relation is time dependent (Okuguchi and Szidarovszky [25]). In this chapter, the most general algorithmic model (1.1) is discussed first, and special cases are then derived from our general convergence theorem.

In order to establish any kind of convergence, X should have some topology. Assume now that X is a Hausdorff topological space that satisfies the first axiom of countability. (For definitions, cf, Tarayama [28]; Hall and Spencer [16]). Let $S \subseteq X$ be the set of desirable points, which are considered to be the solutions to the problem being solved by the algorithm. For example, in the case of an optimization problem, X can be selected as the feasible set and S as the set of the optimal solutions. If a linear or non-linear fixed point problem is solved, X is then the domain of the mapping and S is the set of all fixed points. In analyzing the global asymptotic stability of a discrete dynamic system, set X is the state space and S is the set of equilibrium points.

Definition 1.4. An algorithmic model is said to be convergent if the accumulation points of any iteration sequence $\{x_k\}$ constructed by the algorithm are in S.

Note that the convergence of an algorithm model does not imply that the iteration sequence is convergent. This more general convergence principle was introduced and investigated by many authors (Zangwill [30]; Polak [26]; Hogan [18]; Huard [19]; Klessig [21]). Tishyadhigama et al. presented a comprehensive summary of convergence criteria for algorithmic models. They compared the known convergence criteria and showed that they are all special cases of a very general concept which was first published in that article. The next section further generalizes and analyzes this result.

1.2 CONVERGENCE CRITERIA FOR ALGORITHMIC MODELS

Assume that for $k \geq 0$, functions $c_k : X \rightarrow R^l$ exist and have the following properties:

(A_1) For large k, functions $\{c_k\}$ are uniformly locally bounded below on $X \backslash S$. That is, there is a non-negative integer N_1 such that for all $x \in X \backslash S$ there is a neighborhood U of x and a real $b \in R^l$ (which may depend on x) such that for all $k \geq N_1$ and $x' \in U$,

$$c_k(x') \geq b. \tag{1.4}$$

(A_2) If $k \geq N_1, x' \in f_k(z_1, \ldots, z_k, x)$ $(z, z_i \in X, i = 1, \ldots, k)$, then

$$c_{k+1}(x') \leq c_k(x). \tag{1.5}$$

(A_3) For each $z \in X \backslash S$ if $\{z_i\} \subseteq X$ is any sequence such that $z_i \rightarrow z$ and $\{k_i\}$ is any strictly increasing sequence of non-negative integers such that

$c_{k_i}(z_i) \to c^*$, then for all iteration sequences $\{x_i\}$ such that $x_{k_i} = z_i (i \geq 0)$ an integer N_2 exists such that $k_{N_2} \geq N_1 - 1$ and

$$c_{k_{N_2}+1}(y) < c^* \text{ for all } y \in f_{k_{N_2}}(x_0, x_1, \ldots, x_{K_{N_2}}). \tag{1.6}$$

Theorem 1.1. If conditions $(A_1), (A_2)$, and (A_3) hold, then the algorithmic model (1.1) is convergent.

Proof. Let x^* be an accumulation point of the iteration sequence $\{x_k\}$ constructed by the algorithmic model (1.1), and assume that $x^* \in X \backslash S$. Let $\{k_i\}$ denote the index set such that $\{x_{k_i}\}$ is a subsequence of $\{x_k\}$ converging to x^*. Assumption (A_2) implies that for large k, sequence $\{c_k(x_k)\}$ is decreasing, and from assumption (A_1) we conclude that $\{c_{k_i}(x_{k_i})\}$ is convergent. Therefore, the entire sequence $\{c_k(x_k)\}$ converges to a $c^* \in R^l$. From (1.5) we know that for $k \geq N_1$,

$$c_k(x_k) \geq c^*. \tag{1.7}$$

Use subsequence $\{x_{k_i}\}$ as sequence $\{z_i\}$ in condition (A_3) to see that an N_2 exists such that $k_{N_2} \geq N_1 - 1$ and with the notation $M = k_{N_2} + 1$,

$$c_M(x_M) < c^*,$$

which contradicts relation (1.7) and completes the proof. □

Remark 1.2. Note first that in the special case when (1.1) is a single-step nonstationary process and c_k does not depend on k, this theorem generalizes Theorem 4.3 of Tishyadhigama et al. (1979). If the process is stationary, then this result further specializes to Theorem 3.5 of the same paper.

Remark 1.3. The conditions of the theorem do not imply that sequence $\{x_k\}$ has an accumulation point, as the next example shows.

Example 1.2. Select $X = R^l, S = \{0\}$, and consider the single-step process with $f_k(x) = f(x) = x - 1$, and choose $c_k(x) = x$ for all $x \in X$.

Because functions c_k are continuous and $f_k(x) < x$ for all x, condition (A_1) obviously holds, assumptions (A_2) and (A_3) also hold since for arbitrary $x_0 \in X$, the iteration sequence is strictly decreasing and divergent. (Infinite limit is not considered here as limit point from X.)

Remark 1.4. Even in cases when the iteration sequence has an accumulation point, the sequence does not need to converge, as the following example shows.

Example 1.3. Select $X = R^l, S = \{0,1\}$, and consider the single-step iteration algorithm with function

$$f_k(x) = f(x) = \begin{cases} 1 & \text{if } x = 0 \\ 0 & \text{if } x = 1 \\ x - 1 & \text{if } x \neq S. \end{cases}$$

Choose

$$c_k(x) = c(x) = \begin{cases} 0 & \text{if } x \in S \\ x & \text{otherwise.} \end{cases}$$

On X, function c is continuous, thus assumption (A_1) is satisfied. If $x \neq S$, then $f(x) < x$, which implies that $c(f(x)) < c(x)$. If $x \in S$, then $f(x) \in S$. Therefore, in this case $c(f(x)) = c(x)$. Hence, condition (A_2) also holds. Assumption (A_3) follows from the definition of functions c_k and from the fact that $f(x) < x$ on $X \backslash S$. If x_0 is selected as a non-negative integer, then the iteration sequence has two accumulation points: 0 and 1. If x_0 is selected otherwise, then no accumulation point exists.

Note that Definition 1.4 is considered the definition of global convergence on X, because the initial approximations $x_0, x_1, \ldots, x_{l-1}$ are arbitrary elements of X. Local convergence of algorithmic models can be defined as follows.

Definition 1.5. An algorithmic model is said to be locally convergent if there is a subset X_1 of X such that the accumulation points of any iteration sequence $\{x_k\}$ constructed by the algorithm starting with initial approximations $x_0, x_1, \ldots, x_{l-1}$ from X_1 are in S.

Theorem 1.1 can be modified as a local convergence theorem by substituting X and S by X_1 and $X_1 \cap S$, respectively. The details can be worked out as an easy exercise, so they are omitted.

Notice that the range of function c_k may be a subset of a partially ordered space. The above results can be easily extended to this more general case, and they can be applied in establishing the convergence of algorithms for finding optimal solutions for multi-objective programming problems.

In the case of most applications the iteration process (1.1) is a stationary single-step method. In the following part of this section we discuss how the general conditions of Theorem 1.1 may be simplified for stationary single-step methods. First we reformulate Theorem 1.1 with a stationary function c.

The algorithmic model has now the form

$$x_{k+1} \in f(x_k),$$

where $f : X \rightarrow 2^X$ is the iteration mapping. Assume that a function $c : X \rightarrow R^l$ exists and have the following properties:

$(A_1)^1$ Function c is locally bounded below on $X \backslash S$; that is, for all $x \in X \backslash S$ there is a neighborhood U of x and a real $b \in R^l$ such that for all $x' \in U, c(x') \geq b$.

$(A_2)^1$ $c(x') \leq c(x)$ for all $x \in S$ and $x' \in f(x)$.

$(A_3)^1$ For each $z \in X \backslash S$, if $\{z_i\} \subseteq X$ is any sequence such that $z_i \to z$ and $c(z_i) \to$
       c^*, then an integer N exists such that for all $y \in f(z_N), c(y) < c^*$.

Theorem 1.2. If conditions $(A_1)^1, (A_2)^1$, and $(A_3)^1$ hold, then the above algorithmic model is convergent. This result is the simple restatement of Theorem 1.1, and therefore no proof is needed.

Replace condition $(A_3)^1$ with

$(A_3)^2$ For each $z \in X \backslash S$, if $\{z_i\}, \{y_i\} \subset X$ are such that $z_i \to z, y_i inf(z_i), c(z_i) \to$
       c^*, and $c(y_i) \to c^{**}$, then $c^{**} < c^*$.

Theorem 1.3. If conditions $(A_1)^1, (A_2)^1$, and $(A_3)^2$ hold, then the above algorithmic model is convergent.

Proof. It is sufficient to show that assumption $(A_3)^1$ is implied by the conditions of the theorem. Let $z \in X \backslash S$ and let $\{z_i]\} \subseteq X$ such that $z_i \to z$ and $c(z_i) \to c^*$. Assume that $(A_3)^1$ does not hold. Then there exists $y_i \in f(z_i), (i = 0, 1, \dots)$ such that $c(y_i) \geq c^*$ for all i. Therefore:

$$c(z_i) \geq c(y_i) \geq c^*,$$

which implies that $c(y_i) \to c^*$. This contradicts $(A_3)^2$, because we concluded that $c^{**} = c^*$.   □

A different modification of assumptions $(A_3)^1$ and $(A_3)^2$ can be given as follows:

$(A_3)^3$ For all $x \in X \backslash S$ there exists a $\delta(x) > 0$ (possible depending on x) and a
       neighborhood U of x such that for all $z \in U$ and $z' \in f(z)$,

$$c(z') - c(z) \leq -\delta(x).$$

This condition is sometimes called the locally uniform monotonicity of the pair (c, f).

Theorem 1.4. If conditions $(A_1)^1, (A_2)^1$, and $(A_3)^3$ hold, then the above algorithmic model is convergent.

Proof. We will show that condition $(A_3)^3$ implies $(A_3)^2$. Let $z, \{z_i\}, \{y_i\}, c^*$, and c^{**} be as in condition $(A_3)^2$. Then, $(A_3)^3$ implies that an integer N exists such that $c(y_i) \leq c(z_i) - \delta$ for all $i \geq N$. Therefore, $c^{**} = \lim c(y_i) \leq \lim(c(z_i) - \delta) < c^*$. Hence, condition $(A_3)^2$ holds.   □

Consider the new modifications:

$(A_2)^2$ For all $x \in X$ and $y \in f(x), c(y) - c(x) \geq -\delta(x) \geq 0$ where $\delta : X \to R_+^1$ is
       a given function.
$(A_3)^4$ For each $z \in X \backslash S$, if $\{z_i\} \subseteq X$ is such that $z_i \to z$, then $\sum_{i=0}^{\infty} \delta(z_i) = \infty$.

Theorem 1.5. If conditions $(A_1)^1, (A_2)^2$, and $(A_3)^4$ hold, the above algorithmic model is convergent.

Before proving this theorem, a simple lemma will be formulated and proven.

Lemma 1.1. Assume that $z \in X$ and $\delta : X \to R_+^1$ are such that for all sequences $\{z_i\}$ which converge to z, $\sum_{i=0}^{\infty} \delta(z_i) = \infty$. Then there exists a neighborhood $U(z)$ of z and a $\delta^* > 0$ such that $\delta(z') \geq \delta^*$ for all $z' \in U(z)$.

Proof. Assume that the lemma is false. Then there is a sequence $\{z_i\}$ such that $z_i \to z$ and $\delta(z_i) \to 0$. Define

$$k(i) = \min \left\{ j \mid j \geq i+1, \delta(z_j) \leq \left(\frac{1}{2}\right)^i \right\}.$$

Notice that $k(i)$ is defined for all i, since $\delta(z_i) \to 0$. Define sequence x_i as

$$x_0 = z_0, \ x_1 = z_{k(0)}, \ x_2 = z_{k(k(0))}, \ x_3 = z_{k(k(k(0)))}$$

and so on. Then, by the hypothesis of the lemma, $\sum_{i=0}^{\infty} \delta(x_i) = \infty$; however,

$$\sum_{i=0}^{\infty} \delta(x_i) \leq \sum_{i=0}^{\infty} \left(\frac{1}{2}\right)^i = 2$$

is an obvious contradiction.

Proof of Theorem: Obviously $(A_2)^2$ implies $(A_2)^1$. Therefore, it is sufficient to show that $(A_3)^4$ implies $(A_3)^3$. Let $z \in X \backslash S$, then $(A_3)^4$, and the lemma imply that for all $z' \in U(z)$ and $z'' \in f(z')$,

$$c(z'') - c(z') \leq -\delta(z') \leq -\delta^*.$$

Consequently, condition $(A_3)^3$ holds. $\qquad\qquad\qquad\qquad\qquad\qquad\qquad\qquad\square$

To conclude this section, Theorem 1.1 is reexamined, and we further generalize it in a more general framework, which includes not only the convergence of algorithmic models but also ω-limit points of differential equations as special cases.

Assume again that X is a Hausdorff topological space that satisfies the first axiom of countability. Assume further that I is an unbounded subset of $[0, \infty]$.

Definition 1.6. Any set F of functions f, such that $D(f) = I$ and $R(f) \subseteq X$, is called a dynamic process, and a particular element $f \in F$ is called a trajectory.

Let $S \subset X$ be a given set, which will be called the set of desirable points.

Definition 1.7. The dynamic process F is called convergent, if $f \in F, t_1 < t_2 < \dots (t_i \in I, i = 1, 2, \dots)$, $t_i \to \infty$, and $f(t_i) \to f^*$, then $f^* \in S$.

Assume next that Y is also a Hausdorff topological space that satisfies the first axiom of countability, and $\geq$ is a partial order on Y. It is also assumed that the topology and partial order are interconnected by the following property: (P) If $y_1 \geq y_2 \geq \ldots (y_i \in Y, i = 1, 2, \ldots)$ and $y_i \geq y(i = 1, 2, \ldots)$ with some $y \in Y$, then sequence $\{y_i\}$ converges to an element $y^* \in Y$ such that for all i, $y_i \geq y^*$.

Definition 1.8. A function $c : I \times X \to Y$ is called a generalized Liapunov function if the following conditions are satisfied:

$(A_1)^*$ For large t, function c is uniformly locally bounded below on $X \backslash S$; i.e., there is a non-negative number Q_1 such that for all $z \in X \backslash S$ there is a neighborhood U of z and a real $y \in Y$ (which may depend on z) such that for all $t \geq Q_1$ $(t \in I)$ and $z' \in U, c(t, z') \geq y$.

$(A_2)^*$ If $f \in F$ and $Q_1 \leq t \leq t'(t, t' \in I)$, then $c(t', f(t')) \leq c(t, f(t))$.

$(A_3)^*$ For each $z^* \in X \backslash S$, if $\{z_i\} \subset X$ is such that $z_i \to z^*$ and $\{t_i\} \subset I$ is a strictly increasing sequence such that $t_i \to \infty$ and furthermore $c(t_i, z_i) \to y^*$, then for all trajectories $f \in F$ with the property $f(t_i) = z_i(i = 1, 2, \ldots)$ there exists a $t \in I$ such that $t \geq Q_1$ and $c(t, f(t)) < y^*$.

Theorem 1.6. If the dynamic process F has a generalized Liapunov function, then it is convergent.

Proof. Assume that an $f \in F$ exists such that $f(t_i) \to f^* \in X \backslash S$ with some strictly increasing sequence $\{t_i\} \subset I, t_i \to \infty$. From condition $(A_2)^*$ we know that sequence $\{c(t_i, f(t_i))\}$ is decreasing for large values of i, and assumption $(A_1)^*$ implies that $c(t_i, f(t_i))$ is bounded from below if i is sufficiently large. Therefore, from condition (P) we conclude that sequence $\{c(t_i, f(t_i))\}$ converges to a limit $y^* \in Y$. Furthermore, by again using assumption $(A_2)^*$ we see that

$$c(t, f(t)) \geq y^*.$$

for all $t \geq Q_1$ $(t \in I)$. Use sequence $\{f(t_i)\}$ as $\{z_i\}$ in assumption $(A_3)^*$ to conclude that

$$c(t, f(t)) < y^*.$$

with some $t \geq Q_1$, which contradicts the previous inequality. Thus, the proof is complete. □

Corollary 1.1. Assume that all trajectories are contained in a compact set X. Furthermore, S consists of only one point x^*. Then, under the assumptions of the theorem, for all trajectories $f \in F, f(t_i) \to x^*$ as $t_i \in I(i = 1, 2, \ldots)$ and $t_i \to \infty$; i.e., the dynamic process F is globally asymptotically stable.

Notice first that the special case of $I = \{1, 2, \ldots\}$ corresponds to algorithmic models, which were discussed earlier.

Consider next the case when $I = [0, \infty], X \subset R^n$ and the trajectories are generated by the differential equation

$$\frac{d}{dt} f(t) = A(t, f(t)),$$

where $A : I \times X \to X$ is a given function. Assume that for all $x_0 \in X$ and $t_0 \in I$, the equation has at least one solution such that $f(t_0) = x_0$, and these solutions are defined for all $t \in I$ and remain in X. Let $F(t_0, x_0)$ denote the set of these solutions. Denote

$$F = \bigcup_{t_0 \in I, x_0 \in X} F(t_0, x_0).$$

It is easy to see that condition $(A_3)^*$ can now be reformulated as

$(A_3)^{**}$ For each $z^* \in X \backslash S$, if $\{z_i\} \subset X$ is such that $z_i \to z^*$ and $\{t_i\} \subset I$ is a strictly increasing sequence such that $t_i \to \infty$ further more $c(t_i, z_i) \to y^*$, then for all trajectories $f \in \bigcap_i F(t_i, z_i)$ there is a $t \geq Q_1$ such that $c(t, f(t)) < y^*$.

Assume next that functions A and c do not depend explicitly on t, and for all $x_0 \in X, F(0, x_0)$ has only one trajectory: $f(t, x_0)$. In this special case assumption $(A_3)^*$ is implied by the following condition:

$(A_3)^{***}$ For each $z^* \in X \backslash S$, if $\{z_i\} \subset X$ is such that $z_i \to z^*$ and $c(z_i) \to y^*$, then there exists an integer $j > 0$ such that with some $t > 0, c(f(t, z_j)) < y^*$.

The relation of the above condition to the famous stability result of Uzawa (1961) is next analyzed. Assume again that A and c do not depend explicitly on t; furthermore, for all $x_0 \in X, F(0, x_0)$ has exactly one trajectory, $f(t, x_0)$. Assume also that

1. c is decreasing on S and strictly decreasing on $X \backslash S$.
2. c is continuous on X.
3. $f(t, x_0)$ continuously depends on x_0 for all $t \geq 0$.

We now verify that conditions 1 through 3 imply $(A_1)^*, (A_2)^*$, and $(A_3)^{***}$, i.e., the above results can be considered as the straightforward extensions of the classical theorem of Uzawa (1961). The continuity of c implies $(A_1)^*$, and $(A_2)^*$ is assumed by condition 1 above. Assume next that $y^* \in X \backslash S$, then from condition 1 we conclude that for all $t \geq 0, c(f(t, y^*)) < c(y^*)$. Then the continuity of c and f implies that $c(f(t, z_j)) < c(y^*)$, if z_j is sufficiently close to y^*. Hence, $(A_3)^{***}$ holds.

1.3   APPLICATIONS

1.3.1   FIRST APPLICATION

Our first example is the well-known simplex method. Assume that a linear function $g(x)$ is maximized on a polyhedron $P \subseteq R^n$. The initial solution approximation x_0 is selected as any vertex of P. The process is a single-step algorithmic model, where

$f_k(x_k)$ denotes the set of all vertices that are neighbors of x_k and have higher objective function values than x_k. It is easy to see that if no degeneration is present, the process satisfies the conditions of Theorem 1.1 with $c_k = g$ and S being the set of optimal vertices. Because P has only finite number of vertices, the process terminates in finite number of steps at an optimal vertex.

The algorithm is illustrated by Example 1.4.

Example 1.4.

$$\text{maximize } x_1 + x_2$$
$$\text{subset to } x_1, x_2 \geq 0$$
$$2x_1 + x_2 \leq 4$$
$$x_1 + 2x_2 \leq 4.$$

The vertices of P are (0,0), (2,0), (4/3,4/3), and (0,2). If one starts from the initial vertex $x_0 = (0,0)$, then both neighbor vertices (2,0) and (0,2) give higher objective function values. Therefore,

$$f_0(x_0) = \{(2,0),(0,2)\}.$$

By selecting $x_1 = (2,0)$, the new vertex also has two neighbors: (0,0) and (4/3,4/3). However, only the vertex (4/3,4/3) has higher objective function value. Therefore,

$$f_1(x_1) = \{(4/3,4/3)\}.$$

and we select $x_2 = (4/3,4/3)$. This new vertex also has two neighbors: (2,0) and (0,2), however, both have lower objective function values. Thus, $f_2(x_2)$ is empty, and the process terminates. Then the solution is (4/3,4/3) with objective function value 8/3.

For solving non-linear optimization problems iteration procedures are the most utilized algorithms. A survey of particular methods can be found, for example, in Zangwill (1969) or in Polak (1971).

1.3.2   SECOND APPLICATION

In our second application, let $f : R \rightarrow R$ be a real variable, real valued function, and assume that the fixed point x^* of function f, i.e., the solution of equation $x = f(x)$, has to be determined. An obvious iteration scheme is given as

$$x_{k+1} = f(x_k), \tag{1.8}$$

where x_0 is selected in the neighborhood of the solution. This process is a stationary single-step process.

In the more general case, when $f : R^n \rightarrow R^n$, the above procedure generalizes to

$$x_{k+1} = f(x_k),$$

where the initial approximation x_0 is selected again in the neighborhood of the solution. Algorithm (1.8) is illustrated by Example 1.5.

Example 1.5. We solve the non-linear equation

$$x = \frac{1}{2} \sin x + 1.$$

In this special case, scheme (1.8) has the form

$$x_{k+1} = \frac{1}{2} \sin x_k + 1.$$

Select the initial approximation $x_0 = 0$, then

$$x_1 = \frac{1}{2} \sin 0 + 1 = 1,$$

$$x_2 = \frac{1}{2} \sin 1 + 1 \approx 1.4207,$$

$$x_3 = \frac{1}{2} \sin 1.4207 + 1 \approx 1.4944,$$

$$x_4 = \frac{1}{2} \sin 1.4944 + 1 \approx 1.4985,$$

$$x_5 = \frac{1}{2} \sin 1.4985 + 1 \approx 1.4987,$$

$$x_6 = \frac{1}{2} \sin 1.4987 + 1 \approx 1.4987.$$

Because the solution does not change up to the fifth significant digit, we accept the root approximation $x^* \approx 1.4987$.

The generalized algorithm is now illustrated.

Example 1.6. Consider the system of equations

$$x^{(1)} = \frac{1}{10} \sin(x^{(1)} + x^{(2)}) + 1$$

$$x^{(2)} = \frac{1}{10} \cos(x^{(1)} + x^{(2)}).$$

In this special case

$$x = \begin{pmatrix} x^{(1)} \\ x^{(2)} \end{pmatrix}, f(x) = \begin{pmatrix} \frac{1}{10} \sin(x^{(1)} + x^{(2)}) + 1 \\ \frac{1}{10} \cos(x^{(1)} + x^{(2)}) \end{pmatrix}.$$

and hence, the iteration procedure is as follows:

$$x_{k+1}^{(1)} = \frac{1}{10}\sin(x_k^{(1)} + x_k^{(2)}) + 1$$

$$x_{k+1}^{(2)} = \frac{1}{10}\cos(x_k^{(1)} + x_k^{(2)}).$$

Select the initial approximations $x_0^{(1)} = x_0^{(2)} = 0$, then

$$x_1^{(1)} = 1 \text{ and } x_1^{(2)} = \frac{1}{10}.$$

Simple calculation shows that

$$x_2^{(1)} = \frac{1}{10}\sin(1.1) + 1 \approx 1.08912$$

$$x_2^{(2)} = \frac{1}{10}\cos(1.1) \approx 0.04536,$$

$$x_3^{(1)} = \frac{1}{10}\sin(1.13448) + 1 \approx 1.09063$$

$$x_3^{(2)} = \frac{1}{10}\cos(1.13448) \approx 0.04226,$$

$$x_4^{(1)} = \frac{1}{10}\sin(1.13289) + 1 \approx 1.09056$$

$$x_4^{(2)} = \frac{1}{10}\cos(1.13289) \approx 0.04240,$$

$$x_5^{(1)} = \frac{1}{10}\sin(1.13296) + 1 \approx 1.09057$$

$$x_5^{(2)} = \frac{1}{10}\cos(1.13296) \approx 0.04240.$$

It is easy to see that $x_6^{(1)}$ and $x_6^{(2)}$ are the same as $x_5^{(1)}$ and $x_5^{(2)}$ to the accuracy shown. Therefore, we accept the approximating solutions

$$x^{(1)*} \approx 1.09057 \text{ and } x^{(2)*} \approx 0.04240.$$

In many applications, we can speed up the convergence of the previous method by a simple observation. In calculating the "newest" value $x_{k+2}^{(2)}$ the "old" value $x_k^{(1)}$ of $x^{(1)}$ is still used, however, the "new" value $x_{k+1}^{(1)}$ is already computed by the first equation. Therefore, it seems advantageous that in the second equation $x_{k+1}^{(1)}$ is used instead of $x_k^{(1)}$. This modified process is known as Seidel iteration and in our case has the form

$$x_{k+1}^{(1)} = \frac{1}{10}\sin(x_k^{(1)} + x_k^{(2)}) + 1$$

$$x_{k+1}^{(2)} = \frac{1}{10}\cos(x_k^{(1)} + x_k^{(2)}).$$

Select again the initial approximation $x_0^{(1)} = x_0^{(2)} = 0$, then

$$x_1^{(1)} = \frac{1}{10}\sin(0) + 1 \approx 1$$

$$x_1^{(2)} = \frac{1}{10}\cos(1) \approx 0.05403,$$

$$x_2^{(1)} = \frac{1}{10}\sin(1.05403) + 1 \approx 1.08694$$

$$x_2^{(2)} = \frac{1}{10}\cos(1.14907) \approx 0.04167,$$

$$x_3^{(1)} = \frac{1}{10}\sin(1.12861) + 1 \approx 1.09038$$

$$x_3^{(2)} = \frac{1}{10}\cos(1.13205) \approx 0.04248,$$

$$x_4^{(1)} = \frac{1}{10}\sin(1.13286) + 1 \approx 1.09056$$

$$x_4^{(2)} = \frac{1}{10}\cos(1.13304) \approx 0.04239.$$

It is easy to see that $x_5^{(1)}$ and $x_5^{(2)}$ are the same as in the case of the original scheme. We find no significant improvement by using the Seidel iteration method in this case; we do, however, in many applications.

Notice that a simple observation allows us to rewrite the original problem in the form of a single equation. Introduce the new variable $z = x^{(1)} + x^{(2)}$, and add the two equations:

$$z = \frac{1}{10}(\sin z + \cos z) + 1.$$

The iteration scheme (1.8) for this problem has the form

$$z_{k+1} = \frac{1}{10}(\sin z_k + \cos z_k) + 1$$

Select the initial approximation $z_0 = 0$, then

$$z_1 = \frac{1}{10}(\sin 0 + \cos 0) + 1 = 1.1$$

Similarly,

$$z_2 = \frac{1}{10}(\sin 1.1 + \cos 1.1) + 1 \approx 1.13448,$$

$$z_3 = \frac{1}{10}(\sin 1.13448 + \cos 1.13448) + 1 \approx 1.13289$$

$$z_4 = \frac{1}{10}(\sin 1.13289 + \cos 1.13289) + 1 \approx 1.13296,$$

$$z_5 = \frac{1}{10}(\sin 1.13296 + \cos 1.13296) + 1 = 1.13296,$$

i.e., $z^* \approx 1.13296$, and therefore,

$$x^{(1)*} = \frac{1}{10}\sin(z^*) + 1 \approx 1.09057$$

and

$$x^{(2)*} = \frac{1}{10}\cos(z^*) \approx 0.04240.$$

1.3.3   THIRD APPLICATION

For our third example, assume that a root x^* of a real variable, real valued function g must be determined. One of the most commonly used procedures is the secant method, which is defined as follows. Let x_0 and x_1 be two initial approximations of the root. Approximate function g by the linear segment connecting the points $(x_0, g(x_0)), (x_1, g(x_1))$, and let x_2 be the root of this linear segment. The two-point formula implies that the equation of the linear segment is

$$y - g(x_1) = \frac{g(x_1) - g(x_0)}{x_1 - x_0}(x - x_1).$$

The root of this segment can be obtained by substituting $y = 0$ and solving for x. Therefore,

$$x_2 = x_1 - \frac{g(x_1)(x_1 - x_0)}{g(x_1) - g(x_0)},$$

Continuing the procedure in the same manner we have the general step:

$$x_{k+1} = x_k - \frac{g(x_k)(x_k - x_{k-1})}{g(x_k) - g(x_{k-1})}. \tag{1.9}$$

Notice that this algorithm is a stationary two-step process with iteration function

$$f_k(x_{k-1}, x_k) = x_k - \frac{g(x_k)(x_k - x_{k-1})}{g(x_k) - g(x_{k-1})}. \tag{1.10}$$

which is a point-to-point mapping in this case.

This method can be generalized on n-dimensional vector spaces as follows. Let $g : D \rightarrow R^n$, where $D \subseteq R^n$ is a convex set. Let $x_0, x_1, \ldots, x_n$ be initial approximations of a root of equation $g(x) = 0$. Let $y = Ax + b$ be the linear function passing through the points $(x_1, g(x_l)), l = 0, 1, \ldots, n$. Then, x_{n+1}, can be obtained as the root of this linear function:

$$x_{n+1} = -A^{-1}b.$$

Here, the linear function can be determined as follows. The conditions

$$Ax_l + b = g(x_l), \quad l = 0, 1, \ldots, \tag{1.11}$$

imply that

$$A(x_l - x_0) = g(x_l) - g(x_0), \quad l = 1, 2, \ldots, n. \tag{1.12}$$

Introduces matrices

$$X = (x_1 - x_0, x_2 - x_0, \ldots, x_n - x_0)$$

and

$$G = (g((x_1) - g(x_0)), g(x_2) - g(x_0), \ldots, g(x_n) - g(x_0)),$$

that is, the ith columns of X and G are x_i–x_0 and $g(x_i) - g(x_0)$, respectively. Then equalities (1.12) can be summarized as $AX = G$; i.e., $A = GX^{-1}$. Equation (1.11) with $l = n$ implies that $b = g(x_n) - Ax_n$. Hence,

$$x_{n+1} = x_n - (x_1 - x_0, \ldots, x_n - x_0)(g(x_1) - g(x_0, \ldots, g(x_n) - g(x_0)))^{-1} g(x_n),$$

and in general,

$$x_{K+1} = x_K - (x_{k+1-n} - x_{k-n}, \ldots, x_k - x_{k-n})(g(x_{k+1-n}) - g(x_{k-n}, \ldots, g(x_k) - g(x_{k-n})))^{-1} g(x_k). \tag{1.13}$$

Notice that this is a stationary $(n+1)$-step process, which specializes to the special method (1.9) in the case of $n = 1$. Process (1.9) is next illustrated.

Example 1.7. Consider again the equation of the previous example, which can be rewritten as

$$\frac{1}{2} \sin x + 1 - x = 0,$$

so, we may select

$$g(x) = \frac{1}{2} \sin x + 1 - x.$$

It is easy to see that

$$g(1) \approx 0.4207 > 0$$

and

$$g(2) \approx 0.5454 < 0.$$

therefore, a root is found in the interval $(1, 2)$. By selecting the initial approximations $x_0 = 1$ and $x_1 = 2$, the iteration scheme (1.9) gives the approximations

$$x_2 = x_1 - \frac{g(x_1)(x_1 - x_0)}{g(x_1) - g(x_0)} \approx 2 - \frac{-0.5454(2-1)}{-0.5454 - 0.4307} \approx 1.4355,$$

$$x_3 = x_2 - \frac{g(x_2)(x_2 - x_1)}{g(x_2) - g(x_1)} \approx 1.4355 - \frac{0.05993(1.4355 - 2)}{0.05993 + 0.5454} \approx 1.4914,$$

$$x_4 = x_3 - \frac{g(x_3)(x_3 - x_2)}{g(x_3) - g(x_2)} \approx 1.4914 - \frac{0.007025(1.4914 - 1.4355)}{0.007025 - 0.05993} \approx 1.4988,$$

$$x_5 = x_4 - \frac{g(x_4)(x_4 - x_3)}{g(x_4) - g(x_3)} \approx 1.4988 - \frac{-0.00009531(1.4988 - 1.4914)}{-0.00009531 - 0.007025} \approx 1.4987.$$

Easy calculation shows that up to five decimal places $x_6 = x_5$. Therefore, we accept the root approximation 1.4987.

1.3.4   FOURTH APPLICATION

Assume again, in this fourth example, that a scalar equation $g(x) = 0$ has to be solved, where we assume that g is differentiable in the neighborhood of the root. Approximate function g by its linear Taylor's polynomial

$$p(x) = g(x_0) + g'(x_0)(x - x_0)$$

around an initial approximation x_0, which is the tangent line of the curve $y = g(x)$ at the point $x = x_0$. Define x_1 as the root of this linear function:

$$x_1 = x_0 - \frac{g(x_0)}{g'(x_0)}.$$

Continuing the process in the same manner we have the well-known Newton method:

$$x_{k+1} = x_k - \frac{g(x_k)}{g'(x_k)} \quad (k \geq 0) \tag{1.14}$$

which is a stationary single-step process with iteration function

$$f_k(x_k) = x_k - \frac{g(x_k)}{g'(x_k)}.$$

In the more general case, when $g : D \to R^n$ with D being a convex set in R^n, the method has the form

$$x_{k+1} = x_k - J(x_k)^{-1} g(x_k). \tag{1.15}$$

where J is the Jacobian matrix of function g; i.e., the (i, j) element of J is the partial derivative $\partial g_1(x)/\partial x^{(j)}$. Here, $x^{(j)}$ denotes the jth component of x.

In practical cases equation (1.15) must be written as

$$J(X_k) \cdot \Delta_k = g(x_k),$$

and then x_{k+1}, is obtained by the equation

$$x_{k+1} = x_k - \Delta_k.$$

The advantage of this equation is the fact that the solution of linear equations requires less operations than the inversion of matrices.

Example 1.8. We solve again the equation

$$\frac{1}{2}\sin x + 1 - x = 0,$$

which was the subject of our previous examples. In this case the iteration scheme (1.14) is used. Select the initial approximation $x_0 = 2$. Then

$$x_1 = x_0 - \frac{g(x_0)}{g'(x_0)} \approx 2 - \frac{-0.5454}{-1.2081} \approx 1.5485,$$

$$x_2 = x_1 - \frac{g(x_1)}{g'(x_1)} \approx 1.5485 - \frac{-0.04862}{-0.9889} \approx 1.4993,$$

$$x_3 = x_2 - \frac{g(x_2)}{g'(x_2)} \approx 1.4993 - \frac{-0.0005774}{-0.9643} \approx 1.4987,$$

and it is easy to see that up to five significant decimal figures $x_4 = x_3$. Hence, the solution is accepted to be approximated by 1.4987.

1.3.5 FIFTH APPLICATION

In our fifth example, we mention first that in applying the Newton method at each step we must compute the function value $g(x_k)$ and also the derivative value $g'(x_k)$. We can reduce the number of function calls by 50% by applying the following modification of the Newton method:

$$x_{k+1} = x_k - \frac{g(x_k)}{g'(x_0)} \tag{1.16}$$

where at each step we divide by the derivative at the initial approximation; i.e., the denominator does not have to be updated at each step. If g' is smooth in the neighborhood of the solution, then method (1.16) is a good approximation of process (1.14).

Example 1.9. The modified Newton method is applied in the case of the previous example. Select again $x_0 = 2$, then

$$g'(x_0) = \frac{1}{2} \cos 2 - 1 \approx -1.2081.$$

Because

$$g(x_0) = \frac{1}{2} \sin 2 + 1 - 2 \approx -0.5454,$$

$$x_1 = x_0 - \frac{g(x_0)}{g'(x_0)} \approx 2 - \frac{-0.5454}{1.2081} \approx 1.5485$$

as before. The procedure then continues as follows:

$$x_2 = x_1 - \frac{g(x_1)}{g'(x_0)} \approx 1.5485 - \frac{-0.04862}{-1.2081} \approx 1.5083,$$

$$x_3 = x_2 - \frac{g(x_2)}{g'(x_0)} \approx 1.5083 - \frac{-0.009276}{-1.2081} \approx 1.5006,$$

$$x_4 = x_3 - \frac{g(x_3)}{g'(x_0)} \approx 1.5006 - \frac{-0.001831}{-1.2081} \approx 1.4991,$$

$$x_5 = x_4 - \frac{g(x_4)}{g'(x_0)} \approx 1.4991 - \frac{-0.0003845}{-1.2081} \approx 1.4988,$$

$$x_6 = x_5 - \frac{g(x_5)}{g'(x_0)} \approx 1.4988 - \frac{-0.00009531}{-1.2081} \approx 1.4987.$$

Simple calculation shows that up to five significant figures $x_7 = x_6$. Hence, the accepted solution is 1.4987.

The methods discussed previously are used for solving non-linear equations and systems of non-linear equations. Before turning our attention to other problem areas, some practical applications of non-linear equations are briefly outlined.

Let F be a continuous probability distribution function, and assume that a random number with distribution function F has to be generated. A usual procedure first generates a random, uniformly distributed number x in [0,1], and then solves equation $F(z)-x = 0$. If F is strictly increasing, the equation has a unique solution; otherwise, the smallest solution is selected. Thus, a non-linear equation is solved for z with each given value of X.

Assume next that the global optimum of a differentiable function $f : R^n \to R^l$ has to be determined. It is well known that this problem can be reduced to the solution of the following non-linear equations:

$$\frac{\partial f(x)}{\partial x^{(i)}} = 0 \quad (i = 1, 2, \ldots, n)$$

where $x^{(i)}$ denotes the ith element of vector x. A frequent application of such optimization problems is the non-linear least squares method. Assume that $(x_1, f_1), (x_2, f_2), \ldots, (x_N, f_N)$ are given points, and $f(x, a^{(1)}, \ldots, a^{(n)})$ is a given function, where the parameters $a^{(1)}, a^{(2)}, \ldots, a^{(n)}$ are unknowns, and f is differentiable with respect to these parameters. The least squares method minimizes the sum of the squared errors

$$\sum_{k-1}^{N} [f_k - f(x_k, a^{(1)}, a^{(2)}, \ldots, a^{(n)})]^2.$$

The optimal choice for the unknowns can be obtained either by using a direct optimization method, or by solving the usually non-linear equations

$$\sum_{k-1}^{N} [f_k - f(x_k, a^{(1)}, a^{(2)}), \ldots, a^{(n)}] \frac{\partial f(x_k, a^{(1)}, \ldots, a^{(n)})}{\partial a^{(i)}} = 0$$

for $i = 1, 2, \ldots, n$.

In the special case of $N = n$, we may look for a function that is consistent with the data values; i.e., we may find a function $f(x_k, a^{(1)}, a^{(2)}, \ldots, a^{(n)})$ which passes through the points $(x_1, f_1), \ldots, (x_n, f_n)$. These conditions can be represented by the system of equations

$$f(x_k, a^{(1)}, a^{(2)}, \ldots, a^{(n)}) = f_k \quad (k = 1, 2, \ldots, n).$$

The solution of these equations gives the unknown function parameters. In the further special case when f is a polynomial, f is a linear function of the parameters $a^{(k)}$; therefore, the above system of equations is linear. We note here that the problem of finding a function from a given parametric family which exactly fits the data values is usually called interpolation.

Consider next an n-person game $\Gamma = n, S_1, \ldots, S_n, \varphi_1, \ldots, \varphi_n$ where the strategy sets S_k are finite or infinite intervals in R^l and the payoff functions φ_k are differentiable. A vector $x^* = (x^{(1)^*}, \ldots, x^{(n)^*})$ is an equilibrium point of the game if

(i)   for all $i = 1, 2, \ldots, n, x^{(i)^*} \in S_i$,
(ii) for all $i = 1, 2, \ldots, n$, and $x^{(i)^*} \in S_i$,

$$\varphi_i(x^{(1)^*}, \ldots, x^{(i-1)^*}, x^{(i+1)^*}, \ldots, x^{(n)^*}) \leq \varphi_i(x^{(1)^*}, \ldots, x^{(n)^*})$$

In other words, $x^{(i)'}$, maximizes φ_i with fixed values of

$$x^{(1)^*}, \ldots, x^{(i-1)^*}, x^{(i+1)^*}, \ldots, x^{(n)^*}.$$

Therefore, in the case of interior equilibrium points,

$$\frac{\partial \varphi_i(x^{(1)^*}, \ldots, x^{(n)^*})}{\partial x^{(i)}} = 0 \quad (i = 1, 2, \ldots, n).$$

The problem of finding the percentiles of a probability distribution can also be reduced to the solution of a non-linear equation. The α-percentile of a distribution function F is defined as the smallest value x such that $F(x) \geq \alpha/100$. If F is strictly increasing and continuous, x can be obtained by solving the equation $F(x) = \alpha/100$.

In constructing confidence intervals the smallest intervals with given probabilities are usually selected. Again, let F be a probability distribution function, and let p be a given probability level. Then the corresponding confidence interval is given as $[a, b]$, which is the solution of the following optimization problem:

$$\underset{\text{subject to } F(b) - F(a) = p}{\text{minimize}} b - a$$

This problem can be rewritten as an unconstrained optimization problem:

$$\text{minimize } b - a + \lambda \left[F(b) - F[a] - p \right],$$

where λ is the Lagrangian multiplier. Simple differentiation shows that the values of a, b, and λ can be obtained by solving the non-linear equations:

$$1 + \lambda F'(b) = 0$$
$$1 + \lambda F'(a) = 0$$
$$F(b) - F(a) - p = 0.$$

Notice that the first two equations imply that the probability density function has the same value at a and b.

Consider the special case in which F is strictly decreasing. The third equation implies that

$$b = F'(F(a) + p).$$

Therefore, the first two equations can be combined into a single equation to a single unknown:

$$F'(F^{-1}(F(a) + p)) - F'(a) = 0.$$

This equation can be solved by any previously outlined method.

We note also that certain non-linear differential and integral equations can be reduced to the solutions of non-linear algebraic equations. The details are presented in the later parts of this section.

1.3.6   SIXTH APPLICATION

In the sixth application, assume that a linear equation system of the form $Ax = b$ must be solved, where A is a given $n \times n$ real matrix and b is a given n-element real vector. A possible solution algorithm can be derived by rewriting the equation as a fixed point problem

$$x = (I - A)x + b,$$

where I denotes the $n \times n$ identity matrix. The corresponding iteration procedure is the following:

$$x_{k+1} = (I - A)x_k + b. \tag{1.17}$$

We note that an alternative process is introduced in Section 2.5.

Example 1.10. As a numerical example consider the linear equations

$$\begin{pmatrix} \frac{3}{4} & -\frac{1}{4} \\ -\frac{1}{8} & \frac{7}{8} \end{pmatrix} x = \begin{pmatrix} 1 \\ 1 \end{pmatrix}.$$

In this case, we have

$$A = \begin{pmatrix} \frac{3}{4} & -\frac{1}{4} \\ -\frac{1}{8} & \frac{7}{8} \end{pmatrix}$$

and

$$b = \begin{pmatrix} 1 \\ 1 \end{pmatrix}.$$

Therefore, scheme (1.17) reduces to the following:

$$x^{(k+1)} = \begin{pmatrix} \frac{3}{4} & -\frac{1}{4} \\ -\frac{1}{8} & \frac{7}{8} \end{pmatrix} x^{(k)} + \begin{pmatrix} 1 \\ 1 \end{pmatrix}.$$

Select the initial approximation $x^{(0)} = 0$. Easy calculation shows that

$$x^{(1)} = \begin{pmatrix} 1 \\ 1 \end{pmatrix}, x^{(2)} = \begin{pmatrix} \frac{3}{2} \\ \frac{5}{4} \end{pmatrix}, x^{(3)} = \begin{pmatrix} \frac{27}{16} \\ \frac{43}{32} \end{pmatrix} \approx \begin{pmatrix} 1.6875 \\ 1.3438 \end{pmatrix},$$

$$x^{(4)} \approx \begin{pmatrix} 1.7578 \\ 1.3789 \end{pmatrix}, x^{(5)} \approx \begin{pmatrix} 1.7842 \\ 1.3921 \end{pmatrix}, x^{(6)} \approx \begin{pmatrix} 1.7941 \\ 1.3970 \end{pmatrix},$$

$$x^{(7)} \approx \begin{pmatrix} 1.7978 \\ 1.3989 \end{pmatrix}, x^{(8)} \approx \begin{pmatrix} 1.7992 \\ 1.3996 \end{pmatrix}, x^{(9)} \approx \begin{pmatrix} 1.7997 \\ 1.399 \end{pmatrix},$$

$$x^{(10)} \approx \begin{pmatrix} 1.7999 \\ 1.4000 \end{pmatrix}, x^{(10)} \approx \begin{pmatrix} 1.8000 \\ 1.4000 \end{pmatrix}.$$

Simple substitution shows that $x_1 = 1.8$ and $x_2 = 1.4$ are the exact solutions.

Process (1.17) can be generalized in the following way. Multiply both sides of equation $Ax = b$ by a non-singular matrix H, and decompose matrix HA as $HA = A_1 + A_2$, where A_1 is non-singular. Then the equation is equivalent to the following:

$$A_1 x + A_2 x = Hb,$$

which can be rewritten as

$$x = -A_1^{-1} A_2 x + A_1^{-1} Hb.$$

The corresponding iteration scheme is

$$x_{k+1} = -A_1^{-1} A_2 x_k + A_1^{-1} Hb.$$

Two special cases are well known from the literature. If one selects $H = I$ and $A_1 = diag(a_{11}, a22, \ldots, a_{nn})$, where a_{ii} denotes the ith diagonal element of A, then the Jacobi iteration procedure is obtained. In the other special case when $H = I$ and

$$A_1 = \begin{pmatrix} a_{11} & 0 & 0 & \cdots & 0 \\ a_{21} & a_{22} & 0 & \cdots & 0 \\ a_{31} & a_{32} & a_{33} & \cdots & 0 \\ \cdot & \cdot & \cdot & \cdots & \cdot \\ a_{n1} & a_{n2} & a_{n3} & \cdots & a_{nn} \end{pmatrix}.$$

the Gauss-Seidel method is derived.

We notice here that linear equations have numerous applications in engineering and in economy. For example, linear circuits, input-output systems, and equilibriums of linear systems can be mentioned here.

1.3.7   SEVENTH APPLICATION

In the seventh example, again let A be a real $n \times n$ matrix, furthermore let λ and x be an eigenvalue and an associated eigenvector of A. Then

$$Ax = \lambda x.$$

An iterative method for finding a pair of eigenvalue and eigenvector can be developed as follows.

Consider two vectors being equal when they differ from each other only in a constant multiplier. The eigenvalue equation can be interpreted as the fixed point problem of the mapping Ax. Similar to process (1.8), we can introduce the scheme

$$z_{k+1} = Az_k,$$

and if we require that all vectors in the iteration sequence must have unit length (i.e., normalized), this iteration scheme must be modified as

$$x_{k+1} = \frac{Ax_k}{\|Ax_k\|}, \tag{1.18}$$

where $\|.\|$ is the Euclidean norm. Notice that this method is the same as the popular power method which solves for the largest eigenvalue of real matrices.

Example 1.11. We illustrate the above method for matrix

$$A = \begin{pmatrix} 2 & 3 \\ 1 & 4 \end{pmatrix}.$$

Select the initial vector $x_0 = (1,0)^T$, then

$$Ax_0 = \begin{pmatrix} 2 \\ 1 \end{pmatrix}$$

with

$$\|Ax_0\| = \sqrt{4+1} = \sqrt{5}.$$

Therefore, we obtain

$$x_1 = \frac{1}{\sqrt{5}} \begin{pmatrix} \frac{2}{\sqrt{5}} \\ \frac{1}{\sqrt{5}} \end{pmatrix} \approx \begin{pmatrix} 0.8944 \\ 0.4472 \end{pmatrix}.$$

Continue the process to obtain the sequence

$$x_2 \approx \begin{pmatrix} 0.7593 \\ 0.6508 \end{pmatrix}, \quad x_3 \approx \begin{pmatrix} 0.7182 \\ 0.6958 \end{pmatrix}, \quad x_4 \approx \begin{pmatrix} 0.7094 \\ 0.7048 \end{pmatrix},$$

$$x_5 \approx \begin{pmatrix} 0.7076 \\ 0.7067 \end{pmatrix}, \quad x_6 \approx \begin{pmatrix} 0.7072 \\ 0.7071 \end{pmatrix}, \quad x_7 \approx \begin{pmatrix} 0.7071 \\ 0.7071 \end{pmatrix}.$$

It is easy to check that the last approximation is the exact eigenvector of matrix A, because

$$\begin{pmatrix} 2 & 3 \\ 1 & 4 \end{pmatrix} \begin{pmatrix} 0.7071 \\ 0.7071 \end{pmatrix} \cdot = \begin{pmatrix} 3.5355 \\ 3.5355 \end{pmatrix} \cdot = 5 \cdot \begin{pmatrix} 0.7071 \\ 0.7071 \end{pmatrix}.$$

Hence, the associated eigenvalue is 5.

1.3.8   EIGHTH APPLICATION

Consider again the matrix eigenvalue problem $Ax = \lambda x$ of an $n \times n$ real matrix A in our eighth example. This equation with the additional condition $x \neq 0$ can be rewritten as a system of non-linear equations:

$$Ax - \lambda x = 0$$
$$\bar{x}^T x = 1,$$

where $\bar{x}^T$ denotes the conjugate transpose of x. If λ is real, then x can also be selected as real, and therefore the above system of equations reduces to

$$(x^{(1)})^2 + (x^{(2)})^2 + \cdots + (x^{(n)})^2 - 1 = 0, \tag{1.19}$$

where $x^{(1)}, x^{(2)}, \ldots, x^{(n)}$ denote the components of vector x. This system then can be solved by using the multivariable Newton method (1.15).

The Jacobian of this system has now the special form

$$J(x, \lambda) = \begin{pmatrix} A - \lambda I & x \\ 2x^T & 0 \end{pmatrix},$$

therefore, the method specializes to the scheme

$$\begin{pmatrix} A - \lambda I & x \\ 2x^T & 0 \end{pmatrix} \Delta_k = \begin{pmatrix} Ax_k - \lambda_k x_k \\ x_k^T x_k - 1 \end{pmatrix}$$
$$x_{k+1} = x_k - \Delta_k. \tag{1.20}$$

Example 1.12. In the case of the matrix introduced in the previous example the iteration scheme (1.20) has the special form

$$\begin{pmatrix} 2 - \lambda_k & 3 & x_k^{(1)} \\ 1 & 4 - \lambda_k & x_k^{(2)} \\ 2x_k^{(1)} & 2x_k^{(2)} & 0 \end{pmatrix} \Delta_k = \begin{pmatrix} (2 - \lambda_k)x_k^{(1)} + 3x_k^{(2)} \\ x_k^{(1)} + (4 - \lambda_k)x_k^{(2)} \end{pmatrix}$$
$$x_{k+1} = x_k - \Delta_k.$$

Using a simple computer program and different initial approximations (x_0, λ_0), both eigenvalues 5 and 1 are found with the associated eigenvectors $(0.7071, 0.7071)^T$ and $(-0.9487, 0.3162)^T$, respectively.

The above procedure can be easily generalized for solving the generatized eigen-value problem

$$Ax = \lambda Bx,$$

where A and B are real $n \times n$ matrices. System (1.19) generalizes to equations

$$Ax - \lambda Bx = 0$$
$$(x^{(1)})^2 + (x^{(2)})^2 + \cdots + (x^{(n)})^2 - 1 = 0. \tag{1.21}$$

The Jacobian of this system has the form

$$J(x, \lambda) = \begin{pmatrix} A - \lambda B & -Bx \\ 2x^T & 0 \end{pmatrix},$$

Thus, method (1.16) can be written as

$$\begin{pmatrix} A - \lambda B & -Bx \\ 2x^T & 0 \end{pmatrix} \Delta x = \begin{pmatrix} Ax_k - \lambda Bx_k \\ x_k^T x_k - 1 \end{pmatrix}$$
$$x_{k+1} = x_k - \Delta_k. \tag{1.22}$$

1.3.9 NINTH APPLICATION

For our ninth application, consider the initial value problem

$$\dot{x}(t) = f(t, x(t)), \quad x(t_0) = x_0,$$

where $f : [t_0, T] \times R^n \to R^n$ is a continuous function, and $x_0 \in R^n$ is a given vector. Integrate both sides in the interval $[t_0, t]$ to obtain the fixed point problem

$$x(t) = x_0 + \int_{t_0}^{t} f(\tau, x(\tau)) d\tau.$$

The corresponding iteration scheme can be written as

$$x_0(t) \equiv x_0,$$
$$x_{k+1}(t) = x_0 + \int_{t_0}^{t} f(\tau, x(\tau)) d\tau \tag{1.23}$$

Example 1.13. In the case of the initial value problem

$$\ddot{x}(t) = -x(t), \quad \dot{x}(0) = 1, \quad x(0) = 0$$

we cannot directly use the above procedure, since $\ddot{x}$ is present. However, we can rewrite this second order equation as a system of first order equations by introducing the new variable $y = \dot{x}$. We then have the system

$$\dot{x} = y, \quad x(0) = 0$$
$$\dot{y} = -x, \quad y(0) = 1.$$

The iteration algorithm reduces to the special scheme

$$x_{k+1}(t) = \int_0^t y_k(\tau)d\tau$$

$$(x_0(t) \equiv 0, y_0(t) \equiv 1) \tag{1.24}$$

$$y_{k+1}(t) = 1 - \int_0^t x_k(\tau)d\tau.$$

Simple integration show that

$$x_1(t) = \int_0^t 1 d\tau = t, y_1(t) = 1 - \int_0^t 0 d\tau = 1,$$

$$x_2(t) = \int_0^t 1 d\tau = t, y_2(t) = 1 - \int_0^t \tau d\tau = 1 - \frac{t^2}{2},$$

$$x_3(t) = t - \frac{t^3}{3!}, y_3(t) = 1 - \frac{t^2}{2!},$$

$$x_4(t) = t - \frac{t^3}{3!}, y_4(t) = 1 - \frac{t^2}{2!} + \frac{t^4}{4!},$$

$$x_5(t) = t - \frac{t^3}{3!} + \frac{t^5}{5!}, y_5(t) = 1 - \frac{t^2}{2!} + \frac{t^4}{4!},$$

and so on. Using finite induction one may easily verify that

$$x_k(t) = \begin{cases} t - \frac{t^3}{3!} + \cdot + \pm \frac{t^{k-1}}{(k-1)!} & \text{if } k \text{ is even} \\ t - \frac{t^3}{3!} + \cdot + \pm \frac{t^k}{k!} & \text{if } k \text{ is odd} \end{cases}$$

and

$$y_k(t) = \begin{cases} t - \frac{t^2}{2!} + \cdot + \pm \frac{t^k}{k!} & \text{if } k \text{ is even} \\ t - \frac{t^2}{2!} + \cdot + \pm \frac{t^{k-1}}{(k-1)!} & \text{if } k \text{ is odd} \end{cases}$$

Notice that $x_k(t) \to \sin t$ and $y_k(t) \to \cos t$ as $k \to \infty$. Hence, the solution is

$$x(t) = \sin t, y(t) = \cos t.$$

We can speed up the convergence of the iteration scheme (1.24) by noticing that in computing $y_{k+1}(t)$ only the "old" approximation $x_k(t)$ is used; however, the "new" approximation $x_{k+1}(t)$ is already available. If we replace $x_k(\tau)$ by $x_{k+1}(\tau)$ in the second equation of (1.24), then the modified process

$$x_{k+1}(t) = \int_0^t y_k(\tau)d\tau$$

$$(x_0(t) \equiv 0, y_0(t) \equiv 1) \tag{1.25}$$

$$y_{k+1}(t) = 1 - \int_0^t x_{k+1}(\tau)d\tau.$$

is obtained. It is easy to see that

$$x_1(t) = \int_0^t 1 d\tau = t, y_1(t) = 1 - \int_0^t \tau d\tau = 1 - \frac{t^2}{2},$$

$$x_2(t) = \int_0^t \left(1 - \frac{\tau^2}{2}\right) d\tau = t - \frac{t^3}{3!}, y_2(t) = 1 - \frac{t^2}{2!} + \frac{t^4}{4!},$$

$$x_3(t) = 1 - \frac{t^3}{3!} + \frac{t^5}{5!}, y_3(t) = 1 - \frac{t^2}{2!} + \frac{t^4}{4!} - \frac{t^6}{6!},$$

and so on. Using finite induction one may verify that

$$x_k(t) = t - \frac{t^3}{3!} + \frac{t^5}{5!} + \cdots \pm \frac{t^{2k-1}}{(2k-1)!}$$

and

$$y_k(t) = 1 - \frac{t^2}{2!} + \frac{t^4}{4!} + \cdots \pm \frac{t^{2k}}{(2k)!}.$$

A simple comparison of the two iteration sequences shows that the modified algorithm really converges faster to the solution, as more terms of the Taylor series of $\sin t$ and $\cos t$ are obtained in $x_k(t)$ and $y_k(t)$ in the later cases.

1.3.10   TENTH APPLICATION

Non-linear differential equations are sometimes solved by linearization. This procedure is illustrated next. Consider again the initial value problem

$$\dot{x}(t) = f(t, x(t)), x(t_0) = x_0,$$

where $f : [t_0, T] \times R^n \to R^n$ is continuous and differentiable with respect to x. Furthermore, $x_0 \in R^n$ is a given vector. Let $x_0(t)$ be an approximation of the solution, and linearize the right-hand side of the differential equation around the approximating solution $x_0(t)$:

$$f(t,x(t)) = f(t,x_0(t)) + L(t,x_0(t))(x(t) - x_0(t)),$$

where $J(t,x)$ is the Jacobian of $f(t,x)$ with respect to x. By replacing the right-hand side of the differential equation by the linear approximation, a linear differential equation is obtained:

$$\dot{x}(t) = K(t,x_0(t))x(t) + [f(t,x_0(t)) - J(t,x_0(t))x_0(t)], x(t_0) = x_0.$$

Let the solution of this approximating linearized equation be denoted by $x(t)$. Repeat the above procedure with $x_1(t)$ replacing $x_0(t)$ to obtain $x_2(t)$, and again with $x_2(t)$ replacing $x_0(t)$ to obtain $x_3(t)$, etc. The resulting process can be summarized as:

$X_{k+1}(t)$ is the solution of the initial value problem

$$\dot{x}(t) = J(t,x_k(t))x(t) + [f(t,x_k(t)) - J(t,x_k(t))x_k(t)], x(t_0) = x_0,$$

where $x_0(t)$ is an arbitrary function satisfying the initial condition $x_0(t_0) = x_0$.

Example 1.14. The linearization process is now illustrated in the case of the initial value problem

$$\dot{x}(t) = t + x^2(t), x(0) - 1.$$

Starting from the initial approximation $x_0(t) = 1$ the general step can be formulated as

$$\dot{x}(t) = 2x_k(t)x(t) + [t - x_k^2(t)], x(0) - 1$$

since

$$J(t,x) = 2x.$$

For $k = 0$, this equation reduces to the following:

$$\dot{x}(t) = 2x(t) + (t - 1), x(0) = 1.$$

It is easy to see the solution is

$$x_1(t) = \frac{3}{4}e^{2t} - \frac{t}{2} + \frac{1}{4}.$$

The process then continues for $k = 1$ by again solving a linear initial value problem

$$\dot{x}(t) = \left(\frac{3}{2}e^{2t} - t + \frac{1}{2}\right)x(t) + \left[t - \left(\frac{3}{4}e^{2t} - \frac{t}{2} + \frac{1}{4}\right)^2\right], x(0) = 1$$

The details on the further steps are omitted.

1.3.11   ELEVENTH APPLICATION

In our eleventh application, we show how non-linear boundary value problems can be reduced to the solution of non-linear algebraic equations. Consider the second-order boundary value problem

$$f(t, x(t), \dot{x}(t), \ddot{x}(tk)) = 0, x(a) = x, \text{ and } x(b) = x_b,$$

where $f : R^4 \to R^l$ is continuous, strictly increasing in $\ddot{x}$. The unknown function is $x : [a,b] \to R^l$.

In the first step, divide the solution interval $[a,b]$ into n equal parts by the nodes

$$a = t_0 < t_1 < t_2 \cdots < t_{n-1} < t_n = b,$$

where

$$t_1 - t_0 = t_2 - t_1 = \cdots = t_n - t_{n-1} = h.$$

For $k = 1, 2, \ldots n - 1$, substitute t_k into the differential equation

$$f(t_k, x(t_k), \dot{x}(t_k), \ddot{i_k}(t_k)) = 0,$$

Substitute the derivatives by the numerical differentiation rules

$$\dot{x}(t_k) \approx \frac{x(t_{k+1}) - x(t_{k-1})}{2h}$$

and

$$\ddot{x}(t_k) \approx \frac{x(t_{k+1}) - 2x(t_k) + x(t_{k-1})}{h^2}.$$

The resulting equation has the form

$$f\left(t_k, x(t_k), \frac{x(t_{k+1} - x(t_{k-1}))}{2h}, \frac{x(t_{k+1} - 2x(t_k) + x(t_{k-1}))}{h^2}\right) = 0.$$

Introduce the unknowns

$$x_k = x(t_k) \quad (k = 1, 2, \ldots, n-1)$$

with the additional conditions

$$x_0 = x(t_0) = x(a) = x_a$$

and

$$x_n = x(t_n) = x(b) = x_b,$$

We then have the following system of algebraic equations:

$$f\left(t_1, x_1, \frac{x_2 - x_n}{2h}, \frac{x_2 - 2x_1 + x_a}{h^2}\right) = 0$$

$$f\left(t_k, x_k, \frac{x_{k+1} - x_{k-1}}{2h}, \frac{x_{k+1} - 2x_k + x_{k-1}}{h^2}\right) = 0 (k = 2, \ldots, n-2)$$

$$f\left(t_{n-1}, x_{n-1}, \frac{x_b - x_{n-2}}{2h}, \frac{x_b - 2x_{n-1} + x_{n-2}}{h^2}\right) = 0.$$

The solutions of these equations give the approximations of the solution function at nodes t_k $(k = 0, 1, 2, \ldots, n)$. If a continuous solution function is needed, then any function fitting methods such as interpolation and least squares can be used.

Example 1.15. The above methodology is illustrated for the second order differential equation

$$\ddot{x} = x^2 + \exp(\dot{x})x(0) = x(1) = 0.$$

Select $n = 10$, then $h = 0.1, t_k = k/10$ and

$$x_k = x(k/10) \quad (k = 0, 1, \ldots, 10).$$

The boundary conditions imply that $x_0 = x(0) = 0$ and $x_{10} = x(1) = 0$. Therefore, the resulting systems of equations have the special form:

$$\frac{x_2 - 2x_1}{0.01} = x_1^2 + \exp\left(\frac{x_2}{0.2}\right)$$

$$\frac{x_{k+1} - 2x_k + x_{k-1}}{0.01} = x_k^2 + \exp\left(\frac{x_{k-1} - x_{k-1}}{0.2}\right) \quad \frac{-2x_9 + x_8}{0.01} = x_n^2 + \exp\left(\frac{-x_8}{0.2}\right).$$

for $k = 2, 3, \ldots, 8$.

The solutions of these equations give the approximating solution values at the selected nodes t_k $(k = 0, 1, 2, \ldots, 10)$.

1.3.12   TWELVETH APPLICATION

Consider the Fredholm-type integral equation

$$x(t) = \int_a^b K(s,t)x(s)ds + f(t)$$

where $K : [a,b] \times [a,b] \to R$ and $f : [a,b] \to R$ are continuous functions. Because the equation is given in a fixed-point form, the iteration algorithm

$$x_{k+1}(t) = \int_a^b K(s,t)x_k(s)ds + f(t) \tag{1.26}$$

can be suggested to find the solution.

Example 1.16. As an illustration we present the algorithm for solving the particular problem

$$x(t) = \int_0^1 \frac{t+s}{4}x(s)ds + 1,$$

where $[a,b] = [0,1]$ $k(s,t) = \frac{t+s}{4}$, and $f(t) = 1$. Select the initial approximation $x_0(t) \equiv 1$. Then

$$x_1(t) = \int_0^1 \frac{t+s}{4}ds + 1 = \left[\frac{ts}{4} + \frac{s^2}{8}\right]_{a=0}^1 + 1 = \frac{t}{4} + \frac{9}{8}.$$

Similarly,

$$x_2(t) = \int_0^1 \frac{t+s}{4}\left(\frac{s}{4} + \frac{9}{8}\right)ds + 1 = \int_0^1 \left(\frac{ts}{16} + \frac{s^2}{16} + \frac{9t}{32} + \frac{9s}{32}\right)ds + 1$$

$$= \left[\frac{ts^2}{32} + \frac{s^2}{48} + \frac{9ts}{32} + \frac{9s^2}{64}\right]_{s=0}^1 + 1 = \frac{5t}{16} + \frac{223}{192}.$$

The procedure can be continued in the same way. The details are omitted.

An alternative approach for solving the Fredholm-type integral equation leads to the solution of a system of linear algebraic equations. Divide the interval $[a,b]$ into n equal parts by the nodes

$$a = t_0 < t_1 < t_2 \cdots < t_{n-1} < t_n = b.$$

Substitute $t = t_k$ $(k = 0,1,2,\ldots,n)$ into the original equation to get

$$x_k = \sum_{l=0}^n A_l K(t_l,t_k)x_l + f(t_k) \quad (k = 0,1,\ldots,n)$$

where A_l is the lth integration weight. Because the values of $A_l, K(t_l,t_k)$, and $f(t_k)$ are known, a system of linear equations is obtained for the unknown function values $x_0,x_1,\ldots,x_n$.

Example 1.17. What follows is the above reduction principle for the integral equation of the previous example. Select the nodes $t_k = k/10, k = 0,1,\ldots,10$. By selecting the trapezoidal rule for numerical integration, we have

$$A_0 = \frac{1}{20}, A_1 = \cdots = A_9 = \frac{1}{10}, \text{ and } A_{10} = \frac{1}{20}.$$

Therefore, the resulting linear equations are

$$x_k = \frac{x_0}{20} \cdot \frac{\frac{0}{10} + \frac{k}{10}}{4} + \sum_{l=1}^9 \frac{x_l}{10} \cdot \frac{\frac{l}{10} + \frac{k}{10}}{4} + \frac{x_{10}}{20} \cdot \frac{\frac{10}{10} + \frac{k}{10}}{4} + 1,$$

which can be simplified as

$$x_k = \frac{k}{800}x_0 + \sum_{l=1}^9 \frac{1+k}{400}x_l + \frac{10+k}{800}x_{10} + 1 \text{ for } k = 0,1,2,\ldots,10.$$

Fredholm-type integral equations can be generalized as the non-linear integral equation

$$x(t) = \int_a^b K(t,s,x(s))ds + f(t),$$

where $K : [a,b] \times [a,b] \times R \to R$ and $f : [a,b] \to R$ are continuous functions. The iteration process has the form

$$x_{k+1}(t) = \int_a^b K(t,s,x_k(s))ds + f(t),$$

where the initial approximation $x_0(t)$ is selected as a continuous function. We note also that the method based on numerical integration leads to the system of non-linear equations

$$x_k = \sum_{l=0}^n A_1 K(t_k,t_l,x_l) + f(t_k). \quad (k = 0,1,\dots,n)$$

Finally, in our thirteenth example, consider the Volterra-type integral equation

$$x(t) = \int_a^t C(s,t)x(s)ds + f(t),$$

where $C : [a,b] \times [a,b] \to R$ and $f : [a,b] \to R$ are continuous functions, and the solution is to be determined in the interval $[a,b]$. This equation can be considered a special Fredholm-type integral equation, where

$$K(s,t) = \begin{cases} C(s,t) & \text{if } s \leq t \\ 0 & \text{otherwise.} \end{cases}$$

Therefore, algorithm (1.26) modifies as

$$x_{k+1}(t) = \int_a^t C(s,t)x_k(s)ds + f(s). \qquad (1.27)$$

Example 1.18. We now illustrate the above algorithm in the case of integral equation

$$x(t) = \int_0^t tsx(s)ds + 1.$$

Select the initial approximation $x_0(t) = 1$, then

$$x_1(t) = \int_0^t ts \cdot 1 ds + 1 = \left[\frac{ts^2}{2}\right]_{s=0}^t + 1 = \frac{t^3}{2} + 1.$$

Similarly,

$$x_2(t) = \int_0^t ts\left(\frac{s^3}{2}+1\right)ds + 1 = \left[\frac{ts^5}{10}+\frac{ts^2}{2}\right]_{s=0}^t + 1 = \frac{t^6}{10}+\frac{t^3}{2}+1$$

and so on.

The alternative approach based on numerical integration can be formulated in the following way. Select again the nodes

$$a = t_0 < t_1 < t_2 < \cdots < t_{n-1} < t_m = b$$

with equal step size h, and substitute $t = t_k (k = 0, 1, \ldots, n)$ into the original equation to obtain

$$x(t_k) = \int_a^{t_k} C(s,t_k)x(s)ds + f(t_k).$$

Introduce the notation $x(t_k) = X_k$, and use the numerical integration formula to approximate the integral of the right-hand side:

$$x_k = \sum_{l=0}^k A_{kl}C(t_l,t_k)x_l + f(t_k),$$

where A_{kl} $(l = 0,1,2,\ldots,k)$ are the integration weights based on k subintervals. The solution of these linear equations gives the unknown function values $x_0, x_1, \ldots, x_n$.

Example 1.19. Consider again the Volterra-type integral equation of the previous example. Select the nodes $t_k = k/10, k = 0, 1, \ldots, 10$. By selecting the trapezoidal rule for numerical integration, we have

$$A_{k0} = \frac{1}{20}, A_{kl} = \cdots = A_{k,k-1} = \frac{1}{10}, \text{ and } A_{kk} = \frac{1}{20}.$$

Therefore, the resulting linear equations are as follows:

$$x_0 = 0 + 1$$
$$x_1 = \frac{1}{20}\cdot\frac{1}{20}\cdot 0 \cdot x_0 + \frac{1}{20}\cdot\frac{1}{10}+\frac{1}{10}\cdot x_1 + 1$$

and

$$x_k = \frac{1}{20}\cdot\frac{k}{20}\cdot 0 \cdot x_0 + \sum_{l=1}^k -1\frac{1}{10}+\frac{k}{10}\frac{k}{10}\cdot x_1 + \frac{k}{20}\cdot\frac{k}{10}\cdot\frac{k}{10}\cdot x_k + 1$$

for $k = 2, \ldots, 10$, which can be simplified as

$$x_0 = 1$$
$$x_1 = \frac{1}{2000}\cdot x_1 + 1$$

and for $k = 2, \ldots, 10$,

$$x_k = \sum_{l=1}^{k-1} \frac{k_l}{1000} \cdot x_l + \frac{k^2}{2000} x_k + 1.$$

The solution of these equations gives the unknown function values.

Non-linear Volterra-type integral equations have the general form

$$x(t) = \int_a^t C(s,t,x(s)) ds + f(t),$$

where $C : [a,b] \times [a,b] \times R \to R$ and $f : [a,b] \to R$ are continuous functions. The iteration process has the form

$$x_{k+1}(t) = \int_a^t C(s,t,x_k(s)) ds + f(t),$$

where $X_0(t)$ is selected as a continuous function. We notice also that the method based on numerical integration leads to the system of non-linear equations

$$x_k = \sum_{l=0}^{k} A_{kl} C(t_l, t_k, x_l) + f(t_k) \quad (k = 0, 1, \ldots, n)$$

In engineering applications more complicated integral equations must sometimes be solved. For example, in examining neutron transport, integral equations of the form

$$x(t) = f(t) + \frac{1}{2} \lambda x(t) \int_t^1 k(t,s) x(s) ds$$

are solved, where f is a given function $f : [0,1] \to R^l$ and λ is a given scalar. A similar example is the integral equation:

$$x(t) = 1 + tx(t) \int_0^1 \frac{k(s)x(s)}{t+s} ds,$$

where $k : [0,1] \to R^l$ is a given function. This equation is applied for analyzing radioactive transfer, neutron transport, and in the kinetic theory of gases.

1.4   EXERCISES

1. Extend Theorem 1.3 to the more general case when c depends on k.

2. Extend Theorem 1.4 to the more general case when c depends on k.

3. Extend Theorem 1.5 to the more general case when c depends on k.

4. Maximize $F = x_1 + 2x_2 + x_3$ subject to

$$2x_1 + x_2 - x_3 \leq 2$$
$$-2x_1 + x_2 - 5x^3 \geq -6$$
$$4x_1 + x_2 + x_3 \leq 6$$
$$x_1 \geq 0, i = 1, 2, 3.$$

5. Minimize $F = -400x_1 - 100x_2$ subject to

$$10x_1 + 5x_2 \leq 2500$$
$$4x_1 + 10x_2 \leq 2000$$
$$2x_1 + 3x_2 \leq 900$$
$$x_1 \geq 0, x_2 \geq 0$$

6. Find an interval $[a,b]$ containing a root α of the equation $x = \frac{1}{2}\cos x$ such that for every $x_0 \in [a,b]$ the iteration $x_{n+1} = \frac{1}{2}\cos x_n$ will converge to α.

7. Solve the equation of the previous problem by using the secant method.

8. Repeat Problem 7 by using the Newton method.

9. Three ladies are arguing the proper way to prepare tea for their guests. The first insists that the tea should be made up in at least 17 cup lots by mixing orange pekoe, green, and standard tea in the proportion 4:2:3. The second suggests that the green tea and the standard tea, which they have on hand and which is sufficient to make 10 and 8 cups, respectively, should be adequate for their purpose, although some tea may be left over after the guests have gone. The third points out that no orange pekoe tea is in the house and hence it must be purchased from the market. Formulate this situation in a linear programming format so as to minimize the cost of tea purchased for the party. How many cups of tea of each type will be in evidence at the party?

10. Solve the following non-linear equations by the method of your choice:
    a) $\ln x = x - 2$
    b) $xe^x = 2$
    c) $e^x \cdot \ln x = 1$

11. Use iteration (1.17) to find $(I - A)^{-1}$ for matrix

$$A = \begin{pmatrix} 0.15 & 0.10 \\ 0.20 & 0.05 \end{pmatrix}$$

Perform three steps.

12. Repeat Example 1.11 for matrix

$$A = \begin{pmatrix} 1 & 2 \\ 2 & -1 \end{pmatrix}.$$

13. Solve problem

$$\dot{x} = x + y + 1 \quad x(0) = 0$$
$$\dot{y} = x - y \quad y(0) = 1$$

by the iteration scheme (1.23). Perform three steps.

14. Solve the boundary-value problem

$$\ddot{x} = tx + \dot{x} + 1, x(0) = x(1) = 1$$

by the discretization method. Select $h = 0.1$.

15. Repeat Example 1.16 for problem

$$x(t) = \int_0^1 \frac{(t-s)^2}{10} x(s) ds + 2.$$

16. Repeat Example 1.17 for the integral equation of the previous problem.

17. Repeat Example 1.18 for equation

$$x(t) = \int_0^t (t+s)x(s) ds + 2.$$

18. Repeat Example 1.19 for the integral equation of the previous problem.

REFERENCES

1. Argyros, I.K. 2007. *Computational Theory of Iterative Methods*. eds. Chui, C.K., Wuytack, L. Series: Studies in Computational Mathematics, 15. New York: Elsevier.
2. Argyros, I.K. 2008. *Convergence and Applications of Newton-type Iterations*. New York: Springer-Verlag.
3. Argyros, I.K., and George, S. 2019. *Mathematical Modeling for the Solution of Equations and Systems of Equations with Applications*, Volume III, NY: Nova Science Publisher.
4. Argyros, I.K., George, S., and Magrenan, A.A. 2015. Local convergence for multi-point-parametric Chebyshev-Halley-type methods of high convergence order. *J. Comput. App. Math.*, 282:215–224.
5. Argyros, I.K., George, S., and Thapa, N. 2018. *Mathematical Modeling for the Solution of Equations and Systems of Equations with Applications*, Volume I, NY: Nova Science Publishers.
6. Argyros, I.K., George, S., and Thapa, N. 2018. *Mathematical Modeling for the Solution of Equations and Systems of Equations with Applications*, Volume II, NY: Nova Science Publishers.
7. Argyros, I.K., and Hillout S. 2012. Weaker conditions for the convergence of Newton's method. *J. Complexity*. 28, 3:364–387.
8. Argyros, I.K., and Magreñán, A.A. 2017. *Iterative Methods and Their Dynamics with Applications*, New York: CRC Press, Taylor & Francis.
9. Argyros, I.K., and Magrenan, A.A. 2018. *A Contemporary Study of Iterative Methods*. NY: Academy Press, Elsevier.
10. Argyros, I.K., and Regmi, S. 2019. *Undergraduate Research at Cameron University on Iterative Procedures in Banach and Other Spaces*, NY: Nova Science Publisher.
11. Argyros, M.I., Argyros, I.K., and Regmi, S. 2021. *Hilbert Spaces and Its Applications*, NY: Nova Science Publisher.
12. Brikhoff, G. 1948. *Lattice Theory, Colloq. Publ.*, 25, New York: American Mathematical Society.
13. Brock, W.A. and Scheinkman, J.A. 1975. Some results on global asymptotic stability of difference equations. *J. Econ. Theory*, 10:265–268.
14. Deuflhard, P. and Heindl, G. 1979. Affine invariant convergence theorems for Newton's method and extensions to related methods. *SIAM J. Numer. Anal.*, 16:1–10.
15. Fujimoto, T. 1987. Global asymptotic stability of nonlinear difference equations I. *Econ. Lett.*, 22:247–250.
16. Hall, D.W. and Spencer, G.L. 1955. *Elementary Topology*, New York: John Wiley & Sons.
17. Higle, J.L. and Sen, S. 1989. On the convergence of algorithms with applications to stochastic and nondifferentiable optimization. *SIE Working Paper*, #89-027. University of Arizona, Tucson.
18. Hogan, W.W. 1973. Point to set maps in mathematical programming. *SIAM Rev.*, 15:591–603.
19. Huard, P. 1975. Optimization algorithms and point to set maps. *Math. Programing*, 8:308–331.
20. Kantorovich, L.V. and Akilov, G.P. 1964. *Functional Analysis in Normed Spaces*, New York: Pergamon Press.
21. Klessig, R. 1974. A general theory of convergence for constrained optimization algorithms that use antizigzaging provisions. *SIAM J. Contr.*, 12(4):598–608.
22. Seller, J.P. 1986. *The Stability and Control of Discrete Processes*. New York: Springer-Verlag.

23. Lui, D. and Szidarovszky, F. 1990. Block-M-matrices and their properties. *Pure. Math. Appl.*, 1(Ser.B, 2-3):99–107.
24. Luenberger, D.G. 1979. *Introduction to Dynamics Systems*. New York: John Wiley & Sons.
25. Okuguchi, K. and Szidarovsky, F. 1988. A vote on global asymptotic stability of nonlinear difference equations. *Econ. Lett.*, 26:349–352.
26. Polak, E. 1971. *Computational Methods in Optimization: A Unified Approach*, New York: Academic Press.
27. Szidarovszky, F. and Yakowitz, S. 1978. *Principles and Procedures of Numerical Analysis*, New York: Plenum Press.
28. Tarayama, A. 1974. *Mathematical Economics*, Illinois: Dryden, Hinsdale.
29. Tishyadhigama, S. 1977. General convergence theorems: their relationships and applications. Ph.D. thesis. Department of Electrical Engineering and Computer Sciences. Berkley: University of California.
30. Zangwill, W.I. 1969. *Nonlinear Programming: A Unified Approach*, Englewood, Cliffs, NJ: Prentice Hall.

2 The Convergence of Iteration Sequences

This chapter examines general conditions on the convergence of sequences generated by the general algorithmic model (1.1). Based on the main result (Theorem 1.1) of Chapter 1 a general theorem will be first proven. Then special conditions will be introduced in order to find practical and easily implemented convergence conditions.

2.1 THE GENERAL CONVERGENCE THEOREM

$$x_{k+1} \in f_k(x_0, x_1, \ldots, x_k) \quad (k \geq l-1), \tag{2.1}$$

where for $k \geq l-1, f_k : X^{k+1} \to 2^X$. Here, we assume again that X is a Hausdorff topological space that satisfies the first axiom of countability, and l is a given positive integer. Furthermore, in relation (2.1), any point from the set $f_k(x_0, x_1, \ldots, x_k)$ can be accepted as the successor of x_k. Assume further that the set S of desirable points has only one element s^*. Assume that

(B_1) There is a compact set $C \subseteq X$ such that for all $k, x_k \in C$;
(B_2) Conditions $(A_1), (A_2)$, and (A_3) of Theorem 1.1 are satisfied.

The main result of this section is given as Theorem 2.1.

Theorem 2.1. Under assumptions (B_1) and $(B_2, x_k \to s^*$ as $k \to \infty$ with arbitrary initial points $x_0, x_1, \ldots, x_{l-1} \in X$.

Proof. Because C is compact, sequence $\{x_k\}$ has a convergent subsequence. From Theorem 1.1 we also know that all the limit points of this iteration sequence belong to S, which has only one point s^*. Hence, the iteration sequence has only one limit point s^*, which implies that it converges to s^*. $\qquad\square$

Remark 2.1. The theorem in this formulation can be interpreted as a global convergence result. Assume that the conditions of the theorem hold only in a neighborhood X_1, of s^* such that

$$f_k(x^{(1)}, x^{(2)}, \ldots, x^{(k+1)}) \subseteq X_1 \text{ for all } k \geq l-1 \text{ and } x^{(i)} \in X_1 (i = 1, 2, \ldots, k+1).$$

Then the theorem can be interpreted as a local convergence result.

The speed of convergence of algorithm (2.1) can be estimated as follows. Assume that

(B_3) X is a metric space with distance $d : X \times X \to R^1$;

DOI: 10.1201/9781003128915-2

(B_4)   Non-negative constants a_{ki}, $(k \geq l-1, 0 \leq i \leq k)$ exist such that if $k \geq l-1$
and $x \in f_k(x^{(0)}, x^{(1)}, \ldots, x^{(k)})$,

then

$$d(x,s^*) \leq \sum_{i=0}^{k} a_{ki}(x^{(i)}, s^*).$$

From (2.1) we have

$$\varepsilon_{k+1} \leq \sum_{i=0}^{k} a_{ki}\varepsilon_i,$$

where $\varepsilon_i = d(x_i, s^*)$ for all $i \geq 0$.

Starting from initial values $\delta_i, = \varepsilon_i (i = 0, 1, \ldots, l-1)$ consider the non-stationary
difference equation

$$\delta_{k+1} = \sum_{i=0}^{k} a_{ki}\delta_i. \tag{2.2}$$

Obviously, for all $k \geq 0, \varepsilon_k \geq \delta_k$. In order to obtain a direct expression for δ_k, and
therefore the same for the error bound of $x_k (k > l-1)$, introduce the following addi-
tional notation:

$$d_k = (\delta_0, \delta_1, \ldots, \delta_k)^T$$

$$A_k = \begin{pmatrix} 1 & & & & \\ & 1 & & & \\ & & \cdot & & \\ & & & \cdot & 1 \\ a_{k0} & \cdot & \cdot & \cdot & a_{kk} \end{pmatrix}, \text{ and } a_k^T = (a_{k0}, a_{k1}, \ldots, a_{kk}).$$

Then, from (2.2),

$$d_{k+1} = A_k d_k,$$

and hence, finite induction shows that for all $k \geq 1$,

$$d_k = A_{k-1} A_{k-2} \ldots A_{l-1} d_{l-1}.$$

Note that the components of d_{l-1} are the errors of the initial approximations
$x_0, x_1, \ldots, x_{l-1}$. From (2.2) we have

$$d_{k+1} = a_k^T d_k = (a_k^T A_{k-1} A_{k-2} \ldots A_{l-1}) d_{l-1} = b_k^T d_{l-1}$$

with

$$b_k^T = a_k^T A_{k-1} A_{k-2} \ldots A_{l-1}$$

being an l-dimensional row vector. Finally, introducing the notation

$$b_k^T = (b_{k0}, b_{k1}, b_{k,l-1}),$$

the definition of the numbers δ_k and relation (2.2) imply Theorem 2.2.

Theorem 2.2. Under assumptions (B_3) and (B_4),

$$d(x_{k+l}, s^*) \leq \sum_{i=0}^{l-1} b_{ki} d(x_i, s^*)(k \geq l - 1). \tag{2.3}$$

Corollary 2.1. If for all $i = 0, 1, \ldots, l - 1, b_{ki} \to 0$ as $k \to \infty$, the iteration sequence $\{x_k\}$ generated by algorithm (2.1) converges to s^*. Hence, in this case conditions (B_1) and (B_2) are not needed to establish convergence.

The conditions of Theorem 2.1 are usually difficult to verify in practical cases. Therefore, the next section provides a relaxation of these conditions in order to obtain sufficient convergence conditions which can be easily verified in practical cases.

2.2 CONVERGENCE OF L-STEP METHODS

In this section, the l-step iteration process of the form

$$x_{k+1} \in f_k(x_{k-l+1}, x_{k-l+2}, \ldots, x_k) \tag{2.4}$$

is discussed, where $l \geq 1$ is a given integer, and for all $k, f_k : X^1 \to 2^x$. Assume again that the set S of desirable points has only one element s^*.

Definition 2.1. A function $V : X^1 \to R^1$ is called the Liapunov function of process (2.4), if for arbitrary $x(i) \in X(i = 1, 2, \ldots, l, x(l) \neq s^*)$ and $y \in f_k(x^{(1)}, x^{(2)}, \ldots, x^{(l)})(k \geq l - 1)$,

$$V(x^{(2)}, \ldots, x^{(1)}, y) < V(x^{(1)}, x^{(2)}, \ldots, x^{(1)}). \tag{2.5}$$

Definition 2.2. The Liapunov function V is called closed if it is defined on $\overline{X}_1$, where $\overline{X}_1$ is the closure of X. Furthermore, if $k_i \to \infty, x_i^{(j)} \to x^{(j)}$ as $i \to \infty, (x_i^{(j)}$ for $i \geq 0$ and $j = 1, 2, \ldots, l$ such that $x^{(1)*} \neq s^*)$ and $y_i \in f_{ki}(x^{(1)}, \ldots, x_i^{(1)})$ $(i \geq 0)$ such that $y_i - y^*$ as $i \to \infty$, then

$$V(x^{(2)*}, \ldots, x^{(1)*}, y^*) < V(x^{(1)*}, \ldots, x^{(1)*}). \tag{2.6}$$

Assume that the following conditions hold:

(C_1) For all $k \geq l - 1, f_k(x^{(1)}, \ldots, x^{(l-1)}, s^*) = \{s^*\}$ with arbitrary $x^{(1)}s, \ldots, x^{(l-1)} \in X$;

(C_2) Process (2.4) has a continuous, closed Liapunov function;

(C_3) X is compact

Theorem 2.3. Under assumptions $(C_1), (C_2)$, and $(C_3), x_k \to s^*$ as $k \to \infty$.

Proof. Note first that this process is equivalent to the single step method (1.3), where set X is replaced by $\hat{X} = X^1$, and the set of desirable points is now $\hat{S} = S^1$. Select function c as the Liapunov function V.

We can now easily verify that the conditions of Theorem 2.1 are satisfied, which implies the convergence of the iteration sequence $\{x_k\}$.

Assumption (A_1) follows from (C_3) and the continuity of V. Condition (C_1) and the monotonicity of V imply assumption (A_2). Assumption (A_3) is the consequence of condition (C_2) and relation (2.6). $\square$

Remark 2.2. rem1 Assumption (C_3) can be weakened as follows:

$(C_3)'$ For all $x \in X \setminus S$, there is a compact neighborhood $U \subseteq X$ of x.

Here, we assume that $s^* \in X$, and condition (C_1) is required only if $s^* \in X$.

Remark 2.3. rem2 Assumption $s^* \in X$ is needed in order to obtain s^* as the limit of sequence from X. Assumption (C_1) guarantees that if any iteration step results in the solution s^*, then the process remains at the solution. We may also show that the existence of the Liapunov function is not a very strong assumption. Assume that X is a metric space, and $f : X \to X$. Consider the special iteration process $x_{k+1} = f(x_k)$ and assume that starting from an arbitrary initial point, sequence $\{x_k\}$ converges to the solution s^* of equation $x = f(x)$. Let $V : X \to R^1$ be constructed as follows. For any $x \in X$ select the initial approximation $x_0 = x$, and consider the iteration sequence $x_{k+1} = f(x_k)(k \geq 0)$, and define $V(x) = \max\{d(x_k, s^*), k \geq 0\}$, where d is the distance.

Obviously, $V(f(x)) \geq V(x)$ for all $x \in X$. The need for the continuity-type assumptions in (C_2) is understandable, because without certain continuity assumptions no convergence can be established. Assumption (C_3) requires that the entire sequence $\{x_k\}$ is contained in a compact set. This condition is necessarily satisfied, for example, if X is in a finite dimensional Euclidean space, and either X is bounded or if for every $K \geq 0$ a $Q \geq 0$ exists such that $t^{(1)}, \ldots, t^{(l)} \in X$ and $\|t^{(j)}\| > Q$ (for at least one index j) imply that $V(t^{(1)}, \ldots, t^{(l)}) > K$. In the case of single-step processes, (i.e., if $l = 1$) this last condition can be reformulated as $V(x) \to \infty$ as $\|x\| \to \infty, x \in X$.

Assume next that the iteration process is stationary; i.e., mappings f_k do not depend on k. Replace condition (C_2) by the following pair of conditions:

$(C_2)'$ The process has a continuous Liapunov-function;
$(C_2)''$ Mapping f is closed on X; i.e., if $x_i^{(j)} \to x^{(j)*}$ as $i \to \infty (j = 1, 2, \ldots, l)$ and $\in f(x_i^{(1)}, \ldots, x_i^{(l)})$ such that $y_i \to y^*$, then $y^* \in f(x^{(1)*}, \ldots, x^{(l)*})$.

Theorem 2.4. If process (2.1) is stationary and conditions $(C_1), (C_2)', (C_2)''$, and (C_3) hold, then $x_k \to s^*$ as $k \to \infty$.

Remark 2.4. This result in the special case of $l = 1$ can be considered the discrete-time counterpart of the famous stability theorem of Uzawa [25].

Remark 2.5. Assume that the process is non-stationary, and for all $k \geq l-1$, mapping f_k is closed, and the iteration sequence converges to s^*. Then, for all $k \geq l-1, s^* \in f_k(s^*, \ldots, s^*)$. Hence, s^* is a common fixed point of mappings f_k.

The speed of convergence of process (2.4) is examined next. Two results are introduced. The first is based on the general result presented in Section 2.1, and the second is based on special properties of the Liapunov function.

Note first that in the case of an l-step process assumption (B_4) is modified as

(C_4)　Non-negative constants a_{ki} $(k \geq l-1, k-l+1 \leq i \leq k)$ exist so that for all $k \geq l-1$ and $x \in f_k(x^{(1)}, \ldots, x^{(1)}), (x^{(1)}, \ldots, x^{(l)}) \in X$ are arbitrary),

$$d(x, s^*) \leq \sum_{i=1}^{l} a_{k,k-l+i} d(x^{(i)}, s^*).$$

Then Theorem 2.2 remains valid with the specification that $a_{k,i} = 0$ for all $i \leq k-1$.

In the case of a stationary process constants a_{k+1} do not depend on k. If we introduce the notation $a_i = \bar{a}_{k,k-l+i}$, then (2.2) reduces to

$$\delta_{k+1} = \sum_{i=1}^{l} \bar{a}_i \delta_{k-l+i}. \tag{2.7}$$

Observe that sequence $\{\delta_4\}$ is the solution of this 1th order linear difference equation. Note first that the characteristic polynomial of this equation is

$$\varphi(\lambda) = \lambda^l - \bar{a}_1 \lambda^{l-1} - \bar{a}_{l-1}^{l-2} - \cdots - \bar{a}_2 \lambda - \bar{a}_1.$$

Assume that the roots of φ are $\lambda_1, \lambda_2, \ldots, \lambda_R$ with multiplicities $m_1, m_2, \ldots, m_R$. The general solution of equation (2.7) is given as

$$\delta_{k+1} = \sum_{r=1}^{R} \sum_{s=0}^{m_r-1} c_{rs} k^s \lambda_r^k,$$

where the coefficients c_{rs} are obtained by solving the initial value equations

$$\sum_{r=1}^{R} \sum_{s=0}^{m_r-1} c_{rs} i^s \lambda_r^i = d(x_{i-1}, s^*)(i = 1, 2, \ldots, l).$$

Hence, we proved the following theorem.

Theorem 2.5. Under assumption (C_4) for stationary processes,

$$d(x_{k+1}, s^*) \leq \sum_{r=1}^{R} \sum_{s=0}^{m_r-1} c_{rs} k^s \lambda_r^k (k \geq l-1). \tag{2.8}$$

Corollary 2.2. If for all r, $r = 1, 2, \ldots, R$, $|\lambda_r| < 1$, then $x_k \to s^*$ as $k \to \infty$.

In the rest of the section the speed of convergence of process (2.4) is estimated based on some special properties of the Liapunov function.

Assume now that

(C_5)   Constants $a_i, b_i (i = 1, 2, \ldots, l), a_i > 0$ exists such that

$$\sum i = 1^l a_i d(x^{(i)}, s^*) \le V(x^{(1)}, \ldots, x^{(l)}) \le \sum_{i=1}^{l} b_i d(x^{(i)}, s^*) \text{ for all } x^{(i)} \in x(i = 1, 2, \ldots, l).$$

Theorem 2.6 holds.

Theorem 2.6. Assume that process (2.4) does not terminate in finite steps and has a Liapunov function V, which satisfies condition (C_5). Then for all $k \ge l - 1$,

$$d(x_{k+1}, s^*) \le a_l^{-1} \sum_{i=1}^{l} (b_i - a_{i-1}) d(x_{k-1+i}, s^*) \quad (a_0 = 0).$$

Proof. Condition (C_5) implies that

$$\sum_{i=1}^{l} a_i d(x_{k-l+i+1}, s^*) \le V(x_{k+2-l}, \ldots, x_{k+1})$$

$$\le V(x_{k+1-i}, \ldots, x_k) \le b_i d(x_{k-1+1}, s^*).$$

The assertion is a simple consequence of this inequality.   □

Corollary 2.3. Introduce the notation $\bar{a}_i, = (b_i - a_{i-1})/a, (i = 1, 2, \ldots, l)$, and let sequence $\{\delta_k\}$ denote the solution of difference equation (2.7) with initial conditions of: $= d(x_{i-1}, s^*)(i = 1, 2, \ldots, l)$. Obviously, $d(x_k, s^*) \le \delta_k$ for all $k \ge l - 1$, and with the above coefficients $\bar{a}_i$, inequality (2.7) holds, and thus Theorem 2.5 remains true.

2.3   CONVERGENCE OF SINGLE-STEP METHODS

In this section, single-step processes generated by point-to-set mappings are examined. For the sake of simplicity, we assume that X is a subset of a Banach space B and contains the origin. The iteration process now has the form

$$x_{k+1} \in f_k(x_k) \quad (k \ge 0), \tag{2.9}$$

where $f_k : X \to 2^X$. It is also assumed that 0 is in X and $s^* = 0$. We may make the latter assumption without losing generality, as any solution s* can be transformed into zero by introducing the transformed mappings

$$g_k(x) = \{y - s^*/y \in f_k(x + s^*)\}.$$

It is also assumed that for all k, $f_k(0) = \{0\}$. We begin our analysis with the following useful result.

Theorem 2.7. Assume that X is compact, and there is a real valued continuous function $\alpha : X \backslash \{0\} \to [0,1)$ such that

$$\|y\| \leq \alpha(x)\|x\| \tag{2.10}$$

for all $k \geq 0, x \neq 0$, and $y \in f_k(x)$. Then the iteration sequence (2.9) converges to 0 as $k \to \infty$.

Proof. We now verify that all conditions of Theorem 2.3 are satisfied with the Liapunov function $V(x) = \|x\|$ and $s^* = 0$. Note that (C_1) and (C_3) obviously hold, and condition (C_2) is implied by the facts that the norm is continuous, and $\alpha(x) < 1$ for $x \neq 0$. $\qquad\square$

Remark 2.6. If (2.10) is replaced by the weaker assumption that

$$\|y\| < \|x\|$$

for all $k \geq 0, x \neq 0$, and $y \in f_k(x)$, the result may not hold, as the following example shows.

Example 2.1. Select $B = R^1$ and $X = [0,2]$. For all $k \geq 0$, define the point-to-point iteration mapping

$$f_k(x) = [(k+2)^2 - 1](k+2)^{-2}.$$

If the initial point is chosen as $x_0 = 2$, then finite induction shows that

$$x_k = 1 + (k+1)^{-1} \to 1 \neq 0 \text{ as } k \to \infty.$$

Furthermore, for all $k \geq 0$ and $x \neq 0$,

$$|f_k(x)| < |x|.$$

Corollary 2.4. Recursion (2.9) and inequality (2.10) imply that for $k \geq 0$,

$$\|x_{k+1}\| \leq \alpha(x_k)\|x_k\|,$$

and, therefore finite induction shows that

$$\|x_{k+1}\| \leq \alpha(x_k)\alpha(x_{k-1})\ldots a(x_0)\|x_0\|. \tag{2.11}$$

As a special case assume that $\alpha(x) \leq q < 1$ for all $0 \neq x \in X$. Then, for all $k \geq 0$,

$$\|x_{k+1}\| \leq q^{k+1}\|x_0\|, \tag{2.12}$$

which shows the linear convergence of the process in this special case.

Relation (2.11) serves as the error formula of the algorithm. In addition, it has the following consequence. Assume that (2.10) holds for all $0 \neq x \in X$. Furthermore, $\alpha(x_k)\alpha(x_{k-1}),\ldots,\alpha(x_0) \to 0$ as $k \to \infty$. Then, $x_k \to 0$ for $k \to \infty$. Hence, in this case we may drop the assumptions that $\alpha(x) \in [0,1)(0 \neq x \in X)$ and X is compact.

An alternative approach to Theorem 2.7 is based on the assumption that a function $h : (0,\infty) \to R$ exists such that

$$\|y\| \leq h(r)\|x\| \tag{2.13}$$

for all $k \geq 0, r > 0, \|x\| \leq r, x \in X$ and $y \in f_k(x)$. In this case it is easy to verify that for all k,

$$\|x_k\| \leq q_k,$$

where q_k is the solution of the non-linear difference equation

$$q_{k+1} = h(q_k)q_k, q_0 = \|x_0\|.$$

Hence, the convergence analysis of iteration algorithms defined in a Banach space is reduced to the examination of the solution of a special scalar non-linear difference equation. In deriving further practical convergence conditions, we use the following special result.

Lemma 2.1. Assume that X is convex, and function $h : X \to X$ satisfies the following condition:

$$\|h(x) - h'(x)\| \leq \alpha(\xi)\|x - x'\| \tag{2.14}$$

for all $x, x' \in X$, where ξ is a point on the linear segment between x and x'. Further, $\alpha : X \to R^l$ is a real valued function such that for all fixed x and $x' \in X, \alpha(x' + t(x - x'))$ as the function of the parameter t is Riemann integrable on $[0,1]$. Then, for all x and $x' \in X$,

$$\|h(x) - h(x')\| \leq \int_0^1 \alpha(x' + t(x - x'))dt\|x - x'\|. \tag{2.15}$$

Proof. Let $x, x' \in X$ and define $t_i, = i/N (i = 0,1,2,\ldots,N)$, where N is a positive integer. Then from (2.14),

$$\|h(x) - h(x')\| \leq \sum_{i=1}^{N} \|h(x' + t_i(x - x'))h(x' + t_{i-1}(x - x'))\|$$

$$\leq \sum_{i=1}^{N} \alpha(x' + \tau_i(x - x'))\|t_i - t_{i-1}(x - x')\|,$$

where $\tau_i \in [t_{i-1}, t_i]$, which implies that

$$\|h(x) - h(x')\| \leq \left\{ \sum_{i=1}^{N} \alpha(x' + \tau_i(x - x')(t_i + t_{i-1})) \right\} \|x - x'\|.$$

Observe that the first factor is a Riemann sum of the integral

$$\int_0^1 \alpha(x' + t(x - x'))dt$$

which converges to the integral. Let $N \to \infty$ in the above inequality to obtain the result. □

Remark 2.7. If function α is continuous, then $\alpha(x' + t(x-x'))$ is continuous in t. Therefore, it is Riemann integrable.

Assume next that maps f_k are point-to-point, and process (2.9) satisfies the following conditions:

(D_1) $f_k(0) = 0$ for $k \geq 0$;
(D_2) For all $k \geq 0$,
$$\|f_k(x) - f_k(x')\| \leq \alpha(\xi_k)\|x - x'\| \tag{2.16}$$
for all $x, x' \in X$, where $\alpha : X \to R^l$ is a continuous function, and ξ_k is a point on the linear segment connecting x and x';
(D_3) $\alpha(x) \in [0,1)$ for all $0 \neq x \in X$;
(D_4) X is compact and convex.

Theorem 2.8. Under the above conditions, $x_k \to 0$ as $k \to \infty$.

Proof. Let $0 \neq x \in X$. Relation (2.15) then implies that for all k,
$$\|f_k(x)\| \leq \int_0^1 \alpha(tx)dt\|x\|, \tag{2.17}$$
where we have selected $x' = 0$. Break the integral into two parts to obtain
$$\|f_k(x)\| \leq \left\{ \int_0^\delta \alpha(tx)dt + \int_\delta^1 \alpha(tx)dt \right\} \|x\|.$$

Because a is continuous, $\alpha(0) \leq 1$, and because the interval $[\delta, 1]$ is compact, $\delta(t_x) \leq \beta_\delta(x) < 1$ for all $\delta \geq t \geq 1$, where $\beta_\delta : X\backslash\{0\} \to R^l$ is the real valued function defined as
$$\beta_\delta(x) = \max_{\delta \leq t \leq l}\{a(tx)\}.$$
Therefore,
$$\|f_k(x)\| \leq \{\delta + (1-\delta)\beta_\delta\}\|x\| = \tau_\delta(x)\|x\|,$$
where $\tau_\delta : X\backslash\{0\} \to R^l$ is a continuous function such that for all $x \neq 0$, $\tau_\delta(x) \in [0,1)$.

Hence, the conditions of Theorem 2.7 are satisfied with $\alpha = \tau_\delta$ which implies the assertion. □

Remark 2.8. Replace (2.16) by the following weaker condition. Assume that for all $k \geq 0$ and $x, x' \in X$,
$$\|f_k(x) - f_k(x')\| \leq \alpha_k(\xi_k)\|x - x'\|, \tag{2.18}$$

where for $k \geq 0, \alpha_k : X \to R^l$ is a continuous function, ξ_k is a point on the linear segment connecting x and x', and $\alpha_k(x) \in [0,1)$ for all $k \geq 0$ and $0 \neq x \in X$. Then the assertion of the theorem may not hold. Such a case is illustrated by Example 2.1.

Corollary 2.5. Recursion (2.9) and inequality (2.17) imply that for $k \geq 0$,

$$\|x_{k+1} = \|f_k(x_k)\| \leq \overline{\alpha}(x_k)\|x_k\|, \|$$

where

$$\overline{\alpha}(x) = \int\limits_0^1 \alpha(tx)dt.$$

Hence, by replacing $\alpha(x)$ by $\alpha(x)$, the Corollary of Theorem 2.7 remains valid.

In the previous results no differentiability of functions f_k was assumed. In the special case of Fréchet differentiable functions f_k, the above theorems can be reduced to very practical convergence conditions. These results are presented in Section 2.4.

2.4 CONVERGENCE OF SINGLE-STEP METHODS WITH DIFFERENTIABLE ITERATION FUNCTIONS

Assume now that B is a Banach space, $X \subseteq B$, and functions $f_k : X \to X$ are continuously differentiable on X. It is also assumed that all closed and bounded subsets of X are compact, X is convex, $0 \in X$, and that 0 is a common fixed point of functions f_k. In this special case the following result holds.

Theorem 2.9. Let $f_k'(x)$ denote the Fréchet derivative of f_k at x. Assume that for all $k \geq 0$,

$$\|f_k'(x)\| \leq \beta(x), \tag{2.19}$$

where $\beta : X \to R^l$ is a continuous function such that for all $x \neq 0, \beta(x) \in [0,1)$. Then $x_k \to 0$ as $k \to \infty$.

Proof. Select

$$X_0 = \{x|x \in X \text{ and } \|x\| \leq \|x_0\|\},$$

then X_0 is compact. Select furthermore $\alpha = \beta$. We can easily verify that all conditions of Theorem 2.8 are satisfied with X_0 replacing X, which implies the assertion. Assumptions (D_1) and (D_3) are obviously satisfied. Assumption (D_2) follows from the mean value theorem of derivatives and then from the fact that the linear segment between x and x' is compact and function a is continuous. In order to verify assumption (D_4) we have to show that $x_k \in X_0$ for all $k \geq 0$. From the beginning of the proof of Theorem 2.8 we conclude that for all $0 \neq x \in X, \|f_k(x)\| \leq \|x\|$. Finite induction implies that for all $k \geq 0, \|x_k\| \leq \|x_0\|$, therefore, $x_k \in X_0$ for all $k \geq 0$. Hence, the proof is completed. $\qquad\square$

Remark 2.9. If (2.19) is replaced by the weaker assumption that for all $k \geq 0$ and $x \neq 0$,

$$\|f_k(x)\| \leq 1,$$

the result may not hold, as the case of Example 2.1 illustrates. However, if f_k does not depend on k; i.e., when $f_k = f$, the condition

$$\|f'(x)\| < 1 \text{ for all } x \neq 0,$$

implies that $x_k \to 0$ as $k \to \infty$. To prove this assertion select $\beta(x) = \|f'(x)\|$. Note that this special result was first introduced by Wu and Brown [26].

Corollary 2.6. Note that the Corollary of Theorem 2.8 remains valid with $\alpha(x) = \beta(x)$.

Remark 2.10. Assume that no assumption is made on the derivatives at the fixed point 0.

The conditions of Theorem 2.9 will now be weakened. Assume again that $X \subseteq B$, where B is a Banach space. Furthermore, for all $k \geq 0, f_k$ is Fréchet differentiable at 0, and $\|f'_k(0)\| \leq 1$. As the following example shows, these conditions do not imply even local convergence of the algorithm.

Example 2.2. Select $X = R^1$, and for $k \geq 0$ define

$$f_k(x) = \frac{(k+1)(k+4)x}{(k+2)(k+3)}.$$

It is easy to verify that for all $k \geq 0$,

$$0 \leq f'_k(0) = \frac{(k+1)(k+4)}{(k+2)(k+3)} < 1.$$

If $x_0 \neq 0$ is any initial approximation, then finite induction shows that

$$x_k = \frac{k+3}{3(k+1)}x_0 \to \frac{1}{3}x_0 \neq 0 \text{ as } k \to \infty.$$

Assume now that the process is stationary. Then, the following result holds.

Theorem 2.10. Assume that $f_k = f(k \geq 0)$, 0 is in the interior of X, and f is Fréchet differentiable at 0; furthermore, $\|f'(0) < 1\|$ Then there is a neighborhood U of 0 such that $x_0 \in U$ implies that $x_k \to 0$ as $k \to \infty$.

Proof. Since f is differentiable at 0, we can write $f(x) = L(x) + R(x)$, where L is a bounded linear mapping of X into itself and $\lim \|R(x)\| \|x\|^{-1} = 0$ as $x \to 0$. By assumption, $\|L\| < 1$. Select a number $b > 0$ such that $\|L\| < b < 1$. Then a $d > 0$ exists such that

$$\|R(x)\| < (1-b)\|x\| \text{ for } \|x\| < d.$$

Let $U = \{x \in X \| x \| < d\}$. We shall now prove that U has the required properties. Using the triangle inequality we can easily show that

$$\| f(x) \| < e \| x \|, \text{ if } x \in U,$$

where $e = \| L \| + 1 - b$. Relation $0 < e < 1$ implies that if $x_0 \in U$, the entire sequence of iterates x_k is also contained in U. Notice finally that all conditions of the Corollary of Theorem 2.7 are satisfied with $q = e$, which completes the proof. $\square$

Remark 2.11. Assumption $\| f'(0) \| < 1$ can be weakened by assuming only that the spectral radius of $f'(0)$ is less than one. In this case $\| f'(0)^N \| < 1$ with some $N > 1$, and then apply the theorem for the function

$$f^N(x) = (f \circ f \circ \cdots \circ f)(x).$$

Remark 2.12. Note that no differentiability is assumed for $x \neq 0$.

Remark 2.13. When $X = R^N$, our results can be reduced to those obtained by Ostrowskii [20], Argyros [1–11], Kitchen [17], Rail [21], and Rheinboldt [22].

Assume now that $B = R^N$, and function $f : B \to B$ is continuously differentiable. Let f' denote the Jacobian of f.

Theorem 2.11. If there exists a norm such that

(E_1) $\| f'(0)(x) \| < \| x \|$ for all $x \neq 0$;
(E_2) $\| f'(x)x \| < \| x \|$ when $\| f(x) \| = \| x \| \neq 0$, then $x_n \to 0$ as $k \to \infty$.

Proof. From (E_1), we know that $\| f'(0) \| < 1$. Since f' is continuous, an $r > 0$ exists such that $\| f'(x) \| < 1$, for all x such that $\| x \| < r$. The mean value theorem of derivatives implies that for all such x, $\| f(x) \| < \| x \|$. If this inequality holds for all $x \in B$, then Theorem 2.7 with $\alpha(x) = \frac{\| f(x) \|}{\| x \|}$ implies the assertion. Assume that for some x, $\| f(x) \| > \| x \|$, and define

$$r^* = \min\{\| x \| \| f(x) \| > \| x \|\} \geq r.$$

If for all vectors $\| x \| = r^*, \| f(x) \| > \| x \|$, then the continuity of f implies that the value of r^* can be further reduced. Therefore, at least one x^* is such that

$$\| x^* \| = r^* \text{ and } \| f(x^*) \| = \| x^* \|.$$

Because f is differentiable, we know that for all $\varepsilon > 0$ and sufficiently small $\lambda \in (0,1)$,

$$\| f((1-\lambda)x^*) - f(x^*) - \lambda f'(x^*)x^* \| < \varepsilon \lambda \| x^* \|$$

see Ortega and Rheinboldt [19], which together with (E_2) implies that

$$\| f((1-\lambda)x^*) - f(x^*) - \lambda (\| f'(x^*)x^* \| + \varepsilon \| x^* \|) \| < \lambda \| x^* \| \qquad (2.20)$$

where we selected

$$\varepsilon < \frac{\|x^*\| - \|f(x^*)\| x^*}{\|x^*\|}.$$

Relation (2.20) implies that

$$\|(1-\lambda)x^*\| < \|f((1-\lambda)x^*)\|$$

since $\|f(x^*)\| = \|x^*\|$. Notice that the latter inequality contradicts the selection of r^*, which completes the proof. $\qquad\square$

Remark 2.14. From the proof, we conclude that the selection $\alpha(x) = \|f(x)\|/\|x\|$ satisfies the conditions of Theorem 2.7, and therefore the Corollary of Theorem 2.7 can be applied to estimate the speed of convergence.

Remark 2.15. In the above results the special Liapunov function $V(x) = \|x\|$ was used, where $\|.\|$ is some vector norm. Select the more general Liapunov function $V(x) = \|Px\|$, where P is an $N \times N$ constant non-singular matrix. For the sake of simplicity, we assume that $f_k = f$ for all $k \geq 0$. Then in Theorems 2.7 and 2.11 conditions (2.10) and (2.19) can be substituted by the modified relations

$$\|Pf(x)\| \leq \|Px\|$$

and

$$\|Pf'(x)u\| < \|Pu\| \text{ for all } u \neq 0.$$

If one selects the Euclidean norm $\|x\| = x^T x$, then these conditions are equivalent to the relations

$$f^T(x)P^T Pf(x) < x^T P^T \qquad (2.21)$$

and

$$u^T f'(x)^T P^T Pf'(x)u < u^T P^T Pu. \qquad (2.22)$$

Note that (2.21) holds for all $u \neq 0$ if and only if matrix $f'(x)^T Pf'(x) - P^T P$ is negative definite. This condition has been derived in Fujimoto [16] and it is a generalization of Theorem 1.3.2.3 of Okuguchi [18].

We finally show a further generalization of Theorem 2.11

Theorem 2.12. Assume that a three-times continuously differentiable function V: $R^N \to R_+$ exists to the extent that

$(E_1)'$ $V(x) > 0$ for $x \neq 0, V(0) = 0$ and $\lim_{\|x\| \to \infty} V(x) = \infty$;
$(E_2)'$ If $V(f(x)) = V(x)$ and $x \neq 0$, then $V'(f(x))f'(x)x - V'(x)x < 0$;
$(E_3)'$ $V'(0)f'(0)x - V'(0)x < 0$ if $V'(0) \neq 0$;
(E_4) If $V'(0) = 0$, then for $x \neq 0, x^T f'(0)^T V''(0)f'(0)x - x^T V''(0) < 0$.

Here V' and V'' denote the gradient and Hessian of V. Then $x_k \to 0$ as $k \to \infty$.

Proof. We will prove that for all $x \neq 0$,

$$V(f(x)) < V(x),$$

and therefore Theorem 2.3 implies the assertion. Select an $x \neq 0$. Define the function

$$g(\lambda) = \lambda [V(f(\lambda x)) - V(\lambda x)]$$

for all $\lambda \in [0,2]$. Obviously, $g(0) = 0$, and

$$g'(\lambda) = V(f(\lambda x)) - V(\lambda x) + \lambda [V'(f(\lambda x))f'(\lambda x)x - V'(\lambda x)x].$$

Assumption $(E_2)'$ implies that $g'(\lambda) < 0$ if $V(f(\lambda x)) - V(\lambda x) = 0$. Furthermore, $g'(0) = 0$, and by assumption (E_3),

$$g''(0) = 2[V'(0)f'(0)x - V'(0)x] \text{ if } V'(0) \neq 0.$$

If $V'(0) = 0$, then $g''(0) = 0$ and by (E_4),

$$g'''(0) - 3[x^T f'(0)^T V''(0)f'(0)x - x^T V''(0)x] < 0.$$

The above relations imply that for small $\lambda, g(\lambda) < 0$, and if for some $\lambda > 0, g(\lambda) = 0$, then $g'(\lambda) < 0$. Therefore,

$$g(1) = V(f(x)) - V(x) < 0,$$

which completes the proof. □

Remark 2.16. We will now derive Theorem 2.11 with the norm $\|.\|_2$ from the above result. Select $V(x) = x^T x$, then $V'(x) = 2x$ and $V''(x) = 2I$. Condition $(E_1)'$ is trivially satisfied, and $(E_2)'$ can be shown as follows. Notice that if $\|f(x)\|_2^2 = \|x\|_2^2$ and $x \neq 0$, then $\|f(x)x\| < \|x\|$, and therefore the Cauchy-Schwartz inequality implies that

$$f(x)^T f'(x)x \leq \|x\|_2 \cdot \|f'(x)x\|_2 < \|x\|_2^2$$

Assumption (E_3) also holds, since $V'(0) = 0$. Finally, notice that (E_1) implies (E_4).

2.5   APPLICATIONS

Case Study 2.1. In study number one, consider the continuous linear system

$$\dot{x} = Ax + b,$$

with constant $n \times n$ matrix A and n-vector b, where for the sake of simplicity it is assumed that the input is unity at all times. The equilibrium s^* of this system is the solution of the linear equation $Ax + b = 0$ (see, for example, Szidarovszky and Bahill [23]). Assume that matrix A is strictly diagonally dominant; i.e., for all $i = 1, 2, \ldots, n$,

$$|a_{ij}| > \sum_{j \neq i} |a_{ij}| \qquad (2.23)$$

Rewrite the linear equation as

$$x^{(i)} = -\sum_{j \neq i} \frac{a_{ij}}{a_{ii}} x^{(i)} - \frac{b_i}{a_{ii}} \quad (i = 1, 2, \ldots, n),$$

where the ith component of x is denoted by $x^{(i)}$ $(i = 1, 2, \ldots, n)$. Consider now the iteration scheme

$$x_{k+1}^{(i)} = \sum_{j \neq i} \frac{a_{ij}}{a_{ii}} x_k^{(i)} - \frac{b_i}{a_{ii}} \quad (i = 1, 2, \ldots, n),$$

which is the Jacobi iteration method introduced in Section 1.3. Notice that it is a single-step process having the iteration function with components

$$f^{(i)}(x) = -\sum_{j \neq i} \frac{a_{ij}}{a_{ii}} x^{(j)} - \frac{b_i}{a_{ii}}. \tag{2.24}$$

Select the l_∞-norm of real vectors. Then for all x,

$$\|g(x)\| = \|f(x + s^*) - s^*\| \leq \alpha(x)\|x\|,$$

where for all x,

$$\alpha(x) = \max_i \sum_{j \neq i} \left| \frac{a_{ij}}{a_{ii}} \right| < 1.$$

Thus, all conditions of Theorem 2.7 are satisfied, and the above iteration method converges to the equilibrium of the system. Because the equilibrium is usually nonzero, we had to use the transformation presented at the beginning of the Section 2.3 to reduce the fixed point problem to a similar problem with zero fixed point.

The convergence of the above process can be sped up by using the Seidel-type iteration process, which was introduced in Section 1.3. In this case the modification is as follows:

$$x_{k+1}^{(i)} = -\sum_{j < i} \frac{a_{ij}}{a_{ii}} x_{k+1}^{(j)} - \sum_{j > i} \frac{a_{ij}}{a_{ii}} x_k^{(i)} - \frac{b_i}{a_{ii}} \quad (i = 1, 2, \ldots, n),$$

In many economical applications matrix A has a special block structure:

$$A = \begin{pmatrix} A^{11} & A^{12} & \ldots & A^{1m} \\ A^{21} & A^{22} & \ldots & A^{2m} \\ \ldots & \ldots & \ldots & \ldots \\ A^{m1} & A^{m2} & \ldots & A^{mm} \end{pmatrix},$$

where the blocks are square matrices of the same size and the diagonal blocks are nonsingular, and for all i,

$$\sum_{j \neq i} \|A^{(ij)}\| < \frac{1}{\|A^{ii} - 1\|}.$$

If vectors x and b are divided into blocks according to A, the ith block equation can be written as

$$A^{(ii)}x^{(i)} + \sum_{j \neq i} A^{(ij)}x^{(j)} + b^{(i)} = 0.$$

Solve this equation for $x^{(i)}$,

$$x^{(i)} = -A^{(ii)-1} \sum_{j \neq i} A^{(ij)}x^{(j)} - A^{(ii)-1}b^{(i)}.$$

The corresponding iteration scheme is given as follows:

$$x_{k+1}^{(i)} = -A^{(ii)-1} + \sum_{j \neq i} A^{(ij)}x_k^{(j)} - A^{(ii)-1}b^{(i)} \quad (i = 1, 2, 2, \dots, m).$$

Notice that this is a single-step process having the iteration function with blocks

$$f^{(i)}(x) = -A^{(ii)-1} \sum_{j \neq i} A^{(ij)}x_k^{(j)} - A^{(ii)-1}b^{(i)}.$$

Similar to the case of the iteration function (2.24), we can easily show that

$$\alpha(x) = \max_i \left\{ \|A^{(ii)-1}\| \cdot \sum_{j \neq i} \|A^{(ij)}\| < 1 \right\}.$$

Therefore, Theorem 2.7 implies the convergence of the procedure. Notice that the above norm condition on the blocks is called the block-diagonally dominance of matrix A, and in the special case of 1×1 blocks it is equivalent to condition (2.23).

Consider next a discrete linear system of the form

$$x(t+1) = Ax(t) + b,$$

where A and b are as before, and we assume again that the input is unity at all times. The equilibrium of this system is the solution of the equation

$$x = Ax + b,$$

which can be rewritten as

$$(A - I)x + b = 0.$$

The problem is mathematically the same as for continuous systems. The only difference is that matrix A must be replaced by A–I.

Notice that the global asymptotic stability of the above discrete linear system is equivalent to the convergence of the iteration procedure driven by the same equation.

Theorem 2.13. The iteration procedure $x_{k+1} = Ax_k + b$ converges to the unique solution of equation $x = Ax + b$, starting with arbitrary initial approximation if and only if all eigenvalues of A are inside the unit circle of the complex plane.

Proof. First we show that the necessary and sufficient condition for the convergence with arbitrary x_0 is that $A^t \to 0$ as $t \to \infty$. This assertion follows immediately from the identities

$$x_t = A^t x_0 + (I + A + A^2 + \cdots + A^{t-1})b$$

and

$$(I - A)(I + A + A^2 + \cdots + A^{t-1}) = I - A^t.$$

Assume first that the iteration procedure converges to the unique solution with arbitrary x_0. Then the coefficient of x_0 must converge to zero, because the solution does not depend on x_0. Assume next that $A^t \to 0$ as $t \to \infty$. Then $A^t x_0$ also converges to zero, and the second identity implies that $I + A + A^2 + \cdots + A^t \to (I - A)^{-1}$ as $t \to \infty$. Hence, $x \to (I - A)^{-1}b$, which is the unique solution of equation $x = Ax + b$.

Next we verify that $A^t \to 0$ as $t \to \infty$ if and only if all eigenvalues of A are inside the unit circle. The proof of this assertion is a simple consequence of the Jordan canonical forms of matrices. Assume that the Jordan form J of matrix A has the Jordan blocks $J_1, J_2, \ldots, J_r$, where for $j = 1, 2, \ldots, r$,

$$J_j = \begin{pmatrix} \lambda & 1 & & & \\ & \lambda & 1 & & \\ & & \cdot & \cdot & \\ & & & \cdot & \cdot \\ & & & & \cdot & 1 \\ & & & & & \lambda \end{pmatrix}$$

where λ is an eigenvalue of A. Notice that each eigenvalue of A can be found in at least one Jordan block, and

$$J_j^t = \begin{pmatrix} \lambda^t & \binom{t}{1}\lambda^{t-1} & \cdot & \cdot & \cdot & \binom{t}{s-1}\lambda^{t-s+1} \\ & \lambda^t & & \cdot & \cdot & \cdot & \binom{t}{s-2}\lambda^{t-s+2} \\ & & \cdot & & & \\ & & & \cdot & & \\ & & & & \cdot & \\ & & & & & \lambda^t \end{pmatrix}$$

where $s \times s$ is the size of J_j. Obviously, $A^t \to 0$ if and only if $J_t \to 0$, and this limit relation holds if and only if for all blocks, $J_j^t \to 0$. This last relation for all j is equivalent to the condition that for all eigenvalues λ of A, $|\lambda| < 1$. Thus, the proof is completed.   □

Corollary 2.7. If for some matrix norm, $\|A\| < 1$, then all eigenvalues are inside the unit circle; i.e., the iteration method is convergent. The assertion is the simple consequence of the fact that for all eigenvalues λ of A, $|\lambda| \leq \|A\|$.

In investigating the convergence of iteration algorithms for solving linear equations, some matrix optimization problems must be considered and solved. One particular problem is discussed next. Consider the equation $Ax = b$, where A is an $n \times n$ real matrix and b is a real n-vector. For every $n \times n$ non-singular matrix M consider the splitting

$$A = M + (A - M),$$

which leads to the modified equation

$$x = (I - M^{-1}A)x + M^{-1}b.$$

The corresponding iteration scheme has the form

$$x_{k+1} = (I - M^{-1}A)x_k + M^{-1}b.$$

In order to have the fastest speed of convergence, matrix M must be selected in such a way that the absolute values of the eigenvalues of matrix $I - M^{-1}A$ are as small as possible. Obviously, the selection $M = A$ is the best choice, because in this case $I - M^{-1}A = 0$, and all eigenvalues are equal to zero. However, $M^{-1} = A^{-1}$, and the computation of vector $M^{-1}b$, which is needed to perform the iteration process, is equivalent to the original problem. Hence, this selection has no practical value. In practical applications optimal matrices from certain matrix sets are usually determined and used.

For example, assume that $M = \alpha I$, where I is the identity matrix and α is an unknown constant. Let the eigenvalues of A be denoted by $\lambda_1, \lambda_2, \ldots, \lambda_n$, the eigenvalues of $I - M^{-1}A = I - \frac{1}{\alpha}A$ are then $1 - \frac{\lambda_1}{\alpha}, \ldots, 1\frac{\lambda_n}{\alpha}$. Thus, the optimal selection is defined as

$$\alpha^* = \operatorname{argmin}_\alpha \max \left\{ \left|1 - \frac{1}{\alpha}\lambda_1\right|, \left|1 - \frac{1}{\alpha}\lambda_2\right|, \ldots, \left|1 - \frac{1}{\alpha}\lambda_n\right| \right\}.$$

Again, the application of this formula is difficult because it requires the knowledge of the eigenvalues of A.

An analogous approach is the minimization of the norm of matrix $I - M^{-1}A$. Because the objective function depends on the norm selection, the best selection will also depend on it. For example, consider again the case of $M = \alpha I$, then

$$\|I - M^{-1}A\|_\infty = \max_i \left\{ \left|1 - \frac{1}{\alpha}a_{ii}\right| + \sum_{j \neq i} \left|\frac{1}{\alpha}a_{ij}\right| \right\},$$

$$\|I - M^{-1}A\|_1 = \max_j \left\{ \left|1 - \frac{1}{\alpha}a_{jj}\right| + \sum_{i \neq j} \left|\frac{1}{\alpha}a_{ij}\right| \right\},$$

and

$$\|I - M^{-1}A\|_F = \left\{ \sum_i \left[\left(1 - \frac{1}{\alpha}a_{ii}\right)^2 + \sum_{j \neq i} \frac{1}{\alpha^2}a_{ij}^2 \right] \right\}^{\frac{1}{2}}.$$

Notice that the row norm, the column norm, and the Frobenius norm are used here. By introducing the new variable $\beta = \frac{1}{\alpha}$, the first two norms can be minimized by solving special linear programming problems, in which a new variable is introduced for each absolute value in the objective function, and one additional variable is introduced for the entire objective function. The last norm leads to an unconstrained quadratic optimization problem:

$$\min \sum_i \left[1 - \beta a_{ii}^2 + \sum_{j \neq i} \beta^2 a_{ij}^2 \right].$$

Because the objective function is convex, simple differentiation shows that at the optimal solution

$$\sum_i \left[-2(1 - \beta a_{ii})a_{ii} + 2\beta \sum_{j \neq i} a_{ij}^2 \right] = 0.$$

In other words,

$$\beta^* = \frac{\sum_i a_{ii}}{\sum_i \sum_j a_{ij}^2}.$$

is the optimal choice.

Case Study 2.2. Our second case study considers the linear input–output economic system

$$x = Ax + y, \tag{2.25}$$

where x is the gross production vector, y is the net production of an economy, and A is the technology matrix. Assume that for a given net production vector y, the corresponding gross production vector x must be determined; i.e., for given y, equation (2.25) must be solved for x. A trivial algorithm can be formulated as

$$x_{k+1} = Ax_k + y.$$

From the previous case study, we know that with arbitrary x_0 this iteration scheme converges if and only if all eigenvalues of A are inside the unit circle.

Now consider the non-linear input–output system

$$x = f(x) + y,$$

where $f : R^n \to R^n$ is a differentiable function. Assume that for all $x \in R^n$,

$$\left| \frac{df^{(i)}(x)}{dx^{(j)}} \right| \leq Q_{ij},$$

where Q_{ij} is a constant, and $f^{(i)}$ and $x^{(j)}$ denote the ith element of f and the jth element of x, respectively. Assume in addition that

$$\max_i \sum_{j=1}^{n} Q_{ij} < 1.$$

Then, as in the first case study it is easy to show that the iteration procedure

$$x_{k+1} = f(x_k) + y$$

converges to the unique solution starting with arbitrary initial approximation x_0. In the proof the vector norm $\|.\|_\infty$ is used. We note also that the selection of the vector norms $\|.\|_1$, and $\|.\|_2$ leads to the alternative sufficient convergence conditions

$$\max_j j \sum_{i=1}^{n} Q_{ij} < 1,$$

and

$$\sum_{i=i}^{n} \sum_{j=1}^{n} Q_{ij} < 1.$$

The details are left as an easy exercise.

Case Study 2.3. Consider the circuit, where the numerical data are as follows:

$$R_1 = R_2 = R_3 = 1\Omega, R_4 = R_5 = R_6 = 2\Omega, V_1 = V_2 = 1V.$$

It is well known that the unknown currents solve the linear equations

$$(R_1 + R_2 + R_4)i_1 - R_2 i_2 - R_4 i_3 = 0$$
$$- R_2 i_1 + (R_2 + R_3 + R_6)i_2 - R_3 i_3 = V_1$$
$$- R_4 i_1 - R_3 i_2 + (R_3 + R_4 + R_5)i_3 = V_2.$$

Notice that the matrix of coefficients is strictly diagonally dominant; therefore, the Jacobi iteration procedure can be used for solving these equations as was done in the first case study of this section. The transformed linear equations have the form:

$$i_1 = \frac{R_2}{R_1 + R_2 + R_4} i_2 + \frac{R_4}{R_1 + R_2 + R_4} i_3$$
$$i_2 = \frac{R_2}{R_1 + R_3 + R_6} i_1 + \frac{R_3}{R_2 + R_3 + R_6} i_3 + + \frac{V_1}{R_2 + R_3 + R_6}$$
$$i_3 = \frac{R_4}{R_3 + R_4 + R_5} i_1 + \frac{R_3}{R_3 + R_4 + R_5} i_2 + + \frac{V_2}{R_3 + R_4 + R_5}.$$

By using the numerical data they reduce to equations:

$$i_1 = \frac{1}{4}i_2 + \frac{1}{2}i_3$$

$$i_2 = \frac{1}{4}i_1 + \frac{1}{4}i_3 + \frac{1}{4}$$

$$i_3 = \frac{2}{5}i_1 + \frac{1}{5}i_2 + \frac{1}{5}$$

and the corresponding iteration scheme can be written as follows:

$$i_{1,k+1} = \frac{1}{4}i_{2,k} + \frac{1}{2}i_{3,k}$$

$$i_{2,k+1} = \frac{1}{4}i_{1,k} + \frac{1}{4}i_{3,k} + \frac{1}{4}$$

$$i_{3,k+1} = \frac{2}{5}i_{1,k} + \frac{1}{5}i_{2,k} + \frac{1}{5}.$$

A simple computer program shows that this process converges to the solution

$$i_1 \approx 0.3135859, i_2 \approx 0.4312681, \text{ and } i_3 \approx 0.4116425.$$

Hence, the currents of the three loops are determined.

The Gauss-Seidei iteration process applied to the same system of equations has the form

$$i_{1,k+1} = \frac{1}{4}i_{2,k} + \frac{1}{2}i_{3,k}$$

$$i_{2,k+1} = \frac{1}{4}i_{1,k} + \frac{1}{4}i_{3,k} + \frac{1}{4}$$

$$i_{3,k+1} = \frac{2}{5}i_{1,k} + \frac{1}{5}i_{2,k} + \frac{1}{5}.$$

A simple computer program leads to the same solution as the one obtained previously.

Case Study 2.4. A dynamic economic system is discussed in the fourth case study. Assume that n products are sold on the market, where the unit prices are $P_1, P_2, \ldots, P_n$, respectively. Assume that both the demand and supply functions are linear; i.e.,

$$d_i(P_1, \ldots, P_n) = a_{i1}P_1 + \cdots + a_{in}P_n + a_{i0}$$

and

$$s_i(P_1, \ldots, P_n) = b_{i1}P_1 + \cdots + b_{in}P_n + b_{i0}$$

are the demand and supply of product i, respectively, for $i = 1, 2, \ldots, n$. If the price of any product increases, the producers increase their production of that product in order to enjoy higher profit, and thus switch from other products to the more profitable product.

Mathematically this tendency can be modeled by assuming that

$$\frac{\partial s_i}{\partial P_i} > 0, \text{ and } \frac{\partial s_i}{\partial P_j} \leq 0 \text{ for } j \neq i.$$

Similarly, the demand of any product with increasing price will decrease, as the buyers will switch to other products in order to spend less money:

$$\frac{\partial d_i}{\partial P_i} < 0, \text{ and } \frac{\partial d_i}{\partial P_j} \geq 0 \text{ for } j \neq i.$$

For each product, the difference

$$s_i(P_1, \ldots, P_n) - d_i(P_1, \ldots, P_n)$$

shows the supply-demand balance. If this difference is positive, there is a surplus of this product, which will result in a price decrease. If this difference is negative, then there is a shortage of this product, and therefore the price of this product will increase. This property is modeled by the dynamic equations

$$P_i(t+1) = P_i(t) - K_i[s_i(P_1(t), \ldots, P_n(t)) - d_i(P_1(t), \ldots, P_n(t))]$$

for $i = 1, 2, \ldots, n$. Here, t denotes the time, and $K_i > 0$ is a given constant for all i. We can rewrite this equation as

$$P_i(t+1) = [1 - K_i(b_{ii} - a_{ii})] P_i(t) - \sum_{j \neq i} K_i \cdot (b_{ij} - a_{ij}) P_j(t) - K_i(b_{i0} - a_{i0}). \quad (2.26)$$

Introduce vectors

$$P_t = \begin{pmatrix} P_1(t) \\ P_2(t) \\ \vdots \\ P_n(t) \end{pmatrix} (t \geq 0), b = \begin{pmatrix} -K_1(b_{10} - a_{10}) \\ -K_2(b_{20} - a_{20}) \\ \vdots \\ -K_n(b_{n0} - a_{n0}) \end{pmatrix}$$

and matrix $H = (h_{ij})$ with elements

$$h_{ij} = \begin{cases} 1 - K_i(b_{ii} - a_{ii}) & \text{if } i = j \\ -K_i(b_{ij} - a_{ij}) & \text{if } i \neq j, \end{cases}$$

Equation (2.26) can then be summarized as

$$P_{t+1} = H \cdot P_t + b. \quad (2.27)$$

Notice that the global asymptotical stability of this dynamic process is equivalent to the convergence of process (2.27) as an iteration algorithm starting from arbitrary initial vector P_0. We know from the end of the first case study that with arbitrary P_0, P_t, converges to the equilibrium if and only if all eigenvalues of H are inside the unit circle. A sufficient condition is that $\|H\| < 1$ with some matrix norm. If the row norm is selected, then we have to assume that

$$|1 - K_i(b_{ii} - a_{ii})| + \sum_{j \neq i} |-K_i(b_{ij} - a_{ij})| < 1$$

for all i. Because the derivative conditions imply that

$$b_{ii} > 0, a_{ii} < 0, b_{ij} \leq 0, \text{ and } a_{ij} \geq 0 \quad (j \neq i),$$

the above inequality can be rewritten as

$$\sum_{j \neq i} K_i(a_{ij} - b_{ij}) < K_i(b_{ii} - a_{ii})$$

by assuming that $K_i(b_{ii} - a_{ii}) < 1$ for all i. Dividing by K_i and rearranging the terms the convergence condition can be reduced to the following inequality:

$$\sum_{j=1}^{n} a_{ij} < \sum_{j=1}^{n} b_{ij}.$$

Case Study 2.5. Our fifth model is the classical Cournot oligopoly model, which has a very important role in the theory of dynamic economic systems. Assume that N firms produce a homogeneous product and sell it on the same market. Let $x^{(1)}, x^{(2)}, \ldots, x^{(N)}$ denote the production levels of the firms, and assume that the unit price is a function of the total production of the industry: $P(x^{(1)} + x^{(2)} + \cdots + x^{(N)})$. If $C^{(k)}(x^{(k)})$ denotes the cost of form k, its profit can be determined as

$$\varphi^{(k)}(x^{(1)}, \ldots, x^{(N)}) = x^{(k)} P(x^{(1)} + \cdots + x^{(N)}) - C^{(k)}(x^{(k)}).$$

For the sake of simplicity assume that the price function and all cost functions are linear:

$$P(s) = As + B(s = x^{(1)} + \cdots + x^{(N)}, A < 0)$$

and

$$C^{(k)}(x^{(k)}) = b^{(k)} x^{(k)} + c^{(k)} \quad (b^{(k)} > 0, k = 1, 2, \ldots, N).$$

This static model as an N-person game with sets $[0, \infty)$ of strategies, and payoff functions $\varphi(k)$ is called the Cournot oligopoly.

The dynamic extension of this model can be formulated as follows. At the initial time period $t = 0$, each firm selects a production level $x^{(k)}(0)$. At each further time period $t \geq 1$, each firm k proceeds in two steps. First, each firm estimates the total production level of all other firms. Let the estimator for firm k be denoted by $s_E(t)(t)$.

Then, each firm determines the profit maximizing output based on its production. That is, for each k, function

$$x^{(k)}[As_E^{(k)}(t) + Ax^{(k)} + B] - (b^{(k)}x^{(k)} + c^{(k)})$$

is maximized. Because $A < 0$, the first order optimality conditions are sufficient:

$$2Ax^{(k)} + As_E^{(k)}(t) + B - b^{(k)} = 0,$$

which reduces to

$$x^{(k)} = -\frac{1}{2}s_E^{(k)}(t) + \frac{b^{(k)} - B}{2A}.$$

Assuming that $x(k) \geq 0$ for all k, each firm selects this profit optimizing output $x^{(k)}$ as its production level for the new time period t. Hence, the system is driven by relations

$$x^{(k)} = -\frac{1}{2}s_E^{(k)}(t) + \frac{b^{(k)} - B}{2A} \quad (1 \leq K \leq N) \tag{2.28}$$

Notice that the structure of this dynamic equation depends on the particular prediction schemes of the firms. In the economic literature three major types are discussed.

The first scheme is called Cournot expectation and it is based on the assumption that each firm keeps its previous production level:

$$s_E^{(k)}(t) = \sum_{i \neq k} x^{(i)}(t-1).$$

Substitute this equation into (2.28) to obtain the following first order difference equation:

$$x^{(k)} = -\frac{1}{2}\sum_{i \neq k} x^{(i)} + \frac{b^{(k)} - B}{2A} \quad (1 \leq k \leq N) \tag{2.29}$$

Notice that this scheme is identical to the iteration method of solving the fixed point problem

$$x^{(k)} = -\frac{1}{2}\sum_{i \neq k} x^{(i)}(t-1) + \frac{b^{(k)} - B}{2A} \quad (1 \leq K \leq N),$$

where the iteration steps are replaced by the time periods. We know that scheme (2.29) converges if and only if all eigenvalues of matrix

$$H_c = \begin{pmatrix} 0 & -\frac{1}{2} & -\frac{1}{2} & \cdots & -\frac{1}{2} & -\frac{1}{2} \\ -\frac{1}{2} & 0 & -\frac{1}{2} & \cdots & -\frac{1}{2} & -\frac{1}{2} \\ \cdots & \cdots & \cdots & \cdots & \cdots & \cdots \\ -\frac{1}{2} & -\frac{1}{2} & -\frac{1}{2} & \cdots & -\frac{1}{2} & 0 \end{pmatrix}$$

are inside the unit circle. We will next prove that the eigenvalues of H_c are $1/2$ and $(1-N)/2$. Consider first the eigenvalue problem of matrix 1 with all elements being equal to one. The eigenvalue equation of 1 has the form

$$u_1 + u_2 + \cdots + u_N = \lambda u_1$$
$$u_1 + u_2 + \cdots + u_N = \lambda u_2$$

$$\cdots$$

$$u_1 + u_2 + \cdots + u_N = \lambda u_N.$$

If $\lambda = 0$, then $u_1 + u_2 + \cdots + u_N = 0$ is the only condition on the solution. If $\lambda \neq 0$, then $u_1 = u_2 = \cdots = u_N$, therefore $\lambda = N$. Since

$$H_c = -\frac{1}{2}\mathbf{1} + \frac{1}{2}I,$$

the eigenvalues are $-\frac{1}{2} \cdot 0 + \frac{1}{2} = \frac{1}{2}$ and $-\frac{1}{2} \cdot N + \frac{1}{2} = \frac{1-N}{2}$. Therefore, convergence occurs if and only if

$$-1 < \frac{1-N}{2} < 1,$$

which is equivalent to the relation $N = 2$. Thus, the dynamic oligopoly under Cournot expectations is globally asymptotically stable if and only if only two firms are in the industry.

Adaptive expectations are defined by equations

$$s_E^{(k)}(t) = s_E^{(k)}(t-1) + m^{(k)}\left(\sum_{i \neq k} x^{(i)}(t-1) - s_E^{(k)}(t-1)\right) \qquad (2.30)$$

The first term of the right-hand side is the previous estimation of firm k on the production level of the rest of the industry, and the second term is the $m^{(k)}$-multiple of the prediction error of the previous time period. By substituting this equation into (2.29) the following dynamic equations are obtained:

$$x^{(k)}(t) = -\frac{m^{(k)}}{2}\sum_{l \neq k} x^{(i)}(t-l) - \frac{1-m^k}{2}s_E^{(k)}(t-l) + \frac{b^{(k)} - B}{2A} \qquad (2.31)$$

for $k = 1, 2, \ldots, N$. Equations (2.30) and (2.31) again represent a system of first-order linear difference equations with $2N$ unknown functions, which is globally asymptotically stable if and only if all eigenvalues of matrix

$$H_a = \begin{pmatrix} 0 & -\frac{m^{(1)}}{2} & \cdots & -\frac{m^{(1)}}{2} & -\frac{1-m^{(1)}}{2} & & \\ -\frac{m^2}{2} & 0 & \cdots & \frac{m^{(2)}}{2} & & -\frac{1-m^{(1)}}{2} & \\ & \cdots & & \cdots & & & \cdots \\ -\frac{m^{(N)}}{2} & -\frac{m^{(N)}}{2} & \cdots & 0 & & & -\frac{1-m^{(N)}}{2} \\ 0 & m^{(1)} & \cdots & m^{(1)} & 1-m^{(1)} & & \\ m^{(2)} & 0 & \cdots & m^{(2)} & & 1-m^{(2)} & \\ & \cdots & \cdots & & & & \cdots \\ m^{(N)} & m^{(N)} & \cdots & 0 & & & 1-m^{(N)} \end{pmatrix}$$

are inside the unit circle. Consider the special case, when $m^{(1)} = m^{(2)} = \cdots = m^{(N)} = m$. Notice that matrix H_a. can be rewritten into a block form

$$H_a = \begin{pmatrix} \frac{m}{2}(I-1) & -\frac{1-m}{2}I \\ -m(I-1) & (I-m)I \end{pmatrix},$$

Therefore, its eigenvalue equation can be written as

$$\frac{m}{2}(I-1)u - \frac{I-m}{2}v = \lambda u$$
$$-m(I-1)u - (1-m)v = \lambda v.$$

Multiply the first equation by two and add the resulting equation to the second one,

$$\lambda(2u+v) = 0.$$

If $\lambda = 0$, it is inside the unit circle. If $\lambda \neq 0$, then $v = -2u$, and the first equation implies that

$$\frac{m}{2}(I-1)u + (1-m)u = \lambda u,$$

In other words,

$$\left(\left(1-\frac{m}{2}\right)I - \frac{m}{2}l\right)u = \lambda u.$$

Hence, λ is the eigenvalue of matrix $\left(1-\frac{m}{2}\right)I - \frac{m}{2}l$, which implies that

$$\lambda = \left(1-\frac{m}{2}\right) - \frac{m}{2} \cdot 0 = 1 - \frac{m}{2} \text{ or } \lambda = \left(1-\frac{m}{2}\right) - \frac{m}{2} \cdot N = 1 - m\frac{N+1}{2}.$$

The system is globally asymptotically stable if and only if

$$-1 < 1 < -\frac{m}{2} < 1$$

and

$$-1 < 1 - m\frac{N+1}{2} < 1.$$

Simple algebra shows that these inequalities hold if and only if

$$0 < m < \frac{4}{N+1}.$$

The best selection of the speed of adjustment m can be found by minimizing the largest eigenvalue of H_a; i.e.,

$$m^* = \arg\min\max\left\{\left|1-\frac{m}{2}\right|; \left|1-\frac{m(N+1)}{2}\right|\right\}.$$

A simple graphical analysis shows that

$$m^* = \frac{4}{N+2}.$$

In the case of extrapolative expectations a linear prediction formula is constructed based on the two most recently observed data values:

$$s_E^{(k)}(t) = \alpha^{(k)} \sum_{l \neq k} x^{(i)}(t-l) + (1 - \alpha^{(k)}) \sum_{i \neq k} x^{(i)}(t-2).$$

Substitute this prediction scheme into equation (2.28) to obtain the following second order linear difference equation:

$$x^{(k)}(t) = -\frac{\alpha^{(k)}}{2} \sum_{i \neq k} x^{(i)}(t-1) - \frac{1 - \alpha^{(k)}}{2} \sum_{i \neq k} x^{(i)}(t-2) + \frac{b^{(k)} - B}{2A} \quad (1 \leq K \leq N).$$

$$(2.32)$$

This system can be interpreted as a two-step iteration algorithm. It is easy to see that this process is convergent if and only if all eigenvalues of matrix

$$H_e = \begin{bmatrix} 0 & & & 1 & & \\ & 0 & & & 1 & \\ & & \ddots & & & \ddots \\ & & & 0 & & & 1 \\ 0 & -\frac{1-\alpha^{(1)}}{2} & \cdots & -\frac{1-\alpha^{(1)}}{2} & 0 & -\frac{\alpha^{(1)}}{2} & \cdots & -\frac{\alpha^{(1)}}{2} \\ -\frac{1-\alpha^{(2)}}{2} & 0 & \cdots & -\frac{1-\alpha^{(2)}}{2} & -\frac{\alpha^{(2)}}{2} & 0 & \cdots & -\frac{\alpha^{(2)}}{2} \\ -\frac{1-\alpha^{(N)}}{2} & -\frac{1-\alpha^{(N)}}{2} & \cdots & 0 & -\frac{\alpha^{(N)}}{2} & -\frac{\alpha^{(N)}}{2} & \cdots & 0 \end{bmatrix}$$

are inside the unit circle. Consider again the special case of $\alpha^{(1)} = \alpha^{(2)} \cdots = \alpha^{(N)} = \alpha$. Then, H_e can be rewritten as a block matrix:

$$H_e = \begin{pmatrix} 0 & I \\ \frac{1-\alpha}{2}(I-1) & \frac{\alpha}{2}(I-1) \end{pmatrix}$$

Therefore, its eigenvalue equation is as follows:

$$v = \lambda u$$
$$\frac{1-\alpha}{2}(I-1)u + \frac{\alpha}{2}(I-1)v = \lambda v.$$

Substitute the first equation into the second to obtain the quadratic eigenvalue problem

$$\left(\frac{1-\alpha}{2}(I-1) + \frac{\alpha}{2}(I-1)\lambda = \lambda^2 I \right) v = 0.$$

Hence, the eigenvalues of H_e, are the solutions of the quadratic determinant equation

$$\det \left(\frac{1-\alpha}{2}(I-1) + \frac{\alpha}{2}(I-1)\lambda = \lambda^2 I \right) = 0.$$

A sufficient condition can be derived based on the following idea. Multiply the above quadratic eigenvalue problem by the complex conjugate V^T of v^T. A quadratic equation is then obtained for λ:

$$\lambda^2 - \frac{\alpha}{2}V\lambda - \frac{1-\alpha}{2}V = 0,$$

where

$$V = \frac{V^T(I-1)v}{v^T v}.$$

We now need the following simple result.

Lemma 2.2. Consider the quadratic polynomial equation

$$p(\lambda) = \lambda^2 + b\lambda + c = 0$$

with real coefficients b and c. The roots of the equation are inside the unit circle if and only if

$$b - c - 1 < 0, b + c + 1 > 0, \text{ and } c < 1.$$

Proof. Assume first that the roots are complex. Then

$$b^2 - 4c < 0$$

and

$$\lambda_{12} = \frac{-b \pm \sqrt{bc = b^2}}{2},$$

which implies that

$$|\lambda_{12}|^2 = \frac{b^2}{4} + \frac{4c - b^2}{4} = c.$$

Therefore, in this case the condition is

$$1 > c > \frac{b^2}{4}.$$

Assume next that the roots are real. Then

$$b^2 - 4c \geq 0$$

and

$$\lambda_{12} = \frac{-b + \sqrt{b^2 - 4c}}{2}.$$

The roots are inside the unit circle if and only if

$$-1 < \frac{-b \pm \sqrt{b^2 - 4c}}{2} < 1.$$

This relation is equivalent to the inequalities

$$-1 < \frac{-b - \sqrt{b^2 - 4c}}{2} \quad \text{and} \quad \frac{-b + \sqrt{b^2 - 4c}}{2} < 1.$$

Simple algebra shows that these conditions hold if and only if $b \leq 2, b - c - 1 < 0, b \geq -2$, and $b + c + 1 > 0$. The union of the complex and real cases is summarized in the assertion.

In our case the conditions of the lemma can be reduced to the following relations:

$$-\frac{\alpha}{2}V + \frac{1 - \alpha}{2}V - 1 < 0, \; -\frac{\alpha}{2}V - \frac{1 - \alpha}{2}V + 1 > 0, \; -\frac{1 - \alpha}{2}V < 1.$$

These relations can be rewritten as

$$\left(\frac{1}{2} - \alpha\right)V < 1, \; -\frac{1}{2}V > -1, \; -\frac{1 - \alpha}{2}V < 1.$$

Consider first the case $\alpha > 1$. Then these relations can be solved for V as follows:

$$-\frac{2}{2\alpha - 1} < V < \min\left\{2, \frac{2}{\alpha - 1}\right\}.$$

Since matrix $I - 1$ is symmetric, V is between the largest and smallest eigenvalues of $I - 1$; i.e.,

$$1 - N \leq V \leq 1.$$

Therefore, a sufficient stability condition is given as

$$-\frac{2}{2\alpha - 1} < 1 - N \quad \text{and} \quad \alpha < 3,$$

In other words,

$$\alpha < \frac{N + 1}{2(N - 1)}.$$

Notice that an $\alpha > 1$ satisfies this inequality only if $N = 2$. Assume next that $\alpha = 1$. The conditions are

$$-\frac{V}{2} < 1, \; -\frac{V}{2} > -1, 0 < 1,$$

which can be reduced as

$$2 > V > -2,$$

which will hold if

$$1 - N > -2.$$

That is, $N = 2$ again is the condition.

Assume that $\frac{1}{2} < \alpha < 1$. The conditions are now as follows:

$$\max\left\{-\frac{2}{2\alpha - 1}, \frac{2}{\alpha - 1}\right\} < V < 2,$$

which are necessarily satisfied if

$$\max \left\{ -\frac{2}{2\alpha - 1}, \frac{2}{\alpha - 1} \right\} < 1 - N.$$

Simple algebra shows that this relation is equivalent to

$$\frac{N - 3}{N - 1} < \alpha < \frac{N + 1}{2(N - 1)}.$$

Notice that this inequality has solution in $\left(\frac{1}{2}, 1 \right)$ if only for $N < 7$. Consider next the case of $\alpha = \frac{1}{2}$. Then the conditions reduce to the following inequalities:

$$0 < 1, -\frac{1}{2}V? > -1, -\frac{1}{4}V < 1;$$

i.e.,

$$-4 < V < 2.$$

This relation holds necessarily if $-4 < 1 - N$; that is, when $N < 5$.

Consider now the last special case of $0 < \alpha < \frac{1}{2}$. The conditions are given by the inequalities

$$V < \frac{2}{1 - 2\alpha} V < 2, V > \frac{2}{\alpha - 1}.$$

Because

$$\frac{2}{1 - 2\alpha} > 2,$$

these relations can be reduced as

$$-\frac{2}{1 - \alpha} < V < 2.$$

The system is therefore globally asymptotically stable if

$$-\frac{2}{1 - \alpha} < 1 - N,$$

which can be rewritten as

$$\alpha > \frac{N - 3}{N - 1}.$$

Notice, finally, that this inequality has a solution $0 < \alpha < \frac{1}{2}$ if only for $N < 5$.

The linearity of the above models serve only technical convenience. The non-linear extensions are examined in Okuguchi and Szidarovszky [18], who also discuss other applications of the same mathematical model in engineering.   □

Case Study 2.6. Consider the iteration method (1.8), which we repeat here for the sake of convenience,

$$x_{k+1} = f(x_k).$$

We assume now that $D \leq R^l$ is compact and convex, $f : D \to D$ is a differentiate function, and for all $x \in D, |f'(x)| \leq q < 1$, where q is a fixed real number. The mean value theorem of derivatives implies that this process satisfies the conditions of Theorem 2.8; hence, it converges to a solution of equation $x = f(x)$. It is easy to show that the solution is unique. Assume that s^* and s^{**} are both fixed points in D, then

$$|s^* - s^{**}| = |f(s^*) - f^{s^{**}}| = |f'(\xi)| \cdot |s^* - s^{**}|$$
$$\leq q \cdot |s^* - s^{**}| < |s^* - s^{**}|,$$

which is an obvious contradiction. Notice that Example 1.5 shows a particular application of this method.

The Corollaries of Theorems 2.8 and 2.7 imply that for all $k \geq 1$,

$$|x_k - s^*| < q^k |x_0 - s^*|.$$

This error bound allows us to estimate the necessary number of iteration steps before starting the procedure. Assume that $s^* \in (a, b)$ and x_0 is selected as one of the endpoints of the interval. Then, $|x_0 - s*| < b - a$, and therefore

$$|x_k - s^*| \leq q^k (b - a).$$

The error of x_k is below any given ε, if

$$q^k (b - a) < \varepsilon,$$

that is,

$$k > \frac{\log \frac{\varepsilon}{b-a}}{\log q}.$$

Example 2.3. Assume that in the case of the equation of Example 1.5, we know that $|f'(x)| \leq \frac{1}{2}$ and $s^* \in (1, 2)$. If the solution is needed for five correct significant figures, then we may select $\varepsilon = \frac{1}{2} \cdot 10^{-4}$. Therefore,

$$K > \frac{\log \left(\frac{1}{2} \cdot 10^{-4} \right)}{\log \frac{1}{2}} = \frac{4 + \log 2}{\log 2} \approx 14.29,$$

In other words, 15 iteration steps are sufficient.

Assume next that a single equation $g(x) = 0$ has to be solved. In order to apply the above method, we must rewrite this equation into a fixed point problem. Obviously, the equation is equivalent to the fixed point problem

$$x = x + h(x)g(x),$$

where $h(x) \neq 0$. In order to guarantee the convergence of the corresponding iteration method, we must select function $h(x)$ so that the derivative of function $x + h(x)g(x)$ in the neighborhood of the solution s^* is as small as possible. A logical choice is given by requiring that the derivative at s^* equals zero, because continuity implies that the derivative remains small in the neighborhood of s^*. That is we want to satisfy equation

$$1 + h'(s^*)g(s^*) + h(s^*)g'(s^*) = 0.$$

Because $g(s^*) = 0$, we have

$$h(s^*) = -\frac{1}{g'(s^*)}.$$

Therefore, we may select

$$h(x) = -\frac{1}{g'(x)}$$

for all x. Hence, the fixed point problem has the form

$$x = x - \frac{g(x)}{g'(x)},$$

and the corresponding iteration scheme coincides with the Newton method (1.14). We note here that Example 1.8 presented a numerical case study of the application of the Newton method.

The error of the kth Newton iterate can be given in the following way. Notice first that the Newton method can be rewritten as

$$g(x_k) + g'(x_k)(x_{k+1} - x_k) = 0.$$

The remainder term of the quadratic Taylor's polynomials implies that

$$0 = g(s^*) = g(x_k) + g'(x_k)(s^* - x_k) + \frac{g''(\zeta)}{2}(s^* - x_k)^2,$$

where we assumed that g is twice differentiable and ζ is a point between x_k and s^*. Subtract the two equations to obtain

$$x_{k+1} - s^* = \frac{g''(\zeta)}{2g'(x_k)}(x_k - s^*)^2.$$

Assume that

$$|g''(\zeta)| \leq M \text{ and } |g'(x_k)| \geq m > 0,$$

and introduce the notation

$$\varepsilon_k = |x_k - s^*|, k = \frac{M}{2m}, \delta_k = K\varepsilon_k.$$

Then obviously

$$\varepsilon_{k+1} \leq \frac{M}{2m}\varepsilon_k^2 = K\varepsilon_k^2,$$

which implies that

$$\delta_{k+1} \leq \delta_k^2,$$

and by using finite induction it is easy to see that for all $k \geq 1$,

$$\delta_k \leq \delta_0^{2^k}.$$

This inequality implies that

$$\varepsilon_k = |x_k - s^*| \leq \frac{1}{k}(K|x_0 - s^*|)^{2^k}.$$

If the initial approximation is selected in such a way that it satisfies the relation

$$|x_0 - s^*| < \frac{1}{K},$$

then $\varepsilon_k \to 0$, and the convergence is quadratic.

The secant method (1.9) can be considered an approximation of the Newton method, when at step k the derivative $g'(x_k)$ is approximated by the two-point rule $(g(x_k) - g(x_{k-1}))(x_k - x_{k-1})$. Example 1.7 shows the application of the algorithm.

The error of the secant method can be bounded in the following way. Assume that g is twice differentiable. Notice first that the formula implies that

$$x_{k+1} - s^* = \frac{(x_k - s^*)g(x_{k-1}) - (x_{k-1} - s^*)g(x_k)}{g(x_{k-1}) - g(x_k)},$$

that is,

$$\frac{x_{k+1} - s^*}{(x_k - s^*)(x_{k-1} - s^*)} = \frac{\frac{g(x_{k-1})}{x_{k-1} - s^*} - \frac{g(x_k)}{x_k - s^*}}{g(x_{k-1}) - g(x_k)}.$$

Apply the Cauchy's mean value theorem on the right-hand side, which says that if g and h are differentiable on an interval $[a, b]$, then with some $\zeta \in (a, b)$,

$$\frac{g(b) - g(a)}{h(b) - h(a)} = \frac{g'(\zeta)}{h'(\zeta)}.$$

Because

$$\left(\frac{g(x)}{x - s^*} \right) = \frac{g'(x)(x - s^*) - g(x)}{(x - s^*)^2},$$

we have the equation

$$\frac{x_{k+1} - s^*}{(x_k - s^*)(x_{k-1} - s^*)} = \frac{g'(\zeta)(\zeta - s^*) - g(\zeta)}{(\zeta - s^*)^2 g'(\zeta)}.$$

From the Taylor's formula we know that

$$0 = g(s^*) = g(\zeta)(s^* - \zeta) + \frac{g''(\eta)}{2}(s^* - \zeta)^2.$$

Combine the two relations to conclude that

$$x_{k+1} - s^* = \frac{g''(\eta)}{2g'(\zeta)}(x_k - s^*)(x_{k-1} - s^*).$$

By using the same notation as above in the case of the Newton method, we have

$$\varepsilon_{k+1} \leq \frac{M}{2m}\varepsilon_k\varepsilon_{k-1},$$

and therefore,

$$\delta_{k+1} \leq \delta_k\delta_{k-1}.$$

If $\delta = \max\{\delta_0, \delta_1\}$, then finite induction shows that $\delta_k \leq \delta^{\gamma_k}$, where γ_k is the kth Fibonacci number defined by the recursion $\gamma_k = \gamma_{k-1} + \gamma_{k-2}$ with $\gamma_0 = \gamma_1 = 1$. Hence,

$$\varepsilon_k \leq \frac{1}{K}\delta^{\gamma_K},$$

and if $\delta < 1$, then $x_k \to s^*$. Comparing the error bounds of the original iteration (1.8), secant, and Newton methods we see that Newton's method is the fastest, and the original iteration method is the slowest in the sense of the order of the convergence speed.

Assume next the $D \subseteq R^n$ is compact and convex, and $f : D \to D$ is a differentiable function. Let J denote the Jacobian of f, and assume that for all $x \in D$, $\|J(x)\| \leq q < 1$, where q is a fixed real number. The mean value theorem of Fréchet derivatives (see, for example, Ortega and Rheinboldt [19]) implies that all conditions of Theorem 2.8 are satisfied, which implies the convergence of the iteration process $x_{k+1}, = f(x_k)$. The uniqueness of the fixed point can be shown similarly to the single dimensional case. Example 1.6 presented a numerical example as the application of this iteration procedure.

Consider next the equation $g(x) = 0$, where $g : D \to R^n$. This equation is equivalent to the fixed point problem

$$x = x + H(x)g(x),$$

where matrix $H(x)$ is nonsingular for all $x \in D$. We know that the corresponding iteration process converges if the Jacobian of the right-hand side is sufficiently small in the neighborhood of the solution. We can guarantee this condition by assuming that the Jacobian is zero at the solution s^*, which can be written as follows. For all i and j,

$$0 = \frac{\partial}{\partial x^{(i)}}\left(x(i) + \sum_{i=1}^{n} h_{il}(x)g_i(x)\right)(x - s^*)$$

$$\delta_{ij} + \sum_{i=1}^{n}\left(\frac{\partial h_{ij}}{\partial x^{(j)}}(s^*)g_i(s^*) + h_{il}(s^*)\frac{\partial g_i}{\partial x^{(j)}}(s^*)\right),$$

where

$$\delta_{ij} = \begin{cases} 1 & \text{if } i = j \\ 0 & \text{if } i \neq j, \end{cases}$$

and h_{il} and g_l, denote the (i,l) element of H and the lth element of g, respectively. Because $g_l(s^*) = 0$, the above equation reduces to

$$I + H(s^*)J(s^*) = 0,$$

where J is the Jacobian of g. That is, $H(s^*) = -J(s^*) - 1$, and we may select

$$H(x) = -J(x)^{-1}$$

for all $x \in D$. The fixed point problem is therefore

$$x = x - J(x)^{-1}g(x),$$

and the corresponding iteration scheme coincides with the generalized Newton method (1.15). An application of the multivariable Newton method was presented in Example 1.12.

Case Study 2.7. As was mentioned in Section 1.3, the above methods for solving non-linear algebraic equations can be used to obtain the local optimum of differentiable functions. Let $D \subseteq R^n$ be an arbitrary set with nonempty interior, and assume that an interior point $s^* \in D$ is a local optimum point of a differentiable function g. Then, for $i = 1, 2, \ldots, n$,

$$\frac{\partial}{\partial x^{(i)}}g(x)|_{x=s^*} = 0, \tag{2.33}$$

which is a system of equations for the unknown vector s^*.

Example 2.4. This reduction will be illustrated in the case of minimizing the function

$$g(x^{(1)}, x^{(2)} = 2x^{(1)^2} + 2x^{(2)^2} + 2x^{(1)^2}x^{(2)^2} - 6x^{(1)^2} - 6x^{(2)^2}).$$

In this case, equations (2.33) have the form

$$\frac{\partial g}{\partial x^{(1)}}(x^{(1)}, x^{(2)}) = 4x^{(1)} + 2x^{(2)} - 6 = 0$$

$$\frac{\partial g}{\partial x^{(2)}}(x^{(1)}, x^{(2)}) = 4x^{(2)} + 2x^{(1)} - 6 = 0$$

which has the unique solution $x^{(1)} = x^{(2)} = 1$. This result can be checked easily by noticing that

$$g(x^{(1)}, x^{(2)}) = (x^{(1)} - 1)^2 + (x^{(1)} + x^{(2)} - 2)^2 + (x^{(2)} - 1)^2 - 6.$$

In more complicated cases the solutions of the resulting equations can be determined by using some kind of iteration procedure.

Case Study 2.8. Here we consider the predator-prey population models. These models describe the interaction of a prey population $X(t)$ and a predator population $Y(t)$ at any time $t \geq 0$. Assume that their interaction is modeled by the system of ordinary differential equations

$$\dot{X} = X - \frac{1}{5}X^2 - \frac{1}{20}XY + 1$$

$$\dot{Y} = -\frac{1}{5}Y + \frac{1}{2}XY - \frac{1}{5}.$$

We now determine the equilibrium of this system. At the equilibrium, $\dot{X} = \dot{Y} = 0$, therefore, we must solve the system of algebraic equations

$$X - \frac{1}{5}X^2 - \frac{1}{20}XY + 1 = 0 \qquad -\frac{1}{5}Y + \frac{1}{2}XY - \frac{1}{5} = 0.$$

Before formulating the multivariable Newton method (1.15), the Jacobian must be computed. Simple differentiation shows that

$$J(X,Y) = \begin{pmatrix} 1 - \frac{2}{5}X - \frac{1}{20}Y & -\frac{1}{20}X \\ \frac{1}{2}Y & -\frac{1}{5} + \frac{1}{2}X \end{pmatrix}.$$

Thus, at step k $(k \geq 0)$, the differences

$$\Delta X_k = X_{k+1} - X_k \text{ and } \Delta Y_k = Y_{k+1} - Y_k$$

satisfy the linear equation

$$\begin{pmatrix} 1 - \frac{2}{5}X_k - \frac{1}{20}Y_k & -\frac{1}{20}X_k \\ \frac{1}{2}Y_k & -\frac{1}{5} + \frac{1}{2}X_k \end{pmatrix} \begin{pmatrix} \Delta X_k \\ \Delta Y_k \end{pmatrix} = \begin{pmatrix} -X_k\frac{1}{5}X_k^2 - \frac{1}{20}X_kY_k - 1 \\ \frac{1}{5}Y_k - \frac{1}{2}X_kY_k + \frac{1}{5} \end{pmatrix},$$

which can be obtained by multiplying both sides of equation (1.15) by $J(x_k)$ from the left-hand side. Hence, the application of the Newton method requires the repeated solution of linear equations. A simple computer program gives the results

$$X \approx 5.838060 \text{ and } Y \approx 0.07355648.$$

A discussion of the main properties of predator-prey models are given by Szidarovszky and Bahill [23].

An application of non-linear equations for constructing adaptive control systems is presented in case study 3.7. In recent years increasing attention has been given to systems that are capable of accommodating unpredictable changes, whether these changes arise within the system or externally. This property is called adaptation and is a fundamental characteristic of living organisms, because they attempt to maintain physiological equilibrium in order to survive under changing environmental conditions. In the system theory literature there is no unified definition for adaptive control systems. Therefore, we will consider a system adaptive if it satisfies the following criteria:

1. Continuously and automatically measures the dynamic characteristics of the system;
2. Compares the measurements to the desired dynamic characteristics;
3. Modifies its own parameters in order to maintain desired performance regardless of the environmental changes.

An adaptive control system therefore consists of three blocks: performance index measurement, comparison-decision, and adaptation mechanisms. It is always assumed that there is a closed-loop control on the performance index. An important class of adaptive systems, model reference adaptive systems, is based on replacing the set of given performance indices by a reference model.

The output of this model and that of the adjustable system are continuously compared by a typical feedback comparator, and the difference is used by the adaptation mechanism either to modify the parameters of the adjustable system or to send an auxiliary input signal to minimize the difference between the performance indices of the two systems.

Another often used class of adaptive systems is given by the adaptive model-following control systems. These control systems also use a model that specifies the design objectives.

In this case study the second type of adaptive system is examined. A mathematical model is first presented, and the stability of the resulting adaptive system is investigated. Specifically, we introduce a new necessary and sufficient condition for the existence of globally asymptotically stable adaptive model-following control systems.

The mathematical model is formulated as follows. Assume that the reference model is dimensional and is given as

$$\dot{x} = A_M x + B_M u_M, \tag{2.34}$$

and the plant to be controlled is

$$\dot{y} = A_P y + B_P u_P. \tag{2.35}$$

The plant control input is given by the relation

$$u_P = -K_P y + K_M x + K_U u_M. \tag{2.36}$$

In this formulation A_M, B_M, A_P, B_P are given constant matrices, x and y are the states of the reference model and the plant, and u_M and up are their inputs. The coefficient matrices K_P, K_M, and K_U are unknowns; they are defined so that if the error vector $e = x - y$ is initialized as $e(0) = 0$, then it remains zero for all future time periods. Combining the above equations we have

$$\dot{e} = (A_M - B_P K_M)e + (A_M - A_P + B_P(K_P - K_M))y + (B_M - B_P K_U)u_M. \tag{2.37}$$

Perfect model following requires, therefore, that

$$A_M - A_P + B_P(K_P - K_M) = 0$$
$$B_M - B_P K_U = 0,$$

because these equations imply that for all real vectors y and u_M of appropriate dimensions, equations (2.37) become homogeneous. Thus, the solution of the resulting homogeneous equation with zero initial condition is the zero vector for all $t \geq 0$. We can rewrite the last equation as

$$B_P(K_P - K_M) = A_P - A_M, B_P K_U = B_M. \tag{2.38}$$

The necessary and sufficient condition for the existence of matrices K_P, K_M, and K_U that satisfy the above equations is the following:

$$\text{rank } (B_P) = \text{rank } (B_P, A_P - A_M) = \text{rank } (B_P, B_M). \tag{2.39}$$

These conditions mean that all columns of both matrices $A_P - A_M$ and B_M are in the subspace spanned by the columns of matrix B_P. Note that equations (2.38) can be solved using Gauss elimination (see, for example, Szidarovszky and Yakowitz [24]). The initial condition of the error vector e usually differs from zero. In such cases, we require that $e(t) \rightarrow 0$ as $t \rightarrow \infty$; i.e., equation (2.37) is asymptotically stable. We know that this additional condition holds if and only if all eigenvalues of matrix $A_M - B_P K_M$ have negative real parts.

It is well known from the system theory literature (see, for example, Szidarovszky and Bahill [23]) that a matrix K_M exists such that all eigenvalues of $A_M - B_P K_M$ have negative real parts if the modified controllability matrix $(B_P, A_M B_P, \ldots, A_M^{n-1} B_P)$ has full rank. This condition is sufficient, but not necessary, as the following example illustrates.

Example 2.5. Define $n = 2$,

$$A_P = \begin{pmatrix} 1 & 2 \\ 2 & 1 \end{pmatrix}, \quad B_P = \begin{pmatrix} 1 & 2 \\ 1 & 2 \end{pmatrix},$$
$$A_M = \begin{pmatrix} 0 & 1 \\ 1 & 0 \end{pmatrix}, \quad B_M = \begin{pmatrix} 2 & 1 \\ 2 & 1 \end{pmatrix}.$$

Here, the modified controllability matrix

$$\begin{pmatrix} 1 & 2 & 1 & 2 \\ 1 & 2 & 1 & 2 \end{pmatrix}$$

has unit rank, so the above condition does not hold. However, we can easily find a K_M such that equations (2.38) hold and all eigenvalues of $A_M - B_P K_M$ have negative real parts.

Note first that equations (2.38) can be rewritten as

$$\begin{pmatrix} 1 & 2 \\ 1 & 2 \end{pmatrix} \begin{pmatrix} r_{11} & r_{12} \\ r_{21} & r_{22} \end{pmatrix} = \begin{pmatrix} 1 & 1 \\ 1 & 1 \end{pmatrix},$$

$$\begin{pmatrix} 1 & 2 \\ 1 & 2 \end{pmatrix} \begin{pmatrix} k_{11} & k_{12} \\ k_{21} & k_{22} \end{pmatrix} = \begin{pmatrix} 2 & 1 \\ 2 & 1 \end{pmatrix},$$

where matrix $K_P - K_M$ is denoted by (r_{ij}) and K_U is denoted as (k_{ij}). Expanding the above operations, we obtain the following system of linear equations:

$$r_{11} + 2r_{21} = 1$$
$$r_{12} + 2r_{22} = 1$$
$$k_{11} + 2k_{21} = 2$$
$$r_{12} + 2k_{22} = 1,$$

where the repeated equations are omitted. It is easy to see that $r_{11} = r_{12} = 1, k_{11} = 2, k_{12} = 1, r_{21} = r_{22} = k_{21} = k_{22} = 0$ solve these equations. Hence, we may select

$$K_P - K_M = \begin{pmatrix} 1 & 1 \\ 0 & 0 \end{pmatrix} \text{ and } K_U = \begin{pmatrix} 2 & 1 \\ 0 & 0 \end{pmatrix}.$$

There are still infinite possibilities for selecting matrix K_M, as only $K_P - K_M$ is specified. We wish to make this selection so that matrix $A_M - B_P K_M$ has eigenvalues with only negative real parts. For example, select K_M so that

$$A_M - B_P K_M = -I$$

that is,

$$B_P K_M = A_M + I.$$

If $K_M = (\bar{k}_{ij})$, then this equation has the form

$$\begin{pmatrix} 1 & 2 \\ 1 & 2 \end{pmatrix} \begin{pmatrix} \bar{k}_{11} & \bar{k}_{12} \\ \bar{k}_{21} & \bar{k}_{22} \end{pmatrix} = \begin{pmatrix} 1 & 1 \\ 1 & 1 \end{pmatrix}.$$

It is easy to see that $\bar{k}_{11} = 1, \bar{k}_{12} = 1, \bar{k}_{21} = \bar{k}_{22} = 0$ are solutions. Therefore, the selection of

$$K_U = \begin{pmatrix} 2 & 1 \\ 0 & 0 \end{pmatrix}, K_M = \begin{pmatrix} 1 & 1 \\ 0 & 0 \end{pmatrix}, K_P = \begin{pmatrix} 2 & 2 \\ 0 & 0 \end{pmatrix}$$

is satisfactory in order to construct an asymptotically stable adaptive system.

Next, a necessary and sufficient condition is derived for the existence of a suitable matrix K_M. Assume that the rank condition (2.39) is satisfied. Then equations (2.38) have solutions for $K_P - K_M$ and K_U. Because K_P can be arbitrary, no constraint is needed for K_M. We know that all eigenvalues of $A_M - B_P K_M$ have negative real parts if and only if equation

$$(A_M - B_P K_M)^T Q + Q(A_M - B_P K_M) = -I \qquad (2.40)$$

has positive definite solution Q where I is the identity matrix (see, for example, Szidarovszky and Bahill [19]). This equation is a necessary and sufficient condition, where no restriction is given for K_M, but Q has to be positive definite. Matrix Q is positive definite if and only if it can be decomposed as $Q = LDL^T$, where D is a diagonal matrix with positive diagonal and L is lower triangular with unit diagonal elements. Next, we show that it is sufficient to assume that the diagonal elements of D are only non-negative. This observation follows immediately from the fact that any solution Q of equation (2.40) is necessarily nonsingular. Contrary to the assertion assume that Q is singular. Then there exists a real non-zero vector v such that $Qv = 0$. Then the equation implies that

$$0 > -v^T v = v^T (A_M - B_P K_M)^T Q v + v^T Q(A_M - B_P K_M)v = 0,$$

which is impossible. The non-negativity of the diagonal elements of D can be guaranteed by assuming that they are squares of real numbers. Hence, we proved Theorem 2.14.

Theorem 2.14. There exists an asymptotically stable adaptive model-following control system if and only if the rank condition (2.39) holds and a matrix K_M, a diagonal matrix D, and a lower triangular matrix L with zero diagonal exist such that

$$(A_M - B_P K_M)^T (L+I)D^2(L+I)^T + (L+I)D^2(L+I)^T \times (A_M - B_P K_M) = -I. \quad (2.41)$$

Example 2.6. In the case of the previous example the rank condition obviously holds, and equation (2.41) has the form

$$\begin{aligned}
&\left[\begin{pmatrix} 0 & 1 \\ 1 & 0 \end{pmatrix} - \begin{pmatrix} 1 & 2 \\ 1 & 2 \end{pmatrix} \begin{pmatrix} \bar{k}_{11} & \bar{k}_{12} \\ \bar{k}_{21} & \bar{k}_{22} \end{pmatrix}\right]^T \begin{pmatrix} 1 & 0 \\ l_{21} & 1 \end{pmatrix} \begin{pmatrix} d_{11}^2 & 0 \\ 0 & d_{22}^2 \end{pmatrix} \\
&= \begin{pmatrix} 1 & l_{21} \\ 1 & 1 \end{pmatrix} + \begin{pmatrix} 1 & 0 \\ l_{21} & 1 \end{pmatrix} \begin{pmatrix} d_{11}^2 & 0 \\ 0 & d_{22}^2 \end{pmatrix} \\
&= \begin{pmatrix} 1 & l_{21} \\ 1 & 1 \end{pmatrix} \left[\begin{pmatrix} 0 & 1 \\ 1 & 0 \end{pmatrix} - \begin{pmatrix} 1 & 2 \\ 1 & 2 \end{pmatrix} \begin{pmatrix} \bar{k}_{11} & \bar{k}_{12} \\ \bar{k}_{21} & \bar{k}_{22} \end{pmatrix}\right] \\
&= \begin{pmatrix} -1 & 0 \\ 0 & -1 \end{pmatrix}.
\end{aligned}$$

Simple calculation shows that this matrix equation is equivalent to the following system of algebraic equations:

$$d_{11}^2[l_{21} - (\bar{k}_{11} + 2\bar{k}_{21})l_{21} + 1] = -\frac{1}{2}$$

$$d_{11}^2 + l_{11}^2 d_{11}^2 + d_{22}^2 - d_{11}^2(\bar{k}_{12} + 2\bar{k}_{22})(\bar{l}_{21} + 1)$$
$$- (\bar{k}_{11} + 2\bar{k}_{21})(l_{21}d_{11}^2 + l_{21}^2 d_{11}^2 + d_{22}^2) = 0$$

$$d_{11}^2 l_{21} - (\bar{k}_{12} + 2\bar{k}_{22})(l_{21}d_{11}^2 + l_{22}^2 d_{11}^2 + d_{22}^2) = -\frac{1}{2}.$$

We have three equations for seven unknowns; therefore, infinitely solutions exist. Simple substitution shows that

$$\bar{k}_{11} = \bar{k}_{12} = 1, \quad \bar{k}_{21} = \bar{k}_{22} = 0,$$

$$l_{21} = 0, \quad d_{11}^2 = d_{22}^2 = \frac{1}{2}$$

is a solution; i.e.,

$$K_M = \begin{pmatrix} 1 & 1 \\ 0 & 0 \end{pmatrix}.$$

is satisfactory. Matrices K_P and K_U can be obtained from equations (2.38), as in the previous example.

In the case of large dimensional systems the nonlinear equations (2.41) can be solved by using computer methods such as those discussed earlier in this section.

Case Study 2.9. The solution of linear differential equations with constant coefficients is based on the computation of the fundamental matrix, which is the unique solution of the matrix differential equation

$$\dot{X}(t) = AX(t), \quad X(0) = I. \tag{2.42}$$

Here, $\dot{X}(t)$ denotes the derivative of X at t, and I is the identity matrix of appropriate dimension.

Integrate both sides of equation (2.42) in the interval $[0, t]$ to obtain the fixed point problem

$$X(t) = I + \int_0^t AX(\tau)d\tau, \tag{2.43}$$

which is defined in the set of matrices. An iteration scheme can be constructed as

$$X_{k+l}(t) = I + \int_0^t AX_k(\tau)d\tau. \tag{2.44}$$

Select $X_0(t) = I$. Then from (2.44) we have

$$X_1(t) = I + At,$$

$$X_2(t) = I + At + \frac{A^2 t^2}{2},$$

$$X_3(t) = I + At + \frac{A^2 t^2}{2} + \frac{A^3 t^3}{6}$$

and so on. Using finite induction, it is easy to verify that in general,

$$X_k(t) = I + \frac{At}{1!} + \frac{A^2 t^2}{2!} + \frac{A^3 t^3}{3!} + \cdots + \frac{A^k t^k}{k!},$$

and for arbitrary matrix A, the limit of $X_k(t)$ is the matrix exponential

$$X(t) = e^{At}.$$

Example 2.7. As a particular example consider matrix

$$A = \begin{pmatrix} 0 & w \\ -w & 0 \end{pmatrix} (w > 0)$$

which has an important role in the investigation of harmonic motions. Easy calculation shows that

$$A^2 = \begin{pmatrix} 0 & w \\ -w & 0 \end{pmatrix} \begin{pmatrix} 0 & w \\ -w & 0 \end{pmatrix} = \begin{pmatrix} -w^2 & 0 \\ 0 & -w^2 \end{pmatrix} = -w^2 I$$

therefore,

$$A^3 = -w^2 A,$$
$$A^4 = -w^2 A = w^4 I,$$
$$A^5 = w^4 A,$$
$$A^6 = -w^4 A^2 = -w^6 I,$$

and so on. In general, in other words,

$$A^k = \begin{cases} (-1)^l w^{2l} I & \text{if } k = 2l \\ (-1)^l w^{2l} A & \text{if } k = 2l + 1. \end{cases}$$

Thus,

$$e^{At} = \sum_{i=0}^{\infty} \frac{A^k t^k}{k!} = \sum_{i=0}^{\infty} \frac{(-1)^l w^{2l} t^{2l} I}{(2l)!} + \sum_{i=0}^{\infty} \frac{(-1)^l w^{2l} A t^{2l+i1}}{(2l+1)!}$$

$$= (\cos wt) I + (\frac{l}{w} \sin wt) A = \begin{pmatrix} \cos wt & \sin wt \\ -\sin wt & \cos wt \end{pmatrix}$$

This result can be checked easily by verifying that

$$e^{At}\big|_{t=0} = \begin{pmatrix} \cos 0 & \sin 0 \\ -\sin 0 & \cos 0 \end{pmatrix} = I,$$

and

$$\frac{d}{dt}e^{At} = \begin{pmatrix} -w\sin wt & w\cos wt \\ -w\cos wt & -w\sin wt \end{pmatrix} = \begin{pmatrix} 0 & w \\ -w & 0 \end{pmatrix}\begin{pmatrix} \cos wt & \sin wt \\ -\sin wt & \cos wt \end{pmatrix} = A \cdot e^{At}.$$

The convergence of this algorithm will be discussed more generally in the next case study because matrix exponentials are very important in many fields of applied mathematics, engineering, and economics, we now discuss their main properties.

Assume first that A is a projection; i.e., $A^2 = A$. Then for all $k \geq 1, A^k = A$; therefore,

$$e^{At} = I + \sum_{k=1}^{\infty} \frac{A^k t^k}{k!} = I + \left(\sum_{k=1}^{\infty} \frac{t^k}{k!}\right)A = I + (e^t - 1)A.$$

Assume next that A is a nilpotent matrix; that is, $A^M = 0$ with some M. Then e^{At} is a finite polynomial of A:

$$e^{At} = I + \sum_{k=1}^{M-1} \frac{t^k}{k!}A^k t^k.$$

If A is diagonal, then let $a_1, a_2, \ldots, a_n$ denote the diagonal elements. Since for all $k \geq 1$,

$$A^k = \text{diag}(a_1^k, a_2^k, \ldots, a_n^K),$$

we have

$$e^{At} = \sum_{k=0}^{\infty} \frac{t^k}{k!}\text{diag}a_1^k, a_2^k, \ldots, a_n^k$$

$$= \text{diag}\left(\sum_{k=0}^{\infty} \frac{t^k}{k!}a_1^k, \sum_{k=0}^{\infty} \frac{t^k}{k!}a_2^k, \ldots, \sum_{k=0}^{\infty} \frac{t^k}{k!}a_n^k\right)$$

$$= \text{diag}(e^{a_1 t}, e^{a_2 t}, \ldots, e^{a_n t}).$$

Assume that matrix A can be transformed into diagonal form; i.e., with some nonsingular matrix T and diagonal matrix D, $A = TDT^{-1}$. Therefore, for all $k \geq 1$,

$$A^k = (TDT^{-1})(TDT^{-1})\ldots(TDT^{-1}) = TD^k T^{-1},$$

which implies that

$$e^{At} = \sum_{k=0}^{\infty} \frac{t^k}{k!}A^k = \sum_{k=0}^{\infty} \frac{t^k}{k!}TD^k T^{-1} = T\left(\sum_{k=0}^{\infty} \frac{t^k}{k!}D^k\right)T^{-1} = Te^{Dt}T^{-1},$$

where the second factor can be obtained easily as shown above. Consider next the case of a block diagonal matrix A,

$$A = \text{diag}(A_1, A_2, \ldots, A_n).$$

Then, $A^k = \text{diag}(A_1, A_2, \ldots, A_n)$, and similarly to the diagonal case we have

$$e^{At} = \text{diag}(e^{A_1 t}, e^{A_2 t}, \ldots, e^{A_n t}).$$

Assume now that A is an $n \times n$ Jordan block:

$$A = \begin{pmatrix} \lambda & 1 & & & \\ & \lambda & 1 & & \\ & & \cdot & \cdot & \\ & & & \cdot & \cdot \\ & & & & \cdot & 1 \\ & & & & & \lambda \end{pmatrix}$$

Notice first that $A = \lambda I + N$, where

$$N = \begin{pmatrix} 0 & 1 & & & \\ & 0 & 1 & & \\ & & \cdot & \cdot & \\ & & & \cdot & \cdot \\ & & & & \cdot & 1 \\ & & & & & 0 \end{pmatrix}$$

is a nilpotent matrix, $N^n = 0$. Therefore, for all $k \geq 1$,

$$A^k = (\lambda I + N)^k = \sum_{i=0}^{k} \binom{k}{i} \lambda^{k-i} N^i = \sum_{i=0}^{\min\{k; n-1\}} \binom{k}{i} \lambda^{k-i} N^i$$

$$= \begin{pmatrix} \lambda & \binom{k}{1} \lambda^{k-1} & \binom{k}{1} \lambda^{k-2} & \cdots & \binom{k}{n-1} \lambda \\ & \lambda & \binom{k}{1} \lambda^{k-1} & \cdots & \binom{k}{n-2} \lambda^2 \\ & & \cdot & & \cdot \\ & & & \cdot & \cdot \\ & & & & \cdot \\ & & & & \lambda^k \end{pmatrix}.$$

Consequently, simple calculation shows that

$$e^{At} = \begin{pmatrix} e^{\lambda t} & \frac{\lambda e^{\lambda t}}{1!} & \frac{\lambda^2 e^{\lambda t}}{2!} & \cdots & \frac{\lambda^{n-1} e^{\lambda t}}{(n-1)!} \\ & e^{\lambda t} & \frac{\lambda e^{\lambda t}}{1!} & \cdots & \frac{\lambda^{n-2} e^{\lambda t}}{(n-2)!} \\ & & \cdot & & \cdot \\ & & & \cdot & \cdot \\ & & & & \cdot \\ 0 & & & e^{\lambda t} & \end{pmatrix}$$

Because an arbitrary $n \times n$ real matrix can be transformed into a Jordan canonical form, the previous special cases can be easily combined to compute the matrix exponential for arbitrary real matrices. We know that $A = T \text{diag}(A_1, A_2, \ldots, A_r) T^{-1}$ with some non-singular matrix T, and Jordan blocks $A_1, A_2, \ldots, A_r$. Then,

$$e^{At} = T \text{diag}(e^{A_1 t}, e^{A_2 t}, \ldots, e^{A_r t}) T^{-1}.$$

where all diagonal blocks are easy to determine.

Case Study 2.10. Examined here is the iterative solution of the non-linear initial value problem

$$\dot{X}(t) = f(t, x(t), x(t_0)) = x_0 \tag{2.45}$$

where $f : [t_0, \infty) \times R^n \to R^n$ is a continuous function, which satisfies the Lipschitz condition. That is, for all $x, x' \in R^n$,

$$\|f(t, x) - f(t, x')\| \leq L \cdot \|x - x'\|\|,$$

where $L > 0$ is a fixed constant.

For example, equation (2.42) satisfies the Lipschitz condition, since with arbitrary column x of matrix X,

$$\|Ax - Ax'\| = \|A(x - x')\| \leq \|A\| \cdot \|x - x'\|,$$

therefore, we may select $L = \|A\|$. For solving the initial value problem (2.45) the iteration scheme (1.23) can be applied, which we repeat here for convenience:

$$x_{k+1}(t) = x_0 + \int_0^t f(\tau, x_k(\tau)) d\tau. \tag{2.46}$$

Based on the results of this chapter, it can be shown that by starting with the initial solution $x_0(t) \equiv x_0$, this procedure converges to the unique solution of the initial value problem (2.45).

Example 2.8.

$$\dot{x}(t) = 1 + x(t), \quad x(0) = 0.$$

The iteration procedure (2.46) now reduces to

$$x_{k+1}(t) = \int_0^t (1 + x_k(\tau)) d\tau.$$

Select $x_0(t) = 0$, then

$$x_1(t) = \int_0^t 1 d\tau = \tau,$$

and similarly

$$x_2(t) = \int_0^t (1+\tau)d\tau = t + \frac{t^2}{2},$$

$$x_3(t) = \int_0^t (1+\tau+\frac{\tau^2}{2})dt = t + \frac{t^2}{2} + \frac{t^3}{3!}$$

and so on. Finite induction implies that in general,

$$x_k(t) = t + \frac{t^2}{2} + \frac{t^3}{3!} + \cdots + \frac{t^k}{k!},$$

which converges to the function $x(t) = e^t - 1$. Hence, this is the solution of the original initial value problem.

Higher-order ordinary differential equations can also be solved by the application of the above iteration procedure. In the first step, the higher order equation must be rewritten as a system of first order equations, and then the iteration algorithm can be readily applied.

Example 2.9. Consider the second-order equation

$$\ddot{x} + x = 1, x(0) = 1, \dot{x}(0) = 1.$$

Introduce the new variables

$$x_1 = x \text{ and } x_2 = \dot{x},$$

Then, we have

$$\dot{x}_1 = x_2 \quad x_1(0) = 1$$
$$\dot{x}_2 = -x_1 + 1 \quad x_2(0) = 1.$$

The iteration scheme (2.46) can be written as

$$x_{1,k+1}(t) = 1 + \int_0^t x_{2,k}(\tau)d\tau$$

$$x_{2,k+1}(t)1 + \int_0^t (-x_{1,k}(\tau)+1)d\tau.$$

Select the initial approximations $x_{1,0}(t) = x_{2,0}(t) = 1$. Simple algebra shows that

$$x_{1,1}(t) = 1 + \int_0^t 1 d\tau = 1 + t,$$

$$x_{2,1}(t) = 1 + \int_0^t (-1+1) d\tau = 1,$$

$$x_{1,2}(t) = 1 + \int_0^t 1 d\tau = 1 + t,$$

$$x_{2,2}(t) = 1 + \int_0^t (-\tau) d\tau = 1 - \frac{t^2}{2!},$$

$$x_{1,3}(t) = 1 + \int_0^1 \left(1 - \frac{\tau^2}{2}\right) d\tau = 1 + t - \frac{t^3}{3!},$$

$$x_{2,3}(t) = 1 + \int_0^t (-\tau) d\tau = 1 - \frac{t^2}{2!},$$

$$x_{1,4}(t) = 1 + \int_0^1 \left(1 - \frac{\tau^2}{2}\right) d\tau = 1 + t - \frac{t^3}{3!},$$

$$x_{2,4}(t) = 1 + \int_0^1 \left(\tau - \frac{\tau^2}{3!}\right) d\tau = 1 + \frac{t^2}{2} + \frac{t^4}{4!},$$

By using finite induction, we recognize that $x_1,k(t) \to \sin t + 1$ and $x_2,k(t) \to \cos t$ as $t \to \infty$. Since $x(t) = x_1(t)$, the solution of the original problem is $x(t) = \sin t + 1$.

Finally, we note that a similar second order equation was solved in Example 1.13.

Case Study 2.11. Consider again the Fredholm-type integral equation

$$x(t) = \int_a^b K(s,t)x(s)ds + f(t),$$

where $K : [a,b] \times [a,b] \to R$ and $f : [a,b] \to R$ are continuous functions. Assume furthermore that for all $t \in [a,b]$,

$$\int_a^b |K(s,t)|ds \le q < 1.$$

It is well known that under this additional assumption the Fredholm-type integral equation has exactly one continuous solution, and from the results of Section 2.2 we conclude that starting from arbitrary continuous initial approximation $x_0(t)$, the iteration algorithm

$$x_{k+1}(t) = \int_a^b K(s,t)x_k(s)ds + f(t) \qquad (2.47)$$

converges to the unique solution of the equation. We also note that in the case of the Volterra-type integral equation the continuity of functions C and f is sufficient to guarantee the convergence of algorithm (1.27). Note that Examples 1.16 and 1.18 illustrate these iteration methods.

Case Study 2.12. Assume that $A : D \to D$ is a mapping, with $D \subseteq R^n$ being a closed, convex set. Consider the iteration algorithm

$$x_{k+1} = \alpha_k x_k + (1 - \alpha_k)A(x_k), \qquad (2.48)$$

where $\alpha_k \in [0,1)$ for all $k \geq 0$. Because D is convex, the algorithm is well defined. This is because for all $x \in D, \alpha_k x + (1-\alpha_k)A(x) \in D$.

First, we show that the convergence of the iteration process

$$z_{k+1} = A(z_k) \qquad (2.49)$$

to the unique fixed point of equation $z = A(z)$ does not imply the convergence of algorithm (2.48) to the same limit.

Example 2.10. Define $D = [0,2]$ and

$$A(x) = \begin{cases} 1 & \text{of } 0 \leq x < 1 \\ 2 & \text{of } 1 \leq x \leq 2 \end{cases}$$

Notice that mapping A has a unique fixed point: $z^* = 2$. If $z_0 \in [0,1)$, then $z_1 = A(z_0) = 1$ and $z_2 = A(z_1) = 2 = z^*$. If $z_0 \in [1,2]$, then $z_1 = A(z_0) = 2 = z^*$. Hence, process (2.49) converges to z^*. However, select for all $k, \alpha_k \in (0,1)$, and an arbitrary $x_0 \in [0,1)$. Then, $x_1 = \alpha 1 \cdot x_0 + (1 - \alpha_0) \cdot 1 \in [0,1)$, and finite induction shows that for all $k, x_k \in [0,1)$. Hence, $x_k \nrightarrow z^*$.

Next, we show that if mapping A is a contraction, then the algorithmic mapping (2.48) is not necessarily a contraction even if $\alpha_k = \alpha(x_k)$.

Example 2.11. Let $D = [0,2]$ and define $A(x) = 2$ for all $x \in D$. Set $\alpha(0) = 0.9$ and $\alpha(1) = 0.1$. Then, for the mapping

$$B(x) = \alpha(x)x + (1 - \alpha(x))A(x),$$
$$B(0) = 0.9 \times 0 + 0.1 \times 2 = 0.2$$

and
$$B(1) = 0.1 \times 1 + 0.9 \times 2 = 1.9.$$

Notice that A is obviously a contraction; however,
$$|B(1) - B(0)| = 1.7 \text{ and } |1 - 0| = 1.$$

Hence, B is not a contraction.

Assume now that in algorithm (2.48), mapping A is a contraction. Furthermore, $0 \leq \alpha_k \leq 1 - \varepsilon$ for all k, where $\varepsilon > 0$ is a fixed constant. We can easily show that x_k converges to the unique fixed point of A as $k \to \infty$. Let z^* again denote the fixed point of A, then

$$\begin{aligned}
\|x_{k+1} - z^*\| &= \|\alpha_k x_k + 1(1 - \alpha_k)(x_k) - \alpha_k z^* - (1 - \alpha_k)z^*\| \\
&\leq \|\alpha_k(x_k - z^*)\| + \|(1 - \alpha_k)(A(x_k) - A(z_*))\| \\
&= \alpha_k\|x_k - z^*\| + (1 - \alpha_k)\|A(x_k) - A(z^*)\| \\
&\leq \alpha_k\|(x_k - z^*)\| + (1 - \alpha_k)K\|x_k - z^*\|,
\end{aligned}$$

where $K < 1$ is the contraction constant of mapping A. Therefore,

$$\|(x_{k+1} - z^*)\| \leq (\alpha_k + (1 - \alpha_k)K)\|x_k - z^*\|,$$

where

$$\begin{aligned}
0 \leq \alpha_k + (1 - \alpha_k)K &= \alpha_k(1 - K) + K \\
&\leq (1 - \varepsilon)(1 - K) + K = 1 - \varepsilon(1 - K) < 1.
\end{aligned}$$

Hence,
$$\|x_k - z^*\| \to 0 \text{ as } k \to \infty.$$

The convergence of algorithm (2.48) will now be established under different conditions than above. Let $\|.\|$ denote the usual Euclidean norm, and assume that for all $x \in D$,

$$\|A(x) - z^*\| \leq K\|x - z^*\| \tag{2.50}$$

with some constant K, and

$$(x_k - z^*)^T (A(x) - z^*) \leq \beta(x) \cdot \|x_k - z^*\|^2, \tag{2.51}$$

where $\beta : D \to R$ is some real valued function. The first condition is satisfied if, for example, mapping A is bounded in D, and the second condition is necessarily satisfied if $(-A)$ is monotone (in the sense of Ortega and Rheinboldt [19]). Similar to the previous case we have the inequality

$$\begin{aligned}
\|x_{k+1} - z^*\|^2 &= \|\alpha_k x_k + (1 - \alpha_k)A(x_k) - \alpha_k z^* - (1 - \alpha_k)z^*\|^2 \\
&= \|\alpha_k(x_k - z^*) + (1 - \alpha_k)(A(x_k) - z^*)\|^2 \\
&= \alpha_k^2\|(x_k - z^*)\|^2 + (1 - \alpha_k)^2\|A(x_k) - A(z^*)\|^2 \\
&\quad + 2\alpha_k(1 - \alpha_k)(x_k - z^*)^T (A(x_k) - z^*) \\
&\leq [\alpha_k^2 + (1 - \alpha_k)^2 K^2 + 2\alpha_k(1 - \alpha_k)\beta_k]\|x_k - z^*\|^2 + \|x_k - z^*\|^2,
\end{aligned}$$

where $\beta_k = \beta(x_k)$. The algorithm (2.48) converges to z^*, if with some small positive ε,

$$\alpha_k^2 + (1 - \alpha_k)^2 K^2 + 2\alpha_k(1 - \alpha_k)\beta_k \leq 1 - \varepsilon \quad (k \geq 0).$$

This inequality can be rewritten as

$$\alpha_k^2 + (1 - K^2 - 2\beta_k) + 2\alpha_k(K^2 - \beta_k) + k^2 \leq 1 - \varepsilon.$$

Denote the left-hand side of this inequality by $p(\alpha_k)$. Notice first that the Cauchy-Schwarz inequality implies that for all $x \in D$,

$$(x - z^*)^T (A(x) - z^*) \leq \|x - z^*\| \cdot \|A(x) - z^*\|$$
$$\leq \|x - z^*\| \cdot K\|x - z^*\| = K \cdot \|x - z^*\|^2.$$

Therefore, we may assume that $\beta(x) \leq K$ for all $x \in D$. Observe next that

$$p(1) = 1 + K^2 - 2\beta_k - 2(K^2 - \beta_k) + K^2 = 1$$

and

$$p(0) = k^2.$$

Assume first that $1 + K^2 - 2\beta_k = 0$. Because

$$1 + k^2 - 2\beta_k \geq 1 + K^2 - 2K = (K - 1)^2 \geq 0$$

this is the case only when $K = \beta_k = 1$. Then $p(\alpha_k) = K^2 = 1$, therefore, no satisfactory α_k exists.

Assume next that $1 + K^2 - 2\beta_k > 0$. Then $p(\alpha_k)$ is a convex function. The case of $K < 1$, and α_k is satisfactory, if it is small enough. Satisfactory values of a_k exist if and only if the vertex a^* of the parabola $p(\alpha_k)$ is in the interval $(0, 1)$ and $p(a^*) \leq 1 - \varepsilon$. Simple differentiation shows that

$$\alpha^* = \frac{K^2 - \beta_k}{1 + K^2 - 2\beta_k},$$

and it is in $(0, 1)$ if and only if $\beta_k < 1$. Furthermore,

$$p(\alpha^*) = \frac{K^2 - \beta_k}{1 + k^2 - 2\beta_k}.$$

Hence, satisfactory values of α_k exist for $K > 1$ if and only if $\alpha_k < 1$ and $p(\alpha^*) \leq 1 - \varepsilon$.

We note that algorithm (2.48) has important applications in investigating dynamic economic systems. If the state transition mapping is A and the distance between x_k and $A(x_k)$ is large, then the state of the system cannot be changed from x_k to $A(x_k)$ during a single time period. In such cases the new state at time period $k + 1$ is selected as a point between x_k and $A(x_k)$. In other words, the state moves to the direction of the best choice for the state of the new time period; however, it moves only as far from x_k as possible.

Case Study 2.13. Algorithmic process (2.48) can be further generalized in the following way. Let $A_i : D \to D, i = 1, 2, \ldots, 1$, be bounded mappings on D, where $D \subseteq R^n$ is a closed, convex set. Assume that for all i, z^* is a fixed point of A_i. Further, for all $x \in D$,

$$\|A_i(x) - z^*\| \leq K_i \|x - z^*\|. \tag{2.52}$$

Consider the general algorithm

$$x_{k+1} \sum_{i=1}^{l} \alpha_{ik} A_i(x_k), \tag{2.53}$$

where $\alpha_{ik} \geq 0$ is a constant for all i and k such that $\sum_{i=1}^{l} a)ik = 1$ for all k. Then, similarly to algorithm (2.48),

$$\|x_{k+1} - z^*\| = \|\sum_{i=1}^{l} a_{ik}(A_i(x_k) - z^*)\| \leq \left(\sum_{i=1}^{l} a_{ik} K_i\right) \|x_k - z^*\|$$

If we assume that for all $k \geq 0$,

$$\sum_{i=1}^{l} a_{ik} K_i \leq 1 - \varepsilon$$

with some $\varepsilon > 0$, then algorithm (2.53) converges to z^*.

Notice that this algorithm can be further generalized as

$$x_{k+1} = F_k(A_1(x_k), A_2(x_k), \ldots, A_l(x_k)), \tag{2.54}$$

where $F_k : D^l \to D$ is a mapping such that $F_k(z^*, \ldots, z^*) = z^*$. In addition, for all $x^{(1)}, \ldots, x^{(1)} \in D$,

$$\|F_k(x^{(1)}, \ldots, x^{(l)})\| \leq \sum_{i=1}^{l} \|x^{(i)} - z^*\|.$$

One may easily show that under the same conditions on the coefficients α_{ik} as before, this general algorithm also converges to z^*.

Consider again the linear algorithm (2.53) and assume that for all i, j, and $x \in D$,

$$(A_i(x) - z^*)^T (A_i(x) - z^*) \leq \beta_{ij}(x) \cdot \|x - z^*\|. \tag{2.55}$$

Then with the Euclidean norm, we have

$$\|x_{k+1} - z^*\|^2 = \left\|\sum_{i=1}^{l} a_{ik}(A_i(x_k) - z^*)\right\|^2 \leq \left[\sum_{i=1}^{l} \sum_{j=1}^{l} a_{ik} a_{jk} \beta_{ik}(x_k)\right] \cdot \|x_k - z^*\|^2.$$

Hence, the algorithm converges to z^*, if with some $\varepsilon > 0$,

$$\sum_{i=1}^{l} \sum_{j=1}^{l} \alpha_{jk} \alpha_{jk} \beta_{ik}(x_k) \leq 1 - \varepsilon (k \geq 0).$$

Case Study 2.14. In this case study, let X be any set, and assume that $\rho : X \times X \to R_+$ is the distance function on X, and the metric space (X, ρ) is complete. Assuming that f is a point-to-point mapping, $f : X \to X$. Define

$$V_1 = \{(\alpha, \beta, \gamma) | \alpha, \beta, \gamma \in R, 0 \le \alpha + \beta \le 2(\alpha + \beta) + \gamma < 1, \beta \ge 0\},$$
$$V_2 = \{(\alpha, \beta, \gamma) | \alpha, \beta, \gamma \in R, 0 \le \alpha + \beta \le 2\alpha + \gamma < 1, \beta < 0\},$$

and let $V = V_1 \cup V_2$. Assume that with some $(\alpha, \beta, \gamma) \in V$,

$$\rho(f(x), f(y)) \le \alpha(\rho(x, f(x)) + \rho(y, f(y))) + \beta(\rho(x, f(y)) + \rho(y, f(y))) + \gamma \rho(x, y) \tag{2.56}$$

for all $x, \gamma \in X$.

Theorem 2.15. Under the above assumptions f has at least one fixed point on X. If in addition, $2\beta + \gamma < 1$, then f has exactly one fixed point.

Proof. First we prove the existence of a fixed point. Consider the iteration sequence $x_{k+1} = f(x_k)$. Notice first that assumption (2.56), with $x = x_{n+1}$ and $y = x_n$, implies that

$$\rho(x_{n+2}, x_{n+1}) \le \alpha(\rho(x_{n+1}, x_{n+2}) + \rho(x_n, x_{n+1}))\beta(\rho(x_{n+1}, x_{n+1})$$
$$+ \rho(x_n, x_{n+2})) + \gamma \rho(x_{n+1}, x_n).$$

$\square$

which implies that

$$(I - \alpha)\rho(x_{n+1}, x_{n+2}) \le \beta \rho(x_n, x_{n+2}) + (\alpha + \gamma)\rho(x_n, x_{n+1}). \tag{2.57}$$

Two cases are discussed.

1   If $\beta \ge 0$, then $(\alpha, \beta, \gamma) \in V_1$. The triangle inequality and relation (2.57) imply that

$$(1 - \alpha)\rho(x_{n+1}, x_{n+2}) \le \beta \rho(x_n, x_{n+1}) + \rho(x_{n+1}, x_{n+2})(\alpha + \gamma)\rho(x_n, x_{n+1}).$$

Because $1 - \alpha - \beta > 0$ and $\alpha + \beta + \gamma \ge 0$, simple algebra shows that

$$\rho(x_{n+1}, x_{n+2}) \le \frac{\alpha + \beta \gamma}{1 - \alpha - \beta} \rho(x_n, x_{n+1}).$$

Notice that $q' = \frac{\alpha + \beta + \gamma}{1 - \alpha - \beta} \in [0, 1)$.

2   If $\beta < 0$, then $(\beta, \beta, \gamma) \in V_2$. Relation (2.57) now implies that

$$(1 - \alpha)\rho(x_{n+1}, x_{n+2}) \le \beta - \rho(x_n, x_{n+1}) + \rho(x_{n+1}, x_{n+2})(\alpha + \gamma)\rho(x_n, x_{n+1}).$$

In this case $1 - \alpha - \beta > 0$ and $\alpha + \gamma - \beta \geq 0$; therefore,

$$\rho(x_{n+1}, x_{n+2}) \leq \frac{\alpha + \gamma - \beta}{1 - \alpha - \beta} \rho(x_n, x_{n+1}),$$

and

$$q'' = \frac{\alpha + \gamma - \beta}{1 - \alpha - \beta} \in [0, 1).$$

If one selects $q = \max q', q'' \in [0, 1)$, then

$$\rho(x_n, x_{n+1}) \leq q^n \rho(x_0, x_1).$$

Let $n < m$, then

$$\rho(x_n, x_m) \leq \rho(x_{n+1}, x_{n+2}) \rho(x_{n+1}, x_{n+2}) + \cdots + \rho(x_{m-1}, x_m)$$

$$\leq (q^n + q^{n+1} + \cdots + q^{m-1}) \rho(x_0, x_1) \leq \frac{q^n}{1 - q} \rho(x_0, x_1). \qquad (2.58)$$

Therefore, $\{x_n\}$ is a Cauchy sequence. Since (X, ρ) is complete, $x_n \to x^*$ as $n \to \infty$ with some $x^* \in X$. We will next show that x^* is a fixed point of mapping f, which follows immediately from the relation

$$0 \leq \rho(x^*, f(x^*)) = \lim_{n \to \infty} \rho(f(x_{n-1}), f(x^*))$$

$$\leq \lim_{n \to \infty} \{ \alpha(\rho(x_{n-1}, x_n) \rho(x^*, f(x^*))) + \beta(\rho(x_{n-1}, f(x^*)) + \rho(x^*, x_n)) + \gamma \rho(x_{n-1}, x_*) \}$$

$$= (\alpha + \beta) \rho(x^*, f(x^*)).$$

Because $\alpha + \beta < 1, \rho(x^*, f(x^*)) = 0$. Hence, $x^* = f(x^*)$; i.e., x^* is a fixed point of f.

Assume next that $2\beta + \gamma < 1$, and x^*, x^{**} are two fixed points. Then (2.56) implies that

$$0 \leq \rho(x, x^{**}) \leq (2\beta + \gamma) \rho(x, x^{**}),$$

and therefore $\rho(x^*, x^{**}) \leq (2\beta + \gamma) \rho(x, x^{**})$.

Remark 2.17. If $\alpha = \beta = 0$, and $0 \leq \gamma < 1$, then condition (2.56) means that mapping f is a contraction. Therefore, the theorem generalizes the contraction mapping theorem (see, for example, Szidarovszky and Yakowitz [24]).

Remark 2.18. By letting $m \to \infty$ in inequality (2.58) we conclude that

$$\rho(x_n, x^*) \leq \frac{q^n}{1 - q} \rho(x_0, x_1),$$

which provides a practical error estimator.

Case Study 2.15. Considered here is the two-step process

$$x_{k+1} = f(x_k, x_{k-1}),$$

where $f : D \times D \to D, D \subseteq R^N$ is a closed set. Assume that for all $x, y, z \in D$,

$$\|f(x,y) - f(y,z)\| \le \alpha\|x - y\| + \beta\|y - z\|, \tag{2.59}$$

where $\alpha + \beta < 1$. Notice first that α and β are non-negative, which follows from (2.59) with $y = z$ and $x = y$, respectively.

Theorem 2.16. Under the above assumptions x_k converges to the unique solutions s^* of equation $x = f(x,x)$, and for all $k \ge 0$,

$$\|x_k - s^*\| \le c_1\lambda_1^k + c_2\lambda_2^k, \tag{2.60}$$

where c_1 and c_2 are non-negative constants, and λ_1 and λ_2 are the roots of the quadratic equation $\lambda^2 - \alpha\lambda - \beta = 0$.

Proof. Introduce the notation $\delta_k = \|x_{k+1} - x_k\|$, then

$$\delta_{k+1} = \|x_{k+2} - x_{k+1}\| = \|f(x_{k+1}, x_k) - f(x_k, x_{k-1})\|$$
$$\le \alpha\|x_{k+1} - x_k\| + \beta\|x_k - x_{k-1}\| = \alpha\delta_k + \beta\delta_{k-1}.$$

Define sequence $\{t_k\}$ by the recursion

$$t_{k+1} = \alpha t_k + \beta t_{k-1}, \quad t_0 = \delta_0 \text{ and } t_1 = \delta_1.$$

It is easy to show by finite induction that for all $k \ge 0$,

$$\delta_k \le t_k.$$

Notice that the recursion of the t_k is a second order difference equation with characteristic polynomial $\lambda^2 - \alpha\lambda - \beta$. The roots of this polynomial are

$$\lambda_{1,2} = \frac{-\alpha \pm \sqrt{a^2 + 4\beta}}{2}$$

. If $\alpha = \beta = 0$, then $\lambda_1 = \lambda_2 = 0$, otherwise $-1u\lambda_1 \le 0 \le \lambda_2 < 1$. Therefore,

$$t_k = a_1\lambda_1^k + a_2\lambda_2^k$$

for all $k \ge 0$ with some constants a_1 and a_2. Let $K \ge k$, then

$$\|x_k - x_*\| \le \|x_{k+1} - x_k\| + \|x_{k+2} - x_{k+1}\| + \cdots + \|x_k - x_{k-1}\|$$
$$\le \sum_{i=k}^{\infty} \delta_i \le \sum_{l=k}^{\infty} t_l = \frac{al}{1 - \lambda_1}\lambda_1^k + \frac{al}{1 - \lambda_2}\lambda_2^k. \tag{2.61}$$

Hence, $\{x_k\}$ is a Cauchy sequence. Because D is closed, $x_k \to x^*$ as $k \to \infty$ with some $x^* \in D$. By letting $K \to \infty$ in (2.62), we obtain (2.61) with

$$c_1\frac{a_1}{1 - \lambda_1} \text{ and } c_2\frac{a_2}{1 - \lambda_2}.$$

Next, we show that x^* solves equation $x = f(x,x)$. Assumption (2.59) implies that

$$\|f(x^*,x^*) - x_{k+1}\| \le \|f(x^*,x^*) - f(x^*,x_k)\| + \|f(x^*,x_k) - f(x_k,x_{k-1})\|$$
$$\le \beta(\|x^* - x_k\| + \|x_k - x_{k-1}\|) + \alpha\|x^* - x_k\|, \quad\quad (2.62)$$

where the right-hand side converges to zero as $k \to \infty$. Therefore,

$$x^* = \lim_{k \to \infty} x_{k+1} = f(x^*,x^*).$$

We will now verify that equation $x = f(x,x)$ has a unique solution. Assume that both x^* and x^{**} are solutions. Then

$$\|x^* - x^*\| = \|f(x^*,x^*) - f(x^{**},x^{**})\|$$
$$\le \|f(x^*,x^*) - f(x^*,x^{**})\| + \|f(x^*,x^{**}) - f(x^{**},x^{(**)})\|$$
$$\le \beta\|(x^* - x^{**}) + \alpha(x^* - x^{**})\|$$

which implies that

$$(1 - \alpha - \beta)\|x^* - x^{**}\| \le 0.$$

Because $\alpha + \beta < 1$, $x^* = x^{**}$, which completes the proof. $\quad\quad\quad\quad\square$

Remark 2.19. The theorem can easily be extended to metric spaces and l-step processes for $l > 2$. The details are omitted.

Case Study 2.16. Let $f : X \to X$, where X is a complete metric space. Assume that for all $x \in X$ and with some $q \in [0,1)$,

$$\rho(f(f(x)),f(x)) \le q\rho(f((x)),x). \quad\quad (2.63)$$

Mapping f is called an iterated contraction.

Theorem 2.17. If (2.63) holds, then iteration procedure $X_{k+1} = f(x_k)$ converges to a point $x^* \in X$, and for all $k \ge 0$

$$\rho(x_k,x^*) \le \frac{q^k}{1-q}\rho(x_1,x_0). \quad\quad (2.64)$$

If f is continuous at x^*, then $x^* = f(x^*)$.

Proof. Notice that for all $k \ge 1$,

$$\rho(x_{k+1},x_k) = \rho(f(f(x_{k-1})),f(x_{k-1}))$$
$$\le q\rho(f(x_{k-1}),x_{k-1}) = q\rho(x_k,x_{k-1}).$$

Then similar to the proof of Theorem 2.15, we obtain that for all $K \ge k$,

$$\rho(x_k,x_k) \le \frac{q_k}{1-q}(x_0,x_1).$$

Hence, sequence $\{x_k\}$ is a Cauchy sequence, and thus converges to an $x^* \in X$. Letting $K \to \infty$ leads to (2.64). If f is continuous at x^*, then let $k \to \infty$ in equation $x_{k+1} = f(x_k)$ to conclude that $x^* = f(x^*)$. $\quad\quad\quad\quad\square$

Case Study 2.17. Let X again be a complete metric space, $f : X \to X$, and consider the iteration procedure $X_{k+1} = f(x_k)$. Assume that for all $k \geq 0$,

$$\rho(x_{k+1}, x_k) \leq t_{k+1} - t_k, \tag{2.65}$$

where the real sequence $\{t_k\}$ is non-negative and converges to a finite limit t^*. Sequence $\{t_k\}$ is called a majorizing sequence for $\{x_k\}$. Notice that $\{t_k\}$ is increasing.

Theorem 2.18. Under the above conditions an $x^* \in X$ exists such that $x_k \to x^*$ as $k \to \infty$, and

$$\rho(x_k, x^*) \leq t^* - t_k \text{ for } k \geq 0 \tag{2.66}$$

Proof. Let $K > k$, then

$$\rho(x_k, x_k) \leq \rho(x_k, x_{k+1}) + \rho(x_{k+1}, x_{k+2}) + \cdots + \rho(x_{k-1}, x_k)$$
$$\leq (t_{k+1} - t_k) + (t_{k+2} - t_{k+1}) + \cdots + (t_k - t_{k-1}) = t_k - t_k.$$

Because $\{t_k\}$ is increasing and convergent, $t_k - t_k \to 0$ as $k, K \to \infty$. Therefore, $\{x_k\}$ is a Cauchy sequence, and thus converges to an $x^* \in X$. Letting $K \to \infty$ in the above inequality leads to the error estimator (2.66). $\square$

Case Study 2.18. Outlined here is another generalization of the contraction mapping theorem. Let (X, ρ) be a metric space, and $A \subseteq X$. Define

$$\delta(A) \sup\{\rho(a, b) | a, b \in A\},$$

and for each $x, y \in X$, let

$$\begin{aligned}
\Delta(x, n) &= \{x, f(x), \dots, f^n(x)\}, \\
\Delta(x, \infty) &= \{x, f(x), \dots, f^2(x)\}, \\
\Delta(x, y, n) &= \{x, y, f(x), f(y), \dots, f^{(n)}(x), f^{(n)}(y)\}, \\
\Delta(x, y\infty) &= \{x, y, f(x), f(y), \dots\},
\end{aligned} \tag{2.67}$$

where

$$f^1(x) = f(x) \text{ and } f^n(x) = f(f^{n-1}(x)) \quad (n \geq 2).$$

Let $\{x_k\}$ be an arbitrary sequence in X, and define

$$\Delta(x_i, j) = \{x_i, x_{i+1}, \dots, x_{i+j}\} (j \geq 1 \geq 0)$$

and

$$\Delta(x_i, \infty) = \{x_i, x_{i+1}, \dots\} (i \geq 0).$$

Sequence $\{x_k\}$ is called a Banach sequence if $\delta(\Delta(x_0, \infty)) < \infty$ and a $q \in [0, 1)$ exists such that for all $n \geq 1$,

$$\delta(\Delta(x_n, \infty)) \leq q\delta(\Delta(x_{n-1}, \infty)). \tag{2.68}$$

This condition is obviously equivalent to the following: For all $i \geq n, j \geq n (n \geq 0)$,

$$\rho(x_i, x_j) \leq q\delta(\Delta(x_n, \infty)). \tag{2.69}$$

Let $0 = v_0 < v_1 < v_2 < \ldots$ be a sequence of non-negative integers. We say that $\{x_k\}$ is a generalized Banach sequence if $\delta(\Delta(x_0, \infty)) < \infty$ and a $q \in [0, 1)$ exists such that

$$\delta(\Delta(x_{vn}, \infty)) \leq q\delta(\Delta(x_{v_{n-1}}, \infty)). \tag{2.70}$$

This condition is obviously equivalent to the following: for all $i \geq v_n, j \geq v_n (n \geq 0)$,

$$\rho(x_i, x_j) \leq q\delta(\Delta(x_{v_{n-1}}, \infty)). \tag{2.71}$$

Some important properties of Banach sequences and generalized Banach sequences are worthy of mention.

1   Each Banach sequence (and generalized Banach sequence) is a Cauchy sequence.

   Proof.   Because Banach sequences are special cases of generalized Banach sequences with $v_n = n$ for all $n \geq 0$, it is sufficient to present the proof for the more general case. Let $j \geq i$, and $v_n \leq i$. Then

   $$\rho(x_i, x_j) \leq q\delta(\Delta(x_{v_{n-1}}, \infty)) \leq \cdots \leq q^n \delta(\Delta(x_{v_0}, \infty))$$

   with $q^n \to 0$ as $n \to \infty$.   □

   Corollary 2.8. If $\{x_k\}$ is a (generalized) Banach sequence, and (X, ρ) is a complete metric space, then x_k converges to an $x^* \in X$.

2   Sequence $\{x_k\}$ is a generalized Banach sequence (with parameters q and $\{v_n\}$) if and only if for each $n \geq 0$ and $\beta \in (q, 1)$ an $N = N(n, \beta)$ exists such that for all $m \geq N$,

   $$\delta(\Delta(x_{v_m}, m)) < \beta\delta(\Delta(x_{v_{n-1}}, m)) \tag{2.72}$$

   Proof.   Let $\{x_k\}$ be a generalized Banach sequence. If an N exists such that $x_N = x_{N+1} = \ldots$, then the assertion is obviously true. Otherwise for $n \geq 0$,

   $$\delta(\Delta(x_{v_n}, \infty)) \leq q\delta(\Delta(x_{v_{n-1}}, \infty)) < \beta\delta(\Delta(x_{v_{n-1}}, \infty)).$$

   Because

   $$\delta(\Delta(x_k, \infty)) = \lim_{r \to \infty} \delta(\Delta(x_k, i)), \tag{2.73}$$

   (2.72) holds for sufficiently large m. Assume that a sequence $\{x_k\}$ satisfies (2.72). Since (2.73) holds, let $m \to \infty$ in relation (2.72) to obtain (2.70). Therefore, it is sufficient to show that $\delta(\Delta(x_0, \infty))$ is finite. We shall verify in addition that

   $$\delta(\Delta(x_{v_{n-1}}, \infty)) \leq \frac{1}{1-q}\delta(\Delta(x_{v_{n-1}}, v_n - v_{n-1})) \quad (n \geq 1) \tag{2.74}$$

which implies the assertion. Let $0 < q < \beta < 1$ and let n be arbitrary. For some $m \geq N(n,\beta), m + v_{n-1} > v_n$. Select i and j so that $v_n \leq i \leq v_{n-1} + m$, $v_n \leq j \leq v_{n-1} + m$. Then

$$\rho(x_i, x_j) \leq q\delta(\Delta(x_{v_n})) < \beta\delta(\Delta(x_{v_{n-1}}, m)).$$

Therefore either $\delta(\Delta(x_{\gamma n - 1}, m)) = \delta(\Delta(x_{\gamma n - 1 j}, \gamma_n - \gamma_{n-1}))$ integers k and l exist such that $v_{n-1} \leq k \leq v_n, v_n < 1 \leq v_{n-1} + m$, and

$$\delta(\Delta(x_{v_{n-1}}, m)) = \rho(x_k, x_1),$$

which implies that

$$\begin{aligned}\delta(\Delta(x_{v_{m-1}}, m)) = \rho(x_k, x_1) &\leq \rho(x_k, x_{v_n}) + \rho(x_{v_n}, x_1)\\ &\leq \delta(\Delta(x_{v_{n-1}}, v_n - v_{n-1}))\delta(\Delta(x_{v_n}, m))\\ &< \delta(\Delta(x_{v_{n-1}}, v_n - v_{n-1})) + \beta\delta(\Delta(x_{v_{n-1}}, m)).\end{aligned}$$

Thus,

$$\delta(\Delta(x_{v_{n-1}}, m)) < \frac{1}{1-\beta}\delta(\Delta(x_{v_{n-1}}, v_n - v_{n-1}))$$

Let $m \to \infty$ and notice that $\beta > q$ is arbitrary, which implies that (2.74) holds.   □

3   Let $\{x_k\}$ be a generalized Banach sequence, assume that metric space (X, ρ) is complete, and let $x^* = \lim_{k \to \infty} x_k$. Then

$$\rho(x_{v_n}, x^*) \leq \frac{q^n}{1-q}\delta(\Delta(x_0, v_1)) \quad (n \geq 0) \tag{2.75}$$

and

$$\rho(x_{v_n}, x^*) \leq \frac{q}{1-q}\delta(\Delta(x_{v_{n-1}}, v_n - v_{n-1})) \quad (n \geq 1) \tag{2.76}$$

Proof.   Since for $m > n$,

$$\rho(x_{v_n}, x_{v_m}) \leq \delta(\Delta(x_{v_n}, \infty)) \leq q\delta(\Delta(x_{v_{n-1}}, \infty)) \leq \cdots \leq q^n\delta(\Delta(x_0, \infty)),$$

by using (2.74) with $n = 0$ we have

$$\rho(x_{v_n}, x_{v_m}) \leq \frac{q}{1-q}\delta(\Delta(x_0, v_1)).$$

Letting $m \to \infty$ leads to (2.75). Finally, from (2.74) we see that

$$\rho(x_{v_n}, x_{v_m}) \leq (\Delta(x_{v_n}, \infty)) \leq q\delta(\Delta(x_{v_{n-1}}, \infty)) \leq \frac{q^n}{1-q}\delta(\Delta(x_{v_{n-1}}, v_n - v_{n-1}))$$

which implies (2.76) by letting $m \to \infty$.   □

Corollary 2.9. If $\{x_k\}$ is a Banach sequence, then

$$\rho(x_n, x^*) \le \frac{q^n}{1-q} \rho(\Delta(x_0, x_1)) \tag{2.77}$$

and

$$\rho(x_n, x^*) \le \frac{q}{1-q} \rho(x_{n-1}, x_n), \tag{2.78}$$

since in this case $v_n = n$ for all $n \ge 0$.

Consider next the iteration sequence

$$x_{k+1} = f(x_k), \quad x_0 = x \in X. \tag{2.79}$$

Mapping f is called orbitally continuous if for any x^* such that

$$x^* = \lim_{i \to \infty} f^{v_i}(x) \text{ with some } x \in X, f(x^*) = \lim_{i \to \infty} f^{(v_i)}(x).$$

We say that (X, ρ) is f-orbitally complete, if for every sequence $\{f^{v_i}(x)\}, x \in X$, which is a Cauchy sequence, there is a limit point $x^* \in X$.

An orbitally continuous mapping f is called a generalized Banach contraction, if $\delta(\Delta(x, y, \infty)) < \infty$ for all $x, y \in X$, and a $q \in [0, 1)$ exists such that

$$\rho(f(x), f(y)) \le q\delta(\Delta(x, y, \infty)). \tag{2.80}$$

Notice first that any contraction satisfies these properties; that is, this concept really generalizes the contraction principle. Let f be a contraction, and let $(\overline{X}, \overline{\rho})$ be a complete metric space being the extension of (X, ρ). Then f has exactly one fixed point x^* in $\overline{X}$. Then for all $n, m \ge 0$,

$$\rho(f^n(x), f^m(H)) \le \overline{\rho}(x^*, f^m(x)) \le q^n\overline{\rho}(x, x^*) + q^m\overline{\rho}(x^*, x) \le 2q\overline{\rho}(x, x^*),$$

and similarly

$$\rho(f^n(x), f^m(y)) \le \overline{\rho}(f^n(x), x^*) + \overline{\rho}(x^*, f^m(y)) \le q\overline{\rho}(x, x^*) + q\overline{\rho}(y, x^*)$$

and

$$\rho(f^n(x), f^m(x)) \le \overline{\rho}(f^n(y), x^*) + \overline{\rho}(x^*, f^m(y)) \le 2q\overline{\rho}(y, x^*).$$

Hence,

$$\delta(\Delta(x, y, \infty)) \le 2q \max\{\overline{\rho}(x, x^*), \overline{\rho}(y, x^*)\} < \infty$$

and obviously,

$$\delta(\Delta(x, y, \infty)) \ge \rho(x, y),$$

therefore (2.80) holds.

The concept of generalized Banach contractions can be further generalized. An orbitally continuous mapping f is called a generalized Caccioppoli contraction, if $\delta(\Delta(x,y,\infty)) < \infty$ for all $x,y \in X$, and for all $n \geq 1$ and $x,y \in X$,

$$\rho(f^n(x), f^n(y)) \leq q_n \rho(\Delta(x,y,\infty)), \tag{2.81}$$

where

$$\sum_{n=1}^{\infty} q_n < \infty.$$

Notice that (2.80) implies (2.81) with $q_n = q^n$. Our first generalized fixed point theorem is Theorem 2.18.

Theorem 2.19. Let $f : X \to X$ be a generalized Banach contraction, and assume that (X,ρ) is f-orbitally complete. Then

1. f has a unique fixed point $x^* \in X$;
2. For all $x \in X$, the iteration sequence converges to x^*;
3. $\rho(x_n, x^*) \leq \frac{q^n}{1-q} \rho(x_0, x_1), (n \geq 0)$;
4. $\rho(x_n, x^*) \leq \frac{q}{1-q} \rho(x_{n-1}, x_n), (n \geq 1)$.

Proof. First we show that $\{x_n\}$ is a Banach sequence. Let n be arbitrary and let $j > i \geq n$. Then

$$\rho(x_i, x_j) = \rho(f^i(x), f^i(x)) = \rho(f(f^{i-1}(x)), f(f^{i-1}(x))) \leq q\delta(\Delta(f(f^{i-1}(x),\infty))$$
$$= q\delta(\Delta(f^{i-1}(x),\infty)) \leq q\delta(\Delta(f^{n-1}(x),\infty)),$$

which implies (2.69). Hence, $\{x_k\}$ is a Banach sequence. Consequently, it is a Cauchy sequence, and because X is f-orbitally complete, there is a limit point x^* of $\{x_k\}$. The assumption that f is orbitally continuous implies that x^* is a fixed point of f. We next show that the fixed point is unique. Assume that both x^* and x^{**} are fixed points, then

$$\rho(x^*, x^{**}) = \rho(f(x^*), f(x^{**})) \leq q\delta(\Delta(x^*, x^{**}, \infty)) = q\rho(x^*, x^{**}),$$

which implies that

$$(1-q)\rho(x^*, x^{**}) \leq 0.$$

Hence, $\rho(x^*, x^{**}) = 0$; that is, $x^* = x^{**}$. Notice finally that assertions 3 and 4 above are the same as relations (2.77) and (2.78). □

Remark 2.20. Error bound (3) can be used to estimate the necessary number of iteration steps before starting the computations. Let $\varepsilon > 0$ be given. The iterate x_n will approximate x^* within the absolute error ε if

$$\frac{q^n}{1-q} \rho(x_0, x_1) < \varepsilon.$$

This relation holds for

$$n > \frac{\log \frac{\varepsilon(1-q)}{\rho(x_0,x_1)}}{\log q}$$

The error bound (4) can be used as a practical stopping rule. In other words, for any given $\varepsilon > 0$, the procedure is stopped when

$$\rho(x_n, x_{n-1}) < \frac{\varepsilon(1-q)}{q}.$$

The previous theorem can be further generalized as follows.

Theorem 2.20. Let $f : X \to X$ be a generalized Caccioppoli contraction and assume that (X, ρ) is f-orbitally complete. Then

1. f has a unique fixed point $x^* \in X$;
2. For all $x \in X$, the iteration sequence converges to x^*;
3. $\rho(x_n, x^*) \leq \delta(\Delta(x_k, \infty)) \sum_{i=n-k}^{\infty} q_i$ for all $n \geq 1$ and $0 \leq k \leq n-l$.

Proof. Let $x \in X$, and $0 \leq k \leq n - l$. Then

$$\rho(x_{n+1}, x_n) = \rho(f^{n+1}(x), f^n(x)) = \rho(f^{n-k}(f^{(k+1)}(x)), f^{n-k}(f^k(x)))$$
$$\leq q_{n-k}\delta(\Delta(f^{k+1}(x), f^k(x), \infty)) \leq q_{n-k}\delta(\Delta(f^k(x), \infty)).$$

Let $m > n$. This inequality implies that

$$\rho(x_m, x_n) = \rho(f^m(x), f^n(x)) \leq \rho(f^n(x), f^{n+1}(x)) + \rho(f^{n+1}(x), f^{n+2}(x))$$
$$+ \cdots + \rho(f^{m-1}(x), f^m(x)) \leq \delta(\Delta(f^k(x), \infty)) \sum_{i=n-k}^{\infty} q_i.$$

Because X is f-orbitally complete, sequence $\{x_k\}$ is convergent. Let $x^* = \lim_{k \to \infty} x_k$ and by letting $m \to \infty$ in the above inequality, this leads to 3.

Notice that

$$\rho(f^n(x), f^{n+1}(x)) \leq \rho(f^n(x) f^n(f(x)))$$
$$\leq q_n \delta(\Delta(x, f(x), \infty)) = q_n \delta(\Delta(x), \infty).$$

Because $\sum_{i=1}^{\infty} q_i < \infty, q_n \to \infty$ as $n \to \infty$, which implies that

$$\lim_{n \to \infty} \rho(f^n(x), f^n(f(x))) = 0.$$

We assumed that f is orbitally continuous and therefore inequality

$$\rho(x^*, f(x^*)) \leq \rho(x^*, f^n(x)) + \rho(f^n(x), f^n(f(x))) + \rho(f(f^n(x), f*(x^*)))$$

implies that $x^* = f(x^*)$.

Finally, we show that the fixed point is unique. Assume that both x^* and x^{**} are fixed points. Then

$$\rho(x^*, x^{**}) = \rho(f^n(x^*), f^n * (x^{**})) \leq q_n \delta(\Delta(x^*, x^{**}, \infty)) = q_n \rho(x^*, x^{**}),$$

Thus,

$$(1 - q_n)\rho(x^*, x^{**}) \leq 0.$$

since $q_n < 1$ for large n, $p(x^*, x^{**}) = 0$. Hence, $x^* = x^{**}$, which completes the proof.

$\square$

2.6 EXERCISES

1. Consider a continuous map $T : R^n \to R^n$ such that $T \in C^1$ and $T(0) = 0$. Set
   $A = \{x \mid \|T(x)\| < \|x\|\}$, $B = \{x \mid \|T(x)\| \geq \|x\|\}$. Assume that

   a. A is invariant under T, $T(A) \subseteq A$;
   b. For all $b \in B$, there exists a positive integer $i(b)$ such that $T^{i(b)}(b) \in A$.
       Show that for $x \in R^n$, $T^k(x) \to 0$ as $k \to \infty$.

2. Assume that a function $h : (0, \infty) \to R$ exists such that $\|y\| \leq h(r)\|x\|$ for all
   $k \geq 0, r > 0, \|x\| \leq r, x \in X$ and $y \in f_k(x)$. Show that

   $$\|x_k\| \leq q_k,$$

   where
   $$q_{k+1} = h(q_k)q_k, \quad q_0 = \|x_0\|.$$

   Provide a convergence analysis of iteration (2.9) based on the above estimate.

3. To find a zero for $g(x) = 0$ by iteration, where g is a real function defined
   on $[a, b]$, rewrite the equation as

   $$x = x + c \cdot g(x) \equiv f(x)$$

   for some constant $c \neq 0$. If s^* is a root of $g(x)$ and if $g'(s^*) \neq 0$, how should
   c be chosen in order that the sequence $x_{k+1} = f(x_k)$ converge to s^*?

4. Solve the initial value problem

   $$\dot{x}(t) = 1 + \cos(x(t)), x(0) = 0,$$

   using the approach of case study 2.10.

5. The predator-prey population models describe the interaction of a prey population X and a predator population Y. Assume that their interaction is modeled by the system of ordinary differential equations

   $$\dot{x} = x - \frac{1}{4}x^2 - \frac{1}{10}xy + 1$$

   $$\dot{y} = -\frac{1}{4}y + \frac{1}{7}xy + \frac{2}{3}.$$

   (Assume $x(0) = y(0) = 0$). Solve the system.

6. Repeat the calculations of case study 2.2 if

   $$R_1 = R_2 = R_3 = 3\Omega, R_4 = R_5 = R_6 = 7\Omega$$

   and

   $$V_1 = V_2 = 2V.$$

7. Assume that $f_k = f(k \geq 0)$, 0 is in the interior of X, and f is Fréchet differentiable at 0. Furthermore, the special radius of $f'(0)$ is less than 1. Then show that there is a neighborhood U of 0 such that $x_0 \in U$ implies that $x_k \to 0$ as $k \to \infty$.

8. Let $f : R^n \to R^n$ be a function such that $f(0) - 0, f \in C^0$, and consider the difference equation $x(t+1) = f(x(t))$. If, for some norm, $\|f(x)\| < \|x\|$ for any $x \neq 0$, then show that the origin is globally asymptotically stable equilibrium for the equation.

9. Assume that a strictly increasing function $g : R \to R$ exists such that $g(0) = 0$, and a norm $g(\|f(x)\|) < g(\|x\|)$ for all $x \neq 0$. Then show that 0 is globally symptotically stable equilibrium for equation $x(t+1) = f(x(t))$, where $f(0) = 0$.

10. Consider the following equation in R^2:

$$x_1(t+1) = 0.8 \sin\left(x_1(t) + \frac{\pi}{4}\right) + 0.2x_2(t)$$
$$x_2(t+1) = 0.8x_1(t) + 0.1x_2(t).$$

Do the conclusions of Exercise 8 apply here?

11. Solve the initial value problem

$$\dot{x} = (x(t) + t, x(0) = 1)$$

by the iteration method (2.46). Perform 3 steps.

12. Solve the Fredholm-type integral equation

$$x(t) = \sum_0^1 \frac{t \cdot s}{10} ds + 1$$

by the iteration method (2.47). Perform 3 steps.

13. Solve the Volterra-type integral equation

$$x(t) = \sum_0^t \frac{t \cdot s}{10} ds + 1$$

by the iteration method. Perform 3 steps.

14. Solve equation

$$x = \frac{\sin x}{2} + 1$$

by using algorithm (2.48) with selecting $\alpha_k \equiv \frac{1}{2}$.

15. Repeat the previous problem for equation

$$x = \frac{\sin x + \cos 2x}{10}.$$

REFERENCES

1. Argyros, I.K. 2007. *Computational Theory of Iterative Methods.* eds. Chui, C.K., Wuytack, L. Series: Studies in Computational Mathematics, 15. New York: Elsevier.
2. Argyros, I.K. 2008. *Convergence and Applications of Newton-type Iterations.* New York: Springer-Verlag.
3. Argyros, I.K., and George, S. 2019. *Mathematical Modeling for the Solution of Equations and Systems of Equations with Applications*, Volume III, NY: Nova Science Publisher.
4. Argyros, I.K., George, S., and Magrenan, A.A. 2015. Local convergence for multi-point-parametric Chebyshev-Halley-type methods of high convergence order. *Journal of Computational and Applied Mathematics*, 282:215–224.
5. Argyros, I.K., George, S., and Thapa, N. 2018. *Mathematical Modeling for the Solution of Equations and Systems of Equations with Applications*, Volume I, N.Y: Nova Science Publishers.
6. Argyros, I.K., George, S., and Thapa, N. 2018. *Mathematical Modeling for the Solution of Equations and Systems of Equations with Applications*, Volume II, N.Y: Nova Science Publishers.
7. Argyros, I.K., and Hillout S. 2012. Weaker conditions for the convergence of Newton's method. *J. Complexity.* 28. 3:364–387.
8. Argyros, I.K., and Magreñán, A.A. 2017. *Iterative Methods and Their Dynamics with Applications*, New York: CRC Press. Taylor& Francis.
9. Argyros, I.K., and Magrenan, A.A. 2018. *A Contemporary study of iterative methods.* NY: Academy Press, Elsevier.
10. Argyros, I.K., and Regmi, S. 2019. *Undergraduate Research at Cameron University on Iterative Procedures in Banach and Other Spaces*, N.Y.: Nova Science Publisher.
11. Argyros, M.I., Argyros, I.K., and Regmi, S. 2021. *Hilbert Spaces and Its Applications*, N.Y.: Nova Science Publisher.
12. Balazs, M. and Goldner, G. 1968. On the method of the cord and on a modification of it for the solution of nonlinear operator equations. *Stud. Cerc. Mat.*, 20:981–990.
13. Brikhoff, G. 1948. *Lattice Theory, Colloq. Publ.*, 25, New York: American Mathematical Society.
14. Collatz, L. 1964. *Functional Analysis and Numerical Mathematics*, New York: Academic Press.
15. Dennis, J.E. 1971. Toward a unified convergence theory for Newton-like methods. In *Nonlinear Functional Analysis and Applications (L.B. Rll, ed.)*, 425–472. New York: Academic Press.
16. Fujimoto, T. 1987. Global asymptotic stability of nonlinear difference equations II. *Econ. Lett.*, 23:275–277.
17. Kitchen, J.W. 1966. Concerning the convergence of iterates to fixed points. *Stud. Math.*, XXVII:247–249.
18. Okuguchi, K. and Szidarovsky, F. 1988. A vote on global asymptotic stability of nonlinear difference equations. *Econ. Lett.*, 26:349–352.
19. Ortega, J.M. and Rheinboldt, W.C. 1970. Iterative Solutions of Nonlinear Equations in Several Variables, New York: Academic Press.
20. Ostrowskii, A.M. 1937. Uber die determinanten mit uberwiegender hauptidiagonale. *Comment. Math. Helv.*, 10:69–96.
21. Rall, L.B. 1975. Convergence of Stirling's method in Banach spaces. *Aequationes Math.*, 12:12–20.

22. Rheinboldt, W.C. 1968. A unified convergence theory for a class of iterative processes. *SIAM J. Numer. Anal.*, 5:42–63.

23. Szidarovszky, F. and Bahill, T.A. 1992. *Linear Systems Theory*, Boca Raton, FL: CRC Press.

24. Szidarovszky, F. and Yakowitz, S. 1978. *Principles and Procedures of Numerical Analysis*, New York: Plenum Press.

25. Uzawa, H. 1961. The stability of dynamic processes. *Econometrica*, 29:617–633.

26. Wu, J.W. and Brown, D.P. 1989. Global asymptotic stability in discrete systems. *J. Math. Anal. Appl.*, 140:224–227.

3 Monotone Convergence

In this chapter, we discuss the monotonicity of the iteration sequence generated by the algorithmic model

$$x_{k+l} \in f_k(x_0, x_1, \ldots, x_k) \, (k \geq l-1).$$

(3.1)

Here, we assume again that for $k \geq l-1$, $f_k : X^{k+1} \to 2^X$, where X is a subset of a partially ordered space L, and let s denote the order in L.

3.1 GENERAL RESULTS

Our main results will be based on a certain type of isotony of the algorithmic mappings.

Definition 3.1. The sequence of point-to-set mappings $g_k : X^{k+1} \to 2^L$ is called increasingly isotone from the left on X, if for any sequence $\{x^{(k)}\}$ such that $x^{(k)} \in X (K \geq 0)$ and $x^{(0)} \leq x^{(1)} \leq x^{(2)} \leq \ldots, y_1 \leq y_2$ for all $y_1 \in g_k(x^{(0)}, x^{(1)}, \ldots, x^{(k)})$, $y_2 \in g_{k+1}(x^{(0)}, x^{(1)}, \ldots, x^{(k+1)})$ and $k \geq l-1$.

Consider the special case of single variable mappings g_k. The above properties are now reduced to the following. Assume that $x^{(0)}, x^{(1)} \in X, x(0) \geq g_k(x^{(0)})$, and $y_2 \in g_{k+1}(x^{(1)})$, then $y_1 \leq y_2$.

Similarly, the sequence of point-to-set mappings $g_k : X^{k+1} \to 2^L$ is called increasingly isotone from the right on X, if for any sequence $\{x^{(k)}\}$ such that $x^{(k)} \in X (k \geq 0)$ and $x^{(0)} \leq x^{(1)} \leq x^{(2)} \ldots, y_1 \geq y_2$ for all $y_1 \in g_x(x^{(1)}, x^{(2)}, \ldots, x^{(k)})$, $y_2 \in g_{k+1}(x^{(0)}, x^{(1)}, \ldots, x^{(k)})$ and $k \geq l-1$.

Our main result is the following.

Theorem 3.1. Assume that the sequence of mappings f_k is increasingly isotone from the left; furthermore, $x_0 \leq x1 \leq \cdots \leq x_{l-1} \leq x_l$ ($x_k \in X, 0 \leq k \leq l-1$). Then for all $k \geq 0, X_{k+1} \geq X_k$.

Proof. By induction, assume that for all indices $i (i < k), x_{i+1} \geq x_i$. Then relations

$$x_k \in f_{k-l}(x_0, x_1, \ldots, x_{k-1}) \text{ and } x_{k+1} \in f_k(x_0, x_1, \ldots, x_k)$$

and Definition 3.1 imply that $x_{k+1} \geq x_k$. Because this inequality holds for $k \leq l-1$, the proof is complete. □

Consider next the modified algorithmic model

$$y_{k+1} \in f_k(y_k, y_{k-1}, \ldots, y_1, y_0),$$

(3.2)

DOI: 10.1201/9781003128915-3

which differs from process (3.1) only in the order of the variables in the right-hand side. Using finite induction again similarly to Theorem 3.1, one may easily prove the following theorem.

Theorem 3.2. Assume that the sequence of mappings f_k is increasingly isotone from the right; furthermore, $y_0 \geq y_1 \geq \cdots \geq y_{l-1} \geq y_1 (y_k \in X, 0 \leq k \leq l-1)$. Then for all $k \geq 0, y_{k+1} \leq y_k$.

Corollary 3.1. Assume that $X \subseteq R^n$; furthermore, $x_k \to s^*$ and $y_k \to s^*$ as $k \to \infty$. It is also assumed that $\leq$ is the usual partial order of vectors; i.e., $a = (a^{(i)}) \leq b = (b^{(i)})$ if and only if for all $i, a^{(i)} \leq b^{(i)}$. Under the conditions of Theorems 3.1 and 3.2,

$$x_k \leq s^* \leq y_k.$$

This relation is very useful in constructing an error estimator and a stopping rule for methods (3.1) and (3.2), since for all coordinates of vectors x_k, y_k, and s^*,

$$0 \leq s^{*(i)} - x_k^{(i)} \leq y_k^{(i)} - x_k^{(i)}$$

and

$$0 \leq y_k^{(i)} - s^{*(i)} \leq y_k^{(i)} - x_k^{(i)}.$$

Hence, if each element of $y_k - x_k$ is less than an error tolerance ε, then all elements of x_k and y_k are closer than e to the corresponding elements of the limit vector.

3.2 A GENERAL MODEL IN LINEAR SPACES

Assume now that X is a subset of a partially ordered linear space L. In addition, the partial order satisfies the following conditions:

(F_1) $x \leq y(x, y \in L)$ implies that $-x \geq -y$;
(F_2)   If $x^{(1)}, x^{(2)}, y^{(1)}, y^{(2)} \in L$, $x^{(1)} \leq y^{(1)}$ and $x^{(2)} \leq y^{(2)}$, then $x^{(1)} + x^{(2)} \leq y^{(1)} + y^{(2)}$.

Assume that for $k \geq l-1$ and $p = 1, 2$, mappings $K_k^{(p)}, H_k^{(p)}$ are point-to-set mappings defined on X^{k+1}, and for all $(x^{(1)}, x^{(2)}, \ldots, x^{(k+1)}) \in X^{k+1}, K_k^{(p)}$ and $H_k^{(p)}(x^{(1)}, \ldots, x^{(k+1)})$ are nonempty subsets of L; i.e., $K_k^{(p)}$ and $H_k^{(p)} X^{k+1} \to 2^L$. Consider next the algorithmic model:

$$x_{k+1} = t_{k+1}^{(1)} - s_{k+1}^{(1)} \text{ and } y_{k+1} = t_{k+1}^{(2)} - s_{k+1}^{(2)}, \tag{3.3}$$

where

$$t_{k+1}^{(1)} \in K_k^{(1)}(x_0, x_1, \ldots, x_k), s_{k+1}^{(1)} \in H_k^{(1)}(y_k, y_{k-1}, \ldots, y_0)$$

and

$$t_{k+1}^{(2)} \in K_k^{(2)}(y_k, y_{k-1}, \ldots, y_0), s_{k+1}^{(1)} \in H_k^{(2)}(x_0, x_1, \ldots, x_k)$$

In the formulation of our next theorem, we will need what follows.

Definition 3.2. A point-to-set mapping $g : X^{k+1} \to 2^L$ is called increasingly isotone on X, if for all $x^{(1)} \le x^{(2)} \le \cdots \le x^{(k+2)}$ $(x^{(1)} \in X, i = 1, 2, \ldots, k+2)$, relations $y_1 \in g(x^{(1)}, x^{(2)}, \ldots, x^{(k+1)})$ and $y_2 \in g(x^{(2)}, x^{(3)}, \ldots, x^{(k+2)})$ imply that $y_1 \le y_2$.

Increasingly isotone functions have the following property.

Lemma 3.1. Assume that mapping $g : X^{k+1} \to 2^L$ is increasingly isotone. Then, for all $x^{(i)}$ and $y^{(i)} \in X (i = 1, 2, \ldots, k+1)$ such that

$$x^{(1)} \le x^{(2)} \le \cdots \le x^{(k+1)} \le y^{(1)} \le y^{(2)} \le \cdots \le y^{(k+1)}$$

and for any

$$x \in g(x^{(1)}, x^{(2)}, \ldots, x^{(k+1)}) \text{ and } y \in g(y^{(1)}, y^{(2)}, \ldots, y^{(k+1)}), x \le y.$$

Proof. Let $y_i \in g(x^{i+1}, \ldots, x^{(k+1)}, y^{(1)}, \ldots, y^{(i)})$ be arbitrary for $i = 1, 2, \ldots, k$. Then

$$x \le y_1 \le y_2 \le \cdots \le y_k \le y,$$

which completes the proof. □

Remark 3.1. In the literature a point-to-set mapping $g : X^{k+1} \to 2^L$ is called isotone on X, if for all $x^{(i)}$ and $y^{(i)}$ such that $x^{(i)} \le y^{(i)} (i = 1, 2, \ldots, k+1)$ and for all $x \le g(x^{(1)}, \ldots, x^{(k+1)})$ and $y \in g(y^{(1)}, \ldots, y^{(k+1)}), x \le y$.

It is obvious that an isotone mapping is increasingly isotone, but the reverse is not necessarily true as the following example illustrates.

Example 3.1. Define $L = R^1, X = [0,1], k = 1, \le$ to be the usual order of real numbers, and

$$g(x^{(1)}, x^{(2)}) = \begin{cases} x^{(2)} & \text{if } x^{(1)} \ge 2x^{(2)} - 1 \\ x^{(1)} - x^{(2)} + 1, & \text{if } x^{(1)} < 2x^{(2)} - 1. \end{cases}$$

We will now verify that g is increasingly isotone, but not isotone on X. Select arbitrary $x^{(1)} \le x^{(2)} \le x^{(3)}$ from the unit interval. Note first that $g(x^{(1)}, x^{(2)}) \le x^{(2)}$, since if $x^{(1)} \ge 2x^{(2)} - 1$, then $g(x^{(1)}, x^{(2)}) = x^{(1)}$, and if $x^{(1)} < 2^{x(2)} - 1$, then $g(x^{(1)}, x^{(2)}) = x^{(1)} - x^{(2)} + 1 < 2^{x(2)} - 1 - x^{(2)} + 1 = x^{(2)}$. Note next that $g(x^{(2)}, x^{(3)}) \ge x^{(2)}$, since if $x^{(1)} \ge 2x^{(3)} - 1$, then $g(x^{(2)}, x^{(3)}) = x^{(3)} \ge x^{(2)}$, and if $x^{(2)} < 2x^{(3)} - 1$, then $g(x^{(1)}, x^{(3)}) = x^{(2)} - x^{(3)} + 1 \ge x^{(2)}$. Consequently, $g(x^{(1)}, x^{(2)}) \le x^{(2)} \le g(x^{(2)}, x^{(3)})$. Hence, g is increasingly isotone; however, it is not isotone on X, since for any points $(t, 1)$ and $(t, 1 - \varepsilon)$ (where $t, \varepsilon > 0$ and $t + 2\varepsilon < 1$),

$$g(t, 1) = t - 1 + l = t < g(t, 1 - \varepsilon) = t - (1 - \varepsilon) + 1 = t + \varepsilon,$$

but $1 > 1 - \varepsilon$. Consider now the following assumptions:

(G_1) Sequences of mappings $K_k^{(1)}$ and $H_k^{(2)}$ are increasingly isotone from the left. Furthermore, $K_k^{(}2)$ and $H_k^{(}1)$ are increasingly isotone from the right;

(G_2) For all k and $t^{(0)} \leq t^{(1)} \leq \cdots \leq t^{(k)} \leq \bar{t}^{(k)} \leq \bar{t}^{(1)} \leq \bar{t}^{(0)}$ ($\bar{t}^{(i)}$ and $\bar{t}^{(i)} \in X, i = 0, 1, \ldots, k$),

$$K_k^{(1)}(t^{(0)}, t^{(1)}, \ldots, t^{(k)}) \leq K_k^{(2)}(\bar{t}^{(k)}, \ldots, \bar{t}^{(0)})$$

and

$$H_k^{(1)}(\bar{t}^{(k)}, \ldots, \bar{t}^{(1)}, \bar{t}^{(0)}) \geq H_k^{(2)}(t^{(0)}, t^{(1)}, \ldots, t^{(k)}).$$

(G_3) The initial points are selected so that

$$x_0 \leq x_1 \leq \cdots \leq x_{l-1} \leq x_l \leq y_l \leq y_{l-1}, \cdots \leq y_1 \leq y_0;$$

(G_4) $\langle x_0, y_0 \rangle = \{x | x \in L, x_0 \leq x \leq y_0\} \subseteq X.$

Theorem 3.3. Under assumptions $(F_1), (F_2)$, and (G_1) through (G_4), x_k and y_k are in X for all $k \geq 0$. Furthermore,

$$x_k \leq x_{k+1} \leq y_{k+1} \leq y_k$$

Proof. By induction, assume that $y_0 \geq x_{i+1} \geq x_i$ and $x_0 \leq y_{i+1} \leq y_i$ for all $i < k$. Therefore, points x_i and $y_i (i = 0, 1, \ldots, k)$ are in X. Note next that assumption (G_1) implies that

$$t_{k+1}^{(1)} \geq t_k^{(1)}, t_{k+1}^{(2)} \leq t_k^{(2)}, s_{k+1}^{(1)} \leq s_k^{(1)} \text{ and } s_{k+1}^{(2)} \geq s_k^{(2)},$$

and hence,

$$x_{k+1} = t_{k+1}^{(1)} - s_{k+1}^{(1)} \geq t_k^{(1)} - s_k^{(1)} = x_k$$

and

$$y_{k+1} = t_{k+1}^{(2)} - s_{k+1}^{(2)} \leq t_k^{(2)} - s_k^{(2)} = y_k.$$

Assume next that $x_i \leq y_i (i \leq k)$. Then, assumption (G_2) implies that

$$t_{k+1}^{(1)} \leq t_{k+1}^{(2)} \text{ and } s_{k+1}^{(1)} \leq s_k^{(2)}.$$

Therefore,

$$x_{k+1} = t_{k+1}^{(1)} - s_{k+1}^{(1)} \leq t_{k+1}^{(2)} - s_{k+1}^{(2)} = y_{k+1}.$$

Thus, the proof is complete. □

Introduce next the point-to-set mappings

$$f_k^{(}p)(x^{(1)}, \ldots, x^{(k+1)}) = \{t - s | t \in K_k^{(p)}(x^{(1)}, \ldots, x^{(k+1)}), s \in H_k^{(p)}(x^{(1)}, \ldots, x^{(k+1)})\}.$$

Assume that z is a common fixed point of mappings $f_k^{(1)}$ and $f_k^{(2)}$, i.e., for $p = 1, 2, z \in f_k^{(p)} z, z, \ldots, z$, furthermore (G_5). Mappings $K_k^{(1)}, K_k^{(2)}, H_k^{(1)}, H_k^{(2)}$ increasingly isotone.

Theorem 3.4. Assume that $x_1 \leq z \leq y_1$. Then, under assumptions $(F_1), (F_2)$ and (G_1) through (G_5),

$$x_k \leq z \leq y_k \text{ for all } k \geq 0.$$

Proof. By induction, assume that $x_1 \leq z_i \leq y_1$ for $i \geq k$. Then, for $p = 1, 2$ for all $t^{(p)} \in K_k^{(p)}(z, z, \ldots, z)$ and $s^{(p)} \in H_k^{(p)}(z, z, \ldots, z)$,

$$t_{k+1}^{(1)}) \leq t^{(1)}, t^{(2)} \leq t_{k+1}^{(2)} \text{ and } s_{k+1}^{(1)} \geq s^{(1)} \text{ and } s^{(2)} \geq s_{k+1}^{(2)}.$$

Therefore,

$$x_{k+1} = t_{k+1}^{(1)} - s_{k+1}^{(1)} \leq t^{(p)} - s^{(p)} \leq t_{k+1}^{(2)} - s_{k+1}^{(2)} = y_{k+1}, (p = 1, 2)$$

and since $z = t - s$ with some t and s, the proof is completed. □

Remark 3.2. Two important special cases of algorithm (3.3) can be given:

1. If $K_k^{(1)} = K_k^{(2)} = K_k$ and $H_k^{(1)} = H_k^{(2)} = H_k$, then $f_k^{(1)} = f_k^{(2)} = f_k$, if and so z is a common fixed point of the mapping f_k.
2. Assume next that K_k and H_k and l-variate functions. Then one may select

$$K_k^{(1)}(t^{(0)}, \ldots, t^{(k)}) = K_k(t^{(k-l+1)}, \ldots, t^{(k)}),$$
$$K_k^{(2)}(t^{(0)}, \ldots, t^{(k)}) = K_k(t^{(0)}, \ldots, t^{(l-1)}),$$
$$H_k^{(1)}(t^{(0)}, \ldots, t^{(k)}) = H_k(t^{(0)}, \ldots, t^{(l-1)}),$$

and

$$H_k^{(2)}(t^{(0)}, \ldots, t^{(k)}) = H_k(t^{(k-l+1)}, \ldots, t^{(k)}).$$

Note that the resulting process for $l = 1$ was discussed in Ortega and Rheinboldt (1970, Section 13.2) [13].

Corollary 3.2. Assume that in addition L is a Hausdorff topological space which satisfies the first axiom of countability, and

(F_3) If $X^k \in L(k \geq 0)$ such that $X^k \leq x$ with some $x \in L$ and $X_k \leq X_{k+1}$ for all $k \geq 0$, then the sequence $\{x_k\}$ converges to a limit point $x^* \in L$; furthermore, $x^* \geq x$.

(F_4) If $X_k \in L(k \geq 0)$ and $X_k \to x$, then $-X_k \to -x$.

Note first that assumptions (F_1) through (F_4) imply that any sequence $\{y_k\}$ with the properties $y_k \in L, y_k \geq y_{k+1}$, and $y_k \geq y(k \geq 0)$ is convergent to a limit point $y^* \in L$; furthermore, $y^* \geq y$. Since under the conditions of the theorem $\{x_k\}$ is increasing with an upper bound y_0 and $\{y_k\}$ decreases with a lower bound x_0, both sequences are convergent. If x^* and y^* denote the limit points and z is a common fixed point of mappings f_k such that $x_1 \leq z \leq y_1$, then $x^* \leq z \leq y^*$.

Consider the special l-step algorithm given as special case 2 above. Assume that for all $k \geq l - 1$, mappings K_k and H_k are closed. Then, x^* and y^* are common fixed points of mappings f_k. Note that x^* is the smallest fixed point and y^* is the largest fixed point in $\langle x_1, y_1 \rangle$.

As a further special case we mention the following lemma.

Lemma 3.2. Let B be a partially ordered topological space and let x, y be two points of B such that $x \leq y$. If $f : \langle x, y \rangle \to B$ is a continuous isotone mapping having the property that $x \leq f(x)$ and $y \leq f(y)$, then a point $z \in \langle x, y \rangle$ exists such that $z = f(z)$.

Proof. Select $X = \langle x, y \rangle$, $l = 1$, $H_k = 0$, and $K_k = f$, and consider the iteration sequences $\{x_k\}$ and $\{y_k\}$ starting from $X_0 = x$ and $y_0 = y$. Then, both sequences are convergent, and the limits are fixed points. □

Remark 3.3. This result is known as the famous Kantorovich lemma, (see, for example, Ortega and Rheinboldt [13]).

3.3 APPLICATIONS

Case Study 3.1. Consider the simple iteration method

$$x_{k+1} = f(x_k), \tag{3.4}$$

where $f : [a, b] \to R$ with the additional properties:

$$f(a) > a, f(b) < b, 0 \leq f(x) \leq q < l \text{ for all } x \in [a, b],$$

where q is a given constant. Consider the iteration sequences

$$x_{k+1} = f(x_k), \quad x_0 = a$$

and

$$y_{k+1} = f(y_k), \quad y_0 = b. \tag{3.5}$$

Since f is increasing, Theorems 3.1 and 3.2 apply to both sequences $\{x_k\}$ and $\{y_k\}$ that converge to the unique fixed point s^*; furthermore, sequence $\{x_k\}$ is increasing and $\{y_k\}$ decreases.

Next we drop the assumption $f'(x) \leq q < 1$, but retain all other conditions. In other words, f is continuous, increasing in $[a, b]$,

$$f(a) > a, \text{ and } f(b) < b.$$

Notice that Theorems 3.1 and 3.2 still can be applied for the iteration sequences (3.5). However, the uniqueness of the fixed point is no longer true.

Because f is continuous, there is at least one fixed point in the interval (a,b). Alternatively, Theorems 3.3 and 3.4 may also be applied with $K_k^{(1)} = K_k^{(2)} \equiv f$ and $H_k^{(1)} = H_k^{(2)} = 0$, which has the following additional consequence. Let s^* and s^{**} denote the smallest and largest fixed points between a and b. Because $a < s^*$ and $b > s^{**}, f(a) \leq f(s^*) = s^*$ and $f(b) \geq f(s^{**}) = s^{**}$, thus, s^* and s^{**} are between x_1 and y_1. Therefore, for all $k \geq 1, x_k \leq s^* \leq s^{**} \leq y_k$; furthermore, $x_k \to s^*$ and $y_k \to s^{**}$. If the fixed point is unique, the uniqueness of the fixed point can be established by showing that sequences $\{x_k\}$ and $\{y_k\}$ have the same limit.

Finally, we mention that the above facts can be easily extended to the more general case when $f : R^n \to R^n$.

Example 3.2. We now illustrate procedures (3.5) in the case of the single variable nonlinear equation

$$x = \frac{1}{10}e^x + 1.$$

It is easy to see that

$$\frac{1}{10}e^1 + 1 > 1 \text{ and } \frac{1}{10}e^2 + 1 < 2.$$

Furthermore,

$$\frac{d}{dx}\left(\frac{1}{10}e^x + 1\right) = \frac{1}{10}e^x \in (0, 0.75)$$

Therefore, all conditions are satisfied. Select $x_0 = 1$ and $y_0 = 2$. Then,

$$x_1 = \frac{1}{10}e^1 + 1 \approx 1.27183; \qquad y_1 = \frac{1}{10}e^2 + 1 \approx 1.7183;$$

$$x_2 = \frac{1}{10}e^{1.27183} + 1 \approx 1.35674; \qquad y_2 = \frac{1}{10}e^{1.73891} + 1 \approx 1.56911$$

$$x_3 = \frac{1}{10}e^{1.35674} + 1 \approx 138835; \qquad y_3 = \frac{1}{10}e^{1.56911} + 1 \approx 1.48024$$

$$x_4 = \frac{1}{10}e^{1.38835} + 1 \approx 1.40082; \qquad y_4 = \frac{1}{10}e^{1.48024} + 1 \approx 1.43940$$

$$x_5 = \frac{1}{10}e^{1.40082} + 1 \approx 1.40585; \qquad y_5 = \frac{1}{10}e^{1.43940} + 1 \approx 1.42182,$$

and so on. The monotonicity of both iteration sequences, as expected, is clearly visible.

A nice application of the previous scheme is presented in the second case study. The static emergency vehicle (such as ambulance, fire truck, etc.) location problem is to determine base locations so that some service level objective is optimized. The central problem in such models is to determine vehicle utilization. Goldberg and Szidarovszky [12] summarize alternative models. One particular model can be formulated as follows. Divide the examined area into I zones. In the area there are J

vehicle bases, each with a single vehicle. It is also assumed that for each zone a preference ordering of the vehicles is determined, which is strict and independent of the state of the system. Any call that arrives when all vehicles are busy is assigned to a parallel private system, so we may assume that calls do not queue.

The Hypercube Model (denoted Hypercube), is a descriptive model that is a large Markov chain, and thus requires Markovian assumptions for the call arrival and service processes. Hypercube generates a wide variety of performance statistics such as vehicle utilization, the probability that a vehicle serves calls in a particular area, and the expected number of calls served by a particular vehicle. The model can be computationally burdensome because the number of states in the chain is $2J$. In addition, the model cannot deal with situations in which call service time depends on call location. Each vehicle has a mean service rate which is the parameter of the service time distribution for the particular vehicle.

Several authors have developed models with the probabilistic strengths of Hypercube, but which require less computational effort. Our approach is motivated by two issues. First, we formulate a general descriptive model to approximate vehicle utilizations, and show that many particular models of the approximations and extensions to Hypercube are special cases of the general structure. Second, we give conditions that ensure that the general model has a unique solution. These conditions then will suggest an iteration method which converges to the solution.

The principal issue in an approximation model for Hypercube is to estimate vehicle utilizations. We propose a general model for estimating utilizations which is the common generalization of well-known models from the literature. Once we know the vehicle utilizations, we can estimate the probability that a vehicle serves a particular zone. This in turn leads to estimates of performance measures.

As previously mentioned, we divide an area into I zones, indexed by i. In the area there are J vehicle bases, each with a single vehicle, indexed by j. Given the set of open bases, we can determine a preference ordering of the vehicles for each zone. We assume that the preference ordering is strict, independent of the state of the system, and represents the dispatching order for servicing calls in the zone. We assign calls that arrive when all vehicles are busy to a parallel private system so that we may assume that calls do not queue. Define the following additional notation:

1. ρ_j is the utilization of vehicle j;
2. $r(i, j)$ is the rank of vehicle j in the preference list for zone i;
3. $k(i, 1)$ is the vehicle index with rank 1 in the preference list for zone i;
   $\hat{\rho}$ is an aggregate measure of the ρ_j values (e.g., $\hat{\rho} = \sum_{j=1}^{J} \frac{\rho_j}{J}$);
4. $A_{ij}(\hat{\rho})$ is a function of i, j, and $\hat{\rho}$ which tailors the model to the specific applications.

Using the notation and the above assumptions, we formulate a model to determine the vehicle utilizations. The model is a stationary representation of the call assignment process and the specific application determines the interpretation of the $A_{jj}(\hat{\rho})$ coefficients. The model is

$$\rho_j = \sum_{i=1}^{I}\left[A_{tj}(\hat{\rho})(1-\rho_j) \prod_{l=1}^{r(i,j)-1} \rho_{k(i,i)} \right], \quad \text{for all } j. \tag{3.6}$$

Note that if $r(i,j)=1$, then we set

$$\prod_{l=1}^{r(i,j)-1} \rho_k(i,1) = 1.$$

We construct equation (3.6) by first using an independence assumption. The term

$$(1-\rho_j) \prod_{i=1}^{r(i,j)-1} \rho_{k(i,1)}$$

represents the probability that vehicle j is idle and that the vehicles preferred to j for zone i are all busy. If the vehicles operate independently of each other and for all j,p, represents the probability that j is busy, the term is equal to the probability that vehicle j serves a call in zone i. In general, however, the vehicles do not operate independently. Also, each zone may have demand and service rates different than those of the other zones, and the travel time to the zone depends on the particular vehicle serving the call. Therefore, we multiply the product term by $A_{ij}(\hat{\rho})$ to adjust (approximately) for the independence assumption and include the zone and vehicle specific demand, service, and travel rates. After multiplying by the adjustment factors, the term inside the sum represents the contribution of zone i to the utilization of vehicle j. By summing over all zones, we obtain the total utilization of vehicle j.

The first approximation model to Hypercube consists of the following system of equations:

$$\rho_j = \sum_{i=1}^{i}\left[d_i(1-\rho_j)Q(J,\hat{\rho},r(i,j)-1)\prod)i = 1^{r(i,j)-1}\rho_{k(i,1)} \right] \quad \text{for each } j,$$

where d_i is the call arrival rate for zone i and $Q(J,\hat{\rho},k-1)$) is derived to eliminate the effect of the independence assumption. We can take the joint probability derived from the independence assumption and multiply by the appropriate Q factor. The Q factors can be derived from an $M/M/J/O$ queueing model using the assumptions that all servers have an equal utilization ($\hat{\rho}$), servers are chosen randomly, and each call has an identical mean service time independent of call location and server. The specific value to use for ($r\hat{h}o$) can be computed from the problem data and represents the average utilization of an $M/M/J/O$ queueing model. We note that for fixed J and $r(i,j)-1$, numerical data show that for $\hat{\rho} \in [0,0.75], Q(J,p,r(i,j)-1)$ is a decreasing function in $\hat{\rho}$; and for $\hat{\rho} \in [0.75,0.95], Q(J,\hat{\rho},r(i,j)-1)$ is a decreasing function to three significant digits after the decimal. Therefore, we assume that

$Q(J, \hat{\rho}, r(i, j) - 1)$ is decreasing in $\hat{\rho}$ in the operational range of emergency vehicle systems.

The previous equation is a special case of equation (3.6) where

$$A_{ij}(\hat{\rho}) = d_i Q(J, \hat{\rho}, r(i, j) - 1).$$

Here, d_i is included in $A_{ij}(\hat{\rho})$; however, no differentiation of service time is found among the individual servers or zones. The basic problems with this equation are that it requires the assumption that service time is independent of call location and that the validity of the Q factors is questionable when the vehicles have unequal utilizations and unequal service rates.

To relax the equal service time assumptions made in the above equation we may formulate a general service time model. Define the following additional notation:

1. t_{ij} is the mean travel time from base j to zone i;
2. T_i is the mean service time per call in zone i excluding travel to i.

We can then obtain the following model:

$$\rho_j = \sum_i [d_i(t_{ij} + T_i)(1 - \rho_j)Q(J, \hat{\rho}, r(i, j) - 1) \prod_{i=1}^{r(i,j)-1} \rho_{k(i,1)}] \text{ for each } j.$$

This equation is a special case of equation (3.6) where

$$A_{ij}(\hat{\rho}) = d_i(t_{ij} + T_i)Q(J, \hat{\rho}, r(i, j) - 1).$$

In this instance, $A_{ij}(\hat{\rho})$ includes information on both demand and service characteristics. Note that the validity of the Q factors is suspect because the assumptions used in their derivation do not hold in the cases in which the effect of location dependent service is critical.

Our final model is as follows:

$$\rho_j = \sum_i \left[d_i(t_{ij} + T_i)(1 - \rho_j) \prod_{i=1}^{r(i,j)-1} \rho_{k(i,1)} \right] \text{ for each } j,$$

is identical to the previous model, except that it uses a strict independence assumption and does not include the Q factors. Because this equation is a special case of the previous equation, it must also be a special case of equation (3.6) where:

$$A_{ij} = (\hat{\rho}) = d_i(t_{ij} + T_i).$$

Our first idea in solving these equations is to introduce the iteration process

$$\rho_{j,k+1} = \sum_i d_i(t_{ij} + T_i)(1 - \rho_{j,k}) \prod_{i=1}^{r(i,j)-1} \rho_{k(i,1)}, \text{ for each } k,$$

which is the usual fixed point iteration. The convergence of this procedure depends on the order of magnitude of the model parameters. That is, convergence is guaranteed only under very special conditions. However, a nice trick allows us to use the monotonicity of the iteration sequence, and based on the boundness of the sequence we will be able to establish convergence under general conditions. This idea is based on rewriting equation (3.6) as

$$\rho_j = \frac{Q_j}{1+Q_j} \text{(for all j)}$$

with

$$Q_j = \sum_i \left[d_i(t_{ij} + T_i) \prod_{i=1}^{r(i,j)-1} \rho_{k(i,1)} \right]$$

The corresponding iteration scheme has the form

$$\rho_{j,k+1} = \frac{Q_{j,k}}{1+Q_{j,k}} \tag{3.7}$$

with

$$Q_{j,k} = \sum_i \left[d_i(t_{ii} + T_i) \prod_{i=1}^{r(i,j)-1} \rho k, (i,1), k \right].$$

Observe that this process satisfies the following properties:

1. If $0 \le \rho_{j,k} \le 1$ for all j, then $0 \le \rho_{j,k+1} \le 1$ for all j;
2. The right-hand side of equation (3.7) is increasing in the vector $(\rho1, k, \rho2, k, \ldots, \rho_{J,k})$.

Theorems 3.1 and 3.2 imply that starting with the initial solutions $\rho_{j,0} = 0$ (for all j), a convergent and increasing iteration sequence is obtained, and starting with initial solutions $\rho_{j,0=1}$ (for all j), a convergent and decreasing iteration sequence is constructed. Furthermore, the limit vectors represent the minimal and maximal solutions, respectively. If they are equal, then the solution is unique.

Case Study 3.2. $(I - A)X = I$,

that is,

$$X = AX + I.$$

The corresponding iteration process can be written as

$$X_{k+1} = AX_k + I.$$

Select any matrix X_0. Then the above assumptions imply that this iteration process converges to $(I - A) - 1$. Furthermore, if $AX_0 + I \ge X_0$, then Theorem 3.1 implies that

$$X_0 \le X_1 \le X_2 \le \ldots$$

An obvious selection is $X_0 = 0$. *Then, $X_1 = I$,*

$$X_2 = A + I$$
$$X_3 = A^2 + A + I$$

and so on. In general,

$$X_k = A^{k-1} + A^{k-2} + \cdots + A^2 + A + I.$$

Thus,

$$(I - A)^{-1} = \sum_{i=0}^{\infty} A^i. \tag{3.8}$$

Notice that all terms of the right-hand side are non-negative; therefore,

$$(I - A)^{-1} \geq 0.$$

Example 3.3. The previous iteration scheme is now illustrated in the case of matrix

$$A = \begin{pmatrix} \frac{1}{2} & \frac{1}{4} \\ \frac{1}{4} & \frac{1}{2} \end{pmatrix}$$

It is easy to see that the eigenvalues of A are $1/4$ and $3/4$. Therefore, all conditions are satisfied. Select $X_0 = 0$, then

$$X_1 = AX_0 + I = \begin{pmatrix} 1 & 0 \\ 0 & 1 \end{pmatrix},$$

$$X_2 = AX_1 + I = \begin{pmatrix} \frac{3}{2} & \frac{1}{4} \\ \frac{1}{4} & \frac{3}{2} \end{pmatrix},$$

$$X_3 = AX_2 + I = \begin{pmatrix} \frac{29}{16} & \frac{1}{2} \\ \frac{1}{2} & \frac{29}{16} \end{pmatrix},$$

and so on. The monotonicity of the sequence X_0, X_1, X_2, X_3 is clearly visible. We notice that the true inverse is

$$(I - A)^{-1} = \begin{pmatrix} \frac{1}{2} & -\frac{1}{4} \\ -\frac{1}{4} & \frac{1}{2} \end{pmatrix}^{-1} = \begin{pmatrix} \frac{8}{3} & \frac{4}{3} \\ \frac{4}{3} & \frac{8}{3} \end{pmatrix}$$

Case Study 3.3. The concept of *P*-contractions is based on the above inverse representation. For any vector $x = (x^{(1)}, \ldots, x^{(n)})^T$ denote $|x| = (|x^{(1)}|, \ldots, |x^{(n)}|)^T$. Let $f : D \to D$, where $D \subseteq R^n$ is a closed set. Assume that there is a constant, real non-negative matrix P with all eigenvalues inside the unit circle such that

$$|f(x) - f(y)| \leq P|x - y| \tag{3.9}$$

for all $x, y \in D$.

Theorem 3.5. If f is a P-contraction, then f has a unique fixed point s^* in D, and the iteration sequence $x_{k+1} = f(x_k)$ converges to s^* with arbitrary initial approximation x_0. Additionally, the following error estimates are valid:

$$|x_k - s^*| \le P^k(I-P)^{-1}|x_1 - x_0|$$

and

$$|x_k - s^*| \le P(I-P)^{-1}|x_k - x_{k-1}|.$$

Proof. For all $K > k$,

$$|x_k - x_k| \le |x_{k+1} - x_k| + |x_{k+2} - x_{k+1}| + \cdots + \|x_k - x_{k-1}\|$$

Because for all i,

$$|x_{i+1} - x_i| = |f(x_i) - f(x_{i-1})| \le P|x_i - x_{i-1}| \le \cdots \le P^i|x_1 - x_0|,$$

we have

$$|x_k - x_k| \le \sum_{i=k}^{K-1} P^i|x_1 - x_0| \le \sum_{i=k}^{\infty} P^i|x_1 - x_0| = P^k(I-P)^{-1}|x_1 - x_0|. \qquad (3.10)$$

The assumption that all eigenvalues of P are inside the unit circle implies that $P^k \to 0$ as $k \to 0$ (see Section 2.5). Therefore, $\{x_k\}$ is a Cauchy sequence. By assumption, set D is closed, which implies that $x^k \to s^*$ with some $s^* \in D$.

Next we show that s^* is a fixed point of f. Notice that

$$|s^* - f(s^*)| \le |s^* - x_{k+1}| + |f(x_k) - f(s^*)| \le |s^* - x_{k+1}| + P|x_k - s^*|,$$

which tends to zero as $k \to \infty$. Therefore, $s^* = f(s^*)$. The fixed point is unique. Assume that both s^* and s^{**} are fixed points. Then,

$$|s^* - s^{**}| = |f(s^*) - f(s^{**})| \le P|s^* - s^{**}|;$$

that is, $(I-P)|s^* - s^{**}| \le 0$. Since $(I-P)^{-1} \ge 0, |s^* - s^{**}| \le 0$; i.e., $s^* = s^{**}$. Let $K \to \infty$ in (3.10) to obtain the first error estimate, and the second bound can be obtained from the first bound by selecting $x_0 = x_{k-1}$. $\qquad \square$

Remark 3.4. The first error estimator can be used to predict the number of iteration steps necessary to obtain the solution within a given error bound. The second bound can be used as a stopping rule. Assume that $\varepsilon > 0$ is given, then

$$\|x_k - s^*\|_\infty, \text{ if } \|P^k\|_\infty \|(I-P)^{-1}|x_1 - x_0|\|_\infty < \varepsilon,$$

which can be rewritten as

$$\|P^k\|_\infty < \frac{\varepsilon}{\|(I-P)^{-1}|x_1 - x_0|\|_\infty}$$

Because $P^k \to 0$ as $k \to \infty$, for sufficiently large k, this inequality will certainly hold. For any given $\varepsilon > 0$, the stopping rule can be formulated as

$$\|P_k - x_{k-1}\|_\infty < \frac{\varepsilon}{\|P(I-P)^{-1}\|_\infty}.$$

Notice that in applying the first bound, we do not need to invert matrix $I - P$ because $(I - P) - 1|x_1 - x_0$ is the solution z of the linear equation

$$(I - P)z = |x_1 - x_0|,$$

which is easier to be found than to invert matrices.

Corollary 3.3. Let $f^{(i)}$ denote the ith component of f, and assume that for all i,

$$|f^{(i)}(x^{(1)},\ldots,x^{(n)}) - f^{(i)}(y^{(i)},\ldots,y^{(n)})| \leq \sum_{j=1}^{n} P_{ij}|x^{(j)} - x^{(i)}|,$$

where $P_{i,j} \geq 0$, and all eigenvalues of matrix $P = (p_{i,j})$ are inside the unit circle. All conditions of the theorem hold, and thus the assertion is valid.

The assertion of Theorem 2.5 can be further generalized for non-stationary processes. Consider the iteration procedure

$$x_{k+1} = f_k(x_k), \quad k = 0, 1, 2, \ldots$$

where for all $k, f_k : D \to D$, and $D \subseteq R^n$ is a closed set, and

$$|f_{k+1}(x) - f_k(y)| \leq P_k|x - y|$$

for all $x, y \in D$, where P_k is a constant, real non-negative matrix. Then

$$|x_{i+1} - x_i| = |f_i(x_i) - f_{i-1}(x_{i-1})| \leq P_{i-1}|x_i - x_{i-1}| \leq \ldots \quad \leq P_{i-1}P_{i-2}\ldots|x_1 - x_0|,$$

and therefore for all $K > k$,

$$|x_k - x_k| \leq |x_{k+1} - x_k| + |x_{k+2} - x_{k+1}| + \cdots + |x_k - x_{k-1}|$$

$$\leq \left(\sum_{i=k}^{\infty} P_{i-1}P_{i-2}\ldots P_1 P_0\right)|x_1 - x_0|.$$

If matrix

$$Q_k = \sum_{i=k}^{\infty} P_{i-1}P_{i-2}\ldots P_1 P_0$$

converges to zero as $k \to \infty$, then sequence $\{x_k\}$ is a Cauchy sequence, and because D is closed, $x_k \to s^* \in D$ as $k \to \infty$. By letting $K \to \infty$ we have the error estimate

$$|x_k - s^*| \leq Q_k|x_1 - x_0|,$$

and by selecting x_{k-1} as the initial approximation this inequality reduces to the following:

$$|x_k - s^*| \leq Q_1 |x_k - x_{k-1}|.$$

The first inequality can be used to estimate the necessary number of iteration steps for a given error tolerance. The second relation can be used as a stopping rule similarly to the stationary case.

Notice that for all k,

$$|s^* - f_k(s^*)| \leq |s^* - x_{k+2}| + |f_{k+1}(x_{k+1}) - f_k^{s^*}|$$
$$\leq |s^* - x_{k+2}| + P_k |x_{k+1} - s^*| \leq |s^* - x_{k+2}| + P_k Q_{k+1} |x_1 - x_0|.$$

Therefore, if $P_k Q_{k+1} \to 0$ as $k \to \infty$, then $f_k(s^*) \to s^*$ as $k \to \infty$.

Case Study 3.4. In analyzing the properties of GI/M/1-type positive recurrent Markov chains (for definitions see, for example, [13] the following matrix equation must be solved:

$$R = \sum_{l=0}^{\infty} R^l A_l \qquad (3.11)$$

where matrices A_l are non-negative, and their sum is a stochastic matrix (i.e., the sum of the elements of each row equals one). In the theory of Markov chains the minimal non-negative solution of equation (3.11) has a very important role. The corresponding fixed point iteration scheme has the form

$$R_{k+1} = \sum_{l=0}^{\infty} R_k^l A_l$$

This is a single-step process, therefore, all conditions of Theorem 3.1 are satisfied if the initial approximation is selected so that

$$R_0 = \sum_{l=0}^{\infty} R_0^l A_l$$

The selection $R_0 = 0$ is obviously appropriate. Hence, the iteration sequence $\{R_k\}$ is increasing and converges to the minimal non-negative solution of (3.11).

If A_1 has eigenvalues only in the unit circle, then from the third case study we know that $(I - A_1)^{-1} \geq 0$; therefore, for all $l, A_1(I - A_1)^{-1} \geq 0$. Thus, the corresponding iteration scheme

$$R_{k+1} = \sum_{l=0 l \neq 0}^{\infty} R_k^l (I - A_1)^{-1}$$

has monotone convergence.

A similar scheme can be derived by rewriting equation (3.11) as

$$R = A_0 \left(I - \sum_{l=1}^{\infty} R^{l-1} A_l \right)^{-1}$$

The corresponding iteration process is as follows:

$$R_{k+1} = A_0 \left(I - \sum_{l=1}^{\infty} R_k^{l-1} A_l \right)^{-1}$$

Case Study 3.5. Many iteration procedures for solving linear equations are based on inverses of certain matrices. For example, the general iteration scheme

$$x_{k+1} = -A_1^{-1} A_2 x_k + A_1^{-1} H b$$

introduced in the 6th Application of Section 1.3, and its special cases (such as the Jacobi method) are based on the coefficient matrix $-A_1^{-1} A_2$. In order to apply Theorems 3.1 and 3.2 we need to find conditions which guarantee that mapping $x \to -A_1^{-1} A_2$ is isotone. A sufficient condition is given as $A_1^{-1} \geq 0$ and $A_2 \geq 0$. Hence, the non-negativity of inverse matrices is very important in establishing the monotone convergence of the corresponding iteration sequence. We discuss this problem in detail in Section 4.3; however, we provide here a general scheme that guarantees the non-negativity of the inverses of matrices with special properties. Let $\varphi : C^{n \times n} \to C^{n \times n}$ be a matrix-matrix mapping and let $E : C^{n \times n} \to R^m$ be a matrix-vector mapping, where $m \geq 1$ is a given integer. The following assumptions are made:

1. φ is a projector, i.e., for all $A \in C^{n \times n}$

$$\varphi(\varphi(A)) = \varphi(A);$$

2. φ is linear; i.e., for all $A, B \in C^{n \times n}$ and sclar α and β,

$$\varphi(\alpha A + \beta B) = \alpha \varphi(A) + \beta \varphi(B);$$

3. For all $A \in C^{n \times n}$ and for scalar α such that $|\alpha| \geq 1$,

$$E(\alpha A) \succeq E(A),$$

where $\succ$ is a transitive relation in R^m. We say that a matrix $A \in C^{n \times n}$ is $(\varphi, E, \succ)$-dominant if and only if

$$E(\varphi(A)) \succ E(A - \varphi(A)).$$

Example 3.4. Set $m = n$, and define the relation on R^m as $x \succ y \iff x^{(i)} > y^{(i)}$ for all i. Define

$$E(A) = (E_1(A), \ldots, E_n(A)) \text{ with } E_i(A) = \sum_{j=1}^{n} |a_{ij}| \text{ for } i = 1, 2, \ldots, n.$$

Let $\varphi(A) = \mathrm{diag}(a^{11},\ldots,a^{nn})$. A is then $(\varphi,E,\succ)$-dominant if and only if it is strictly diagonally dominant.

Let $M \subset C^{n \times n}$ be a class of $n \times n$ matrices. We say that $(\varphi,E,\succ)$ is regular with respect to M if and only if every $(\varphi,E,\succ)$-dominant matrix in M is necessarily nonsingular.

Example 3.5. It is well known that $(\varphi,E,\succ)$, defined in Example 3.4, is regular with respect to $M = C^{n \times n}$.

Define φ and E as in Example 3.4 and let $x \succ y \iff x^{(i)}$ as $y^{(i)}$ for all i and for at least one $i, x^{(i)} > x^{(i)}$. Define M as the set of all $n \times n$ irreducible matrices. Then $(\varphi,E,\succ)$ is regular with respect to M.

We shall assume that

4. For all $A \in M$ and for real $\alpha, A + \alpha \varphi(A) \in M$.

Our main result will be based on the following lemma.

Lemma 3.3. Assume that $(\varphi,E,\succ)$ is regular with respect to M. Further, matrix $A \in M$ is $(\varphi,E,\succ)$-dominant, and $\varphi(A)$ is nonsingular. Then all eigenvalues of matrix $I - \varphi(A)^{-}A$ are inside the unit circle.

Proof. Assume in contrary to the assertion that for an eigenvalue λ of $I - \varphi(A)^{-1}A, |\lambda| \geq 1$. Then

$$E(\lambda \varphi(A)) \succeq E(\varphi(A)) \succ E(A - \varphi(A)). \qquad (3.12)$$

Introduce matrix

$$C = A + (\lambda - 1)\varphi(A).$$

Simple calculation shows that

$$\varphi(C) = \varphi(A) + (\lambda - 1)\varphi(\varphi(A)) = \lambda \varphi(A),$$

and

$$C - \varphi(C) = A - \varphi(A).$$

Hence (3.12) implies that matrix C is $(\varphi,E,\succ)$-dominant. Since $C \in M, C$ is nonsingular. Therefore, no $x \neq 0$ exists such that

$$0 = Cx = (A + (\lambda - 1)\varphi(A))x;$$

that is, equation

$$(\varphi(A) - A)x = \lambda \varphi(A)x$$

cannot be satisfied with $x \neq 0$. Since $\varphi(A)$ is invertible, λ is not an eigenvalue of matrix

$$\varphi(A)^{-1}(\varphi(A) - A) = I - \varphi(A)^{-1}A,$$

which contradicts the definition of λ. $\qquad \square$

The general characterization theorem is formulated as follows.

Theorem 3.6. Assume that for a matrix $A \in M, \varphi(A)$-exists and is non-negative, and $I - \varphi(A)^{-1}A \geq 0$. Assume further that A is $(\varphi, E, \succ)$-dominant and $(\varphi, E, \succ)$ is regular with respect to M. Then A^{-1} exists and is non-negative.

Proof. Notice that relation (3.8) implies that matrix

$$I - (I - \varphi(A)^{-1}A)$$

is invertible, and its inverse is non-negative. Since $\varphi(A)^{-1} \geq 0$,

$$0 \leq [I - (I - \varphi(A)^{-1}A)]^{-1}\varphi(A)^{-1} = (\varphi(A)^{-1}A)^{-1}\varphi(A)^{-1}$$
$$= A^{-1}\varphi(A)\varphi(A)^{-1} = A^{-1},$$

which completes the proof. $\square$

We finally note that further special cases of this theorem are investigated in Section 4.3.

Case Study 3.6. Consider the single variable Newton's method (1.14),

$$x_{k+1} = x_k - \frac{g(x_k)}{g'(x_k)}.$$

Assume that s^* is a root of a single variable function g. It is also assumed that g is twice differentiable in the interval $[s^*, x_0]$. In addition, for $x \in [s^*, x_0], g'(x) > 0$ and $g''(x) \geq 0$. These assumptions imply that g increases in $[s^*, x_0]$, and therefore $g(x) > 0$ for all $x \in (s^*, x_0]$. Introduce function

$$f(x) = x - \frac{g(x)}{g'(x)},$$

then

$$f'(x) = 1 - \frac{g'(x)^2 - g(x)g''(x)}{g'(x)^2} = \frac{g(x)g''(x)}{g'(x)^2} \geq 0.$$

Thus, function f increases in $[s^*, x_0]$, and since $x_1 < x_0$, Theorem 3.2 implies that the iteration sequence $\{x_k\}$ is decreasing. By using finite induction it is easy to show that for all $k \geq 0, x_k \geq s^*$. For $k = 0$ this is obvious. Assume that for some $k \geq 0, x_k \geq s^*$. Then,

$$x_{k+1} = f(x_k) \geq f(s^*) = s^*.$$

Hence, sequence $\{x_k\}$ is decreasing with a lower bound s^*, and therefore converges. Let x^* denote the limit of $\{x_k\}$. The definition of the Newton method with $k \to \infty$ implies that

$$x^* = x^* - \frac{g(x^*)}{g'(x^*)},$$

i.e., $g(x^*) = 0$. We know from the above discussion that $g(x) > 0$ for $x > s^*$, therefore $x^* = s^*$. In summary, under the above conditions the Newton method converges to s^*, and the iteration sequence decreases.

The above property of the Newton method is used mainly in selecting the initial approximation. If g increases and is convex in the neighborhood of the root, then an $x_0 > s^*$ in this neighborhood is the appropriate choice, because the method then monotonically converges to s^*. Similar derivation gives the appropriate initial approximation selection in the four main cases depending on the signs of g' and g''. If g increases and is concave, then $x_0 < s^*$ is the selection. If g decreases and is convex, then an $x_0 < s^*$ must be reselected. If g decreases and is concave, then an $x_0 > s^*$ is the appropriate choice.

Example 3.6. In Example 1.8, the Newton method was applied for equation

$$\frac{1}{2}\sin x + 1 - x = 0,$$

and it is shown that there is a unique root in $(1, 2)$. Let

$$g(x) = \frac{1}{2}\sin x + 1 - x,$$

then

$$g'(x) = \frac{1}{2}\cos x - 1 < 0$$

and

$$g''(x) = -\frac{1}{2}\sin x - 1 < 0 \text{ for } x \in (1,2).$$

Hence, $x_0 > s^*$ is the appropriate selection. If $x_0 = 2$, then the iteration sequence is decreasing, as illustrated by actual computation in Example 1.8.

The case of the multivariable Newton method can be discussed in a similar way.

Case Study 3.7. Examine the linear fixed point problem

$$x = Ax + b,$$

where A is a non-negative $n \times n$ constant matrix and $b \geq 0$ is an n-dimensional constant vector. Assume that there is a unique non-negative fixed point s^*. Select vectors $x_0 \leq s^*$ and $y_0 \geq s^*$ as initial approximations. For example, $x_0 = 0$ and any upper bound of s^* is a suitable choice for y_0. If $x_1 \geq x_0$ and $y_1 \leq y_0$, then Theorems 3.1 and 3.2 are applicable. That is, the iteration sequences

$$x_{k+1} = Ax_k + b$$

and

$$y_{k+1} = Ay_k + b$$

converge to s^*. Furthermore, for all k,

$$x_k \leq s^* \leq y_k.$$

Example 3.7. As an illustration of the above procedure, consider the linear fixed point problem:

$$x^{(1)} = \frac{1}{4}x^{(1)} + \frac{1}{4}x^{(2)} + \frac{1}{2}$$
$$x^{(2)} = \frac{1}{4}x^{(1)} + \frac{1}{8}x^{(2)} + \frac{5}{8}.$$

Since the row norm of the coefficient matrix

$$A = \begin{pmatrix} \frac{1}{4} & \frac{1}{4} \\ \frac{1}{4} & \frac{1}{8} \end{pmatrix}$$

is less than one, the iteration procedure

$$x_{k+1}^{(1)} = \frac{1}{4}x_k^{(1)} + \frac{1}{4}x_k^{(2)} + \frac{1}{2}$$
$$x_{k+1}^{(2)} = \frac{1}{4}x_k^{(1)} + \frac{1}{8}x_k^{(2)} + \frac{5}{8}.$$

converges to the unique fixed point. Select $x_0 = 0$, then

$$x_1 = \begin{pmatrix} \frac{1}{2} \\ \frac{5}{8} \end{pmatrix} \geq x_0.$$

Alternatively, select

$$y_0 = \begin{pmatrix} 2 \\ 2 \end{pmatrix}, \text{ then } y_1 = \begin{pmatrix} \frac{3}{2} \\ \frac{11}{8} \end{pmatrix} \leq y_0.$$

Hence, sequence $\{x_k\}$ increases and sequence $\{y_k\}$ decreases. Some initial terms are given as follows:

$$x_2 = \begin{pmatrix} \frac{1}{4} & \frac{1}{4} \\ \frac{1}{4} & \frac{1}{8} \end{pmatrix} \begin{pmatrix} \frac{1}{2} \\ \frac{5}{8} \end{pmatrix} + \begin{pmatrix} \frac{1}{2} \\ \frac{5}{8} \end{pmatrix} = \begin{pmatrix} \frac{25}{32} \\ \frac{53}{64} \end{pmatrix}$$

$$x_3 = \begin{pmatrix} \frac{1}{4} & \frac{1}{4} \\ \frac{1}{4} & \frac{1}{8} \end{pmatrix} \begin{pmatrix} \frac{25}{32} \\ \frac{53}{64} \end{pmatrix} + \begin{pmatrix} \frac{1}{2} \\ \frac{5}{8} \end{pmatrix} = \begin{pmatrix} \frac{231}{256} \\ \frac{473}{512} \end{pmatrix}$$

and so on, and

$$y_2 = \begin{pmatrix} \frac{1}{4} & \frac{1}{4} \\ \frac{1}{4} & \frac{1}{8} \end{pmatrix} \begin{pmatrix} \frac{3}{2} \\ \frac{11}{8} \end{pmatrix} + \begin{pmatrix} \frac{1}{2} \\ \frac{5}{8} \end{pmatrix} = \begin{pmatrix} \frac{39}{32} \\ \frac{75}{64} \end{pmatrix}$$

$$x_3 = \begin{pmatrix} \frac{1}{4} & \frac{1}{4} \\ \frac{1}{4} & \frac{1}{8} \end{pmatrix} \begin{pmatrix} \frac{39}{32} \\ \frac{75}{64} \end{pmatrix} + \begin{pmatrix} \frac{1}{2} \\ \frac{5}{8} \end{pmatrix} = \begin{pmatrix} \frac{281}{256} \\ \frac{551}{512} \end{pmatrix}$$

etc. Notice that the true solution is $s^* = (1,1)^T$, and really

$$x_0 \leq x_1 \leq x_2 \leq x_3 \leq s^* \leq y_3 \leq y_3 \leq y_2 \leq y_1 \leq y_0.$$

Consider again the above linear fixed point problem, but drop the special condition that matrix A is non-negative. In this general case, we can rewrite A as a difference of two non-negative matrices B and C; i.e., $A = B - C$. Obviously an infinite number of such decompositions exist. For example, the selection

$$b_{ij} = \max\{0, a_{ij}\}, c_{ij} = \max\{0, -a_{ij}\} \quad \text{(for all } i, j)$$

is appropriate. Construct the iteration process (3.3) with point-to-point mappings

$$K_k^{(1)}(x_0, x_1, \ldots, x_k) = Bx_k + b,$$
$$H_k^{(1)}(y_k, y_{k-1}, \ldots, y_0) = Cy_k,$$
$$K_k^{(2)}(y_k, y_{k-1}, \ldots, y_0) = By_k + b,$$
$$H_k^{(2)}(x_0, x_1, \ldots, x_k) = Cx_k.$$

The non-negativity of matrices B and C imply that conditions (G_1) and (G_2) hold. In order to satisfy condition (G_3), we must select the initial approximations x_0 and y_0 such that

$$x_0 \le Bx_0 - Cy_0 + b \le By_0 - Cx_0 + b \le y_0 \tag{3.13}$$

Under these conditions sequence $\{x_k\}$ increases and sequence $\{y_k\}$ decreases. Because both sequences are bounded, they are convergent. Let the limit points be denoted by x^* and y^*, respectively. Then allowing $k \to \infty$ in the iteration equations gives the equations

$$x^* = Bx^* + b - Cy^*$$

and

$$y^* = By^* + b - Cx^*.$$

Rewrite these equations as

$$\begin{pmatrix} I - B & C \\ C & I - B \end{pmatrix} \begin{pmatrix} x^* \\ y^* \end{pmatrix} = \begin{pmatrix} b \\ b \end{pmatrix}$$

Notice that from the definition of s^* we obtain

$$\begin{pmatrix} I - B & C \\ C & I - B \end{pmatrix} \begin{pmatrix} s^* \\ s^* \end{pmatrix} = \begin{pmatrix} b \\ b \end{pmatrix}$$

which implies the following important assertion. If matrix

$$\begin{pmatrix} I - B & C \\ C & I - B \end{pmatrix}$$

is nonsingular, then $x^* = y^* = s^*$. In other words, both iteration sequences $\{x_x\}$ and $\{y_x\}$ converge to the same fixed point s^*.

Case Study 3.8. We have seen in Section 1.3 that matrix eigenvalue problems can be solved by applying the multivariable Newton method, which could be rewritten in the form (1.20). The monotonicity of the resulting iteration sequence can be discussed based on the general properties of Newton methods.

Case Study 3.9. For solving initial value problems of ordinary differential equations a simple iteration scheme (1.23) was introduced, which we repeat here for the sake of convenience:

$$x_{k+1}(t) = x_0 + \int_{t_0}^{t} f(\tau, x_k(\tau)) d\tau, \quad x_0(t) \equiv x_0$$

In order to apply the methodology of this chapter, we must find conditions that guarantee that the mapping on the right-hand side of the above iteration equation is increasing in x. Further, for all $t, x_1(t) \geq x_0(t)$ (or alternatively, $X_1(t) \leq x_0(t)$). Obviously, if for all $t, f(t, x0(t)) \geq 0$ (or alternatively, $f(t, x0(t)) \leq 0$), then the second condition is satisfied. It is also easy to see that the first condition holds if f increases in x (when $\bar{x} \geq x$ implies that $f(t, \bar{x}) \geq f(t, x)$ for all t). Assume that f is differentiable. Then this condition is necessarily satisfied if all elements of the Jacobian of f are non-negative.

Example 3.8. As an illustration of the above conditions, consider the initial value problem

$$\dot{x} = t + x, \quad x(0) = 1.$$

For all $t \geq 0, t + x$ increases in x. Furthermore, by selecting $x_0(t) = 1, t + x_0(t) = t + 1 \leq 0$. Thus, the general conditions are satisfied, and the iteration sequence is increasing. In our case the iteration procedure has the form

$$x_{k+1}(t) = 1 + \int_{0}^{t} (\tau + x_k(\tau)) d\tau.$$

That is,

$$x_1(t) = 1 + \int_{0}^{t} (\tau + 1) d\tau = 1 + t + \frac{t^2}{2},$$

$$x_2(t) = 1 + \int_{0}^{t} (\tau + 1 + \tau + \frac{\tau^2}{2}) d\tau = 1 + t + t^2 + \frac{t^3}{6},$$

$$x_3(t) = 1 + \int_{0}^{t} (\tau + 1 + \tau + \tau^2 + \frac{\tau^3}{6}) d\tau = 1 + t + t^2 + \frac{t^3}{6} + \frac{t^4}{24},$$

and so on. The monotonicity of this sequence is clear.

Case Study 3.10. Non-linear differential equations are often solved by linearization. The resulting process was illustrated in Section 1.3, Application 10. In order to guarantee the monotonicity of the iteration sequence $\{x_k(t)\}$ we need to find conditions in which the solution $x(t)$ of the linear differential equation

$$\dot{x}(t) = J(t,z(t))x(t) + [f(t,z(t)) - J(t,z(t))z(t)], \quad x(t_0) = x_0$$

increases in z(t).

Case Study 3.11. In this case study we consider the iteration algorithms

$$x_{k+1(t)} = \int_a^b K(s,t,x_k(s))ds + f(t)$$

and

$$x_{k+1}(t) = \int_a^t C(s,t)x_k(s)ds + f(t)$$

for solving Fredholm and Volterra integral equations. These procedures were discussed in Section 1.3.

In order to apply Theorem 3.1 we must guarantee that $x,(t) \geq x_0(t)$ for all $t \in [a,b]$, and the right-hand sides of the above iteration equations increase in $x_k(s)$. If we assume that $K(s,t) \geq 0$ and $C(s,t) \geq 0, f(t) \geq 0$ for all t and s from the interval $[a,b]$, then, be selecting $X_0(t) = f(t)$, both conditions are satisfied.

It is easy to see that Theorem 3.2 is applicable with $x_0(t) = 0$ if $K(s,t), C(s,t)$, and $f(t)$ are nonpositive for all $t,s \in [a,b]$.

Example 3.9. Consider again the Fredholm-type integral equation of Example 1.16:

$$x(t) = \int_a^1 \frac{t+s}{4}x(s)ds + 1/$$

By selecting $x_0(t) = 1$, we have shown that

$$x_1(t) = \frac{t}{4} + \frac{9}{8}$$

and

$$x_2(t) = \frac{5t}{16} + \frac{223}{192}.$$

Obviously $x_0(t) \leq x_1(t) \leq x_2(t)$ for all $t \in [0,1]$. This monotonicity continues for all $k > 2$, because in this case $K(s,t) = \frac{t+s}{4} \geq 0$ and $f(t) = 1 > 0$.

Case Study 3.12. A dynamic economic system was introduced in the previous chapter. If we assume that the demand and supply functions are linear, the resulting equations can be rewritten as presented in equation (2.27), which we repeat here for the sake of convenience:

$$P_i(t+1) = [1 - K_i(b_{ii} - a_{ii})]P_i(t) - \sum_{j \neq i} K_i(b_{ij} - a_{ij})P_j(t) - L_i(b_{i0} - a_{i0}),$$

We assume here that $K_i > 0, b_{ii} > 0, a_{ii} < 0, b_{ij} \leq 0$, and $a_{ij} \geq 0$ for all i and $j = i$.

Assume that the following additional conditions also hold:

1. $K_j \cdot (b_{ii} - a_{ii}) \leq 1$ for all i;
2. $P_i(1) \geq P_i(0)$ for all i.

Then Theorem 3.1 can be applied to show that for all products, the prices are always increasing; that is, for all i and t, $P_i(t+1) \geq P_i(t)$.

Case Study 3.13. A classical oligopoly model was introduced in Chapter 2. For the sake of convenience, we repeat here the resulting dynamic equations:

$$x^{(t)}(t) = -\frac{1}{2} \sum_{l \neq k} x^{(l)}(t-l) + \frac{b^{(k)} - B}{2A} (1 \leq K \leq N),$$

where $x^{(k)}(t)$ denotes the production level of firm k at time period t. Since the coefficient of $x^{(1)}(t-1)$ is negative, the right-hand side of this equation is decreasing in all $x^{(1)}(t-1)$. Therefore, no direct monotone convergence can be established, since if for some $x \geq 0, x(x) \leq x(x-1)$, then $x(x+1) \geq x(x)$.

Notice, however, that this scheme is a special case of the iteration procedure

$$x_{k+1} = -Ax_k + b$$

where $A \geq 0$ and $b \geq 0$. Since

$$x_{k+2} = -Ax_{k+1} + b = -A(-Ax_k + b) = A^2 x_k + (I - A)b,$$

where $A^2 \geq 0$, under certain conditions we are able to establish the monotonicity of the sequences $x_0, x_2, x_4, \ldots$ and $x_1, x_3, x_5, \ldots$.

Assume that equation $x = -Ax + b$ has a unique non-negative fixed point s^*. Then s^* is a non-negative fixed point of mapping $x \to A^2 x_k + (I - A)b$. As the following example shows this mapping may have other non-negative fixed points than s^*.

Example 3.10. Select $A = I$ and $b = 1$, where all components of 1 are equal to one. Then equation

$$x = -Ax + b$$

has a unique fixed point $x = 1/2b$; however, equation $x = A^2x + (I - A)b$ is equivalent to the trivial equation $x = x$, which has infinite by many fixed points.

Select an x_0 such that $x_2 = A^2x_0 + (I - A)b \geq X_0$. Theorem 3.1 implies that

$$x_0 \leq x_2 \leq x_4 \leq \dots.$$

Notice that the selection $x_0 = 0$ is not always satisfactory, since $(I - A)b$ is not non-negative. If $(I - A)b \geq 0$, then $x_0 = 0$ is an appropriate selection. In the general case it is rather difficult to find a feasible non-negative solution of inequality

$$(A^2 - I)x_0 \geq (A - I)b.$$

For example, the first phase of the simplex method might be used. If x_0 is selected in such a way that $x_3 = A^2x_1 + (I - A)b \geq x_1$ then $x_1 \leq x_3 \leq x_5 \leq \dots.$

If $A^2x_0 + (I - A)b \leq x_0$, then similarly $x_0 \geq x_2 \geq x_4 \geq \dots.$ Finally, of x_0 selected so that $A^2x_1 + (I - A)b \leq x_1, x_1 \geq x_3 \geq x_5 \geq \dots$, and if $A^2x_1 + (I - A)b \geq x_1$, then $x_1 \geq x_3 \geq x_5 \geq \dots.$

Case Study 3.14. Consider in this case study the modified iteration algorithm

$$x_{k+1} = \alpha_k x_+ (1 - \alpha_k)A(x_k), \quad (\alpha_k \in [0,1))$$

where $A : D \to D$ with $D \subseteq R^n$ being a closed, convex set. This process was examined in Chapter 2. Notice first that if A is increasing (or decreasing) in x_k, the entire right-hand side of the iteration equation increases (or decreases) in x_k. Furthermore, if x_0 selected so that $A(x_0) \geq x_0$, (or $A(x_0) \geq x_0$), then $x_1 \geq x_0$ (or $x_1 \leq x_0$). Hence, under these assumptions Theorem 3.1 (or 3.2) can be applied, and the resulting iteration sequence is monotonic.

3.4   EXERCISES

1.   Consider the problem of approximating a locally unique zero s^* of the equation

$$g(x) = 0,$$

where g is a given real function define on a closed interval $[a, b]$. The Newton iterates $\{x_n\}, \{y_n\}$ $n \geq 0$ are defined as

$$g(x_n) + g'(x_n)(x_{n+1} - x_n) = 0, \quad x_0 = a$$

and

$$g(y_n) + g'(y_n)(y_{n+1} - y_n) = 0, \quad y_0 = b.$$

Use Theorems 3.1 and 3.2 to find sufficient conditions for the monotone convergence of the above sequences to $s^* \in [a, b]$.

2. Examine the above problem, but using the secant iterations $\{x_n\}, \{y_n\} n \geq -1$ given by

$$g(x_n) + \frac{g(x_n) - g(x_{n-1})}{x_n - x_{n-1}} (x_{n+1} - x_n) = 0, \quad x_1 = a, x_0 = \text{given}$$

and

$$g(y_n) + \frac{g(y_n) - g(y_{n-1})}{y_n - y_{n-1}} (y_{n+1} - y_n) = 0, \quad y_1 = b, y_0 = \text{given}$$

3. Examine the above problem, but using the Newton-like iterations, where $\{x_n\}, \{y_n\}, n \geq 0$ are given by

$$g(x_n) + h(x_n)(x_{n+1} - x_n) = 0, \quad x_0 = a$$

and

$$g(y_n) + h(y_n)(y_{n+1} - y_n) = 0, \quad y_0 = b$$

with a suitably chosen function h.

4. Examine the above problem, but using the iterations $\{x_n\}, \{y_n\}, n \geq 0$, given by

$$g(x_n) + c_n(x_{n+1} - x_n) = 0, \quad x_0 = a$$

and

$$g(y_n) + d_n(y_{n+1} - y_n) = 0, \quad x_0 = b$$

with suitably chosen real sequences $\{c_n\}$ and $\{d_n\}$.

5. Generalize Problems 1 to 4 to R^n.

6. Generalize Problems 1 to 4 to R^n by choosing g to be defined on a given subset of a partially ordered topological space.

7. To find a zero s^* for $g(x) = 0$, where g is a real function defined on $[a,b]$, rewrite the equation as

$$x = x_c g(x) = f(x)$$

for some constant $c \neq 0$. Consider the sequences

$$x_{n+1} = f(x_n), \quad x_0 = a$$

and

$$y_{n+1} = f(y_n), \quad y_0 = b$$

Use Theorems 3.1 and 3.2 to find sufficient conditions for the monotone convergence of the above sequences to a locally unique fixed point $s^* \in [a,b]$ of the equation $x = f(x)$.

8. Many engineering problems can be formulated as the system of differential equations

$$\ddot{x}_i = f_i(t, x_1, x_2) \, i = 1, 2, 0 \le t \le 1$$

subject to the boundary condition

$$x_i(0) = \alpha_i, \quad x_i(1) = \beta_i, 1 = 1, 2.$$

By using a uniform discretization mesh, construct the corresponding system of real equations to be solved. Find sufficient conditions for the monotone convergence of the multivariable Newton's iterations (1.15) to a zero of the system of the discretized equations.

9. Apply the result of Problem 9 to solve the differential equations, when

$$f_1(t, x_1, x_2) = x_1^2 + x_1 + 0.1x_2^2 - 1.2$$
$$f_2(t, x_1, x_2) = 0.2x_1^2 + x_2^2 + 2x_2 - 0.6,$$
$$\alpha_1 = \alpha_2 \qquad\qquad\qquad = \beta_1 = \beta_2 = 0.$$

10. Let $A \ge 0$ and $b \ge 0$, and investigate the solvability of the inequality

$$(A_2 - I)x \ge (A - I)b,$$

which has an important role in case study 3.13 of Section 3.3.

11. Repeat Example 3.2 for problem

$$x = (x^2 + 1)e^x - 1$$

12. Repeat Example 3.7 for problem

$$x^{(1)} = \frac{x^{(1)}}{10} + \frac{x^{(2)}}{4} + 2$$
$$x^{(2)} = \frac{x^{(1)}}{8} + \frac{x^{(2)}}{16} + 1$$

13. Repeat Example 3.8 for problem

$$\dot{x} = t^2 + x - 1, \quad x(0) = 0.$$

REFERENCES

1. Argyros, I.K. 2007. *Computational Theory of Iterative Methods*. eds. Chui, C.K., Wuytack, L. Series: Studies in Computational Mathematics, 15. New York: Elsevier.
2. Argyros, I.K. 2008. *Convergence and Applications of Newton-type Iterations*. New York: Springer-Verlag.
3. Argyros, I.K., and George, S. 2019. *Mathematical Modeling for the Solution of Equations and Systems of Equations with Applications*, Volume III, NY: Nova Science Publisher.
4. Argyros, I.K., and Hillout S. 2012. Weaker conditions for the convergence of Newton's method. *J. Complexity*. 28. 3:364–387.
5. Argyros, I.K., and Magreñán, A.A. 2017. *Iterative Methods and Their Dynamics with Applications*, New York: CRC Press. Taylor& Francis.
6. Argyros, I.K., and Magrenan, A.A. 2018. *A Contemporary Study of Iterative Methods*. NY: Academy Press, Elsevier.
7. Argyros, I.K., and Regmi, S. 2019. *Undergraduate Research at Cameron University on Iterative Procedures in Banach and Other Spaces*, NY: Nova Science Publisher.
8. Argyros, M.I., Argyros, I.K., and Regmi, S. 2021. *Hilbert Spaces and Its Applications*, NY: Nova Science Publisher.
9. Bramble, J. and Hubbard, B. 1964. On a finite difference analogue of an elliptic boundary value problem with is neither diagonally dominant nor of negative type. *J. Math. Phys.*, 40:117–132.
10. Brock, W.A. and Scheinkman, J.A. 1975. Some results on global asymptotic stability of difference equations. *J. Econ. Theory*, 10:265–268.
11. Deuflhard, P. and Heindl, G. 1979. Affine invariant convergence theorems for Newton's method and extensions to related methods. *SIAM J. Numer. Anal.*, 16:1–10.
12. Goldberg, J. and Szidarovsky, F. 1991. A general model and convergence results for determining vehicle utilization emergence system. *Stoch. Models*, 7(1):137–160.
13. Ortega, J.M. and Rheinboldt, W.C. 1970. Iterative Solutions of Nonlinear Equations in Several Variables, New York: Academic Press.
14. Szidarovszky, F. and Bahill, T.A. 1992. *Linear Systems Theory*, Boca Raton, FL: CRC Press.
15. Tarayama, A. 1974. *Mathematical Economics*, Illinois: Dryden, Hinsdale.
16. Zangwill, W.I. 1969. Convergence conditions for nonlinear programming algorithms. *Manage. Sci.*, 16(1):1–13.

4 Applications in:

In this chapter, we present more applications of the theory developed previously in the areas of Neural Networks, Reliability Engineering, and Economical Models.

4.1 NEURAL NETWORKS

Artificial Neural Networks (ANN) are often called connectionist systems. They are inspired by biological neural networks, how information is distributed in human and animal brain. They are mainly used in modeling input-output transformation if physical models are not available, but a large set of simultaneous input-output past data exists. An ANN consists of an input layer, an output layer and hidden layer. Each layer consists of nodes called artificial neurons. In the input layer the nodes correspond to the input parameter, those of the output layer correspond to be output parameter, and the nodes in the hidden layers serve as model parameters. A sample ANN with M input, N output parameters and with one hidden layer with K nodes can be used to study the details further [1–22].

Each input node x_m is connected to all hidden nodes z_r $(r = 1, 2, \ldots, K)$ and an unknown weight w_{mr} is attached to each arc. Similarly each hidden node z_r is connected to all output nodes y_n $(n = 1, 2, \ldots, N)$ with corresponding weights $\overline{w}_{rn}$. In this ANN models usually linear relation is assumed between the input and hidden nodes as well as between the hidden and output nodes with the weights w_{mr} and $\overline{w}_{rn}$:

$$z_r = \sum_{m=1}^{M} w_{mr} x_m \text{ and } y_n = \sum_{r=1}^{K} \overline{w}_{rn} z_r, \tag{4.1}$$

so a nonlinear input-output model is created,

$$y_n = \sum_{r=1}^{K} \overline{w}_{rn} \sum_{m=1}^{M} w_{mr} x_m, \tag{4.2}$$

where the weights w_{mr} and $\overline{w}_{rn}$ are determined based on known past data. Let $(\underline{x}^{(i)}, \underline{y}^{(i)})$ $(i = 1, 2, \ldots, I)$ denote the simultaneous input-output pairs with components $x_m^{(i)}, y_n^{(i)}$ as $\underline{x}^{(i)} = (x_1^{(i)}, x_2^{(i)}, \ldots, x_M^{(i)})$ and $\underline{y}^{(i)} = (y_1^{(i)}, y_2^{(i)}, \ldots, y_N^{(i)})$.

For each input vector $\underline{x}^{(i)}$, relation (4.2) provides an estimate $\underline{\overline{y}}^{(i)}$ of the components of the corresponding output vector:

$$\overline{y}_k^{(i)} = \sum_{r=1}^{K} \sum_{m=1}^{M} \overline{w}_{rn} w_{mr} x_m^{(i)}, \tag{4.3}$$

DOI: 10.1201/9781003128915-4

however the corresponding "true" output component $y_n^{(i)}$ is known. The error of the estimate (4.3) can be characterized by the sequence of the difference of $\bar{y}_n^{(i)}$ and $y_n^{(i)}$. This error can be computed for all components n and measurements i leading to the overall squared error

$$Q = \sum_{n=1}^{N} \sum_{i=1}^{I} \left[\sum_{r=1}^{K} \sum_{m=1}^{M} \bar{w}_{rn} w_{mr} x_m^{(i)} - y_n^{(i)} \right]^2. \qquad (4.4)$$

The "training" of the ANN is done by minimizing Q with respect to unknowns $\bar{w}_{rn}$ and w_{mr}, which is a non-linear optimization problem. This is usually solved by an iteration procedure like the Newton or the gradient methods. Then the output components to all future input vector can be easily predicted by using equation (4.2).

I the practical applications, the input and output parameters are transformed to have the same order of magnitude. One often used transformation function is

$$f(x) = \frac{e^x}{1 + e^x} \text{ and } f(y) = \frac{e^y}{1 + e^y},$$

since function f is strictly increasing with zero limit of negative infinity and unit limit at positive infinity.

The number of the hidden layers are the number of nodes on each of them are used selected parameters, which have to be tested. for this purpose only about half of the data pairs $(\underline{x}^{(i)}, \underline{y}^{(i)})$ is used for training, and for the other half the overall error $Q^{(i)}$ is computed. If its value is below a user selected threshold, then the ANN structure with the optimal weights is accepted, otherwise the user might add additional hidden layers and/or additional nodes to the hidden layers. If more than one hidden layer is selected, then each node of each hidden layer is connected to each node of the next hidden layer by introducing additional weights, so relations (4.4) or (4.2) become much more complicated.

4.2   RELIABILITY ENGINEERING

Consider an equipment which is subject to random failures. The predicted time of the first failure of a new equipment has special interest. From. recorded past failures of identical equipment, the cumulative distribution function (CDF) of the time to first failure can be addressed by statistical method. The maximum likelihood method is the most popular. It is assumed that the type of the distribution is known and only its parameters have to be estimated [1–22]. let $F(t, \theta)$ denote the CDF, where θ is an unknown parameter vector. Its derivative

$$f(t, \theta) = \frac{d}{dt} F(t, \theta) \qquad (4.5)$$

is called the Probability Density Function (PDF). Let $t_1, t_2, \ldots, t_N$ denote the past failure data, then the likelihood function is defined as

$$\Pi_{n=1}^{N} f(t_n, \boldsymbol{\theta}), \tag{4.6}$$

which is maximized to have the optimal parameter vector. In many practical cases the PDF has exponential term, therefore in such cases its logarithm

$$L(\boldsymbol{\theta}) = \sum_{n=1}^{N} \log f(t_n, \boldsymbol{\theta}), \tag{4.7}$$

is considered, which is called the likelihood function. This function is then maximized. Any textbook on mathematical statistics offers example when the optimal parameter values can be derived analytically, however for several practical distribution type this is not the case if analytic solution is not available, then function (4.6) or (4.7) is maximized which is a non-linear optimization problem.

As an example consider the case of Weibull distribution, the PDF of Weibull is given as

$$f(t, \beta, \eta) = \frac{\beta}{\eta^{\beta}} t^{\beta-1} e^{-\left(\frac{t}{\eta}\right)^{\beta}} \quad (t \geq 0). \tag{4.8}$$

Function (4.6) has now the form,

$$\Pi_{n=1}^{N} \frac{\beta}{\eta^{\beta}} t_n^{\beta-1} e^{-\left(\frac{t}{\eta}\right)^{\beta}} = \frac{\beta N}{\eta^{\beta N}} \left(\Pi_{n=1}^{N} t_n\right)^{\beta-1} e^{-\Sigma_{n=1}^{N}\left(\frac{t}{\eta}\right)^{\beta}},$$

and so

$$L(\beta, \eta) = N \log \beta - N\beta \log \eta + (\beta - 1) \log \Pi_{n=1}^{N} t_n - \sum_{n=1}^{N} \left(\frac{t_n}{\eta}\right)^{\beta}. \tag{4.9}$$

Maximizing this bi-variable function can be iteration process, like the gradient method, or by considering the first order conditions and solving the resulting system of nonlinear equations. Notice first the

$$\frac{\partial L(\beta, \eta)}{\partial \eta} = -\frac{N\beta}{\eta} - \sum_{n=1}^{N} \beta \left(\frac{t_n}{\eta}\right)^{\beta-1} \left(-\frac{t_n}{\eta^2}\right) = 0,$$

which can be simplified as

$$\eta = \left(\frac{1}{N} \sum_{n=1}^{N} t_n^{\beta}\right)^{\frac{1}{\beta}}. \tag{4.10}$$

Similarly,

$$\frac{\partial L(\beta, \eta)}{\partial p} = \frac{N}{p} - N \log \eta + \log \Pi_{n=1}^{N} t_n - \sum_{n=1}^{N} \left(\frac{t_n}{\eta}\right)^{\beta} \log \left(\frac{t_n}{\eta}\right) = 0. \tag{4.11}$$

If parameter β is known, then the value of η can be obtained from (4.10), and if both parameters are unknown, then the system (4.10) - (4.11) has to be solved. A difference approach is offered by substituting (4.10) into objective function (4.9), which then becomes a single variable optimum problem for the only unknown β.

In additional to the CDF and PDFT of the distribution of time to failure, the reliability function if often used. It is defined as

$$R(t) = 1 - F(t) \tag{4.12}$$

giving the probability that no failure occurs in interval $[0,t]$. The manufacturer of an equipment is interested to find out the time when the reliability of the equipment drops under a certain threshold. This problem arises, for example, in designing warranty contracts. In such cases equation

$$R(t) = p \tag{4.13}$$

has to be solved. There are cases, when no analytic solutions are available, such cases include gamma distribution:

$$R(t) = \int_t^\infty \frac{x^{t-1}e^{-\frac{x}{\theta}}}{\theta\Gamma(r)}\,dx \quad (t \geq 0)$$

normal distribution:

$$R(t) = \int_t^\infty \frac{1}{\sigma\sqrt{2\pi}}e^{-1\frac{(x-\mu)^2}{2\sigma^2}}\,dx$$

log-gamma distribution:

$$R(t) = \int_t^\infty \frac{\lambda(\lambda\ln x)^{\alpha-1}e^{-\lambda\ln x}}{\Gamma(\alpha)x}\,dx$$

or log-normal distribution:

$$R(t) = \int_t^\infty \frac{1}{\sigma\sqrt{2\pi}}e^{-\frac{(\ln x-\mu)^2}{2\sigma^2}}\,dx$$

and many others. In such case equation (4.13) has to be solved numerically.

Assume next that an equipment was in working condition in interval $[0,x]$ and we need to know the probability that it will break down during additional y time periods:

$$P(t < x+y | t \geq x) = \frac{P(x \leq t < x+y)}{P(t \geq x)} = \frac{F(x+y) - F(x)}{1 - F(x)} \tag{4.14}$$

which can be expressed via the reliability function as follows:

$$R(x+y|x) = \frac{R(x) - P(x+y)}{R(x)} \tag{4.15}$$

which is called the conditional relatability function. We can ask the same questions as before: when will the reliability of a working equipment of age x be dropped under a certain threshold. In this case (4.13) is modified as

$$R(x+y|x) = P \tag{4.16}$$

where x is known and the additional time length y is unknown.

If a breakdown occurs, then there are three major probabilities to avoid the problem. Minimal repair is used, when the state of the equipment becomes the same as it was before break down. If repair is not possible or economically not advised, then failure replacement is done, when the equipment is replaced with a new one. In performing particular repairs, then depending on the type of repair, the expected number of future failures can decrease, on the age of the equipment is decreased. To allocated skilled manpower and spare parts for future repairs, it is important to estimate the expected number of failures in any future time intervals. Let $M(t)$ denote the expected number of failures in interval $[0,t]$, then in any future interval $[t_1,t_2]$ it is $M(t_2) - M(t_1)$. So function $M(t)$ has to be determined. It will also be assumed that at each repair the equipment is idle for T time periods, when no failure may occur and the equipment is not aged.

Assume first minimal repair. Let X_t denote the actual number of failures in interval $[0,t]$ and assume that the first failure occurs at time S_1. Then

$$M(t) = E(X_t) = E_{S_1}[E(X_t|S_1)] \tag{4.17}$$

where expectation by conditioning is used. Notice that

$$E(X_t|S_1) = \begin{cases} 0 & \text{if } t < S_1 \\ 1 & \text{if } S_1 < t < S_1 + T \\ 1 + M(t-T) - M(S_1) & \text{if } t > S_1 + T, \end{cases} \tag{4.18}$$

since before first failure no other failure may occur, no additional failure may occur during repair and after repair at time t the effective age of the equipment becomes $t - T$. Therefore from (4.17)

$$M(t) = \int_{t-T}^{t} f(s)ds + \int_{0}^{t-T} [1 + M(t-T) - M(s)]f(s)ds, \tag{4.19}$$

where $f(s)$ is the PDF of time to failure. So

$$M(t) = F(t) - F(t-T) + F(t-T)(1 + M(t-T)) - \int_{0}^{t-T} M(s)f(s)ds$$

and after simplification,

$$M(t) = M(t-T)F(t-T) + F(t) - \int_0^{t-T} M(s)f(s)ds \quad (t > T),\tag{4.20}$$

which is an integral equation for the unknown function $M(t)$. Notice that if $t \leq T$, then at most one failure can occur, so

$$M(t) = F(t) \quad (t \leq T).\tag{4.21}$$

In the special case, when repairs are instantaneous, equation (4.20) becomes

$$M(t) = M(t)F(t) + F(t) - \int_0^t M(s)f(s)ds\tag{4.22}$$

with $M(0) = 0$. This equation is a bit more complicated than a Volterma-type interval equation discussed earlier in this book.

Assume next failure replacement, then equation (4.20) is modified as

$$E(X_t|S_1) = \begin{cases} 0 & \text{if } t < S_1 \\ 1 & \text{if } S_1 < t < S_1 + t \\ 1 + E(X_t - S_1 - T) & \text{if } t > S_1 + T, \end{cases}\tag{4.23}$$

since at time $t(t > S_1 + T)$ the effective age is $t - S_1 - T$, because at time $S_1 + T$ the equipment is replaced by a new one. Then (4.19) is modified as follows:

$$M(t) = \int_{t-T}^t f(s)ds + \int_0^{t-T} [1 + M(t-s-T)]f(s)ds,$$

which is simplified as

$$M(t) = F(t) + \int_0^t t - TM(t-s-T)f(s)ds \quad (t > T)\tag{4.24}$$

with

$$M(t) = F(t) \quad (t \leq T).\tag{4.25}$$

If $T = 0$, then we have

$$M(t) = F(t) + \int_0^t M(t-s)f(s)ds \quad (t > T)\tag{4.26}$$

$$M(t)F(t) \quad (t \leq T).\tag{4.27}$$

By introducing the new iteration variable $z = t - s$ in equation (4.26), it becomes

$$M(t) = F(t) + \int_0^t M(z)f(t-z)dz,$$

which a Volterra-type integral equation.

In the case of one type of partial repairs assume that after repair is done, the expected number of future failure decreases by a factor $1 - \alpha (\alpha < 1)$. Then equation (4.18) modifies as

$$E(X_t|S_1) = \begin{cases} 0 & \text{if } t < S_1 \\ 1 & \text{if } S_1 < t < S_1 + T \\ 1 + \alpha(M(t-T) - M(S_1)) & \text{if } t > S_1 + T \end{cases} \qquad (4.28)$$

implying that for $t > T$,

$$M(t) = \int_{t-T}^t f(s)ds + \int_0^{T-t} [1 + \alpha(M(t-T) - M(S_1))]f(s)ds$$

resulting in the simplified equations

$$M(t) = F(t) + \alpha M(t-T)F(t-T) - \alpha \int_0^{t-T} M(s)f(s)ds \quad (t > T) \qquad (4.29)$$

$$M(t) = F(t) \quad (t < T). \qquad (4.30)$$

In the special case when $T = 0$, this equation becomes

$$M(t) = F(t) + \alpha M(t)F(t) - \alpha \int_0^M (s)f(s)ds. \qquad (4.31)$$

In the other type of partial repair, the effective age of the equipment decreased by a factor $1 - \alpha (\alpha < 1)$, in which case equation (4.18) becomes

$$E(X_t|S_1) = \begin{cases} 0 & \text{if } t < S_1 \\ 1 & \text{if } S_1 < t < S_1 + T \\ 1 + M(t - S_1 - T + \alpha S_1) - M(\alpha S_1) & \text{if } t > S_1 + T. \end{cases} \qquad (4.32)$$

The first two cases are the same as in the previous models, however in the third case we have to notice that after repair the effective age of the equipment becomes αS_1 and after repair is done, $t - S_1 - T$ time intervals are needed to reach time t. Similarly to equations (4.20), (4.24), and (4.29), we have that for $t > T$,

$$M(t) = \int_{t-T}^t f(s)ds + \int_0^{t-T} [1 + M(t - s - T + \alpha s) - M(\alpha s)]f(s)ds$$

or

$$M(t) = F(t) + \int_0^{t-T} [M(t-s-T+\alpha s) - M(\alpha s)]f(s)ds, \qquad (4.33)$$

and for $t \geq T$,

$$M(t) = F(t). \qquad (4.34)$$

In the special case when $T = 0$, equation (4.33) simplifies as

$$M(t) = F(t) + \int_0^t [M(t-s+\alpha s) - M(\alpha s)]f(s)ds. \qquad (4.35)$$

In summary, we can see that in case of minimal partial repairs and failure replacements the expected numbers of break-downs as functions of time can be determined by solving the corresponding interval equations.

(4.22) Convergence:

$$M_{n+1}(t) = M_n(t)F(t) + F(t) - \int_0^t M_n(s)f(s)ds \qquad \text{iteration}$$

$$M(t) = M(t)F(t) + F(t) - \int_0^t M(s)f(s)ds \qquad \text{true solution}$$

- -

$$M_{n+1}(t) - M(t) = (M_n(t) - M(t))F(t) - \int_0^t (M_n(s) - M(s))f(s)ds$$

with maximum norm,

$$\|M_{n+1} - M\| \leq \|M_n - M\|F(t) + \int_0^t \|M_n' - r'\|f(s)ds$$

$$= |M_n - r|2F(t).$$

So if $F(t) < \frac{1}{2}$, that is for small t, iteration converges.

(4.26) Convergence:

$$M_{n+1}(t) = F(t) + \int_0^t M_n(z)f(t-z)dz \qquad \text{iteration}$$

$$M(t) = F(t) + \int_0^t M(z)f(t-z)dz \qquad \text{true solution}$$

$$M_{n+1}(t) - M(t) = \int_0^t [M_n(z) - M(z)]f(t-z)dz$$

$$\|M_{n+1} - M\| \le \|M_n - M\| \int_0^t f(t-z)dz.$$

Let $u = t - z$, then we have

$$\|M_{n+1} - M\| \le \|M_n - M\| \int_0^t f(u)du$$

$$\|M_{n+1} - M\| \le \|M_n - M\|F(t).$$

If $F(t) < 1$, process converges.

(4.11) Convergence:

$$M_{n+1}(t) = F(t) + \alpha M_n(t)F(t) - \alpha \int_0^t M_n(s)f(s)ds \qquad \text{iteration}$$

$$M(t) = F(t) + \alpha M(t)F(t) - \alpha \int_0^t M(s)f(s)ds \qquad \text{true solution}$$

$$M_{n+1}(t) - M(t) = \alpha[M_n(t) - M(t)]F(t) - \alpha \int_0^t (M_n(s) - M(s))f(s)ds$$

$$\|M_{n+1} - M\| \le \alpha\|M_n - M\|F(t) + \alpha\|M_n - M\|F(t)$$
$$= \|M_n - M\|2\alpha F(t).$$

If $F(t) < \frac{1}{2\alpha}$, process converges.

(4.35) Convergence:

$$M_{n+1}(t) - M(t) = \int_0^t \{(M_n(t-s+\alpha s) - M(t-s+\alpha s)) - (M_n(\alpha s) - M(\alpha s))\} f(s)ds$$

$$\|M_{n+1} - M\| \le 2\|M_n - M\| F(t)$$

so if $F(t) < \frac{1}{2}$, process converges.

4.3   ECONOMIC MODELS

1. First example is an economy of multiple products. Let $q = (q_1, q_2, \ldots, q_n)$
    denote the production vector and $p = (p_1, p_2, \ldots, p_n)$ the prices of the differ-
    ence products [1–22]. Clearly the produced quantities depend on the prices
    for how much the products can be sold for and similarly the demand of each
    product depends on the prices the products are bought. The demand func-
    tions is denoted by

    $$\underline{D}(\underline{p}) = \begin{pmatrix} d_1(p_1, p_2, \ldots, p_n) \\ d_2(p_1, p_2, \ldots, p_n) \\ \vdots \\ d_n(p_1, p_2, \ldots, p_n) \end{pmatrix}$$

    and similarly the price function is denoted by

    $$\underline{P}(\underline{q}) = \begin{pmatrix} p_1(q_1, q_2, \ldots, q_n) \\ p_2(q_1, q_2, \ldots, q_n) \\ \vdots \\ p_n(q_1, q_2, \ldots, q_n) \end{pmatrix}$$

    Economic equilibrium occurs if all produced items are sold meaning that

    $$q = \underline{D}(\underline{p}) \text{ and } p = \underline{P}(\underline{q}) \tag{4.36}$$

    This is a system of non-linear equation of $2n$ variables. However, the number
    of equations and unknown can be reduced if we combine the two equations:

    $$\underline{q} = \underline{D}(\underline{p}(\underline{q})), \tag{4.37}$$

    where only vector $\underline{q}$ is the unknown with n components.

2. We consider next a simple equilibrium model in macroeconomics. Let
    $C(t), I(t)$ and $Y(t)$ denote consumptions, investment and national income,

then a single dynamic model can be constructed as

$$C(t) = \alpha Y(t - \eta) \tag{4.38}$$
$$I(t) = \varphi(\dot{Y}(t - \delta)) \tag{4.39}$$

$$Y(t) = \int\limits_0^t \frac{1}{2} e^{-\frac{t-\tau}{\varepsilon}} E(\tau) d\tau \tag{4.40}$$

where $\eta > 0$ and $\delta > 0$ are the consumption and investment delays, φ is the investment acceleration function, $E(\tau) = C(\tau) + I(\tau)$ is the total expenditure. The last equation indicated that the national income lags behind the expenditure and the delay is continuously distributed with an exponential kernel function. If the last equation is differentiated with respect to t, and substitute the delayed consumption and investment into the resultant equation an ordinary differential equation is obtained with two fixed delays:

$$\varepsilon \frac{dY(t)}{dt} - \phi\left(\frac{dY(t-\delta)}{dt}\right) + Y(t) - \alpha Y(t-\eta) = 0. \tag{4.41}$$

The unique stationary equilibrium is $Y^e = 0$ with $\frac{dY^e(t)}{dt} = 0$. In examining the local asymptotical stability of the system, we first linearize equation (4.41) around the equation:

$$\varepsilon \frac{dY(t)}{dt} - \phi'(0)\frac{dY(t-\delta)}{dt} + Y(t) - \alpha Y(t-\eta) = 0$$

and with the notation $X(t) = Y(t), a = \frac{1}{\varepsilon}, b = \frac{\phi'(0)}{\varepsilon}, c = \frac{\alpha}{\varepsilon}$ we have

$$\dot{x}(t) + ax(t) - b\dot{x}(t-\sigma) - cx(t-\eta) = 0. \tag{4.42}$$

Then the eigenvalues have to be located by assuming exponential solution $x(t) = e^{\lambda t} u$ to get

$$\lambda + a - b\lambda e^{-\delta\lambda} - ce^{-\eta\lambda} = 0 \tag{4.43}$$

We need to locate all real and complex solution of this equation. Let a solution be $\lambda = r + is$, then (4.43) can be rewritten as

$$r + is + a - b(r + is)e^{-\lambda(r+is)} - ce^{-\eta(r+is)} = 0 \tag{4.44}$$

Notice that

$$e^{-\delta(r+is)} = e^{-\delta r} \cdot e^{-i\delta s} = e^{-\delta r}(\cos \delta s - i \sin \delta s)$$

and

$$e^{-\eta(r+is)} = e^{-\eta r} e^{-i\eta s} = e^{-\eta r}(\cos \eta s - i \sin \eta s)$$

so (4.44) becomes

$$r + is + a - b(r + is)e^{-\delta r}(\cos \delta s - i \sin \delta s) - ce^{-\eta r}(\cos \eta s - i \sin \eta s)$$

Separating the real and imaginary parts two nonlinear equations are obtained for the two real unknowns r and sL

$$r + a - (br\cos\delta s + b\sin\delta s)e^{-\delta s} - ce^{-\eta r}\cos\eta s = 0 \qquad (4.45)$$

$$s + (br\sin\delta s - bs\cos\delta s)e^{-\delta r} + ce^{-\eta r}\sin\eta s = 0 \qquad (4.46)$$

For each solution (r,s) the corresponding eigenvalues is $\lambda = r + is$. $\lambda_2(t)$ and $y(t)$ denote the capital stock and natural income, let $\alpha(0 < \alpha < 1)$ is the marginal propensity to consume and ε a positive adjustment coefficient. Then the dynamic model without delay is formulated by the following two equations:

$$\varepsilon\dot{y}(t) = \dot{r}(t) - (1-\alpha)y(t) \qquad (4.47)$$

$$\dot{r}(t) = \varphi(\dot{y}(t)), \qquad (4.48)$$

where $\varphi(\dot{y})$ is the induced investment. Combining these equation implies that

$$\varepsilon\dot{y}(t) - \varphi(\dot{y}(t)) + (1-\alpha)y(t) = 0$$

with linearized version

$$\varepsilon\dot{y}(t) - \varphi'(0)\dot{y}(t) + (1-\alpha)y(t) = 0. \qquad (4.49)$$

If there is investment delay, then the equation modifies as

$$\varepsilon\dot{y}(t) - \varphi'(0)\dot{y}(t-\theta) + (1-\alpha)y(t) = 0 \qquad (4.50)$$

where θ is the delay. It can be examined similarly to previous case.

3. Our next example is an n-firm oligopoly with product differentiation. Let $q_1, q_2, \ldots, q_n$ be the production levels of the firm which produce different items. The price function of firm r is assumed to be $p_r(q_1, q_2, \ldots, q_n)$ and its cost function is $C_r(a_r)$. Then the profit of this form is the difference of its revenue and cost:

$$\Pi_r(q_1, q_2, \ldots, q_n) = q_1 p_r(q_1, q_2, \ldots, q_n) - C_r(a_r) \qquad (4.51)$$

Assuming interior optimum, the first order conditions are the equations

$$\frac{\partial\Pi_r}{dq_r} = p_r(q_1, q_2, \ldots, q_n) - q_r\frac{\partial p_r(q_1, q_2, \ldots, q_n)}{\partial q_r} - C_r'(q_r) = 0 \qquad (4.52)$$

for $r = 1, 2, \ldots, n$. The equilibrium state of this economic situation is a production vector $q^* = (q_1^*, \ldots, q_n^*)$ such that for each firm q_r^* provides maximal profit with given $q_l^*(l \neq r)$ values. If we consider equation (4.52) for $r = 1, 2, \ldots, n$, a system of n nonlinear equations is obtained, the solution of which provides equilibrium.

4. Consider n-agents who are competing to earn a certain prize values as A. Let x_r be the effort of agent r, then the probability that it will win the prize is

$$P_r = \frac{x_r}{\sum_{e=1}^{n} x_e}. \tag{4.53}$$

If $C_r(x_r)$ denotes its cost to produce effort x_r, then its net gain is

$$G_r(x_1,x_2,\ldots,x_n) = \frac{Ax_r}{\sum_{e=1}^{n} x_r} - C_r(x_r). \tag{4.54}$$

The equilibrium of this economic situation is a vector $\underline{x}^* = (x_1^*,x_2^*,\ldots,x_n^*)$ of efforts which provides maximal profit for all agents. Assuming interior optimum the first order conditions are satisfied of for $r = 1,2,\ldots,n$,

$$\frac{\partial G_r}{\partial x_r} = \frac{A\sum_{e=1}^{n} x_e - Ax_r}{(\sum_{e=1}^{n} x_e)^2} - C_r'(x_r) = 0. \tag{4.55}$$

Let $s = \sum_{e=1}^{n} x_e$, then this equation can be rewritten as

$$As - Ax_r - C_r'(x_r)s^2 = 0$$

or

$$Ax_r + C_r'(x_r)s^2 = As. \tag{4.56}$$

Assume that the cost function is strictly convex, so the left hand side is strictly increasing. Therefore there is at most one solution for x_r with any fixed $s > 0$ value. At $x_r = 0$, the left hand side is below the right hand side if $C_r'(0)s^2 < As$ or $S < \frac{A}{C_r'(0)}$, and the left hand side is always large than the right hand side at $x_r = s$, since $C_r(x_r)$ strictly increases in x_r. So in interval $r \in \left[0, \min_r \frac{A}{C_r'(0)}\right]$ there is a unique $x_r(s)$ value for all r satisfying equation (4.56). And finally conclude the single-variable equation

$$\sum_{r=1}^{n} x_r(s) - s = 0 \tag{4.57}$$

the solution of which gives the equilibrium total effort of all agents, and the corresponding $x_r(s)$ values provides the equilibrium efforts of all agents. In the first step function $x_r(s)$ have to be determined by solving equation (4.56) for a dense grid of the s values, and then in the second step equation (4.57) has to be solved.

5. Two competing species are modeled in the following way. Let $x(t)$ and $y(t)$ be the population density of two complexity species at time t. In addition let $\varepsilon_1 > 0$ and $\varepsilon_2 > 0$ be the intrinsic growth rates, υ_{ij} delays and a_{ij} constant parameters. The dynamic behavior of the population are usually modeled by the following system of nonlinear ordinary differential equations:

$$\dot{x}(t) = x(t)\left[\varepsilon_1 - a_{11}x(t-\upsilon_{11}) - a_{12}y(t-\upsilon_{12})\right] \tag{4.58}$$
$$\dot{y}(t) = y(t)\left[\varepsilon_1 - a_{21}x(t-\upsilon_{21}) - a_{22}y(t-\upsilon_{22})\right] \tag{4.59}$$

The state trajectories of this system can be solved by using Picard iteration which was discussed earlier. For the existence of unique solution, we need the initial function values $x(-t)$ for $t \in [0, \max\{v_{11}, v_{21}\}]$ and the values $y(-t)$ for $t \in [0, \max\{v_{12}, v_{22}\}]$. The iteration scheme has the following form:

$$x_{n+1}(t) = x_n(t) + \int_0^t x(v)[\varepsilon_1 - a_{11}x_n(v - v_{11}) - a_{12}y_n(v - v_{12})]dv \quad (4.60)$$

$$y_{n+1}(t) = y_n(t) + \int_0^t y(v)[\varepsilon_2 - a_{21}x_n(v - v_{21}) - a_{22}y_n(v - v_{22})]d \quad (4.61)$$

6. An interesting extension of the classical oligopoly models are the labor managed models. In the n-firm case, let $x_1, x_2, \ldots, x_n$ be the production levels of identical items of the firms, then $s = x_1 + x_2 + \cdots + x_n$ is the industry output. The price function is given as $p(s)$ and the cost firm r is $C_r(x_r) = \omega h_r(x_r) + c_r$ where ω is the competitive wage rate, C_r is the fixed cost and $h_r(x_r)$ is amount of labor needed to produce the x_r amount of product. It is also assume that the firms are owned by their employees, so instead of maximizing total profits, the firm maximize their profits per unit labor:

$$F(x_1, x_2, \ldots, x_n) = \frac{x_r p(s) - \omega h(x_1) - c_r}{h_r(x_r)} \quad (4.62)$$

$$\left[p\left(\sum_{e=1}^n x_e \right) + x_r p'\left(\sum_{e=1}^n x_e \right) - \omega h'(x_2) \right] h_r(x_r) - \left(x_r p\left(\sum_{e=1}^n x_e \right) \right.$$

$$\left. - \omega h(x_r) - C_r \right) h_r'(x_r) = 0$$

for $r = 1, 2, \ldots, n$, which provides a system of non-linear equations for the equilibrium production levels $x_1, x_2, \ldots, x_n$.

We can use same idea as in 4, let $\sum_{e=1}^n x_e - s$, then for each s we can solve equation

$$[p(s) + x_r p'(s) - \omega h'(x_r)]h_r(x_r) - (x_r p(s) - \omega h(x_r) - c_r)h'(x_e) = 0$$

for $x_r = x_r(s)$, and then solve equation

$$\sum_{r=1}^n x_r(s) - s = 0.$$

4.4   EXERCISES

1. Prove claims made in this chapter.
2. Provide more examples of the theory developed.

REFERENCES

1. Argyros, I.K. 2007. *Computational Theory of Iterative Methods*. eds. Chui, C.K., Wuytack, L. Series: Studies in Computational Mathematics, 15. New York: Elsevier.
2. Argyros, I.K. 2008. *Convergence and Applications of Newton-type Iterations*. New York: Springer-Verlag.
3. Argyros, I.K., and Hillout S. 2012. Weaker conditions for the convergence of Newton's method. *J. Complexity*. 28. 3:364–387.
4. Argyros, I.K., and Magreñán, A.A. 2017. *Iterative Methods and Their Dynamics with Applications*, New York: CRC Press. Taylor& Francis.
5. Argyros, I.K., and Magrenan, A.A. 2018. *A Contemporary Study of Iterative Methods*. NY: Academy Press, Elsevier.
6. Argyros, I.K., and Regmi, S. 2019. *Undergraduate Research at Cameron University on Iterative Procedures in Banach and Other Spaces*, NY: Nova Science Publisher.
7. Argyros, M.I., Argyros, I.K., and Regmi, S. 2021. *Hilbert Spaces and Its Applications*, NY: Nova Science Publisher.
8. Brikhoff, G. 1948. *Lattice Theory, Colloq. Publ.*, 25, New York: American Mathematical Society.
9. Chandler, P. 1983. The nonlinear input-output model. *J. Econ. Theory*, 30:219–229.
10. Davis, A.C. 1955. A characterization of complete lattices. *Pacific. J. Math.*, 5:313–319.
11. Fujimoto, T. 1987. Global asymptotic stability of nonlinear difference equations II. *Econ. Lett.*, 23:275–277.
12. Higle, J.L. and Sen, S. 1989. On the convergence of algorithms with applications to stochastic and nondifferentiable optimization. *SIE Working Paper*, #89–027. University of Arizona, Tucson.
13. Seller, J.P. 1986. *The Stability and Control of Discrete Processes*. New York: Springer-Verlag.
14. Luenberger, D.G. 1979. *Introduction to Dynamics Systems*. New York: John Wiley & Sons.
15. Okuguchi, K. 1977. *Mathematical Foundation of Economics Analysis*, Tokyo: McGraw-Hill.
16. Ortega, J.M. and Rheinboldt, W.C. 1970. Iterative Solutions of Nonlinear Equations in Several Variables, New York: Academic Press.
17. Polak, E. 1970. On the implementation of conceptual algorithms. In *Nonlinear Programming* (J.B. Rosen, O.L. Mangasarian, and K. Ritter, eds.), 275–291. New York: Academic Press.
18. Szidarovszky, F. and Okuguchi, K. 1986. Stability of the Cournot equilibrium for Oligopoly with multi-product firms. *Proc. 5th IFAC/FORS Conf. Dynamic Modelling and Control of National Economics*, Budapest. 17–20.
19. Szidarovszky, F. and Yakowitz, S. 1978. *Principles and Procedures of Numerical Analysis*, New York: Plenum Press.
20. Tarayama, A. 1974. *Mathematical Economics*, Illinois: Dryden, Hinsdale.
21. Tishyadhigama, S., Polak, E., and Klessig, R. 1979. A comparative study of several general convergence conditions for algorithms modeled by point-to-set maps. *Math. Programming Study*, 10:172–190.
22. Zangwill, W.I. 1969. Convergence conditions for nonlinear programming algorithms. *Manage. Sci.*, 16(1):1–13.

5 A Novel Scheme Free from Derivatives

The objective of this chapter is the expansions of the utilization for a fifth convergence order scheme without derivatives for finding solutions of of Banach space valued equations. Conditions of the first order divided difference of the operator involved are only imposed. This way the utilization of the scheme is expanded, since in earlier articles derivatives until order four are required for convergence. Our technique also provides bounds on error distances as well as information about the location of the solutions not given in earlier articles. Experiments with concrete problems complete this chapter.

A plethora of applications from Mathematics as well as scientific disciplines require finding a simple solution x_* of nonlinear equation

$$F(x) = 0, \tag{5.1}$$

where for $\mathbb{X}$ being a Banach space and $\Delta \subset \mathbb{X}$ standing for an open set, $F : \Delta \to \mathbb{X}$ denotes a continuous operator. To solve $F(x) = 0$, we study the local convergence of the following multi-step methods defined for $\sigma = 0, 1, 2, \ldots$ as

$$
\begin{aligned}
u_\sigma &= x_\sigma + \beta F(x_\sigma), \ v_\sigma = x_\sigma - \beta F(x_\sigma), \\
y_\sigma &= x_\sigma - A_\sigma^{-1} F(x_\sigma) \\
z_\sigma &= y_\sigma - \frac{9}{5} A_\sigma^{-1} F(y_\sigma) \\
h_\sigma &= y_\sigma - \frac{16}{5} A_\sigma^{-1} F(y_\sigma) \\
x_{\sigma+1} &= z_\sigma - \frac{1}{5} A_\sigma^{-1} F(h_\sigma)
\end{aligned}
\tag{5.2}
$$

where $A_\sigma = [u_\sigma, v_\sigma; F]$, $A : \Delta \times \Delta \to \ell(\mathbb{X}, \mathbb{X})$ and $\beta \in \mathbb{R}$. The conclusions were obtained for the special case when $\mathbb{X} = \mathbb{Y} = \mathbb{R}^i$. It is a fifth order scheme using up to the fourth derivatives in the local convergence order [6]. But, it is important to note that the scheme (5.2) is derivative free. So, these hypotheses restrict the applicability of the methods. Let us consider a motivational example. We assume the following function F on $\mathbb{X} = \mathbb{R}$ and $\mathbb{D} = [-\frac{1}{2}, \frac{3}{2}]$ such as:

$$
F(\kappa) = \begin{cases} \kappa^3 \ln \kappa^2 + \kappa^5 - \kappa^4, & \kappa \neq 0 \\ 0, & \kappa = 0 \end{cases}, \tag{5.3}
$$

that leads to

$$F'(\kappa) = 3\kappa^2 \ln \kappa^2 + 5\kappa^4 - 4\kappa^3 + 2\kappa^2,$$

DOI: 10.1201/9781003128915-5

$$F''(\kappa) = 6\kappa \ln \kappa^2 + 20\kappa^3 - 12\kappa^2 + 10\kappa,$$
$$F'''(\kappa) = 6\ln \kappa^2 + 60\kappa^2 - 12\kappa + 22.$$

We note that $F'''(\kappa)$ is not bounded in $\mathbb{D}$. Therefore, results requiring the existence of $F''(\kappa)$ or higher cannot be applied for studying the convergence of (5.2). Moreover, no computable error bounds $\|x_\sigma - x_*\|$, where x_* solves the equation $F(x) = 0$, or any information regarding the uniqueness of the solution are provided using Lipschitz-type functions. Furthermore, the convergence criteria can not be compared, since they are based on different hypotheses. We address all these problems by using only the first derivative. Moreover, we rely on the computational order of convergence (COC) or approximated computational order of convergence ($ACOC$) to determine the c-order (Computational order of convergence) not requiring derivatives of order higher than one. The new technique uses the same set of conditions for the three methods. Furthermore, it can also be used to extend the applicability of other methods along the same lines.

5.1 CONVERGENCE

It is convenient to define functions on the real line given $\beta \in \mathbb{R}, \alpha \geq 0|$ and $\gamma \geq 0$.Set $I = [0, \infty)$. Consider the existence of a continuous and increasing function $\Omega_0 : I \times I \to I$ such that

$$\Omega_0(\gamma\zeta, \delta\zeta) = 1, \tag{5.4}$$

has a smallest positive solution r, with $\Omega_0(0, 0) = 0$.

Let $\Omega : I_0 \times I_0 \to I$ be a continuous and increasing function with with $\Omega(0, 0) = 0$, and $I = [0, \infty)$. Define functions $\phi_m, \psi_m, m = 1, 2, 3, 4$ to be on the interval I_0 by

$$\phi_1(\zeta) = \frac{\Omega\big((1 + \gamma)\zeta, \delta\zeta\big)}{1 - \Omega_0(\gamma\zeta, \delta\zeta)},$$

$$\phi_2(\zeta) = \frac{\left[\Omega\big((\gamma + \phi_1(\zeta))\zeta, \delta\zeta\big) + \frac{4}{5}\alpha\right]\phi_1(\zeta)}{1 - \Omega_0(\gamma\zeta, \delta\zeta)},$$

$$\phi_3(\zeta) = \phi_2(\zeta) + \frac{16\alpha\phi_1(\zeta)}{5\big(1 - \Omega_0(\gamma\zeta, \delta\zeta)\big)},$$

$$\phi_4(\zeta) = \phi_2(\zeta) + \frac{\alpha\phi_3(\zeta)}{5\big(1 - \Omega_0(\gamma\zeta, \delta\zeta)\big)},$$

and

$$\psi_m(\zeta) = \phi_m(\zeta) - 1.$$

Using these definitions, we have $\psi_m(0) = -1$ and $\psi_m(\zeta) \to \infty$ as $\zeta \to r^-$. Denote by R_m the least solutions of equations $\psi_m(\zeta) = 0, m = 1, 2, 34$ in the interval $(0, r)$ assured to exist by the intermediate value theorem. Define a radius of convergence R by

$$R = \min\{R_m\}. \tag{5.5}$$

Then, if $\zeta \in [0, R)$

$$0 \le \Omega_0(\gamma\zeta, \delta\zeta) < 1, \tag{5.6}$$

and

$$0 \le \phi_m(\zeta) < 1. \tag{5.7}$$

Set $U(x_*, \rho) = \left\{ x \in \Delta : \|x - x_*\| < \rho \right\}$ and let $\bar{U}(x_*, \rho)$ stand for the closure of $U(x_*, \rho)$.

Let us introduce conditions (C):

(C_1) $F : \Delta \to \mathbb{X}$, is Fréchet differentiable at a simple solution x_* of equation $F(x_*) = 0$ with $[\cdot, \cdot; F] : \mathbb{D} \times \mathbb{D} \to \ell(\mathbb{X}, \mathbb{X})$ a standard divided difference of order one, and for $\beta \in \mathbb{R}$

$$\left\| F'(x_*)^{-1} [x, x_*; F] \right\| \le \alpha$$
$$\| I + \beta[x, x_*; F] \| \le \gamma$$
and
$$\| I - \beta[x, x_*; F] \| \le \delta$$

holds for all $x \in \Delta$ and some $\gamma, \delta \ge 0$.

(C_2) There exists a continuous and increasing function $\Omega_0 : I \times I \to I$ with $\Omega_0(0, 0) = 0$ such that for each $x, y \in \Delta$

$$\left\| F'(x_*)^{-1} \left([x, x_*; F] - F'(x_*) \right) \right\| \le \Omega_0(\|x - x_*\|, \|y - x_*\|).$$

Set $\Delta_0 = \Delta \cap U(x_*, r)$.

(C_3) There exists a continuous and increasing function $= w : I_0 \times I_0 \to I$ with $\Omega_0(0, 0) = 0$ such that for each $x, y, z \in \Delta$

$$\left\| F'(x_*)^{-1} \left([x, y; F] - [z, x_*; F] \right) \right\| \le \Omega(\|x - x_*\|, \|y - x_*\|).$$

(C_4) The ball $\bar{U}(x_*, \tilde{R}) \subset \Delta$, where R is defined in (5.5), and r exists, is given in (5.4) and $\tilde{R} = \max\{R, \gamma R, \delta R\}$.

(C_5) There exists $R_* \ge R$ such that

$$\Omega_0(0, R_*) < 1 \text{ or } \Omega_0(R_*, 0) < 1.$$

Set $\Delta_1 = \Delta \cap U^*(x_*, R_*)$.

Next, we base the convergence analysis on the (C) conditions, and the developed notations.

Theorem 5.1. Suppose the (C) conditions are satisfied. Then, if we choose $x_0 \in U(x_*, R) - \{x_*\}$, the following assertions hold:

$$\{x_\sigma\} \subset \Delta, \tag{5.8}$$

$$\lim_{\sigma \to \infty} x_\sigma = x_*, \tag{5.9}$$

$$\|y_\sigma - x_*\| \le \phi_1(\|x_\sigma - x_*\|)\|x_\sigma - x_*\| \le \|x_\sigma - x_*\| < r, \tag{5.10}$$

$$\|z_\sigma - x_*\| \le \phi_2(\|x_\sigma - x_*\|)\|x_\sigma - x_*\| \le \|x_\sigma - x_*\|, \tag{5.11}$$

$$\|h_\sigma - x_*\| \le \phi_3(\|x_\sigma - x_*\|)\|x_\sigma - x_*\| \le \|x_\sigma - x_*\|, \tag{5.12}$$

$$\|x_{\sigma+1} - x_*\| \le \phi_4(\|x_\sigma - x_*\|)\|x_\sigma - x_*\| \le \|x_\sigma - x_*\|, \tag{5.13}$$

and the only solution of equation $F(x) = 0$ in the set Δ_1 given below (C_4) is x_*.

Proof. By conditions $x_0 \in U(x_*, R) - \{x_*\}$ and (C_1), we have

$$
\begin{aligned}
\|u_0 - x_*\| &= \|x_0 - x_* + \beta F(x_0)\| \\
&= \|(I + \beta[x_0, x_*; F])(x_0 - x_*)\| \\
&\le \|I + \beta[x_0, x_*; F]\|\|x_0 - x_*\| \\
&\le \gamma\|x_0 - x_*\| \le \gamma R,
\end{aligned}
$$

$$
\begin{aligned}
\|v_0 - x_*\| &= \|x_0 - x_* - \beta F(x_0)\| \\
&= \|(I - \beta[x_0, x_*; F])(x_0 - x_*)\| \\
&\le \delta\|x_0 - x_*\| \le \delta R,
\end{aligned}
$$

and

$$\left\|F'(x_*)^{-1}[x_0, x_*; F]\right\| \le \alpha.$$

Then, we have using (5.5), (5.6), and (C_2) that

$$
\begin{aligned}
\left\|F'(x)^{-1}\left(A_0 - F'(x_*)\right)\right\| &\le \Omega_0(\|u_0 - x_*\|, \|v - x_*\|) \\
&\le \Omega_0(\gamma\|x_0 - x_*\|, \delta\|x_0 - x_*\|) \le \Omega_0(\gamma R, \delta R) < 1.
\end{aligned}
$$

which together with a Lemma by Banach operators that are invertible [1, 6, 7], leads to A_0 is invertible,

$$\left\|A_0^{-1}F'(x)\right\| \le \frac{1}{\Omega_0(\gamma\|x_0 - x_*\|, \delta\|x_0 - x_*\|)}, \tag{5.14}$$

and y_0, z_0, h_0, x_1 exist by C conditions for method (5.2) for $\sigma = 0$.

Using (5.5), (5.7) (for $\sigma = 1$), (C_3) and (5.14), we get

$$
\begin{aligned}
\|y_0 - x_*\| &= \left\|x_0 - x_* - A_0^{-1}F(x_0)\right\| \\
&= \left\|\left(A_0^{-1}F'(x_*)\right)F'(x_*)^{-1}\left(A_0 - [x_0, x_*; F]\right)(x_0 - x_*)\right\| \\
&\le \left\|A_0^{-1}F'(x_*)\right\|\left\|F'(x_*)^{-1}\left(A_0 - [x_0, x_*; F]\right)\right\|\|x_0 - x_*\| \\
&\le \frac{\Omega\left(\|(u_0 - x_*) + (x_* - v_0)\|, \|v_0 - x_*\|\right)\|x_0 - x_*\|}{1 - \Omega_0(\gamma\|x_0 - x_*\|, \delta\|x_0 - x_*\|)} \\
&\le \frac{\Omega\left((1 + \gamma)\|x_0 - x_*\|, \delta\|x_0 - x_*\|\right)\|x_0 - x_*\|}{1 - \Omega_0(\gamma\|x_0 - x_*\|, \delta\|x_0 - x_*\|)} \\
&= \phi_1(\|x_0 - x_*\|)\|x_0 - x_*\| \le \|x_0 - x_*\| < R,
\end{aligned} \tag{5.15}
$$

$$\|z_0 - x_*\| = \left\| \left(y_0 - x_* - A_0^{-1}F(y_0) \right) - \frac{4}{5}A_0^{-1}F(y_0) \right\|$$

$$\leq \left\| y_0 - x_* - A_0^{-1}F(y_0) \right\| + \frac{4}{5} \left\| A_0^{-1}F'(x_*) \right\| \left\| F'(x_*)^{-1}F(y_0) \right\|$$

$$\leq \left\| A_0^{-1}F'(x_*) \right\| \left\| F'(x_*)^{-1}\left(A_0 - [y_0, x_*; F] \right) \right\| \| y_0 - x_* \|$$

$$+ \frac{4}{5} \left\| A_0^{-1}F'(x_*) \right\| \left\| F'(x_*)^{-1}F(y_0) \right\| \tag{5.16}$$

$$\leq \frac{\left[\Omega\left(\| (u_0 - x_*) + (x_* - y_0) \|, \|v_0 - x_*\| \right) + \frac{4}{5}\alpha \right] \|y_0 - x_*\|}{1 - \Omega_0(\gamma\|x_0 - x_*\|, \delta\|x_0 - x_*\|)}$$

$$\leq \frac{\left[\Omega\left(\|x_0 - x_*\|, \|x_0 - x_*\| \right) + \frac{4}{5}\alpha \right] \phi_1(\|x_0 - x_*\|)\|x_0 - x_*\|}{1 - \Omega_0(\gamma\|x_0 - x_*\|, \delta\|x_0 - x_*\|)}$$

$$= \phi_2(\|x_0 - x_*\|)\|x_0 - x_*\| \leq \|x_0 - x_*\|,$$

$$\|h_0 - x_*\| = \left\| (z_0 - x_*) - \frac{16}{5}A_0^{-1}F(y_0) \right\|$$

$$\leq \|z_0 - x_*\| + \frac{16}{5} \left\| A_0^{-1}F'(x_*) \right\| \left\| F'(x_*)^{-1}F(y_0) \right\|$$

$$\leq \left[\phi_2(\|x_0 - x_*\|) + \frac{16\alpha\phi_1(\|x_0 - x_*\|)}{5\left(1 - \Omega_0(\gamma\|x_0 - x_*\|, \delta\|x_0 - x_*\|) \right)} \right] \|x_0 - x_*\|$$

$$= \phi_3(\|x_0 - x_*\|)\|x_0 - x_*\| \leq \|x_0 - x_*\|,$$

$$\tag{5.17}$$

and

$$\|x_1 - x_*\| = \left\| z_0 - x_* - \frac{1}{5}A_0^{-1}F(h_0) \right\|$$

$$\leq \|z_0 - x_*\| + \frac{1}{5} \left\| A_0^{-1}F'(x_*) \right\| \left\| F'(x_*)^{-1}F(h_0) \right\|$$

$$\leq \left[\phi_2(\|x_0 - x_*\|) + \frac{\alpha\phi_3(\|x_0 - x_*\|)}{5\left(1 - \Omega_0(\gamma\|x_0 - x_*\|, \delta\|x_0 - x_*\|) \right)} \right] \|x_0 - x_*\|$$

$$= \phi_4(\|x_0 - x_*\|)\|x_0 - x_*\| \leq \|x_0 - x_*\|,$$

$$\tag{5.18}$$

deducing that $y_0, z_0, h_0, x_1 \in U(x_*, R)$ and (5.10)–(5.13) hold for $\sigma = 0$. Next, by the inequation

$$\|x_{k+1} - x_*\| \leq p\|x_k - x_*\| < R, \tag{5.19}$$

where $p = \phi_4(\|x_0 - x_*\|) \in [0, 1)$, we obtain $\lim_{k \to \infty} x_k = x_*$, with $x_{k+1} \in U(x_*, R)$. It remains to show the uniqueness of the x_* in the set Δ_1 with $F(y_*) = 0$. Set $T =$

$[x_*, y_*; F]$. In view of (C_2) and (C_5)

$$\left\| F'(x_*)^{-1}\left(T - F'(x_*)\right)\right\| \leq \Omega_0(0, \|x_* - y_*\|) \leq \Omega_0(0,R) < 1,$$

so T is invertible. Finally, we get $x_* = y_*$ from the approximation $0 = F(y_*) - F(x_*) = T(y_* - x_*)$.

$\square$

5.2 APPLICATIONS

The theoretical results developed in the previous sections are illustrated numerically in this section. We consider two real life problems and two standard non-linear problems that are illustrated in examples 5.1–5.3. The results are listed in Tables 5.1, 5.3 and 5.4. Additionally, we obtain the COC approximated by means of

$$\xi = \frac{\ln \frac{\|x_{\sigma+1} - x_*\|}{\|x_\sigma - x_*\|}}{\ln \frac{\|x_\sigma - x_*\|}{\|x_{\sigma-1} - x_*\|}}, \quad \text{for } \sigma = 1, 2, \ldots \tag{5.20}$$

or $ACOC$ [21] by:

$$\xi^* = \frac{\ln \frac{\|x_{\sigma+1} - x_\sigma\|}{\|x_\sigma - x_{\sigma-1}\|}}{\ln \frac{\|x_\sigma - x_{\sigma-1}\|}{\|x_{\sigma-1} - x_{\sigma-2}\|}}, \quad \text{for } \sigma = 2, 3, \ldots \tag{5.21}$$

We adopt $\varepsilon = 10^{-200}$ as the error tolerance and the terminating criteria to solve nonlinear system or scalar equations are: (i) $\|x_{\sigma+1} - x_\sigma\| < \varepsilon$, and (ii) $\|F(x_\sigma)\| < \varepsilon$.

The computations are performed with the package *Mathematica* 9 with multiple precision arithmetic. The divided difference in all examples is given by $[x, y; F] = \int_0^1 F'\left(y + \theta(x - y)\right)d\theta$. We also consider $\beta = 0$ in all examples 5.1–5.3.

Example 5.1. Let us consider the following system of nonlinear equations (chosen from Grau-Sánchez et al. [5])

$$F(x_1, x_2, \ldots, x_\sigma) = \sum_{j=1, j\neq i}^{\sigma} x_j - e^{-x_i}, \quad 1 \leq i \leq \sigma. \tag{5.22}$$

We choose $\sigma = 10$ in order to check the theoretical results mentioned with a large size system. The obtained solution of this problem is

$$x_* = (0.5671433\ldots, 0.5671433\ldots, 0.5671433\ldots, \cdots, 0.5671433\ldots(10times))^T.$$

Choose $\mathbb{X} = \mathbb{R}^{10}$ and $\Delta = U\left(x_*, \frac{1}{2}\right)$. Then, we get

$$\Omega_0(t, s) = \frac{t+s}{2}, \quad \Omega(t, s) = t + s, \quad \alpha = 7, \quad \gamma = 1, \quad \text{and } \delta = 1.$$

The obtained results can be observed in Table 5.1.

Method	(5.2)
R_1	0.076923
R_2	0.053984
R_3	0.022085
R_4	0.279501
R	0.022085
x_0	$(0.55, 0.55, \ldots, 0.55(10\, times))^T$
σ	3
ξ	5.0000

Table 5.1: Radii for Example 5.1

i	ψ_i	φ_i
0	1.3954170041747090114	1.7461756494150842271
1	1.7444828545735749268	2.0364691127919609051
2	2.0656234369405315689	2.2390977868265978920
3	2.4600678478912500533	2.4600678409809344550

Table 5.2: Values of ψ_i and φ_i (in radians) for Example 5.2.

Example 5.2. The kinematic synthesis problem for steering [4, 13], is given as

$$[E_i (x_2 \sin(\psi_i) - x_3) - F_i (x_2 \sin(\varphi_i) - x_3)]^2$$
$$+ [F_i (x_2 \cos(\varphi_i) + 1) - F_i (x_2 \cos(\psi_i) - 1)]^2$$
$$- [x_1 (x_2 \sin(\psi_i) - x_3)(x_2 \cos(\varphi_i) + 1) - x_1 (x_2 \cos(\psi_i) - x_3)(x_2 \sin(\varphi_i) - x_3)]^2$$
$$= 0, \text{ for } i = 1, 2, 3,$$

where

$$E_i = -x_3 x_2 (\sin(\varphi_i) - \sin(\varphi_0)) - x_1 (x_2 \sin(\varphi_i) - x_3) + x_2 (\cos(\varphi_i) - \cos(\varphi_0)),$$

and

$$F_i = -x_3 x_2 \sin(\psi_i) + (-x_2) \cos(\psi_i) + (x_3 - x_1) x_2 \sin(\psi_0) + x_2 \cos(\psi_0) + x_1 x_3,$$

where

$$i = 1, 2, 3.$$

In Table 5.2, we present the values of ψ_i and φ_i (in radians).
The approximated solution is for $\Delta = U(x_*, 1)$

$$x_* = (0.9051567\ldots, 0.6977417\ldots, 0.6508335\ldots)^T.$$

Choose $\mathbb{X} = \mathbb{R}^3$ and $\Delta = U(x_*, 1) \times U(x_*, 1) \times U(x_*, 1)$.
Then, we get

$$\Omega_0(t, s) = \frac{t+s}{4}, \quad \Omega(t, s) = \frac{t+s}{2} \quad \alpha = 11, \ \gamma = 1, \ \text{and} \ \delta = 1.$$

We provide the radii of convergence for Example 5.2 in Table 5.3.

Method	(5.2)
R_1	0.10526
R_2	0.073747
R_3	0.030365
R_4	0.051422
R	0.030365
x_0	(0.88,0.67,0.63)
σ	5
ξ	4.0000

Table 5.3: Radii for Example 5.2

Example 5.3. We choose a prominent 2D Bratu problem [16, 23], which is given by

$$
\begin{aligned}
&u_{xx} + u_{tt} + Ce^u = 0, \text{on}\\
&A : (x,t) \in 0 \le x \le 1,\ 0 \le t \le 1,\\
&\text{along boundary hypothesis } u = 0 \text{ on } A.
\end{aligned}
\tag{5.23}
$$

Let us assume that $\Theta_{i,j} = u(x_i, t_j)$ is a numerical result over the grid points of the mesh. In addition, we consider that τ_1 and τ_2 are the number of steps in the direction of x and t, respectively. Moreover, we choose that h and k are the respective step sizes in the direction of x and y, respectively. In order to find the solution of PDE (5.23), we adopt the following approach

$$
u_{xx}(x_i, t_j) = \frac{\Theta_{i+1,j} - 2\Theta_{i,j} + \Theta_{i-1,j}}{h^2}, \quad C = 0.1,\ t \in [0,1],
\tag{5.24}
$$

which further yields the succeeding SNE

$$
\Theta_{i,j+1} + \Theta_{i,j-1} - \Theta_{i,j} + \Theta_{i+1,j} + \Theta_{i-1,j} + h^2 C \exp\left(\Theta_{i,j}\right)
\tag{5.25}
$$

where

$$
i = 1,2,3,\ldots,\tau_1 \text{ and } j = 1,2,3,\ldots,\tau_2.
$$

By choosing $\tau_1 = \tau_2 = 11$, $h = \frac{1}{11}$, and $C = 0.1$, we get a large SNE of order 100×100 and converges to the following required root

$$
x_* = \begin{pmatrix}
0.0011\ldots,0.0018\ldots,0.0022\ldots,0.0025\ldots,0.0026\ldots,0.0026\ldots,0.0025\ldots,0.0022\ldots,0.0018\ldots,0.0011\ldots\\
0.0018\ldots,0.0030\ldots,0.0038\ldots,0.0043\ldots,0.0046\ldots,0.0046\ldots,0.0043\ldots,0.0038\ldots,0.0030\ldots,0.0018\ldots\\
0.0022\ldots,0.0038\ldots,0.0049\ldots,0.0056\ldots,0.0059\ldots,0.0059\ldots,0.0056\ldots,0.0049\ldots,0.0038\ldots,0.0022\ldots\\
0.0025\ldots,0.0043\ldots,0.0056\ldots,0.0064\ldots,0.0068\ldots,0.0068\ldots,0.0064\ldots,0.0056\ldots,0.0043\ldots,0.0025\ldots\\
0.0026\ldots,0.0046\ldots,0.0059\ldots,0.0068\ldots,0.0072\ldots,0.0072\ldots,0.0068\ldots,0.0059\ldots,0.0046\ldots,0.0026\ldots\\
0.0026\ldots,0.0046\ldots,0.0059\ldots,0.0068\ldots,0.0072\ldots,0.0072\ldots,0.0068\ldots,0.0059\ldots,0.0046\ldots,0.0026\ldots\\
0.0025\ldots,0.0043\ldots,0.0056\ldots,0.0064\ldots,0.0068\ldots,0.0068\ldots,0.0064\ldots,0.0056\ldots,0.0043\ldots,0.0025\ldots\\
0.0022\ldots,0.0038\ldots,0.0049\ldots,0.0056\ldots,0.0059\ldots,0.0059\ldots,0.0056\ldots,0.0049\ldots,0.0038\ldots,0.0022\ldots\\
0.0018\ldots,0.0030\ldots,0.0038\ldots,0.0043\ldots,0.0046\ldots,0.0046\ldots,0.0043\ldots,0.0038\ldots,0.0030\ldots,0.0018\ldots\\
0.0011\ldots,0.0018\ldots,0.0022\ldots,0.0025\ldots,0.0026\ldots,0.0026\ldots,0.0025\ldots,0.0022\ldots,0.0018\ldots,0.0011\ldots
\end{pmatrix}^T
$$

a the column vector. Choose $\mathbb{X} = \mathbb{R}^{100}$ and $\Delta = U(x_*, 0.006)$. Then, we have
Then, we get

$$\Omega_0(t,s) = \Omega(t,s) = 6(t+s), \quad \alpha = 12, \quad \gamma = 1, \quad \text{and} \quad \delta = 1.$$

The obtained results are depicted in Table 5.4 with the following initial approximation (the column vector)

$$x_0 = \begin{pmatrix} 0.001113, 0.001812, 0.002260, 0.002530, 0.002657, 0.002657, 0.002530, 0.002260, 0.001812, 0.001113, \\ 0.001812, 0.003048, 0.003870, 0.004374, 0.004613, 0.004613, 0.004374, 0.003870, 0.003048, 0.001812, \\ 0.002260, 0.003870, 0.004968, 0.005652, 0.005979, 0.005979, 0.005652, 0.004968, 0.003870, 0.002260, \\ 0.002530, 0.004374, 0.005652, 0.006454, 0.006841, 0.006841, 0.006454, 0.005652, 0.004374, 0.002530, \\ 0.002657, 0.004613, 0.005979, 0.006841, 0.007257, 0.007257, 0.006841, 0.005979, 0.004613, 0.002657, \\ 0.002657, 0.004613, 0.005979, 0.006841, 0.007257, 0.007257, 0.006841, 0.005979, 0.004613, 0.002657, \\ 0.002530, 0.004374, 0.005652, 0.006454, 0.006841, 0.006841, 0.006454, 0.005652, 0.004374, 0.002530, \\ 0.002260, 0.003870, 0.004968, 0.005652, 0.005979, 0.005979, 0.005652, 0.004968, 0.003870, 0.002260, \\ 0.001812, 0.003048, 0.003870, 0.004374, 0.004613, 0.004613, 0.004374, 0.003870, 0.003048, 0.001812, \\ 0.001113, 0.0018125, 0.002260, 0.002530, 0.002657, 0.002657, 0.002530, 0.002260, 0.001812, 0.001113 \end{pmatrix}^T.$$

Method	(5.2)
R_1	0.033333
R_2	0.0050705
R_3	0.000080359
R_4	0.0846490
R	0.000080359
x_0	above the matrix
σ	3
ξ	3.9742

Table 5.4: Radii of convergence for example 5.3

5.3 EXERCISES

1. Consider a continuous map $T : R^n \rightarrow R^n$ such that $T \in C^1$ and $T(0) = 0$. Set $A = \{x \,|\, \|T(x)\| < \|x\|\}$, $B = \{x \,|\, \|T(x)\| \geq \|x\|\}$. Assume that

   a. A is invariant under $T, T(A) \subseteq A$;
   b. For all $b \in B$, there exists a positive integer $i(b)$ such that $T^{i(b)}(b) \in A$. Show that for $x \in R^n, T^k(x) \rightarrow 0$ as $k \rightarrow \infty$.

2. To find a zero for $g(x) = 0$ by iteration, where g is a real function defined on $[a,b]$, rewrite the equation as

$$x = x + c \cdot g(x) \equiv f(x)$$

for some constant $c \neq 0$. If s^* is a root of $g(x)$ and if $g'(s^*) \neq 0$, how should c be chosen in order that the sequence $x_{k+1} = f(x_k)$ converge to s^*?

3. Assume that $f_k = f(k \geq 0)$, 0 is in the interior of X, and f is Fréchet differentiable at 0. Furthermore, the special radius of $f'(0)$ is less than 1. Then show that there is a neighborhood U of 0 such that $x_0 \in U$ implies that $x_k \to 0$ as $k \to \infty$.

4. Let $f : R^n \to R^n$ be a function such that $f(0) - 0, f \in C^0$, and consider the difference equation $x(t+1) = f(x(t))$. If, for some norm, $\|f(x)\| < \|x\|$ for any $x \neq 0$, then show that the origin is globally asymptotically stable equilibrium for the equation.

5. Assume that a strictly increasing function $g : R \to R$ exists such that $g(0) = 0$, and a norm $g(\|f(x)\|) < g(\|x\|)$ for all $x \neq 0$. Then show that 0 is globally symptotically stable equilibrium for equation $x(t+1) = f(x(t))$, where $f(0) = 0$.

REFERENCES

1. Argyros, I.K., and Regmi, S. 2019. *Undergraduate Research at Cameron University on Iterative Procedures in Banach and Other Spaces*. Nova Science Publishers. New York.
2. Argyros, I.K., and Regmi, S. 2020. *Contemporary Algorithms for Solving Problems in Economics and Other Disciplines*. Nova Science Publishers. New York.
3. Awawdeh, F. 2010. On new iterative method for solving systems of nonlinear equations. *Numer. Algor.* 54:395–409.
4. Fousse, L., Hanrot, G., Lefevre, V., Pélissier, P., and Zimmermann, P. 2007. MPFR: a mulitple-precision binary floating-point library with correct rounding. *ACM Trans. Math. Softw.*, 15. Art. 13. 33 (2).
5. Grau-Sánchez, M., Noguera, Á., and Amat, S. 2013. On the approximation of derivatives using divided difference operators preserving the local convergence order of iterative methods. *J. Comput. Appl. Math.*, 237:363–372.
6. Kapania, R.K. 1990. A pseudo-spectral solution of 2-parameter Bratu's equation. *Comput. Mech.*, 6:55–63.
7. Liu, Z., Zheng, Q., and Zhao P. 2012. A variants of Steffensen's method of fourth-order convergence and its application. *Appl. Math. Comput.*, 216:1978–1983.
8. Magrenan, A.A., and Argyros, I.K. 2018. *A Contemporary Study of Iterative Methods: Convergence, Dynamics and Applications*. Elsevier.
9. Moré J.J. 1990. A collections of nonlinear model problems, in: *Computational solution of nonlinear systems of equations lectures in applied mathematics*. ed. E.L. Allgower, K. Georg, vol. 26, 723–762. American Mathematical Society, Providence, RI,
10. Ortega, J.M., and Rheinboldt, W.C. 1970. *Iterative solution of nonlinear equations in several variables*. Academic Press. New York.
11. Rheinboldt, W.C. 1978. An adaptive continuation process for solving systems of nonlinear equations, Polish Academy of Science. *Banach Ctr. Publ.*, 3:129–142.
12. Sharma, J.R., and Kumar, D. 2020. On a reduced cost derivative-free higher order numerical algorithm for nonlinear systems. *Computational and Applied Mathematics* 39:202.
13. Sharma, J.R., and Arora, H. 2016. Efficient derivative free numerical methods for solving systems of nonlinear equations. *Comput. Appl. Math.*, 35:269–284.
14. Sharma, J.R., and Arora, H. 2013. An efficient derivative free iterative method for solving systems of nonlinear equations. *Appl. Anal. Discrete Math.*, 7:390–403.
15. Simpson, R.B. 1975. A method for the numerical determination of bifurcation states of nonlinear systems of equations. *SIAM J. Numer. Anal.*, 12:439–451.
16. Steffensen, J.F. 1993. Remarks on iteration. *Skand. Aktuar Tidskr.*, 16:64–72.
17. Traub, J.F. 1964. *Iterative methods for the solution of equations*. Englewood Cliffs, N.J: Prentice Hall Series in Automatic Computation.
18. Tsoulos, I.G., and Stavrakoudis, A. 2010. On locating all roots of systems of nonlinear equations inside bounded domain using global optimization methods. *Nonlinear Anal. Real World Appl.*, 11:2465–2471.
19. Wang, X., and Zhang, T. 2013. A family of Steffensen type methods with seventh-order convergence. *Numer. Algor.*, 62:429–444.
20. Wang, X., Zhang, T., Qian, W., and Teng, M. 2015. Seventh-order derivative free method for solving nonlinear systems. *Numer. Algor.*, 70:545–558.
21. Weerakoon, S., and Fernando, T.G.I. 2000. A variant of Newton's method with accelerated third order convergence. *Appl. Math. Lett.*, 13:87–93.
22. Wolfram, S. 2003. The Mathematica Book, fifth ed., Wolfram Media.

23. Xiao, X.Y., and Yin, H.W. 2016. Increasing the order of convergence for iterative methods to solve nonlinear systems. *Calcolo*, 53:285–300.

6 Efficient Sixth Convergence Order Method

In this chapter, we present a sixth order Steffensen-type method with one parameter for solving system of equations. The novelty of this chapter lies in the fact that the local convergence is established under weak conditions including computable errors bounds and uniqueness of the solution results in contrast to earlier works using the seventh derivative. The efficiency of the method is discussed and compared to the other using simpler information. Finally large systems of equations are solved to test the theoretical results

Let $\mathbb{B}$, be a Banach space and $\Omega \subseteq \mathbb{B}$ be a nonempty, and open set. By $\mathbb{S}(x,r)$, $\bar{\mathbb{S}}(x,r)$, we denote the open and closed balls in $\mathbb{B}$ and of radius $r > 0$. Define $\ell(\mathbb{B},\mathbb{B}) := \{M : \mathbb{B} \to \mathbb{B}, \, M \text{ is linear and bounded}\}$.

We solve equation

$$F(x) = 0, \tag{6.1}$$

where $F : \Omega \to \mathbb{B}$ is continuous operator using iterative solvers approximating a solution x_*. The task of finding x_* is very challenging and extreme importance in analysis, since problems from diverse disciplines such as Mathematical, Biology, Chemistry, Economics, Physics, Engineering (see also numerical examples) are converted to an equation of the type (6.1) by utilizing mathematical modeling [1–5]. Then, suitable closed form of solutions can rarely be found. Recently, in order to increase the convergence order authors develop $k-$step iterative solvers which converge to x_* under certain assumptions which involve a Fréchet-differentiable operator defined on a subset Ω of a Banach space $\mathbb{B}$ with values in a Banach space $\mathbb{B}$. Taylor series expansions, and derivatives of high order are used in the convergence analysis although these derivatives do not appear in the methods. Moreover, no computable error bounds or any information on the uniqueness of the solution are provided based on Lipschitz-type functions. Furthermore, the convergence criteria can not be compared, since they are based on different hypotheses. These problems limit the applicability of the method.

We suggest a Steffensen-like method (SLM) defined for all $n = 0, 1, 2, \ldots$

$$y_n = x_n - [x_n + F(x_n), x_n - F(x_n); \, F]^{-1} F(x_n)$$

$$z_n = y_n - \left(2[y_n, x_n; F] - [x_n + F(x_n), x_n - F(x_n); \, F]\right)^{-1} F(y_n) \tag{6.2}$$

$$x_{n+1} = z_n - \left(2[y_n, x_n; F] - [x_n + F(x_n), x_n - F(x_n); \, F]\right)^{-1} F(z_n),$$

where $x_0 \in \Omega$ is an initial point, $[\cdot,\cdot;F] : \Omega \times \Omega \to \ell(\mathbb{B},\mathbb{B})$ divided difference of order one satisfying $[x,y;F](x-y) = F(x) - F(y)$ and for all $x,y \in \Omega$ with $x \neq y$, and $[x,x;F] = F'(x)$ if F is differentiable. Method (6.2) studied in [6] (in the special

DOI: 10.1201/9781003128915-6

case $\mathbb{B} = \mathbb{R}^i$) and found to be of order six by using upto the seventh order derivatives limiting the applicability.

As a motivational example, we define function ϕ on $\mathbb{B} = \mathbb{R}$, $\Omega = [-\frac{1}{2}, \frac{3}{2}]$ such as:

$$\phi(t) = \begin{cases} t^3 \ln t^2 + t^5 - t^4, & t \neq 0 \\ 0, & t = 0 \end{cases}. \tag{6.3}$$

We have that

$$\phi'(t) = 3t^2 \ln t^2 + 5t^4 - 4t^3 + 2t^2,$$

$$\phi''(t) = 6t \ln t^2 + 20t^3 - 12t^2 + 10t,$$

$$\phi'''(t) = 6 \ln t^2 + 60t^2 - 12t + 22.$$

It can easily be seen that $\phi'''(t)$ is unbounded on Ω. Hence, there is no guarantee that $\lim_{n \to \infty} x_n = x_*$, since the hypothesis in [6] requires the existence of the seventh derivative. Moreover no computable uniqueness results of bounds on $\|x_n - x_*\|$ were given in [6]. We do find all these using lipschitz like conditions only on the first divided difference. Hence, we extend the applicability of method (6.2). The convergence order is found using COC or $ACOC$ that do not require higher order derivatives or even the knowledge of x_* in the case of $ACOC$. Furthermore, local convergence results are important, since they show the degree of difficulty in choosing initial guesses.

In this chapter, we expand the applicability of method (6.2) using only hypothesis on the continuity of function F. Furthermore, we also propose a scheme for deriving the radii of convergence and the error bounds based on Lipschitz constants. We address the range of initial guesses x_* that gives information about how close it must be to guarantee the convergence of (6.2).

The chapter is organized as follows. In sections 6.1, the local convergence analysis of method $G_k SM$ is presented. In section 6.2, several applications are discussed.

6.1 LOCAL CONVERGENCE

It is convenient for the local convergence analysis of method (6.2) that follows to introduce some scalars and real functions. Let $\alpha > 0, \beta > 0$, and $\gamma > 0$ be given parameters. Consider function $w_0 : [0, \infty) \times [0, \infty) \to [0, \infty)$ to be continuous and increasing with $w_0(0, 0) = 0$.

Assume equation

$$p_0(t) - 1 = 0, \tag{6.4}$$

has a minimal positive solution ρ_0, where $p_0(t) = w_0(\alpha t, \beta t)$.

Consider also function $w : [0, \rho_0) \times [0, \rho_0) \to [0, \rho_0)$ be continuous and increasing with $w(0, 0) = 0$. Define function g_1 and h_1 on the interval $[0, \rho_0)$ by

$$g_1(t) = \frac{w(\gamma t, \beta t)}{1 - w_0(\alpha t, \beta t)},$$

$$h_1(t) = g_1(t) - 1.$$

Assume that

$$h_1(t) = 0 \tag{6.5}$$

has a minimal solution $r_1 \in (0, \rho_0]$.

In addition, we consider that equations

$$p_0\left(g_1(t)t\right) - 1 = 0, \text{ and } p_1(t) - 1 = 0, \tag{6.6}$$

have minimal positive solutions ρ_1 and ρ_2, respectively, where $p_1(t) = 2w_0\left(g_1(t)t, t\right) + w_0(\alpha t, \beta t)$. Let $\rho_3 = \min\{\rho_0, \rho_1, \rho_2\}$.

Define functions g_2 and h_2 on $[0, \rho_3)$ by

$$g_2(t) = \left[\frac{w\left(\gamma g_1(t)t, \beta g_1(t)t\right)}{1 - w_0\left(\alpha g_1(t)t, \beta g_1(t)t\right)} + \frac{2w\left((1+\gamma+g_1(t))t, \gamma t\right)}{\left(1 - p_0(g_1(t)t)\right)\left(1 - p_1(t)\right)} \right] g_1(t),$$

and $h_2(t) = g_2(t) - 1$. Suppose that $h_2(t) = 0$, has a minimal solution $r_2 \in (0, \rho_3]$.

Assume equation

$$p_0\left(g_2(t)t\right) - 1 = 0, \tag{6.7}$$

has a minimal positive solution ρ_4. Let $\rho = \min\{\rho_2, \rho_4\}$.

Define functions g_3 and h_3 on $[0, \rho)$ by

$$g_3(t) = \left[\frac{w\left(\gamma g_2(t)t, \beta g_2(t)t\right)}{1 - p_0\left(g_2(t)t\right)} + w\left((g_1(t) + (1+\gamma)g_2(t))t, (1+(1+\gamma)g_2(t))t\right) \right.$$

$$\left. + \frac{w\left((1+\gamma+g_1(t))t, \gamma t\right)}{\left(1 - p_0(g_2(t)t)\right)\left(1 - p_1(t)\right)} \right] g_2(t),$$

and $h_3(t) = g_3(t) - 1$. Let us assume that $h_3(t) = 0$, has a minimal solution $r_3 \in (0, \rho]$. Define a radius of convergence r by

$$r = \min\{r_i\}, \quad i = 1, 2, 3. \tag{6.8}$$

We shall show that r is radius of convergence for SLM.

Then, we have

$$0 \le p_0(t) < 1, \tag{6.9}$$

$$0 \le p_0\left(g_1(t)t\right) < 1, \tag{6.10}$$

$$0 \le p_0\left(g_2(t)t\right) < 1, \tag{6.11}$$

$$0 \le p_1(t) < 1, \tag{6.12}$$

and

$$0 \le g_i(t) < 1, \text{ for all } t \in [0, r). \tag{6.13}$$

The $S(\gamma, r)$, stands for the open ball centered at $\gamma \in \mathbb{B}$, and of radius $r > 0$. By $\bar{S}(\gamma, r)$, we denote the closure of $S(\gamma, r)$.

In the local convergence analysis of method (6.2) that follows we use conditions (A):

(a_1) $F : \Omega \rightarrow \mathbb{B}$, is differentiable in the Fréchet sense, $[\cdot, \cdot; F] : \Omega \times \Omega \rightarrow \ell(\mathbb{B}, \mathbb{B})$ is divided difference of order one, and there exists $x_* \in \Omega$ with $F(x_*) = 0$, and $F'(x_*)^{-1} \in \ell(\mathbb{B}, \mathbb{B})$.

(a_2) There exist function $w_0 : [0, \infty) \times [0, \infty) \rightarrow [0, \infty)$ continuous and increasing with parameters $\alpha > 0, \beta > 0$ such that for each $x, y \in \Omega$

$$\left\| F'(x_*)^{-1} \left([x, y; F] - F'(x_*) \right) \right\| \leq w_0 \left(\|x - x_*\|, \|y - x_*\| \right),$$

$$\|I + [x, x_*; F]\| \leq \alpha$$

$$\|I - [x, x_*; F]\| \leq \beta$$

and

$$\|[x, x_*; F]\| \leq \gamma.$$

Set $\Omega_0 = \Omega \cap S(x_*, \rho)$, where ρ exists and is given by (6.4).

(a_3) There exist function $w : [0, \rho_1) \times [0, \rho_1) \rightarrow [0, \infty)$ continuous and increasing such that for each $x, y, u, v \in \Omega_0$

$$\left\| F'(x_*)^{-1} \left([x, y; F] - [u, v; F] \right) \right\| \leq w \left(\|x - u\|, \|y - v\| \right).$$

(a_4) $\bar{S}(x_*, R) \subseteq \Omega$, where $R = \max\{r, \alpha r, \beta r, \gamma r\}$, r is defined in (6.8).

(a_5) There exists $\bar{r} \geq r$ such that

$$w_0(0, \bar{r}) < 1 \text{ or } w_0(\bar{r}, 0) < 1.$$

Set $\Omega_1 = \Omega \cap \bar{S}(x_*, \bar{r})$.

Theorem 6.1. Under the conditions (A), further suppose that $x_0 \in S(x_*, r) - \{x_*\}$. Then, the follows items hold

$$\{x_n\} \subseteq S(x_*, r), \tag{6.14}$$

$$\lim_{n \to \infty} x_n = x_*, \tag{6.15}$$

$$\|y_n - x^*\| \leq g_1(\|x_n - x^*\|) \|x_n - x^*\| \leq \|x_n - x^*\| < r, \tag{6.16}$$

$$\|z_n - x^*\| \leq g_2(\|x_n - x^*\|) \|x_n - x^*\| < \|x_n - x^*\|, \tag{6.17}$$

and

$$\|x_{n+1} - x^*\| \leq g_3(\|x_n - x^*\|) \|x_n - x^*\| < \|x_n - x^*\|, \tag{6.18}$$

where the functions g_i are given previously and r is defined by (6.8). Moreover, x_* is the only solution of equation $F(x) = 0$ in the set Ω_1 given in (a_5).

Proof. We use mathematical induction to show items (6.16)–(6.18). Let $V \in S(x_*, r) - \{x_*\}$. By using hypotheses (a_2), we get

$$\|v + F(v) - x_*\| = \left\|\left(I + [v, x_*; F]\right)(v - x_*)\right\|$$
$$\leq \|I + [v, x_*; F]\| \|v - x_*\| \leq \alpha \|v - x_*\|$$

and

$$\|F(v)\| = \|F(v) - F(x_*)\| = \left\|[v, x_*; F](v - x_*)\right\|$$
$$\leq \|[v, x_*; F]\| \|v - x_*\| \leq \beta \|v - x_*\|$$

so $v + F(v) - x_*$ and $F(v)$ belong in $\bar{S}(x_*, r)$. We also have for $x, y \in S(x_*, r) - \{x_*\}$ such that

$$\left\|F'(x_*)^{-1}\left([x, y; F] - F'(x_*)\right)\right\| \leq w_0\left(\|x - x_*\|, \|y - x_*\|\right)$$
$$\leq w_0(r, r) < 1,$$

so the Banach Lemma on invertible operators [4, 5] that $[x, y; F]^{-1} \in \ell(\mathbb{B}, \mathbb{B})$ and

$$\left\|[x, y; F]^{-1} F'(x^*)\right\| \leq \frac{1}{1 - w_0\left(\|x - x^*\|, \|y - x^*\|\right)}. \tag{6.19}$$

It follows that y_0 is well defined by the first substep of method (6.2) for $n = 0$. Using (6.8), (6.9) (for $i = 1$), hypotheses (a_2), (a_3) (6.14) and the definition of y_0, we have in turn

$$\|y_0 - x^*\| \leq \|x_0 - x^* - [x_0 + F(x_0), x_0; F]^{-1} F(x_0)\|$$
$$= \left\|[x_0 + F(x_0), x_0; F]^{-1}\left([x_0 + F(x_0), x_0; F] - [x_0, x_*; F]\right)(x_0 - x_*)\right\|$$
$$\leq \left\|[x_0 + F(x_0), x_0; F]^{-1} F'(x_*)\right\| \left\|F'(x_*)\left([x_0 + F(x_0), x_0; F] - [x_0, x_*; F]\right)\right\|$$
$$\times \left\|\|(x_0 - x_*)\|\right\|$$
$$\leq \frac{w\left(\|F(x_0)\|, \|(x_0 - x_*)\|\right)\|(x_0 - x_*)\|}{1 - w_0\left(\alpha\|(x_0 - x_*)\|, \beta\|(x_0 - x_*)\|\right)}$$
$$\leq \frac{w\left(\beta\|(x_0 - x_*)\|, \|(x_0 - x_*)\|\|\right)\|(x_0 - x_*)\|}{1 - w_0\left(\alpha\|(x_0 - x_*)\|, \beta\|(x_0 - x_*)\|\right)}$$
$$= g_1(\|(x_0 - x_*)\|)\|(x_0 - x_*)\| \leq \|(x_0 - x_*)\| < r, \tag{6.20}$$

so $y_0 \in S(x_*, r)$ (for $y_0 \neq x_*$) and (6.16) hold for $n = 0$.

As in (6.19) but using (6.10) and (6.20), we have $[y_0 + F(y_0), y_0 - F(y_0); F]^{-1} \in \ell(\mathbb{B}, \mathbb{B})$ and

$$
\begin{aligned}
\left\| [y_0 + F(y_0), y_0 - F(y_0); F]^{-1} F'(x_*) \right\| \\
\leq \frac{1}{1 - w_0(\alpha \|y_0 - x_*\|, \beta \|y_0 - x_*\|)} \\
\leq \frac{1}{1 - w_0\Big(\alpha g_1(\|x_0 - x_*\|) \|x_0 - x_*\|, \beta g_1(\|x_0 - x_*\|) \|x_0 - x_*\|\Big)} \qquad (6.21) \\
\leq \frac{1}{1 - p_0\Big(g_1(\|x_0 - x_*\|) \|x_0 - x_*\|\Big)}.
\end{aligned}
$$

Moreover, by (6.11), (a_2) and (6.21), we get in turn that

$$
\begin{aligned}
\left\| F'(x_*)^{-1} \Big[2\Big([y_0, x_0; F] - F'(x_*)\Big) - \Big([x_0 + F(x_0), x_0 + F(x_0); F] - F'(x_*)\Big) \Big] \right\| \\
\leq 2w_0(\|y_0 - x_*\|, \|x_0 - x_*\|) + w_0(\alpha\|x_0 - x_*\|, \beta\|x_0 - x_*\|) \\
\leq 2w_0\Big(g_1(\|x_0 - x_*\|) \|x_0 - x_*\|, \|x_0 - x_*\|\Big) + w_0(\alpha\|x_0 - x_*\|, \beta\|x_0 - x_*\|) \\
= p_1(\|x_0 - x_*\|) < p_1(r) < 1,
\end{aligned}
$$

so $\Big(2[y_0, x_0; F] - [x_0 + F(x_0), x_0 + F(x_0); F]\Big)^{-1} \in \ell(\mathbb{B}, \mathbb{B})$ and

$$
\left\| \Big(2[y_0, x_0; F] - [x_0 + F(x_0), x_0 + F(x_0); F]\Big)^{-1} F'(x_*) \right\| \leq \frac{1}{p_1(\|x_0 - x_*\|)}. \qquad (6.22)
$$

Then, z_0 and x_1 are well defined. By the second substep of method (6.2) for $n = 0$, we can also write

$$
\begin{aligned}
z_0 - x_* \\
= \Big(y_0 - x_* - [y_0 + F(y_0), y_0 - F(y_0); F]^{-1} F(y_0)\Big) \\
+ \Big[[y_0 + F(y_0), y_0 - F(y_0); F]^{-1} - \Big(2[y_0, x_0; F] - [x_0 + F(x_0), x_0 - F(x_0); F]\Big)^{-1} \Big] \\
\times \Big(2[y_0, x_0; F] - [x_0 + F(x_0), x_0 - F(x_0); F]\Big)^{-1} \\
= B_1 + B_2,
\end{aligned}
$$

$$
\qquad (6.23)
$$

for simplicity. Then, by (B_1), (6.8), (a_3), (6.21), and (6.23), we get

$$\|B_1\|$$
$$\leq \frac{w(\gamma\|y_0 - x_*\|, \beta\|y_0 - x_*\|)}{1 - w_0(\alpha\|y_0 - x_*\|, \beta\|y_0 - x_*\|)}\|y_0 - x_*\|$$
$$\leq \frac{w\left(\gamma g_1(\|x_0 - x_*\|)\|x_0 - x_*\|, \beta g_1(\|x_0 - x_*\|\|x_0 - x_*\|)\right)}{1 - p_0\left(g_1(\|x_0 - x_*\|\|x_0 - x_*\|)\right)}g_1(\|x_0 - x_*\|)\|x_0 - x_*\|$$

$$(6.24)$$

and

$$\|B_2\| \leq \frac{2w\left(\|y_0 - x_0 - F(x_0)\|, \|F(x_0)\|\right)}{\left(1 - p_0(g_1(\|x_0 - x_*\|)\|x_0 - x_*\|)\right)\left(1 - p_1(\|x_0 - x_*\|)\right)}. \qquad (6.25)$$

Then, summing B_1 and B_2 and using (6.22), (6.23), and (6.8) (for $i = 2$), we have

$$\|z_0 - x_*\| \leq g_2(\|x_0 - x_*\|)\|x_0 - x_*\| \leq \|x_0 - x_*\| < r, \qquad (6.26)$$

so (6.17) holds for $n = 0$ and $z_0 \in S(x_*, r)$. Furthermore, we can write by the third substep of method (6.2) for $n = 0$

$$x_1 - x_*$$
$$= \left(z_0 - x_* - [z_0 + F(z_0), z_0 - F(z_0); F]^{-1}F(z_0)\right) + \left[[z_0 + F(z_0), z_0 - F(z_0); F]^{-1}\right.$$
$$- \left(2[y_0, x_0; F] - [x_0 + F(x_0), x_0 - F(x_0); F]^{-1}\right)\right]$$
$$\times \left(2[y_0, x_0; F] - [x_0 + F(x_0), x_0 - F(x_0); F]^{-1}\right)F(z_0)$$
$$= B_3 + B_4.$$

$$(6.27)$$

Then, we have as above

$$\|B_3\| \leq \frac{w\left(\gamma\|z_0 - x_*\|, \beta\|z_0 - x_*\|\right)\|z_0 - x_*\|}{1 - w_0\left(\alpha\|z_0 - x_*\|, \beta\|z_0 - x_*\|\right)}$$
$$\leq \frac{w\left(\gamma g_2(\|x_0 - x_*\|)\|x_0 - x_*\|, \beta g_2(\|x_0 - x_*\|\|x_0 - x_*\|)\right)\|z_0 - x_*\|}{1 - p_0\left(g_2(\|x_0 - x_*\|)\|x_0 - x_*\|\right)}$$

$$(6.28)$$

and

$\|B_4\|$

$$\leq \frac{w\left(\left\|F'(x_*)^{-1}[y_0,x_0;F] - [z_0 + F(z_0), z_0 - F(z_0);F]\right\|\right)}{\left(1 - p_0\big(g_2(\|x_0 - x_*\|)\|x_0 - x_*\|\big)\right)\left(1 - p_1(\|x_0 - x_*\|)\right)}$$

$$\leq \frac{w\left(\left\|(y_0 - x_*) + (x_* - z_0) - F(z_0)\right\|, \left\|(x_0 - x_*) + (x_* - z_0) + F(z_0)\right\|\right)\|z_0 - x_*\|}{\left(1 - p_0\big(g_2(\|x_0 - x_*\|)\|x_0 - x_*\|\big)\right)\left(1 - p_1(\|x_0 - x_*\|)\right)}$$

$$\leq \frac{a_0}{\left(1 - p_0\big(g_2(\|x_0 - x_*\|)\|x_0 - x_*\|\big)\right)\left(1 - p_1(\|x_0 - x_*\|)\right)}$$

$$\leq g_3(\|x_0 - x_*\|)\|x_0 - x_*\| \leq \|x_0 - x_*\| < r,$$

$$(6.29)$$

where $a_0 = w\Big(g_1(\|x_0 - x_*\|)\|x_0 - x_*\| + g_2(\|x_0 - x_*\|)\|x_0 - x_*\| + \gamma g_2(\|x_0 -$

$x_*\|)\|x_0 - x_*\|, \|x_0 - x_*\| + g_2(\|x_0 - x_*\|)\|x_0 - x_*\| + \gamma g_2(\|x_0 - x_*\|)\|x_0 - x_*\|\Big)$, so

(6.18) holds for $n = 0$ and $x_1 \in S(x_*, r)$. Simply replace x_0, y_0, z_0 and x_1 by x_m, y_m, z_m and x_{m+1} in the proceding estimations to finish the induction for (6.16)–(6.18). Then, it follows from the estimation

$$\|x_{m+1} - x^*\| \leq c\|x_m - x^*\| < r, \text{ where } c = g_3(\|x_0 - x_*\|) \in [0,1), \qquad (6.30)$$

that $\lim_{m \to \infty} x_m = x^*$ and $x_{m+1} \in S(x^*, r)$.

Finally, to show the uniqueness part, set $T = [x_*, y_*; F]$ for $y_* \in \Omega_1$ with $F(y_*) = 0$. Then, by hypotheses (a_2) and (a_5), we get

$$\left\|F'(x^*)^{-1}(T - F'(x^*))\right\| \leq \|w_0(0, \|x_* - y_*\|)\| \leq w_0(0, \bar{r}) < 1,$$

so $T^{-1} \in \ell(\mathbb{B}, \mathbb{B})$. Finally, $x^* = y^*$ is deduced from $0 = F(x_*) - F(y_*) = T(x_* - y_*)$. $\qquad \square$

6.2 APPLICATIONS

In this section, we illustrate numerically the theoretical results proposed in the previous sections. We consider three real life problems and three standard nonlinear problems which are illustrated in Examples 6.1–6.5 for the case of scalar equations as well as non-linear systems and corresponding initial guesses mentioned in these examples. We display in Tables 6.1–6.2, radius of convergence and the minimum number of iterations required to get the desire accuracy to the corresponding zeros of $F(x)$, absolute residual error at the corresponding iteration and the theoretical order of convergence.

We adopt $\varepsilon = 10^{-100}$ as a tolerance error and the following stopping criteria are used for computer programs to solve nonlinear system: $(i)\|x_{n+1} - x_n\| < \varepsilon$ and

$(ii) \|F(x_n)\| < \varepsilon.$

The computations are performed with the package *Mathematica* 9 with multiple precision arithmetic. Moreover, we used $[x,y;F] = \int_0^1 F'\left(y + \theta(x-y)\right) d\theta$ in the examples.

Example 6.1. Let $\mathbb{B}_1 = \mathbb{B}_2 = \mathbb{R}^3$, $\Omega = \mathbb{S}(0,\ 1)$, $x^* = (0,\ 0,\ 0)^T$. Define F on Ω by means of

$$F(u) = F(u_1, u_2, u_3) = \left(e^{u_1} - 1, \frac{e-1}{2} u_2^2 + u_2,\ u_3 \right)^T,\qquad (6.31)$$

where, $u = (u_1, u_2, u_3)^T$. The Fréchet-derivative is defined by

$$F'(u) = \begin{bmatrix} e^{u_1} & 0 & 0 \\ 0 & (e-1)u_2 + 1 & 0 \\ 0 & 0 & 1 \end{bmatrix}.$$

Then, $F'(x_*) = F'(x_*)^{-1} = diag\{1,\ 1,\ 1\}$, so, we have

$$w_0(s,t) = \frac{e-1}{2}(s+t),\ \ w(s,t) = \frac{1}{2} e^{\frac{1}{e-1}}(s+t),\ \ \alpha = \frac{e+3}{2},\ \ \beta = \frac{e-1}{2},\ \ \text{and}\ \ \gamma = \frac{e+1}{2}.$$

Hence, we obtain radii of convergence for example 6.1 that are mentioned in Table 6.1.

method	SLM
r_1	0.1542
r_2	0.1033
r_3	0.09615
r	0.09615
x_0	$(0.08, 0.08, 0.08)^T$
n	3
$|f(x_3)|$	3.2×10^{-245}
ξ	6.0000

Table 6.1: Radii of convergence for example 6.1

Example 6.2. Here, we choose a well known Fisher's equation

$$\theta_t = D\theta_{xx} + \theta(1-\theta) = 0,$$

with homogeneous Neumann's boundary conditions

$$\theta(x,0) = 1.5 + 0.5 cos(\pi x),\ 0 \le x \le 1, \qquad (6.32)$$
$$\theta_x(0,t) = 0, \forall t \ge 0,$$
$$\theta_x(1,t) = 0, \forall t \ge 0,$$

where D is the diffusion parameter. We adopt finite difference discretization technique, in order to convert the above differential equation (6.32) in to a system of nonlinear equations. So, we choose $w_{i,j} = \theta(x_i, t_j)$ as the required solution at the grid points of the mesh. In addition, x and t are the numbers of steps in the direction of M and N, respectively. Moreover, h and k, respectively are corresponding step size of M and N. By adopting central, backward and forward difference, we have

$$\theta_{xx}(x_i, t_j) = (w_{i+1,j} - 2w_{i,j} + w_{i-1,j})/h^2,$$
$$\theta_t(x_i, t_j) = (w_{i,j} - w_{i,j-1})/k,$$

and

$$\theta_x(x_i, t_j) = (w_{i+1,j} - w_{i,j})/(h), \quad t \in [0, 1],$$

leading to

$$\frac{w_{1,j} - w_{i,j-1}}{k} - w_{i,j}\left(1 - w_{i,j}\right) - D\frac{w_{i+1,j} - 2w_{i,j} + w_{i-1,j}}{h^2}, \tag{6.33}$$

where,

$$h = \frac{1}{M}, k = \frac{1}{N}, i = 1, 2, 3, \ldots, \text{ and } M, j = 1, 2, 3, \ldots, N,.$$

For particular values of $M = 9$, $N = 9$, $h = \frac{1}{9}, k = \frac{1}{9}$ and $D = 1$ that leads us to a nonlinear system of size 81×81, convergence towards the following column vector solution (not a matrix)

$$x_* = u(x_i, t_j) = \begin{pmatrix} 1.6017\ldots, 1.4277\ldots, 1.3328\ldots, 1.2740\ldots, 1.2331\ldots, 1.2022\ldots, 1.1772\ldots, \\ 1.1563\ldots, 1.1385\ldots, 1.5726\ldots, 1.4159\ldots, 1.3277\ldots, 1.2717\ldots, 1.2322\ldots, \\ 1.2017\ldots, 1.1770\ldots, 1.1563\ldots, 1.1384\ldots, 1.5203\ldots, 1.3940\ldots, 1.3182\ldots, \\ 1.2676\ldots, 1.2303\ldots, 1.2009\ldots, 1.1767\ldots, 1.1561\ldots, 1.1384\ldots, 1.4521\ldots, \\ 1.3648\ldots, 1.3055\ldots, 1.2619\ldots, 1.2278\ldots, 1.1998\ldots, 1.1762\ldots, 1.1559\ldots, \\ 1.1383\ldots, 1.3771\ldots, 1.3321\ldots, 1.2911\ldots, 1.2556\ldots, 1.2250\ldots, 1.1985\ldots, \\ 1.1756\ldots, 1.1556\ldots, 1.1381\ldots, 1.3045\ldots, 1.2998\ldots, 1.2768\ldots, 1.2492\ldots, \\ 1.2221\ldots, 1.1973\ldots, 1.1750\ldots, 1.1554\ldots, 1.1380\ldots, 1.2429\ldots, 1.2719\ldots, \\ 1.2642\ldots, 1.2436\ldots, 1.2196\ldots, 1.1961\ldots, 1.1745\ldots, 1.1551\ldots, 1.1379\ldots \\ 1.1990\ldots, 1.2514\ldots, 1.2550\ldots, 1.2395\ldots, 1.2178\ldots, 1.1953\ldots, 1.1742\ldots, \\ 1.1550\ldots, 1.1379\ldots, 1.1768\ldots, 1.2406\ldots, 1.2501\ldots, 1.2373\ldots, 1.2168\ldots, \\ 1.1949\ldots, 1.1740\ldots, 1.1549\ldots, 1.1378\ldots \end{pmatrix}^T .$$

Further, we have on $S(x_*, 1.6)$

$$w_0(s, t) = w(s, t) = 4(s + t), \quad \alpha = 5, \beta = 5, \text{ and } \gamma = 4.$$

Example 6.3. Bratu 2D Problem:

The well known 2D Bratu problem [16] is defined in $\Omega = \{(x, t) : 0 \leq x \leq 1, 0 \leq t \leq 1\}$, as

$$u_{xx} + u_{tt} + Ce^u = 0, \text{on}$$
$$\textit{with boundary conditions } u = 0 \textit{ on } \partial\Omega. \tag{6.34}$$

The approximated solution of a nonlinear partial differential equation can be found using finite difference discretization which reduces to solving a system of nonlinear equations. Let $u_{i,j} = u(x_i, t_j)$ be its approximate solution at the grid points of the mesh. Let M and N be the number of steps in x and t directions, h and k be the respective step size. To solve the given PDE, apply central difference to u_{xx} and u_{tt} i.e $u_{xx}(x_i, t_j) = (u_{i+1,j} - 2u_{i,j} + u_{i-1,j})/h^2$, $C = 0.1, t \in [0,1]$. We look for the solution of the system for $M = 11$ and $N = 11$ of size 100×100, and converges to the following column vector (not a matrix)

$$x_* = \begin{pmatrix} 0.00111\ldots,0.00181\ldots,0.00226\ldots,0.00253\ldots,0.00266\ldots,0.00266\ldots,0.00253\ldots,0.00226\ldots, \\ 0.00181\ldots,0.00111\ldots,0.00181\ldots,0.00305\ldots,0.00387\ldots,0.00437\ldots,0.00461\ldots,0.00461\ldots, \\ 0.00437\ldots,0.00387\ldots,0.00305\ldots,0.00181\ldots,0.00226\ldots,0.00387\ldots,0.00497\ldots,0.00565\ldots, \\ 0.00598\ldots,0.00598\ldots,0.00565\ldots,0.00497\ldots,0.00387\ldots,0.00226\ldots,0.00253\ldots,0.00437\ldots, \\ 0.00565\ldots,0.00645\ldots,0.00684\ldots,0.00684\ldots,0.00645\ldots,0.00565\ldots,0.00437\ldots,0.00253\ldots, \\ 0.00266\ldots,0.00461\ldots,0.00598\ldots,0.00684\ldots,0.00726\ldots,0.00726\ldots,0.00684\ldots,0.00598\ldots, \\ 0.00461\ldots,0.00266\ldots,0.00266\ldots,0.00461\ldots,0.00598\ldots,0.00684\ldots,0.00726\ldots,0.00726\ldots, \\ 0.00684\ldots,0.00598\ldots,0.00461\ldots,0.00266\ldots,0.00253\ldots,0.00437\ldots,0.00565\ldots,0.00645\ldots, \\ 0.00684\ldots,0.00684\ldots,0.00645\ldots,0.00565\ldots,0.00437\ldots,0.00253\ldots,0.00226\ldots,0.00387\ldots, \\ 0.00497\ldots,0.00565\ldots,0.00598\ldots,0.00598\ldots,0.00565\ldots,0.00497\ldots,0.00387\ldots,0.00226\ldots, \\ 0.00181\ldots,0.00305\ldots,0.00387\ldots,0.00437\ldots,0.00461\ldots,0.00461\ldots,0.00437\ldots,0.00387\ldots, \\ 0.00305\ldots,0.00181\ldots,0.00111\ldots,0.00181\ldots,0.00226\ldots,0.00253\ldots,0.00266\ldots,0.00266\ldots, \\ 0.00253\ldots,0.00226\ldots,0.00181\ldots,0.00111\ldots, \end{pmatrix}^T.$$

Further, on $S(x_*, 0.007)$, we have

$$w_0(t,s) = w(s,t) = 0.1(s+t), \quad \alpha = 2, \ \beta = 2, \quad \text{and} \quad \gamma = 1.$$
$(i, j = 1, 2, 3, \cdots, 10.)$

Example 6.4. Finally, we choose

$$F(x) = \begin{cases} x_j^2 x_{j+1} - 1 = 0, \ 1 \le j \le \zeta, \\ x_\zeta^2 x_1 - 1 = 0. \end{cases} \tag{6.35}$$

We pick $\zeta = 110$ in order to deduce a huge system of 110×110. In addition, we select the starting guess $x_0 = (1.25, \ 1.25, \ 1.25, \ \cdots, \ (110\,times))^T$ that converges to $x_* = (1, \ 1, \ 1, \ \cdots, \ (110\,times))^T$. Further, we have on $S(x_*, 0.26)$

$$w_0(t,s) = w(s,t) = 4(s+t), \quad \alpha = 4, \ \beta = 4, \quad \text{and} \quad \gamma = 3.$$

Example 6.5. Returning back to the counter example which were mentioned in the introduction. We have the following values for the $x_* = 1$

$$w_0(t,s) = w(s,t) = \frac{97}{2}(s+t), \quad \alpha = 2, \ \beta = 2, \quad \text{and} \quad \gamma = 1.$$

The distinct radii are stated in Table 6.2.

method	r_1	r_2	r_3	r	x_0	n	$|f(x_3)|$	ξ
SLM	0.002946	0.001912	0.001768	0.001768	1.001	3	1.4×10^{-519}	6.0000

Table 6.2: Radii of convergence for example 6.5

6.3   EXERCISES

1. Solve the initial value problem

$$\dot{x}(t) = 1 + \cos(x(t)), x(0) = 0.$$

2. The predator-prey population models describe the interaction of a prey population X and a predator population Y. Assume that their interaction is modeled by the system of ordinary differential equations

$$\dot{x} = x - \frac{1}{4}x^2 - \frac{1}{10}xy + 1$$

$$\dot{y} = -\frac{1}{4}y + \frac{1}{7}xy + \frac{2}{3}.$$

   (Assume $x(0) = y(0) = 0$). Solve the system.

3. Solve the initial value problem

$$\dot{x} = (x(t) + t, x(0) = 1)$$

   by the iteration method (6.2). Perform 3 steps.

4. Solve the Fredholm-type integral equation

$$x(t) = \sum_{0}^{1} \frac{t \cdot s}{10} ds + 1$$

   by the iteration method (6.2). Perform 3 steps.

5. Solve the Volterra-type integral equation

$$x(t) = \sum_{0}^{t} \frac{t \cdot s}{10} ds + 1$$

   by the iteration method. Perform 3 steps.

6. Solve equation

$$x = \frac{\sin x}{2} + 1$$

   by using algorithm (6.2).

7. Repeat the previous problem for equation

$$x = \frac{\sin x + \cos 2x}{10}.$$

REFERENCES

1. Amat, S., Berñudez, C., Herñandez-Veñon, M.A., and Martínez, E. 2016. On an efficient $k-$step iterative method for nonlinear equations. *J. Comput. Appl. Math.*, 302:258–271.
2. Amat, S., Busquier, S., Berñudez, C., and Plaza, S. 2012. On two families of high order Newton type methods, *Appl. Math. Lett.*, 25. 12:2209–2217.
3. Argyros, I.K., and Magrenan, A.A., 2017. *A Contemporary Study of Iterative Methods.* Academic Press, Elsevier.
4. Argyros I.K., and George, S. 2019. *Mathematical Modeling for the Solutions with Application, vol. III.* New York: Nova publisher.
5. Argyros, I.K., and Hilout, S., Weaker conditions for the convergence of Newon's method. *J. Complexity*, 28. 3:364–387.
6. Cordero, A., Hueso, J.L., Martínez, E., and Torregrosa, J.R. 2012. Steffensen type methods for solving nonlinear equations. *J. Comput. Appl. Math.*, 236, 12:3058–3064.
7. Burden, R.L., and Faires, J.D., 2001. *Numerical Analysis.* Boston: PWS Publishing Company.
8. Ezquerro, J.A., and Hernández, M.A. 2014. How to improve the domain of starting points for Steffensen's method. *Stud. Appl. Math.*, 132. 4:354–380.
9. Potra, F.A., and Pták, V. 1984. *Nondiscrete Induction and Iterative Process.* vol. 103. Pitman Advanced Publishing Program.
10. Simpson, R.B. 1975. A method for the numerical determination of bifurcation states of non-linear systems of equations. *SIAM J. Numer. Anal.*, 12:439-451.
11. Kapania, R.K. 1990. A pseudo-spectral solution of 2-parameter Bratu's equation. *Comput. Mech.*, 6:55-63.
12. Sauer, T., 2012. *Numerical Analysis.* 2nd edn. USA: Pearson.
13. Ortega, J.M., and Rheinboldt, W.C. 1970. *Iterative solution of nonlinear equations in several variables*, New-York: Academic Press.
14. Ostrowski, A.M. 1960. Solution of equations and systems of equations. *Pure and Applied Mathematics* Vol. IX. New York-London: Academic Press.
15. Ostrowski, A.M., 1973. Solution of equations in Euclidean and Banach spaces, *Pure and Applied Mathematics*. Vol. 9. New York-London: Academic Press.
16. Petkovic, M.S., Neta, B., Petkovic, L., and Džunič, J. 2013. *Multipoint methods for solving nonlinear equations.* Elsevier.
17. Rheinboldt, W.C. 1978. An adaptive continuation process for solving systems of nonlinear equations. *Polish Academy of Science, Banach Ctr. Publ.*, 3:129–142.
18. Traub, J.F. 1964. *Iterative methods for the solution of equations.* Englewood Cliffs, N.J.: Prentice Hall Series in Automatic Computation.
19. Tsoulos, I.G., and Stavrakoudis, A. 2010. On locating all roots of systems of nonlinear equations inside bounded domain using global optimization methods. *Nonlinear Anal. Real World Appl.*, 11:2465–2471.

7 High-Order Iterative Methods

In this chapter, we develop a sixth order Steffensen-type method with one parameter to solve systems of equations. Our study's novelty lies in the fact that two types of local convergence are established under weak conditions including computable error bounds and uniqueness of the results. The performance of our methods is discussed and compared to other schemes using similar information. Finally, very large systems of equations (100×100 and 200×200) are solved to test the theoretical results and compare them favorably to earlier works.

Numerous problems from the Biology, Chemistry, Economics, Engineering, Mathematics, Physics are converted to a mathematical expression of the following form

$$F(u) = 0. \tag{7.1}$$

Here, $F : \Omega \subset \mathbb{B} \to \mathbb{B}$, is differentiable, $\mathbb{B}$ is a Banach space and Ω is nonempty and open. Closed form solutions are rarely found, so iterative methods [1–8, 15] are used converging to the solution u_*.

In particular, we propose the following new scheme

$$y_p = u_p - \left[u_p + F(u_p), u_p; F\right]^{-1} F(u_p)$$

$$z_p = u_p - \lambda \left[u_p + F(u_p), u_p; F\right]^{-1} \left(F(u_p) + F(y_p)\right) - (1 - \lambda)\left[u_p, y_p; F\right]^{-1} F(u_p)$$

$$u_{p+1} = z_p - \left[z_p + F(z_p), z_p; F\right]^{-1} F(z_p), \tag{7.2}$$

$u_0 \in \Omega$ is an initial point and $\lambda \in \mathbb{R}$ is a free parameter. In addition to this, $[\cdot, \cdot; F] : \Omega \times \Omega \to \ell(\mathbb{B}, \mathbb{B})$ is a divided difference of order one.

We shall present two convergence analyses. Later we present the advantages over other methods using similar information.

7.1 LOCAL CONVERGENCE ANALYSIS I

We assume that $\mathbb{B} = \mathbb{R}$. We use method (7.2) with standard Taylor expansions [11] for studying local convergence.

Theorem 7.1. Suppose mapping F is s sufficient differentiable on Ω, with $u_* \in \Omega$, a simple zero of F. We also consider that the inverse of F, $F'(u_*)^{-1} \in \ell(\mathbb{B}, \mathbb{B})$. Then, $\lim_{p \to \infty} u_p = u_*$ provided that u_0 is close enough to u_*. Moreover, the convergence order is six.

DOI: 10.1201/9781003128915-7

Proof. Set $\varepsilon_p = u_p - u_*$ and $Q_p = \frac{F'(u_*)}{p!}$, where $(\varepsilon_p)^\gamma = (\varepsilon_1, \varepsilon_2, \ldots, \varepsilon_k)^\gamma$, $\varepsilon_p \in \mathbb{R}^p$.

We shall use some Taylor series expansions, first for $F(u_p)$ and $F\left(u_p + F(u_p)\right)$:

$$F(u_p) = Q_1 \varepsilon_p + Q_2 \varepsilon_p^2 + O(\varepsilon_p^3) \tag{7.3}$$

and

$$F\left(u_p + F(u_p)\right) = (Q_1 + Q_1^2)\varepsilon_p + (3Q_1 Q_2 + Q_2 + Q_1^2 Q_2)\varepsilon_p^2 + O(\varepsilon_p^3), \tag{7.4}$$

respectively.

By using the expressions (7.3) and (7.4) in the first substep of scheme (7.2), we have

$$\widehat{\varepsilon}_p = y_p - u_* = b_1 \varepsilon_p^2 + b_2 \varepsilon_p^3 + b_3 \varepsilon_p^4 + O(e_p^5), \tag{7.5}$$

where

$$b_1 = \frac{Q_2}{Q_1} + Q_2,$$

$$b_2 = \frac{2Q_3}{Q_1} - \frac{2Q_2^2}{Q_1} - \frac{2Q_2^3}{Q_1^2} + Q_1 Q_3 + 3Q_3 - Q_2^2$$

and

$$b_3 = \frac{3Q_2^3}{Q_1} - 2Q_1 Q_2 Q_3 + Q_3^2 + Q_1^2 Q_4 + 4Q_1 Q_4 + 6Q_4 - 7Q_2 Q_3$$

$$+ \frac{3Q_4}{Q_1} + \frac{4Q_2^3}{Q_1^2} + \frac{5Q_3^3}{Q_1^2} - \frac{10Q_2 Q_3}{Q_1}.$$

Secondly, we expand $F(y_p)$

$$F(y_p) = Q_1 \widehat{\varepsilon}_p + Q_2 \widehat{\varepsilon}_p^2 + O(\widehat{\varepsilon}_p^3). \tag{7.6}$$

In view of (7.3)– (7.6), we get in the second substep of scheme (7.2)

$$\bar{\varepsilon}_p = u_{p+1} - u_* = z_p - u_* = b_4 \varepsilon_p^3 + O(\varepsilon_p^4), \tag{7.7}$$

where

$$b_4 = \frac{3Q_2^2}{Q_1} + \frac{2Q_2^2}{Q_1^2} + Q_2^2 - \lambda \left(\frac{4Q_2^2}{Q_1} + Q_2^2 + \frac{3Q_2^2}{Q_1^2} \right).$$

Thirdly, we need the expansions for $F(z_p)$ and $F\left(z_p + F(z_p)\right)$

$$F(z_p) = Q_1 \bar{e}_p + Q_2 \bar{e}_p^2 + O(\bar{e}_p^3), \tag{7.8}$$

Hence, by (7.5) and (7.8), we get

$$F(z_p + F(z_p)) = b_5 \bar{e}_p + b_6 \bar{e}_p^2 + O(\bar{e}_p^3), \tag{7.9}$$

leading together with the third substep of method (7.2) to

$$e_{p+1} = u_{p+1} - u_* = b_7 e_p^6 + O(e_p^7), \tag{7.10}$$

where

$$b_5 = Q_1 + Q_1^2,$$
$$b_6 = 3Q_1 Q_2 + Q_2 + Q_1^2 Q_2$$

and

$$b_7 = \left(Q_2 + \frac{Q_2}{Q_1} \right) \left[\frac{3Q_2^2}{Q_1} + \frac{2Q_2^2}{Q_1} + Q_2^2 - \lambda \left(\frac{4Q_2^2}{Q_1} + \frac{3Q_2^2}{Q_1} + Q_2^2 \right) \right].$$

$\square$

According to Theorem 7.1 the applicability of method (7.2) is limited to mappings F with derivatives up to the seventh order.

Next, we choose $\mathbb{B} = \mathbb{R}$, $\Omega = [-\frac{3}{2}, \frac{1}{2}]$. Define a function f as follows:

$$f(\xi) = \begin{cases} \xi^3 \ln \xi^2 + \xi^5 - \xi^4, & \xi \neq 0 \\ 0, & \xi = 0 \end{cases}. \tag{7.11}$$

We have the following derivatives of function f

$$f'(\xi) = 3\xi^2 \ln \xi^2 + 5\xi^4 - 4\xi^3 + 2\xi^2,$$
$$f''(\xi) = 12\xi \ln \xi^2 + 20\xi^3 - 12\xi^2 + 10\xi,$$
$$f'''(\xi) = 12 \ln \xi^2 + 60\xi^2 - 12\xi + 22.$$

But, $f'''(\xi)$ is not bounded on Ω, so section 7.2, cannot be used. In this case, we have a more general alternative given in the up coming section.

7.2   LOCAL CONVERGENCE ANALYSIS II

Consider $a \geq 0$ and $b > 0$. Let $w_0 : [0, \infty) \times [0, \infty) \to [0, \infty)$ be a increasingly continuous map with $w_0(0,0) = 0$.

Suppose equation

$$w_0(at, t) = 1 \tag{7.12}$$

has ρ_1 as the smallest positive zero. In addition, we assume that $w : [0, \rho_1) \times [0, \rho_1) \to [0, \infty)$ is a increasingly continuous map with $w(0,0) = 0$.

Consider functions g_1 and h_1 defined on semi open interval $[0, \rho_1)$ as follow:

$$g_1(t) = \frac{w(bt, t)}{1 - w_0(at, t)},$$

and

$$h_1(t) = g_1(t) - 1.$$

By these definitions, we have $h_1(0) = -1$ and $h_1(t) \to \infty$ as $t \to \rho_1^-$. Then, the intermediate value theorem assures that function h_1 has minimum one solution in $(0, \rho_1)$. Let r_1 be the minimal such zero.

The expression

$$w_0\left(t, g_1(t)t\right) = 1 \qquad (7.13)$$

has the smallest positive zero ρ_2. Set $\rho_3 = \min\{\rho_1, \rho_2\}$.

We construe the functions g_2 and h_2 on interval $[0, \rho_3)$ in the following way

$$g_2(t) = g_1(t) + \frac{b|1 - \lambda|w\left(bt, \left(1 + g_1(t)\right)t\right)g_1(t)}{\left(1 - w_0(at, t)\right)\left(1 - w_0\left(t, g_1(t)t\right)\right)},$$

and

$$h_2(t) = g_2(t) - 1.$$

We yield $h_2(0) = -1$ and $h_2(t) \to \infty$ since $t \to \rho_3^-$. The r_2 stand for the minimal such zero of function h_2 on $(0, \rho_3)$.

The equation

$$w_0\left(ag_2(t)t, g_2(t)t\right) = 1 \qquad (7.14)$$

has ρ_4 as the smallest positive solution. Set $\rho = \min\{\rho_3, \rho_4\}$. Define functions g_3 and h_3 on $[0, \rho)$ as

$$g_3(t) = \frac{w\left(bg_2(t)t, g_2(t)t\right)g_2(t)}{1 - w_0\left(ag_2(t)t, g_2(t)t\right)},$$

and

$$h_3(t) = g_3(t) - 1.$$

We obtain $h_3(0) = -1$ and $h_3(t) \to \infty$ as $t \to \rho^-$. The r_3 imply the minimal zero of h_3 on $(0, \rho)$. Moreover, define

$$r = \min\{r_i\}, \text{ for } i = 1, 2, 3. \qquad (7.15)$$

So, we have

$$0 \le w_0(at, t) < 1, \qquad (7.16)$$

$$0 \le w_0\left(t, g_1(t)t\right) < 1, \qquad (7.17)$$

$$0 \le w_0\left(ag_2(t)t, g_2(t)t\right) < 1, \qquad (7.18)$$

and

$$0 \le g_i(t) < 1, \qquad (7.19)$$

for all $t \in [0, r)$.

$S(v,c)$ denotes the open ball centered at $v \in \mathbb{B}$ and of radius $c > 0$. By $\bar{S}(v,c)$, we denote the closure of $S(v,c)$

We use the following conditions (A) in order to study the local convergence:

(a_1) $F : \Omega \to \mathbb{B}$ is a differentiable operator in the Fréchet sense, $[\cdot,\cdot;F] : \Omega \times \Omega \to \ell(\mathbb{B},\mathbb{B})$ is a divided difference of order one. In addition to this, we assume that $u_* \in \Omega$ is a simple zero of F. At last, the inverse of operator F, $F'(u_*)^{-1} \in \ell(\mathbb{B},\mathbb{B})$.

(a_2) Let $w_0 : [0,\infty] \times [0,\infty) \to [0,\infty)$ be a increasingly continuous function with $w_0(0,0) = 0$, parameters $a \geq 0$ and $b > 0$ such that for each $u,y \in \Omega$

$$\left\| F'(u_*)^{-1} \left([u,y;F] - F'(u_*)\right) \right\| \leq w_0(\|u - u_*\|, \|y - u_*\|),$$
$$\|I + [u,u_*;F]\| \leq a,$$

and

$$\|[u,u_*;F]\| \leq b.$$

Set $\Omega_0 = \Omega \cap S(u_*,\rho_1)$, where ρ_1 exists and is given by (7.12).

(a_3) We assume that $w : [0,\rho_1) \times [0,\rho_1) \to [0,\infty)$ is a increasingly continuous $\forall x,y,\zeta,\eta \in \Omega_0$

$$\left\| F'(u_*)^{-1} \left([u,y;F] - [\zeta,\eta;F]\right) \right\| \leq w(\|u - \zeta\|, \|y - \eta\|) = \rho_3.$$

(a_4) $\bar{S}(u_*,R) \subseteq \Omega$, where $R = \max\{r, ar, br\}$, r is defined in (7.15) and ρ_2, ρ_4 exist and are given by (7.13) and (7.13), respectively.

(a_5) There exists $\bar{r} \geq r$ such that

$$w_0(0,\bar{r}) < 1 \text{ or } w_0(\bar{r},0) < 1.$$

Set $\Omega_1 = \Omega \cap \bar{S}(u_*,\bar{r})$.

Theorem 7.2. Under the hypotheses (A) further consider that $u_0 \in S(u_*, r) - \{u_*\}$. So, the proceeding assertions hold

$$\{u_p\} \subseteq S(u_*,r), \tag{7.20}$$

$$\lim_{p \to \infty} \{u_p\} = u_*, \tag{7.21}$$

$$\|y_p - u_*\| \leq g_1(\|u_p - u_*\|)\|u_p - u_*\| \leq \|u_p - u_*\| < r, \tag{7.22}$$

$$\|z_p - u_*\| \leq g_2(\|u_p - u_*\|)\|u_p - u_*\| \leq \|u_p - u_*\|, \tag{7.23}$$

and

$$\|u_{p+1} - u_*\| \leq g_3(\|u_p - u_*\|)\|u_p - u_*\| \leq \|u_p - u_*\|. \tag{7.24}$$

In addition, the u_* is the unique solution of $F(u) = 0$ in the set Ω_1 mentioned in hypothesis (a_5).

Proof. We first show items (7.20) – (7.24) by adopting mathematical induction. Since $p \in S(u_*, r) - \{u_*\}$ hold and by condition (a_2), we have

$$\|p + F(p) - u_*\| = \|(I + [p, u_*; F])(p - u_*)\|$$
$$\leq \|I + [p, u_*; F]\| \|p - u_*\|$$
$$\leq a\|p - u_*\|$$

and

$$\|F(p)\| = \|F(p) - F(u_*)\|$$
$$\leq \|[p, u_*; F](p - u_*)\|$$
$$\leq \|[p, u_*; F]\| \|p - u_*\|$$
$$\leq b\|p - u_*\|$$

so $p + F(p) - u_*$ and $F(p)$ belong in $\bar{S}(u_*, R)$. Then, for $u, y, \in S(u_*, r) - \{u_*\}$, and

$$\left\| F'(u_*)^{-1}([u, y; F] - F'(u_*)) \right\| \leq w_0\left(\|u - u_*\|, \|y - u_*\|\right)$$
$$\leq w_0(r, r) < 1,$$

so the Banach lemma on invertible operators [2–4,9] gives $[u, y; F]^{-1} \subset \ell(\mathbb{B}, \mathbb{B})$, and

$$\left\| [u, y; F]^{-1} F'(u_*) \right\| \leq \frac{1}{1 - w_0\left(\|u_0 - u_*\|, \|y - u_*\|\right)}. \tag{7.25}$$

It also follows that y_0 is defined.

Adopting (7.15), (7.16), (7.19) (for $i = 1$), (a_2), (a_3), (7.25) and y_0, we get

$$\|y_0 - u_*\|$$
$$= \left\| u_0 - u_* - [u_0 + F(u_0), u_0; F]^{-1} F(u_0) \right\|$$
$$= \left\| [u_0 + F(u_0), u_0; F]^{-1} \left([u_0 + F(u_0), u_0; F] - [u_0, u_*; F](u_0 - u_*)\right) \right\|$$
$$\leq \left\| [u_0 + F(u_0), u_0; F]^{-1} F'(u_*) \right\|$$
$$\quad \times \left\| F'(u_*)^{-1}\left([u_0 + F(u_0), u_0; F] - [u_0, u_*; F]\right) \right\| \|u_0 - u_*\| \tag{7.26}$$
$$\leq \frac{w\left(\|F(u_*)\|, \|u_0 - u_*\|\right) \|u_0 - u_*\|}{1 - w_0(a\|u_0 - u_*\|, \|u_0 - u_*\|)}$$
$$\leq \frac{w\left(b\|u_0 - u_*\|, \|u_0 - u_*\|\right) \|u_0 - u_*\|}{1 - w_0(a\|u_0 - u_*\|, \|u_0 - u_*\|)}$$
$$= g_1(\|u_0 - u_*\|) \|u_0 - u_*\| < \|u_0 - u_*\| < r,$$

so $y_0 \in S(u_*, r)$ (for $y_0 \neq u_*$) and (7.22) holds for $n = 0$.

$$\left\| F'(u_*) \left([u_0, y_0; F] - F'(u_*) \right) \right\| \leq w_0 \left(\|u_0 - u_*\|, \|y_0 - u_*\| \right)$$

$$\leq w_0 \left(\|u_0 - u_*\| \right), g_1(\|u_0 - u_*\|) \|u_0 - u_*\| \right)$$

$$\leq w_0 \left(r, g_1(r)r \right) < 1,$$

so $[u_0, y_0; F]^{-1} \in \ell(\mathbb{B}, \mathbb{B})$ and

$$\left\| [u_0, y_0; F]^{-1} F'(u_*) \right\| \leq \frac{1}{1 - w_0 \left(\|u_0 - u_*\|, g_1(\|u_0 - u_*\|) \|u_0 - u_*\| \right)}. \qquad (7.27)$$

It also follows that z_0 is well defined by the second substep of method (7.2) for $n = 0$. In particular, we have

$$z_0 = u_0 - \lambda [u_0 + F(u_0), u_0; F]^{-1} (F(u_0) + F(y_0)) - (1 - \lambda) [u_0 + F(u_0), u_0; F]^{-1} F(u_0)$$

$$= \left(u_0 - [u_0 + F(u_0), u_0; F]^{-1} F(u_0) \right) + [u_0 + F(u_0), u_0; F]^{-1} F(u_0)$$

$$- \lambda [u_0 + F(u_0), u_0; F]^{-1} F(u_0)$$

$$- (1 - \lambda) [u_0, y_0; F]^{-1} F(u_0) - \lambda [u_0 + F(u_0), u_0; F]^{-1} F(y_0)$$

$$= y_0 + (1 - \lambda) \left([u_0 + F(u_0), u_0; F]^{-1} - [u_0, y_0, F]^{-1} \right) F(u_0)$$

$$- \lambda [u_0 + F(u_0), u_0; F]^{-1} F(y_0)$$

$$(7.28)$$

Next, by (7.15), (7.19) (for $i = 2$) and (7.25)–(7.28), we get in turn that

$$\|z_0 - u_0\|$$

$$\leq \|y_0 - u_*\| + |1 - \lambda| \left\| [u_0 + F(u_0), u_0; F]^{-1} F'(u_*) \right\|$$

$$\times \left\| F'(u_*)^{-1} \left([u_0, y_0; F] - [u_0 + F(u_0), u_0; F] \right) \right\|$$

$$\times \frac{\left\| [u_0, y_0; F]^{-1} F'(u_*) \right\| \left\| F'(u_*)^{-1} [u_0, u_*; F] \right\| \Lambda}{\left(1 - w_0(a\Lambda, \Lambda) \right) \left(1 - w_0 \left(\Lambda, g_1(\Lambda)\Lambda \right) \right)}$$

$$+ |\lambda| \left\| [u_0 + F(u_0), u_0; F]^{-1} F'(u_*) \right\| \left\| F'(u_*)^{-1} [y_0, u_*; F] \right\| \|y_0 - u_*\|$$

$$\leq \left[g_1(\Lambda) + \frac{b|1 - \lambda| w \left(b\Lambda, (1 + g_1(\Lambda))\Lambda \right)}{\left(1 - w_0(a\Lambda, \Lambda) \right) \left(1 - w_0 \left(\Lambda, g_1(\Lambda)\Lambda \right) \right)} + |\lambda| \frac{bg_1(\Lambda)}{1 - w_0(a\Lambda, \Lambda)} \right] \Lambda$$

$$= g_2(\Lambda)\Lambda \leq \Lambda,$$

$$(7.29)$$

where

$$\Lambda = \|u_0 - u_*\|$$

so $z_0 \in S(u_*, r)$ (for $z_0 \neq u_*$) and (7.23) holds for $p = 0$.

We have by (7.15), (7.18), and (7.29)

$$\left\| F'(u_*)^{-1} \left([z_0 + F(u_0), z_0; F] - F'(u_*) \right) \right\| \leq w_0 \Big(b\|z_0 - u_*\|, \|z_0 - u_*\| \Big)$$

$$\leq w_0 \Big(bg_2(\Lambda)\Lambda, g_2(\Lambda)\Lambda \Big)$$

$$\leq w_0 \Big(bg_2(r)r, g_2(r)r \Big) < 1$$

So $[z_0 + F(z_0), z_0; F]^{-1} \in \ell(\mathbb{B}, \mathbb{B})$ and

$$\left\| [z_0 + F(z_0), z_0; F]^{-1} F'(u_*) \right\| \leq \frac{1}{1 - w_0 \Big(bg_2(\|u_0 - u_*\|)\Lambda, g_2(\Lambda)\Lambda \Big)}. \qquad (7.30)$$

It also follows that u_1 is well defined by (7.30) and the last substep of method (7.2) for $n = 0$. Then, as in (7.25) and (7.26) (for $z = 3$), and (7.30), we obtain in turn

$$\|u_1 - u_*\| \leq \frac{w(b\|z_0 - u_*\|, \|z_0 - u_*\|)\|z_0 - u_*\|}{1 - w_0(a\|z_0 - u_*\|, \|z_0 - u_*\|)}$$

$$\leq \frac{w\Big(bg_2(\Lambda)\Lambda, g_2(\Lambda)\Lambda \Big) g_2(\Lambda)\Lambda}{1 - w_0 \Big(ag_2(\Lambda)\Lambda, g_2(\Lambda)\Lambda \Big)} \qquad (7.31)$$

$$= g_3(\Lambda)\Lambda \leq \Lambda,$$

where, Λ is defined previously. So, $u_1 \in S(u_*, r)$ (for $u_1 \neq u_*$) and (7.24) holds for $n = 0$. Then, substituting u_0, y_0, z_0, u_1 by u_m, y_m, z_m, u_{m+1}, respectively. Hence, the induction for (7.30) and (7.22)–(7.24) is complete. Using the estimation

$$\|u_{m+1} - u_*\| < \alpha\|u_m - u_*\| < r, \qquad (7.32)$$

where $\alpha = g_3(\|u_0 - u_*\|) \in [0, 1]$, we deduce that $\lim_{m \to \infty} u_m = u_*$ and $u_{m+1} \in S(u_*, r)$.

Finally, we want to illustrate that the required solution is unique. Therefore, let $T = [u_*, y_*; F]$ for $y_* \in \Omega_1$ so that $F(y_*) = 0$. Then, by (a_2) and (a_5), we get

$$\|F'(u_*)^{-1}(T - F'(u_*))\| \leq w_0(0, \|u_* - y_*\|)$$

$$\leq w_0(0, \bar{r}) < 1,$$

so $T^{-1} \in \ell(\mathbb{B}, \mathbb{B})$. Finally, $u_* = y_*$ is deduced from $0 = F(u_*) - F(y_*) = T(u_* - y_*)$. □

Remark 7.1. Another way of defining functions g_i, h_i and radii r_i is as follows:

Let $\alpha = \max\{1, a\}$, $i = 1, 2, 3$. Then, as in (7.12)–(7.18), we shall have instead: Suppose that equation

$$w_0(\alpha t, t) = 1 \qquad (7.33)$$

has a smallest positive solution $\bar{\rho}_1$. Let $\bar{w} : [0, \bar{\rho}_1] \times [0, \bar{\rho}_1]$ be a increasingly continuous function with $\bar{w}(0,0) = 0$.

Let functions $\bar{g}_1$ and $\bar{h}_1$ be defined in the interval $[0, \bar{\rho}_1]$ by

$$\bar{g}_1(t) = \frac{\bar{w}(bt,t)}{1 - w_0(\alpha t, t)} \quad \text{and} \quad \bar{h}_1(t) = \bar{g}_1(t) - 1.$$

The $\bar{r}_1$ stands for the smallest positive root of $\bar{h}_1(t) = 0$ in $(0, \bar{\rho}_1)$. Moreover, define functions $\bar{g}_2$, $\bar{g}_3, \bar{h}_2$ and $\bar{h}_3$ on the closed interval $[0, \bar{\rho}_1]$ as follow:

$$\bar{g}_2(t) = \bar{g}_1(t) + \frac{b|1 - \lambda|w\left(bt, \left(1 + \bar{g}_1(t)\right)t\right)\bar{g}_1(t)}{\left(1 - w_0(\alpha t, t)\right)^2},$$

$$\bar{g}_3(t) = \frac{w\left(b\bar{g}_2(t)t, \bar{g}_2(t)t\right)\bar{g}_2(t)}{1 - w_0(\alpha t, t)},$$

$$\bar{h}_2(t) = \bar{g}_2(t) - 1$$

and

$$\bar{h}_3(t) = \bar{g}_3(t) - 1.$$

The $\bar{r}_2$ and $\bar{r}_3$ are serve as the minimal positive roots of $\bar{h}_2(t) = 0$ and $\bar{h}_3(t) = 0$ on closed interval $[0, \bar{\rho}_1]$, respectively. Then, Theorem 7.2 can be written by using the "bar" conditions and functions, with $\bar{r} = \min\{\bar{r}_i\}$.

Remark 7.2. The convergence of method (7.2) to u_* is established under the conditions of Theorem 7.1. But, the order convergence under the conditions of Theorem 7.2 can be established by using (COC) and (ACOC) (for the details, please see section 5).

7.3 APPLICATIONS

Here, we monitor the convergence conditions on three problems 7.1–7.3. We choose $[u, y; F] = \int_0^1 F'(y + \theta(u - y))d\theta$ in the examples. We can confirm the verification the hypotheses of Theorem 7.2 for the given choices of the "w" functions and parameters a and b.

Example 7.1. Here, we investigate the application of our results on Hammerstein integral equations (see [11, pp. 19–20]) for $\mathbb{B} = C[0,1]$ as follows:

$$F(u(s_1)) = u(s_1) - \frac{1}{5}\int_0^1 S(s_1, s_2)u(s_2)^3 ds_2 = 0, \quad u \in C[0,1], \ s_1, s_2 \in [0,1], \quad (7.34)$$

where

$$S(s_1, s_2) = \begin{cases} s(1 - s_2), & s \leq s_2, \\ (1 - s)s_2, & s_2 \leq s. \end{cases}$$

We use $\int_0^1 \phi(t)dt \simeq \sum_{k=1}^8 w_k \phi(t_k)$ in (7.34), where t_k and w_k are the abscissas and weights, respectively. Using $u(t_j)$ for u_j ($j = 1,2,3,...,8$), leads to

$$5u_i - 5 - \sum_{k=1}^8 a_{jk}u_k^3 = 0, \; j = 1,2,3...,8,$$

$$a_{jk} = \begin{cases} w_k t_k (1 - t_j), & k \leq j, \\ w_k t_j (1 - t_k), & j < k. \end{cases}$$

The values of t_k and w_k when $k = 8$, are illustrate in Table 7.1. Then, we have

$$u_* = (1.002096\ldots, \; 1.009900\ldots, \; 1.019727\ldots, \; 1.026436\ldots, \; 1.026436\ldots,$$
$$1.019727\ldots, \; 1.009900\ldots, \; 1.002096\ldots)^T.$$

So, we set $w_0(s_1, s_2) = w(s_1, s_2) = \frac{3}{80}(s_1 + s_2)$, $a = \frac{163}{80}$ and $b = \frac{83}{80}$. The radii for Example 7.1 are listed in Tables 7.2 and 7.3:

j	t_j	w_j
1	0.0198550717512318841582195 7...	0.0506142681451881295762656 7...
2	0.1016667612931866302042230 3...	0.1111905172266872352721780 0...
3	0.2372337950418355070911304 7...	0.1568533229389436436689811 0...
4	0.4082826787521750975302619 3...	0.1813418916891809914825752 2...
5	0.5917173212478249024697380 7...	0.1813418916891809914825752 2...
6	0.7627662049581644929088695 2...	0.1568533229389436436689811 0...
7	0.8983332338706813369795776 96...	0.1111905172266872352721780 0...
8	0.9801449282487681158417804 3...	0.0506142681451881295762656 7...

Table 7.1: Abscissas and weights for $k = 8$

Example 7.2. Here, we choose as integral equation [12, 13], for $\mathbb{B} = C[0, 1]$ as

$$\left[F(\mu)\right](\gamma_1) = \mu(\gamma_1) - \int_0^1 G(\gamma_1, \; \gamma_2)\left(\mu(\gamma_2)^{\frac{3}{2}} + \frac{\mu(\gamma_2)^2}{2}\right)d\gamma_2 = 0, \qquad (7.35)$$

λ	r_1	r_2	r_3	r
0	5.25452	3.87208	4.09301	3.87208
0.5	5.25452	4.26006	4.42602	4.26006
1	5.25452	5.25452	5.25452	5.25452

Table 7.2: Convergence radii for Example 7.1

λ	r_1	r_2	r_3	r
0	5.25452	3.67748	3.87626	3.67748
0.5	5.25452	4.07351	4.17413	4.07351
1	5.25452	5.25452	4.89162	4.89162

Table 7.3: Convergence radii for Example 7.1 with bar functions

λ	r_1	r_2	r_3	r
0	1.03137	0.502403	0.61211	0.502403
0.5	1.03137	0.61199	0.70738	0.61199
1	1.03137	1.03137	1.03137	1.03137

Table 7.4: Convergence radii for Example 7.2 with bar functions

where

$$G(\gamma_1, \gamma_2) = \begin{cases} (1-\gamma_2)\gamma_2, & \gamma_2 \le \gamma_1, \\ \gamma_1(1-\gamma_2), & \gamma_1 \le \gamma_2. \end{cases} \tag{7.36}$$

Since $\mathbb{B} = C[0,1]$ so, $F : C[0,1] \to C[0,1]$ is given as

$$\left[F(\mu)\right](\gamma_1) = \mu(\gamma_1) - \int_0^{\gamma_1} G(\gamma_1, \gamma_2) \left(\mu(\gamma_2)^{\frac{3}{2}} + \frac{\mu(\gamma_2)^2}{2} \right) d\gamma_2. \tag{7.37}$$

We get

$$\left\| \int_0^{\gamma_1} G(\gamma_1, \gamma_2) d\gamma_2 \right\| \le \frac{1}{8}. \tag{7.38}$$

Moreover,

$$\left[F'(\mu)\eta\right](\gamma_1) = \eta(\gamma_1) - \int_0^{\gamma_1} G(\gamma_1, \gamma_2) \left(\frac{3}{2}\mu(\gamma_2)^{\frac{1}{2}} + \mu(\gamma_2) \right) \eta(\gamma_2) d\gamma_2,$$

so $\mu_*(\gamma_1) = 0$, since $F'(\mu_*(\gamma_1)) = I$,

$$\left\| F'(\mu_*)^{-1} \left(F'(\mu) - F'(\eta) \right) \right\| \le \frac{1}{8} \left(\frac{3}{2}\|\mu - \eta\|^{\frac{1}{2}} + \|\mu - \eta\| \right). \tag{7.39}$$

Hence, we have

$$w_0(s,t) = w(s,t) = \frac{1}{16} \left[\frac{3}{2}(\sqrt{s} + \sqrt{t}) + s + t \right], \ a = \frac{53}{16}, \text{ and } b = \frac{37}{16}.$$

Therefore, our results can be utilized even though F' is not bounded on Ω. The radii for Example 7.2 are given in Table 7.4.

Example 7.3. We assume the following differential equations

$$
\begin{aligned}
q_1'(\mu) - q_1(\mu) - 1 &= 0 \\
q_2'(\eta) - (e-1)\eta - 1 &= 0 \\
q_3'(\theta) - 1 &= 0
\end{aligned}
\tag{7.40}
$$

characterizes the progress/movement of a molecule in 3D with $(\mu,\ \eta,\ \theta) \in \Omega$ for $q_1(0) = q_2(0) = q_3(0) = 0$. The required solution $v = (\mu,\ \eta,\ \theta)^T$ describes to $K :=$ $(q_1, q_2, q_3) : \Omega \to \mathbb{R}^3$ given as

$$
K(v) = \left(e^\mu - 1,\ \frac{e-1}{2}\eta^2 + \eta,\ \theta \right)^T = 0.
\tag{7.41}
$$

It follows from (7.41) that

$$
K'(v) = \begin{bmatrix} e^\mu & 0 & 0 \\ 0 & (e-1)\eta + 1 & 0 \\ 0 & 0 & 1 \end{bmatrix},
$$

which yields

$$
w_0(s,t) = \frac{1}{2}(e-1)(s+t),\ w(s,t) = \frac{1}{2}e(s+t),\ a = \frac{1}{2}(e+3),\ \text{and } b = \frac{1}{2}(e+1).
$$

We depicted the radii of Example 7.3 in Tables 7.5 and 7.6.

λ	r_1	r_2	r_3	r
0	0.1388596	0.921375	0.083356	0.083356
0.5	0.1388596	0.921375	0.086297	0.086297
1	0.1388596	0.1388596	0.1388596	0.1388596

Table 7.5: Convergence radii for Example 7.3

λ	r_1	r_2	r_3	r
0	0.1388596	0.0487471	0.1229551	0.0487471
0.5	0.1388596	0.0487471	0.1377815	0.0487471
1	0.1388596	0.1388596	0.1380780	0.1380780

Table 7.6: Convergence radii for Example 7.3 with bar functions

Example 7.4. By the example of section 2, for $\Omega = \mathbb{B} = \mathbb{R}$, $f(\xi) = 0$, we get

$$
w_0(s,t) = w(s,t) = \frac{96.66297}{2}(s+t),\ a = \frac{5}{2},\ \text{and } b = \frac{3}{2}.
$$

The radii of method (7.2) for Example 7.4 are listed in Tables 7.7 and 7.8.

λ	r_1	r_2	r_3	r
0	0.00344841	0.00239612	0.00256623	0.00239612
0.5	0.00344841	0.00267769	0.00280807	0.00267769
1	0.00344841	0.00344841	0.00344841	0.00344841

Table 7.7: Convergence radii for Example 7.4

λ	r_1	r_2	r_3	r
0	0.00344841	0.00225955	0.00246765	0.00225955
0.5	0.00344841	0.00225955	0.00246765	0.00225955
1	0.00344841	0.00344841	0.00334891	0.00344841

Table 7.8: Convergence radii for Example 7.4 with bar functions

7.4 APPLICATIONS WITH LARGE SYSTEMS

We choose $\lambda = 0$, $\lambda = 0.5$ and $\lambda = 1$ in our scheme (7.2), called by $(PS1)$, $(PS2)$ and $(PS3)$, respectively. Now, we compare our schemes with a 6th-order iterative methods suggested by Abbasbandy et al. [1] and Hueso et al. [14], among them we picked the methods $\left(\text{for } t_1 = -\frac{9}{4} \text{ and } s_2 = \frac{9}{8}\right)$, respectively, known as (AS) and (HS). Moreover, comparison of them has been done with the 6th-order iterative methods given by Wang and Li [25], among their method we chose expression, denoted by and (WS). At the last, we contrast (7.2) with sixth-order scheme given by Sharma and Arora [21], we pick expression known as (SM). The details of all the iterative expressions are given as follows:

Method AS:

$$y_j = u_j - \frac{2}{3} F'(u_j)^{-1} F(u_j),$$

$$z_j = u_j - \left[I + \frac{21}{8} F'(u_j)^{-1} F'(y_j) - \frac{9}{2} \left(F'(u_j)^{-1} F'(y_j) \right)^2 + \frac{15}{8} \left(F'(u_j)^{-1} F'(y_j) \right)^3 \right]$$
$$\times F'(u_j)^{-1} F(u_j),$$

$$u_{j+1} = z_j - \left[3I - \frac{5}{2} F'(u_j)^{-1} F'(y_j) + \frac{1}{2} \left(F'(u_j)^{-1} F'(y_j) \right)^2 \right] F'(u_j)^{-1} F(z_j).$$
$$\tag{7.42}$$

Scheme HS:

$$y_j = u_j - F'(u_j)^{-1} F(u_j),$$
$$H(u_j, y_j) = F'(u_j)^{-1} F(y_j), \ H(y_j, u_j) = F'(y_j)^{-1} F(u_j),$$
$$G_s(u_j, y_j) = s_1 I + s_2 H(y_j, u_j) + s_3 H(u_j, y_j) + s_4 H(y_j, u_j)^2, \tag{7.43}$$
$$z_j = u_j - G_s(u_j, y_j) F'(u_j)^{-1} F(u_j),$$
$$u_{j+1} = z_j.$$

where $s_1\ s_2, s_3$ and s_4 are real numbers.

Iterative method WS:

$$y_j = u_j - F'(u_j)^{-1}F(u_j),$$

$$z_j = y_j - \left[2I - F'(u_j)^{-1}F'(y_j)\right]F'(u_j)^{-1}F(y_j),$$ (7.44)

$$u_{j+1} = z_j - \left[2I - F'(u_j)^{-1}F'(y_j)\right]F'(u_j)^{-1}F(z_j).$$

Scheme SM:

$$y_j = u_j - \frac{2}{3}F'(u_j)^{-1}F(u_j),$$

$$z_j = u_j - \left[pI + F'(u_j)^{-1}F'(y_j)\left(qI + rF'(u_j)^{-1}F'(y_j)\right)\right]F'(u_j)^{-1}F(u_j),$$ (7.45)

$$u_{j+1} = z_j - \left[\frac{5}{2}I - \frac{3}{2}F'(u_j)^{-1}F'(y_j)\right]F'(u_j)^{-1}F(z_j),$$

where $p = \frac{23}{8}$, $q = -3$ and $r = \frac{9}{8}$.

The (j), $(\|F(u_j)\|)$, $\|u_{j+1} - u_j\|$, and $\rho^* \approx \dfrac{\log\left[\|u_{j+1}-u_j\|/\|u_j-u_{j-1}\|\right]}{\log\left[\|u_j-u_{j-1}\|/\|u_{j-1}-u_{j-2}\|\right]}$ stands for index of iteration, absolute residual errors in the function F, error between two successive iterations and computational convergence order, receptively. There values are listed in the Tables. Moreover, the quantity η is the final obtained value of $\dfrac{\|u_{j+1}-u_j\|}{\|u_j-u_{j-1}\|^6}$.

The estimation of all the above parameters have been calculated by Mathematica-9. For minimizing the round-off errors, we have chosen multiple precision arithmetic with 1000 digits of mantissa. The term b_1 $(\pm b_2)$ symbolizes the $b_1 \times 10^{(\pm b_2)}$ in all mentioned tables. We adopted the command "AbsoluteTiming[]" in order to calculate the CPU time. We run our programs three times and depicted the average CPU time in Table 7.9, also one can observe the times used for each iterative method, where, we want to point out that for big size problems the method $PS1$ uses the minimum time, so it is being very competitive. The configuration of the used computer is given below:

Processor: Intel(R) Core(TM) i7-4790 CPU @ 3.60GHz
Made: HP
RAM: 8:00GB
System type: 64-bit-Operating System, x64-based processor.

Example 7.5. Here, we deal with a boundary value problem from Ortega and Rheinboldt [11], given by

$$y'' = \frac{y^3 + 6y' + 1}{2} - \frac{3}{2-x}, \quad y(0) = 0,\ y(1) = 1.$$ (7.46)

We assume

$$u_0 = 0 < u_1 < u_2 < u_3 < \cdots < u_p, \text{ where } u_{p+1} = u_p + h,\ h = \frac{1}{p},$$ (7.47)

partition of the interval $[0, 1]$ and $y_0 = y(u_0) = 0$, $y_1 = y(u_1)$, ..., $y_{n-1} = y(u_{n-1})$, $y_p = y(u_p) = 1$.

Next, we discretize expression (7.46) by adopting following numerical formula for derivatives

$$y'_j = \frac{y_{j+1} - y_{j-1}}{2h}, \quad y''_j = \frac{y_{j-1} - 2y_j + y_{j+1}}{h^2}, \quad j = 1, 2, \ldots, p-1,$$

which leads to

$$y_{j+1} - 2y_j + y_{j-1} - \frac{h^2}{2}y_j'^3 - \frac{3}{2}h(y_{k+1} - y_{k-1}) - \frac{3}{2-u_j}h^2 - \frac{1}{h^2} = 0, \quad j = 1, 2, \ldots, p-1,$$

$(p-1) \times (p-1)$ system of nonlinear equations.

For specific value of $p = 7$, we have a 6×6 system and the required solution is

$$u_* = \left(0.07654\ldots, 0.165873\ldots, 0.27152\ldots, 0.39845\ldots, 0.55388\ldots, 0.74868\ldots\right)^T.$$

The computational estimations are on the basis of initial approximation

$$y_j^{(0)} = \left(\frac{3}{2}, \frac{3}{2}, \frac{3}{2}, \frac{3}{2}, \frac{3}{2}, \frac{3}{2}\right)^T$$

.

Example 7.6. The classical 2D Bratu problem [16] is given by

$$u_{\mu\mu} + u_{\theta\theta} + Ce^u = 0,$$

$$\Omega = \left\{(\mu, \theta) \in 0 \le \mu \le 1, 0 \le \theta \le 1\right\}, \text{ with boundary hypotheses } u = 0 \text{ on } \Omega.$$
$$(7.48)$$

By adopting finite difference discretization, we can reduce the above PDE (7.48) to a nonlinear system. For this purpose, we denote $\Delta_{i,j} = u(\mu_i, \theta_j)$ as numerical solution at the grid points of the mesh. In addition to this, M_1 and M_2 stand for the number of steps in the directions of μ and θ, respectively. The h and k called as the respective step sizes in the directions of μ and θ. Adopt the following central difference formula to $u_{\mu\mu}$ and $u_{\theta\theta}$

$$u_{\mu\mu}(u_i, \theta_j) = \frac{\Delta_{i+1,j} - 2\Delta_{i,j} + \Delta_{i-1,j}}{h^2}, \quad C = 0.1, \ \theta \in [0,1], \quad (7.49)$$

leads to us

$$\Delta_{i,j+1} + \Delta_{i,j-1} - \Delta_{i,j} + \Delta_{i+1,j} + \Delta_{i-1,j} + h^2 C \exp\left(\Delta_{i,j}\right) \quad (7.50)$$

where,

$$i = 1, 2, 3, \ldots, \text{ and } M_1, j = 1, 2, 3, \ldots, M_2$$

For obtaining a large system of 100×100, we choose $M_1 = M_2 = 11$, $C = 0.1$ and $h = \frac{1}{11}$. The numerical results are listed in the Table based on the initial guess $u_0 = 0.1\left(\sin(\pi h i)\sin(\pi h j)\right)^T$, $i = j = 10$.

Example 7.7. Let us consider the following nonlinear system

$$F(x) = \begin{cases} u_j^2 u_{j+1} - 1 = 0, \ 1 \le j \le p-1, \\ x_p^2 u_1 - 1 = 0. \end{cases} \tag{7.51}$$

For specific value $p = 200$, we have 200×200 system, and chose the following starting point

$$x^{(0)} = (1.25, \ 1.25, \ 1.25, \ \ldots, \ 1.25)^T.$$

The $u_* = (1, \ 1, \ 1, \ \ldots, \ 1)^T$ is the required solution of system 7.51. The numerical results are given in the Table.

Methods	Ex. 7.5	Ex. 7.6	Ex. 7.7	Total time	Average time
AS	0.465330	210.079553	356.906591	567.451474	189.1504913
HS	0.583412	189.541919	366.511753	556.637084	185.5456947
WS	0.274193	128.377322	182.956711	311.608226	103.8694087
SM	1.130812	126.641140	401.627979	529.399931	176.4666437
$PS1$	0.101071	120.094370	52.204957	172.400398	57.46679933
$PS2$	0.100071	117.901198	52.146903	170.148172	56.71605733
$PS3$	0.100083	117.923227	51.972773	169.996083	56.665361

Table 7.9: CPU time of different methods on Examples 7.5–7.7

(According to CPU time, method $PS3$ is taking the lowest time for executing the results. All the other schemes AS, HS, SM and SM consuming at least double CPU timing as compare to our methods namely $PS1, PS2$ and $PS3$. So, we conclude that our methods provide results faster than the other existing methods.)

Remark 7.3. On the basis of the tables, we conclude that our methods namely $PS1, PS2$ and $PS3$ perform better in the contrast of existing schemes AS, HS, SM and SM on the basis of residual errors, errors between two consecutive iterations and asymptotic error constant. In addition, our methods also demonstrate the stable computational order of convergence. Finally, we concluded that our methods not only perform better than existing methods in numerical results but also take half of the CPU time in contrast to other existing methods (results can be easily found in Table 7.9).

7.5 EXERCISES

1. Solve the initial value problem

$$\dot{x}(t) = 1 + \cos(x(t)), x(0) = 0.$$

2. The predator-prey population models describe the interaction of a prey population X and a predator population Y. Assume that their interaction is modeled by the system of ordinary differential equations

$$\dot{x} = x - \frac{1}{4}x^2 - \frac{1}{10}xy + 1$$

$$\dot{y} = -\frac{1}{4}y + \frac{1}{7}xy + \frac{2}{3}.$$

(Assume $x(0) = y(0) = 0$). Solve the system.

3. Solve the initial value problem

$$\dot{x} = (x(t) + t, x(0) = 1)$$

by the iteration method (7.2). Perform 3 steps.

4. Solve the Fredholm-type integral equation

$$x(t) = \sum_0^1 \frac{t \cdot s}{10} ds + 1$$

by the iteration method (7.2). Perform 3 steps.

5. Solve the Volterra-type integral equation

$$x(t) = \sum_0^t \frac{t \cdot s}{10} ds + 1$$

by the iteration method. Perform 3 steps.

6. Solve equation

$$x = \frac{\sin x}{2} + 1$$

by using algorithm (7.2).

7. Repeat the previous problem for equation

$$x = \frac{\sin x + \cos 2x}{10}.$$

REFERENCES

1. Abbasbandy, S., Bakhtiari, P., Cordero, A., Torregrosa, J.R., and Lotfi, T. 2016. New efficient methods for solving nonlinear systems of equations with arbitrary even order. *Appl. Math. Comput.*, 287:287–288.
2. Alarcón, V., Amat, S., Busquier, S., and López, D.J. 2008. A Steffensens type method in Banach spaces with applications on boundary-value problems. *J. Comput. Appl. Math.*, 216. 1:243–250.
3. Amat, S., Bermudez, C., Hernández-Verón, M.A., and Martínez, E. 2016. On an efficient k-step iterative method for nonlinear equations. *J. Comput. Appl. Math.*, 302:258–271.
4. Argyros, I.K. 2008. *Convergence and Applications of Newton-type Iterations*. New York: Springer-Verlag.
5. Argyros, I.K., and George, S. 2019. *Mathematical Modeling for the Solution of Equations and Systems of Equations with Applications*. Volume III. NewYork: Nova Publishers.
6. Argyros, I.K., and Hillout S. 2012. Weaker conditions for the convergence of Newton's method. *J. Complexity*. 28. 3:364–387.
7. Argyros, I.K., and Magrenan, A.A. 2018. *A Contemporary Study of Iterative Methods*. NY: Academy Press, Elsevier.
8. Behl, R., Argyros, I.K., and and Machado, J.A.T. 2020. Ball comparison between three sixth order methods for Banach space valued operators. *Math.*, 8. 5. *https://doi.org/*10.3390/*math*8050667
9. Cordero, A., and Torregrosa, J.R. 2015. Low-complexity root finding iteration functions with no derivatives of any order of convergence. *J. Comput. Appl. Math.*, 275: 502–515.
10. Džuníc, J., and Petkovíc, M.S. 2012. A cubically convergent Steffensen-like method for solving nonlinear equations. *Appl. Math. Let.*, 25. 11:1881–1886.
11. Ezquerro, J.A., and Hernández, M.A. 2014. How to improve the domain of starting points for Steffensen's method. *Stud. Appl. Math*,132. 4:354–380.
12. Ezquerro, J.A., and Hernández, M.A. 2009. New iterations of R-order four with reduced computational cost. *BIT Numer. Math.*, 49: 325– 342.
13. Hernández, M.A., and Martinez, E. 2015. On the semilocal convergence of a three steps Newton-type process under mild convergence conditions. *Numer. Algor.*, 70: 377–392.
14. Hueso, J.L., Martínez, E., and Teruel, C. 2015. Convergence, efficiency and dynamics of new fourth and sixth order families of iterative methods for nonlinear systems. *J. Comput. Appl. Math.*, 275:412–420.
15. Iliev, A., and Kyurkchiev, N. 2010. *Nontrivial Methods in Numerical Analysis: Selected Topics in Numerical Analysis*. Saarbrucken: LAP LAMBERT Academic Publishing.
16. Kapania, R.K. 1990. A pseudo-spectral solution of 2-parameter Bratu's equation. *Comput. Mech.*, 6:55–63.
17. Ortega, J.M., and Rheinboldt, W.C. 1970. *Iterative Solution of Nonlinear Equations in Several Variables*. New York: Academic Press.
18. Potra, F.A., and Pták, V. 1984. *Nondiscrete Induction and Iterative Processes*. Volume 103, Pitman Advanced Publishing program.
19. Rheindoldt, W. C. 1978. An adaptive continuation process for solving systems of equations. *Polish Academy of Science, Banach Center Publications*, 3. 1:129–142.
20. Sauer, T. 2012. *Numerical Analysis, second edition*. USA: Pearson.
21. Sharma, J.R., and Arora, H. 2014. Efficient Jarratt-like methods for solving systems of nonlinear equations. *Calcolo*, 51:193–210.
22. Sharma, J.R., Ghua, R.K., and Sharma, R. 2013. An efficient fourth-order weighted Newton method for systems of nonlinear equations. *Numer. Algo.*, 62. 2:307–325.

23. Simpson, R.B. 1975. A method for the numerical determination of bifurcation states of nonlinear systems of equations. *SIAM J. Numer. Anal.*, 12:439–451.
24. Traub, J.F. 1982. *Iterative Methods for the Solutions of Equations*. Amer. Math. Soc.
25. Wang, X., and Li, Y. 2017. An Efficient Sixth Order Newton Type Method for Solving Nonlinear Systems. *Algorithms*, 10. 45:1–9.

8 Unified Local Convergence of $k-$Step Solvers

There is a plethora of $k-$step solvers for equations involving operators on Banach spaces. Their convergence is estimated by adopting hypotheses in high order derivatives which are not on these iterative solvers. In addition, no computable error bounds or information on the uniqueness of the solution based on Lipschitz-type functions are given. Moreover, the choice of the initial guess is like shooting in dark. Finally, the criteria of convergence differ from solver to solver, so no comparison can be made between their convergence domains. The novelty of our work that we address these problems by introducing a generalized $k-$step solver containing all previous $k-$step solvers. Moreover, we address all previously stated, problems utilizing weaker conditions needing only the continuity of the operator involved. Applications are also presented where we test the convergence criteria.

Let $\mathbb{B}_1$, $\mathbb{B}_2$ be two Banach spaces and $\Omega \subset \mathbb{B}_1$ be an open and non void set. The $\mathbb{S}(x,\lambda)$, $\bar{\mathbb{S}}(x,\lambda)$, stand for open and closed balls in $\mathbb{B}_1$ and of radius $\lambda > 0$. Define $\ell(\mathbb{B}_1,\mathbb{B}_2) := \{ M : \mathbb{B}_1 \to \mathbb{B}_2, M \text{ is linear and bounded} \}$.

We solve

$$F(x) = 0, \tag{8.1}$$

provided, $F : \Omega \to \mathbb{B}_2$ is continuous using iterative solvers converging a solution x_*. The task of finding x_* is very challenging and extreme importance in analysis, since applications from different disciplines such as Mathematical, Biology, Chemistry, Economics, Physics, Engineering (see the numerical examples) are converted to an equation of the type (8.1) by utilizing mathematical modeling [1–21]. Then, suitable closed form of solutions can rarely be found. Recently, to elevate the convergence order authors develop $k-$step iterative solvers which converge to x_* under certain assumptions which involve a Fréchet-differentiable operator defined on a subset Ω of a Banach space $\mathbb{X}$ with values in a Banach space $\mathbb{Y}$. Taylor series expansions, and derivatives of high order are used in the convergence analysis although these derivatives do not occur in the expressions. Moreover, no computable error bounds or any information on the uniqueness of the solution are provided using Lipschitz-type functions. Furthermore, the convergence criteria can not be compared, since they are based on different hypotheses. These problems limit the applicability of the method. Let us consider a motivational example, we assume the following function F on $\mathbb{B}_1 = \mathbb{B}_2 = \mathbb{R}$, $\mathbb{D} = [-\frac{5}{2}, \frac{1}{2}]$ such as:

$$F(\kappa) = \begin{cases} \kappa^3 \ln \kappa^2 + \kappa^5 - \kappa^4, & \kappa \neq 0 \\ 0, & \kappa = 0 \end{cases}. \tag{8.2}$$

We yield

$$F'(\kappa) = 3\kappa^2 \ln \kappa^2 + 5\kappa^4 - 4\kappa^3 + 2\kappa^2,$$

DOI: 10.1201/9781003128915-8

$$F''(\kappa) = 6\kappa \ln \kappa^2 + 20\kappa^3 - 12\kappa^2 + 10\kappa,$$

$$F'''(\kappa) = 6\ln \kappa^2 + 60\kappa^2 - 12\kappa + 22.$$

We identify that $F'''(\kappa)$ is not bounded in $\mathbb{D}$. Therefore, results requiring the existence of $F''(\kappa)$ or higher cannot apply for studying the convergence of (8.3).

To address all these problems, we unify $k-$step methods by studying generalized $k-$step method $G_k SM$ given as

$$x_\sigma^{(0)} = g(x_\sigma),$$
$$x_\sigma^{(1)} = x_\sigma^{(0)} - L_\sigma F(x_\sigma^{(0)}),$$

$$\cdot$$
$$\cdot$$
$$\cdot$$

$$x_\sigma^{(j)} = x_\sigma^{(j-1)} - L_\sigma x_\sigma^{(j-1)}, \tag{8.3}$$

$$\cdot$$
$$\cdot$$
$$\cdot$$

$$x_\sigma^{(k-1)} = x_\sigma^{(k-2)} - L_\sigma F(x_\sigma^{(k-2)}),$$
$$x_{\sigma+1} = x_\sigma^{(k)} = x_\sigma^{(k-1)} - L_\sigma F(x_\sigma^{(k-1)}),$$

where $L : \Omega \to \Omega \to \ell(\mathbb{B}_2, \mathbb{B}_1)$, and $g, \varphi : \mathbb{B}_1 \to \mathbb{B}_1$ are iteration functions.

$G_k SM$ specializes to existing $k-$step iterative methods by choosing L, g and φ approximately:

1.$k-$step Newton's method

$$y_\sigma = g(x_\sigma) = x_\sigma - L_\sigma F(x_\sigma),$$
$$\varphi(x_\sigma) = y_\sigma$$
and
$$L_\sigma = L(x_\sigma, y_\sigma) = F'(x_\sigma)^{-1}.$$

2.$k-$step secant method:

$$g(x_\sigma) = x_\sigma - [x_{\sigma-1}, x_\sigma; F]^{-1} F(x_\sigma),$$
$$L_\sigma = [x_{\sigma-1}, x_\sigma; F]^{-1}$$

3.$k-$step Steffensen-type methods

$$g(x_\sigma) = x_\sigma - [x_\sigma - \lambda_\sigma F(x_\sigma), x_\sigma + \mu_\sigma F(x_\sigma); F]^{-1} F(x_\sigma),$$
$$\{\lambda_\sigma\}, \{\mu_\sigma\} \subseteq \mathbb{R}, \mathbb{B}_1 = \mathbb{B}_2$$

4.$k-$step Stirling's method

$$g(x_\sigma) = x_\sigma - F'\left(x_\sigma - F(x_\sigma)\right)^{-1} F(x_\sigma),$$
$$\text{and } \mathbb{B}_1 = \mathbb{B}_2,$$

5.$k-$step Newton-like method:

$$g(x_\sigma) = x_\sigma - A_\sigma^{-1} F(x_\sigma),$$
$$L_\sigma = A_\sigma^{-1},$$
$$\text{and } A = A(\cdot) : \Omega \to \ell(\mathbb{B}_1, \mathbb{B}_2).$$

Therefore, it is extremely important to look at the local convergence of (8.3). The semi-local convergence can also be given along the same lines.

Notice that, we have a huge amount of iteration functions [1–19] used for the solutions of equations. It is mentioned in these articles that x_0 should be enough close to x_* for convergence to be realized. But, nothing is said about how close x_0 should be to x_* for convergence. Hence, the radius of the ball convergence is needed. The counter example (8.2) can also be used to other methods which were proposed in [1–21]. Local convergence results are important, since they show how difficult it is to chose the starting guesses.

We use only hypothesis on the continuity of function F that employs method (8.3). We also develop a scheme based on Lipschitz constants to find convergence radii as well as the error estimations. We talk the range of starting guess x_* to guarantee the convergence of (8.3).

We coordinated the work as pursue. The section 8.1 is concerned to analysis of the local convergence of scheme (8.3). In section 8.2, several experiments are discussed.

8.1 LOCAL CONVERGENCE

Let us develop some maps and parameters to be used later as follows. Set $M = [0, \infty)$. Consider, $h : M \to M$ is a continuous and increasing function.

Assume equation

$$h(\zeta) = 1, \tag{8.4}$$

has a minimal positive solution ρ_h. Set $M_h = [0, \rho_h)$. Let $\psi : M_h \to M$ be continuous, and increasing. Define real functions α, and γ_m, $m = 0, 1, 2, \ldots, k$ on the interval M_h by

$$\alpha(\zeta) = \psi(h(\zeta)\zeta),$$

and

$$\gamma_m(\zeta) = \alpha^m(\zeta) h(\zeta).$$

Assume equation

$$\gamma_m(\zeta) = 1, \tag{8.5}$$

has a minimal positive solution ρ_m. Set $\bar{\rho}_m = \min\{\rho_m\}$.

Define a parameter ρ as

$$\rho = \min\{\rho_h, \bar{\rho}_m\}. \tag{8.6}$$

It follows

$$0 \le h(\zeta) < 1 \tag{8.7}$$

and

$$0 \le \gamma_m(\zeta) < 1 \tag{8.8}$$

for all $t \in [0, \rho)$.

The following conditions (A) shall be used

(a_1) $F : \Omega \to \mathbb{B}_2$, $g : \Omega \to \mathbb{B}_2$, $\varphi : \Omega \to \mathbb{B}_1$ are continuous operators. In addition, $L : [\cdot, \cdot; F] \to \ell(\mathbb{B}_2, \mathbb{B}_1)$, and $F(x_*) = 0$ for some $x_* \in \Omega$.

(a_2) Parameters ρ_h, ρ_m given previously exist.

(a_3) Function $h : M \to M$ is continuous, increasing with

$$\|g(x) - x_*\| \le h(\|x - x_*\|)\|x - x_*\|, \text{ for all } x \in \Omega.$$

Let $\Omega_0 = \Omega \cap U(x_*, \rho_h)$.

(a_4) Function $\psi : M_h \to M$ is continuous, increasing with

$$\|x - x_* - L(x, \varphi(x))F(x)\| \le \psi(\|x - x_*\|)\|x - x_*\|, \text{ for all } x \in \Omega_0.$$

(a_5) $\bar{U}(x_*, \rho) \subset \Omega$ with ρ defined in expression (8.6).

Theorem 8.1. Assume condition (A), and choose $x_0 \in U(x_*, \rho) - \{x_*\}$. Then, the obtained sequence $\{x_\sigma\}$ by scheme (8.3) is well defined in $U(x_*, \rho)$ for each $\sigma = 0, 1, 2, 3, \ldots$ and $\lim_{n \to \infty} x_n = x_*$ solving $F(x) = 0$ so that

$$\|x_\sigma^{(0)} - x_*\| \le h(\|x_\sigma - x_*\|)\|x_\sigma - x_*\| \le \|x_\sigma - x_*\| < \rho, \tag{8.9}$$

$$\begin{aligned}
\|x_\sigma^{(m)} - x_*\| &\le \alpha^m(\|x_\sigma - x_*\|)\|x_\sigma^{(0)} - x_*\| \\
&\le \alpha^m(\|x_\sigma - x_*\|)h(\|x_\sigma - x_*\|)\|x_\sigma - x_*\| \\
&= \gamma_m(\|x_\sigma - x_*\|) \le \|x_\sigma - x_*\|
\end{aligned} \tag{8.10}$$

and

$$\begin{aligned}
\|x_{\sigma+1} - x_*\| &= \|x_\sigma^{(k)} - x_*\| \le \alpha^k(\|x_\sigma - x_*\|)\|x_\sigma^{(0)} - x_*\| \\
&= \gamma_k(\|x_\sigma - x_*\|)\|x_\sigma - x_*\| \le \|x_\sigma - x_*\|,
\end{aligned} \tag{8.11}$$

where α and γ_m are given previously.

Proof. We use mathematical induction. First, we show (8.9). Indeed, using the choice $x_0 \in U(x_*, \rho) - \{x_*\}$, (8.3), (8.6), (8.7), and (a_1)–(a_3) to obtain

$$\|x_\sigma^{(0)} - x_*\| = \|g(x_\sigma) - x_*\| \le h(\|x_\sigma - x_*\|)\|x_\sigma - x_*\| \le \|x_\sigma - x_*\| < \rho, \tag{8.12}$$

which shows (8.9) and $x_\sigma^{(0)} \in U(x_*,\rho)$. By (8.3), (a_4), (8.6), (8.8), and (8.12), we get for $x = x_\sigma^{(0)}$

$$
\begin{aligned}
\|x_\sigma^{(1)} - x_*\| &\leq \psi(\|x_\sigma^{(0)} - x_*\|)\|x_\sigma^{(0)} - x_*\|, \\
&\leq \psi\left(h(\|x_\sigma - x_*\|)\|x_\sigma - x_*\|\right)\|x_\sigma^{(0)} - x_*\|, \\
&\leq \alpha(\|x_\sigma - x_*\|)h(\|x_\sigma - x_*\|)\|x_\sigma - x_*\|, \\
&= \gamma_1(\|x_\sigma - x_*\|)\|x_\sigma - x_*\| \leq \|x_\sigma - x_*\|.
\end{aligned}
$$

Similarly, we have

$$
\begin{aligned}
\|x_\sigma^{(2)} - x_*\| &\leq \alpha(\|x_\sigma - x_*\|)\|x_\sigma^{(1)} - x_*\|, \\
&\leq \alpha^2(\|x_\sigma - x_*\|)\|x_\sigma^{(0)} - x_*\|, \\
&= \gamma_2(\|x_\sigma - x_*\|)\|x_\sigma - x_*\| \leq \|x_\sigma - x_*\|,
\end{aligned}
$$

$$\cdots$$

$$
\|x_\sigma^{(m)} - x_*\| \leq \alpha^m(\|x_\sigma - x_*\|)\|x_\sigma^{(m)} - x_*\| = \gamma(\|x_\sigma - x_*\|)\|x_\sigma - x_*\| \leq \|x_\sigma - x_*\|,
$$

and

$$
\begin{aligned}
\|x_{\sigma+1} - x_*\| = \|x_\sigma^{(k)} - x_*\| &\leq \alpha^k(\|x_\sigma - x_*\|)\|x_\sigma^{(0)} - x_*\| \\
&\leq \gamma_k(\|x_\sigma - x_*\|)\|x_\sigma - x_*\| \leq \|x_\sigma - x_*\|,
\end{aligned}
\tag{8.13}
$$

which show (8.10), (8.11) and $x_{\sigma+1} \in U(x_*,\rho)$. The induction for (8.9)–(8.11) is complete. Then, by

$$
\|x_{\sigma+1} - x_*\| = a\|x_\sigma - x_*\| < \rho,
\tag{8.14}
$$

where $a = \gamma_k(\|x_0 - x_*\|) \in [0,1)$, so $\lim_{\sigma \to \infty} x_\sigma = x_*$, with $x_{\sigma+1} \in U(x_*,\rho)$. $\qquad\square$

Concerning the uniqueness of the solution:

Proposition 8.1. Under the condition of Theorem 8.1 further assume:
(a_6) There exist $D \in \ell(\mathbb{B}_1,\mathbb{B}_2), E : \Omega \times \Omega \to \ell(\mathbb{B}_1,\mathbb{B}_2)$ such that $D^{-1} \in \ell(\mathbb{B}_2,\mathbb{B}_1)$, and

$$
\|D^{-1}(D - E)\| < 1,
$$

where $E(x,y)(x - y) = F(x) - F(y)$ for each $x,y \in U(x_*,\rho)$, and for some $\bar{\rho} \geq \rho_0$. Hence, x_* is the only solution of $F(x) = 0$ in $U(x_*,\bar{\rho})$.

Proof. Assume that there exists $y_* \in U(x_*,\bar{\rho})$. Then, by (a_6), we have that

$$
E^{-1} \in \ell(\mathbb{B}_2,\mathbb{B}_1).
$$

Then, from

$$
0 = F(x_*) - F(y_*) = E(x_*,y_*)(x_* - y_*),
$$

we deduce $x_* = y_*$. $\qquad\square$

Remark 8.1. We show how we select real functions in the case of choice 5 given the introduction.

Assume:

(a_7) there exist $B : \Omega \times \Omega \to \ell(\mathbb{B}_1, \mathbb{B}_2)$ with $B^{-1} \in \ell(\mathbb{B}_2, \mathbb{B}_1)$, function $q : [0, \infty) \to [0, \infty)$ with

$$q(\zeta) = 1$$

has a minimal positive solution ρ_q,

$$\left\| B^{-1} \left(B - A(x, \varphi(x)) \right) \right\| \leq q(\|x - x_*\|)$$

and

$$\left\| B^{-1} \left[A(x, \varphi(x))(x - x_*) - F(x) \right] \right\| \leq \psi(\|x - x_*\|)\|x - x_*\|,$$

for all $x \in U(x_*, \rho_q)$.

Notice that it follows from these conditions that $A^{-1} \in \ell(\mathbb{B}_2, \mathbb{B}_1)$, and

$$\|A^{-1}B\| \leq \frac{1}{1 - q(\|x - x_*\|)}. \tag{8.15}$$

Moreover, from the estimate

$$\begin{aligned}
x_\sigma^{(m)} - x_* &= x_\sigma^{(m-1)} - x_* - A_\sigma^{-1} F\left(x_\sigma^{m-1}\right) \\
&= (A_\sigma^{-1} B_\sigma) \left[B_\sigma^{-1} \left(A_\sigma(x_\sigma^{(m-1)}) - x_* \right) - F\left(x_\sigma^{(m-1)}\right) \right]
\end{aligned} \tag{8.16}$$

the conclusion of the Theorem 8.1 in this stronger setting, if we replace (a_4) by (a_7),

$$\psi(\zeta) \text{ by } \psi_1(\zeta) = \frac{\psi(\zeta)}{1 - q(\zeta)},$$

and use $\rho = \min\{\rho_p, \rho_q, \rho_m\}$.

By further specializing A and g and φ (see choices (1)–(4) in the introduction) the choices of real functions h, ψ and q can be determined.

Hence, we now have uniform way of comparing the aforementioned five and other methods. Results for single step methods are immediately derived from (8.3), if we take $k = 0$ (i.e. restrict only in the first substep of (8.3)), and set $\psi = 0$ (see also Applications).

8.2 APPLICATIONS

The theoretical results developed in the previous sections are tested in this section. Considering three real life problems and three standard non-linear problems which are illustrated in the examples. We display our finding in the tables. Additionally, we obtain the computational order of convergence approximated by adopting the pursuing techniques

$$\xi = \frac{\ln \frac{\|x_{\sigma+1} - x^*\|}{\|x_\sigma - x^*\|}}{\ln \frac{\|x_\sigma - x^*\|}{\|x_{\sigma-1} - x^*\|}}, \quad \text{for each } \sigma = 1, 2, \ldots \tag{8.17}$$

or the approximate computational order of convergence (ACOC) [16]

$$\xi^* = \frac{\ln \frac{\|x_{\sigma+1}-x_\sigma\|}{\|x_\sigma-x_{\sigma-1}\|}}{\ln \frac{\|x_\sigma-x_{\sigma-1}\|}{\|x_{\sigma-1}-x_{\sigma-2}\|}}, \qquad \text{for each } \sigma = 2,3,\ldots \qquad (8.18)$$

We adopt $\varepsilon = 10^{-100}$ as the error tolerance and the terminating criteria to solve non-linear system or scalar equations are: $(i)\|x_{\sigma+1}-x_\sigma\| < \varepsilon$ and $(ii)\|F(x_{\sigma+1})\| < \varepsilon$. Computations are performed with the package *Mathematica* 9 with multiple precision arithmetic.

For simplicity, we apply the results for k−step Newton's method with choice provided in (8.1). Moreover, if for some $L_0 > 0, L > 0$, we have

$$\left\|F'(x_*)^{-1}\left(F'(x)-F'(x^*)\right)\right\| \le L_0\|x-x_*\| \text{ for each } x \in \Omega$$

and

$$\left\|F'(x_*)^{-1}\left(F'(y)-F'(x^*)\right)\right\| \le L_0\|y-x_*\| \text{ for each } x,y \in \Omega_0.$$

It then follows from (a_3) and (a_4) that we can choose $h = \psi$. Then, we can choose $h(\zeta) = \psi(\zeta) = \frac{Lt}{2(1-L_0 t)}$ in all examples 8.1–8.6. Notice that in this case, and if $h(\zeta) = 1$, for some $t \ge 0$ then $\alpha(\zeta) = \psi(h(\zeta)\zeta) = \psi(\zeta)$, so $\rho = \rho_h = \bar{\rho}_m$ provided that $h(\zeta) = 1, (\zeta \ge 0)$. It is worth noticing that in this case the radii are independent of m.

Example 8.1. The trajectory of an electron moving between two parallel plates is

$$y(\zeta) = y_0 + \left(v_0 + e\frac{E_0}{m\omega}\sin(\omega\zeta_0+\alpha)\right)(\zeta-\zeta_0)$$

$$+ e\frac{E_0}{m\omega^2}\left(\cos(\omega\zeta+\alpha)+\sin(\omega+\alpha)\right), \qquad (8.19)$$

where e and m are the charge and the mass of the electron at rest, y_0 and v_0 are the position and velocity of the electron at time ζ_0 and $E_0\sin(\omega t + \alpha)$ is the RF electric field between the plates. We select the specific values in the expression (8.19) that ahead us to:

$$f(x) = -\frac{1}{2}\cos(x)+x+\frac{1}{4}\pi. \qquad (8.20)$$

The zero of expression (8.20) is $x_* = -0.30909327154179495274198 6808924$.

Then, we get $L_0 = L = 1.523542095$. The radii of convergence for expression 8.1 are presented in Table 8.1.

Cases	ρ	x_0	n	ξ
1	0.437577	−0.6	7	2.0000
2	0.437577	{−0.6, −0.5}	10	1.6187
3	0.437577	−0.6	7	2.0000
4	0.437577	−0.6	6	2.0000

Table 8.1: Radii for Example 8.1

Example 8.2. Let $\mathbb{B}_1 = \mathbb{B}_2 = \mathbb{R}$ and $\Omega = [-1,1]$. Consider function F on Ω as

$$F(x) = \sin(x). \tag{8.21}$$

We yield $L_0 = L = 1$ by adopting $x_* = 0$. In Table 8.2, we report on the radii for the example 8.2.

Cases	ρ	x_0	n	ξ
1	0.666667	0.5	5	3.0000
2	0.666667	{0.5,0.4}	8	2.0199
3	0.666667	0.5	5	3.0000
4	0.666667	0.5	5	3.0000

Table 8.2: Radii for Example 8.2

Example 8.3. By the example in introduction for $x_* = 1$, we can set $L = L_0 = 96.662907$. Next in Table 8.3, We obtain radii for example 8.3.

Cases	ρ	x_0	n	ξ
1	0.006897	1.004	6	2.0000
2	0.006897	{1.004,1.001}	9	1.6169
3	0.006897	1.004	6	2.0000
4	0.006897	1.004	7	2.0000

Table 8.3: Radii for Example 8.3

Example 8.4. Let $\mathbb{B}_1 = \mathbb{B}_2 = \mathbb{R}^3$ and $\Omega = \mathbb{S}(0, 1)$. Assume F on Ω with $v = (x, y, z)^T$ as

$$F(u) = F(u_1, u_2, u_3) = \left(e^{u_1} - 1, \frac{e-1}{2}u_2^2 + u_2, u_3 \right)^T, \tag{8.22}$$

where, $u = (u_1, u_2, u_3)$. Defined as

$$F'(u) = \begin{bmatrix} e^{u_1} & 0 & 0 \\ 0 & (e-1)u_2 + 1 & 0 \\ 0 & 0 & 1 \end{bmatrix},$$

the Fréchet-derivative. Then, for $x_* = (0, 0, 0)^T$, $F'(x_*) = F'(x_*)^{-1} = diag\{1, 1, 1\}$, we have $L_0 = e - 1$, $L = e^{\frac{1}{e-1}}$. Hence, we obtain convergence radii that are mentioned in Table 8.4.

i	ψ_i	φ_i
0	1.3954170041747090114	1.7461756494150842271
1	1.7444828545735749268	2.0364691127919609051
2	2.0656234369405315689	2.2390977868265978920
3	2.4600678478912500533	2.4600678409809344550

Table 8.5: Values of ψ_i and φ_i (in radians) for Example 8.5.

Cases	ρ	x_0	n	ξ
1	0.382692	$(0.26, 0.26, 0.26)$	8	2.0000
2*2	2*0.382692	$\{(0.26, 0.26, 0.26),$	2*11	2*1.6177
		$(0.26, 0.26, 0.26)\}$		
3	0.382692	$(0.26, 0.26, 0.26)$	8	2.0000
4	0.382692	$(0.26, 0.26, 0.26)$	8	2.0000

Table 8.4: Radii for example 8.4

Example 8.5. The kinematic synthesis problem for steering [12, 21], is given as

$$[E_i (x_2 \sin (\psi_i) - x_3) - F_i (x_2 \sin (\varphi_i) - x_3)]^2$$
$$+ [F_i (x_2 \cos (\varphi_i) + 1) - F_i (x_2 \cos (\psi_i) - 1)]^2$$
$$- [x_1 (x_2 \sin (\psi_i) - x_3) (x_2 \cos (\varphi_i) + 1) - x_1 (x_2 \cos (\psi_i) - x_3) (x_2 \sin (\varphi_i) - x_3)]^2$$
$$= 0,$$

where

$$E_i = -x_3 x_2 (\sin (\varphi_i) - \sin (\varphi_0)) - x_1 (x_2 \sin (\varphi_i) - x_3) + x_2 (\cos (\varphi_i) - \cos (\varphi_0)),$$

$$F_i = -x_3 x_2 \sin (\psi_i) + (-x_2) \cos (\psi_i) + (x_3 - x_1) x_2 \sin (\psi_0) + x_2 \cos (\psi_0) + x_1 x_3,$$

and,

$$i = 1, 2, 3$$

In Table 8.5, we present the values of ψ_i and φ_i (in radians).
The approximated solution is

$$x_* = (0.9051567\ldots, 0.6977417\ldots, 0.6508335\ldots)^T.$$

Then, we get $L_0 = L = 19.1237412$. We provide the radii of convergence for Example 8.5 in Table 8.6.

Cases	ρ	x_0	n	ξ
1	0.034861	$(0.88,0.66,0.62)$	8	2.0102
2*2	2*0.034861	$\{(0.88,0.66,0.62),$ $(0.87,0.67,0.63)\}$	2*13	2*1.6272
3	0.034861	$(0.88,0.66,0.62)$	12	2.0003
4	0.034861	$(0.88,0.66,0.62)$	8	2.0054

Table 8.6: Radii for Example 8.5

Example 8.6. Let us consider that $\mathbb{B}_1 = \mathbb{B}_2 = C[0, 1]$, $\Omega = \bar{S}(0, 1)$ and introduce the space of maps continuous in $[0, 1]$ having the max norm. We consider the following function φ on $\mathbb{A}$:

$$\Psi(\phi)(x) = \Psi(x) - 5\int_0^1 x\tau\phi(\tau)^3 d\tau, \tag{8.23}$$

which further yields:

$$\Psi'(\phi(\mu))(x) = \mu(x) - 15\int_0^1 x\tau\phi(\tau)^2\mu(\tau)d\tau, \text{ for each } \mu \in \mathbb{A}.$$

We have $x_* = 0$, and $L = L_0 = 15$. We record the radii of convergence for example 8.6 in Table 8.7.

Cases	ρ
1	0.044444
2	0.044444
3	0.044444
4	0.044444

Table 8.7: Radii of convergence for example 8.6

8.3 EXERCISES

1. Solve the initial value problem

$$\dot{x}(t) = 1 + \cos(x(t)), x(0) = 0.$$

2. The predator-prey population models describe the interaction of a prey population X and a predator population Y. Assume that their interaction is modeled by the system of ordinary differential equations

$$\dot{x} = x - \frac{1}{4}x^2 - \frac{1}{10}xy + 1$$

$$\dot{y} = -\frac{1}{4}y + \frac{1}{7}xy + \frac{2}{3}.$$

(Assume $x(0) = y(0) = 0$). Solve the system.

3. Solve the initial value problem

$$\dot{x} = (x(t) + t, x(0) = 1)$$

by the iteration method (8.3). Perform 3 steps.

4. Solve the Fredholm-type integral equation

$$x(t) = \sum_0^1 \frac{t \cdot s}{10} ds + 1$$

by the iteration method (8.3). Perform 3 steps.

5. Solve the Volterra-type integral equation

$$x(t) = \sum_0^t \frac{t \cdot s}{10} ds + 1$$

by the iteration method. Perform 3 steps.

6. Solve equation

$$x = \frac{\sin x}{2} + 1$$

by using algorithm (8.3) with selecting $\alpha_k \equiv \frac{1}{2}$.

7. Repeat the previous problem for equation

$$x = \frac{\sin x + \cos 2x}{10}.$$

REFERENCES

1. Amat, S., Argyros, I.K., Busquier, S., and Hernández-Verón, M.A. 2018. On two high-order families of frozen Newton-type methods, *Numerical Linear Algebra with Applications*. 25. 1.

2. Amat, S., Bermúdez, C., Hernández-Verón, M.A., and Martínez, E. 2016. On an efficient $k-$step iterative method for nonlinear equations, *J. Comput. Appl. Math.*, 302: 258–271.

3. Amat, S., Busquier, S., Bermúdez, C., and Plaza. S. 2012. On two families of high order Newton type methods. *Appl. Math. Lett.*, 25. 12:2209–2217.

4. Magrenan, A.A., and Argyros I.K. 2017. *A Contemporary Study of Iterative Methods: Convergence, Dynamics and Applications*. Academic Press. Elsevier.

5. Argyros, I.K., and Magreñán, A.A. 2017. *Iterative Methods and Their Dynamics with Applications*, New York: CRC Press. Taylor & Francis.

6. Artidiello, S., Cordero, A., Torregrosa, J.R., and Vassileva, M.P. 2015. Multidimensional generalization of iterative methods for solving nonlinear problems by means of weigh-function procedure. *Appl. Math. Comput.*, 268:1064–1071.

7. Behl, R., and Motsa, S.S. 2015. Geometric construction of eighth-order optimal families of Ostrowski's method. *T. Wor. Sci. J.* Article ID: 614612. 11 pages.

8. Burden, R.L., and Faires, J.D. 2001. *Numerical Analysis*. Boston: PWS Publishing Company.

9. Cordero, A., and Torregrosa, J.R. 2007. Variants of Newton's method using fifth-order quadrature formulas. *Appl. Math. Comput.*, 190. 1:686–698.

10. Kanwar, V., Behl, R., and Sharma, K.K. 2011. Simply constructed family of a Ostrowski's method with optimal order of convergence. *Comput. Math. Appli.*, 62. 11:4021–4027.

11. Ezquerro, J.A., and Hernández, M.A. 2009. New iterations of R-order four with reduced computational cost. *BIT Numer. Math.*, 49:325–342 .

12. Awawdeh, F. 2010. On new iterative method for solving systems of nonlinear equations, *Numer. Algor.*, 54:395–409.

13. Magreñán, Á.A. 2014. Different anomalies in a Jarratt family of iterative root-finding methods, *Appl. Math. Comput.*, 233: 29–38.

14. Magreñán, Á.A. 2014. A new tool to study real dynamics: The convergence plane. *Appl. Math. Comput.*, 248:215–224.

15. Ortega, J.M., and Rheinboldt, W.C. 1970. *Iterative solution of nonlinear equations in several variables*, New-York: Academic Press.

16. Ostrowski, A.M. 1960. *Solution of equations and systems of equation*. Pure and Applied Mathematics, Vol. IX. New York-London: Academic Press.

17. Ostrowski, A.M. 1973. *Solution of equations in Euclidean and Banach spaces*. Pure and Applied Mathematics, Vol. 9, New York-London: Academic Press.

18. Petkovic, M.S., Neta, B., Petkovic, L., and Džunič, J. 2013. *Multipoint methods for solving nonlinear equations*. Elsevier.

19. Rheinboldt, W.C. 1978. An adaptive continuation process for solving systems of nonlinear equations. *Polish Academy of Science, Banach Ctr. Publ*, 3:129–142.

20. Traub, J.F. 1964. *Iterative methods for the solution of equations*. Englewood Cliffs, N.J.: Prentice Hall Series in Automatic Computation.

21. Tsoulos, I.G., Stavrakoudis, A. 2010. On locating all roots of systems of nonlinear equations inside bounded domain using global optimization methods. *Nonlinear Anal. Real World Appl.*, 11:2465–2471.

22. Weerakoon, S., and Fernando, T.G.I. 2000. A variant of Newton's method with accelerated third order convergence. *Appl. Math. Lett.*, 13:87–93.

9 Ball Comparison Between Three Sixth-Order Methods

Three methods of sixth order convergence are tackled for approximating the solution of an equation defined on the finitely dimensional Euclidean space. This convergence requires the existence of derivatives of, at least, order seven. However, only derivatives of order one are involved in such methods. Moreover, we have no estimates on the error distances, conclusions about the uniqueness of the solution in any domain and the convergence domain is not sufficiently large. Hence, these methods have limited usage. This chapter introduces a new technique on a general Banach space setting based only the first derivative and Lipschitz type conditions that allow the study of the convergence. In addition, we find usable error distances as well as uniqueness of the solution. A comparison between the convergence balls of three methods, not possible to drive with the previous approaches, is also given. The technique is possible to use with methods available in literature to improving, consequently, their applicability. Several numerical examples compare these methods and illustrate the convergence criteria.

Let $F : \Omega \subset \mathbb{X} \to \mathbb{Y}$ be Fréchet differentiable operator, $\mathbb{X}$, $\mathbb{Y}$ be two Banach spaces and $\Omega \subset \mathbb{X}$ be open, convex and non void. To solve $F(x) = 0$, we study the local convergence of the following three step methods defined for $\sigma = 0, 1, 2, \ldots$ as

$$y_\sigma = x_\sigma - \frac{2}{3} F'(x_\sigma)^{-1} F(x_\sigma)$$

$$z_\sigma = x_\sigma - \frac{1}{2}\left[I + 2F'(x_\sigma)\left(3F'(y_\sigma) - F'(x_\sigma)\right)^{-1}\right] F'(x_\sigma)^{-1} F(x_\sigma) \qquad (9.1)$$

$$x_{\sigma+1} = z_\sigma - 2\left(3F'(y_\sigma) - F'(x_\sigma)^{-1}\right) F(z_\sigma),$$

$$y_\sigma = x_\sigma - \frac{2}{3} F'(x_\sigma)^{-1} F(x_\sigma)$$

$$z_\sigma = x_\sigma - \frac{1}{2}\left[I + 2F'(x_\sigma)\left(3F'(y_\sigma) - F'(x_\sigma)\right)^{-1}\right] F'(x_\sigma)^{-1} F(x_\sigma) \qquad (9.2)$$

$$x_{\sigma+1} = z_\sigma - \frac{1}{4}\left[I + 2F'(x_\sigma)\left(3F'(y_\sigma) - F'(x_\sigma)\right)^{-1}\right]^2 F'(x_\sigma)^{-1} F(z_\sigma),$$

and

$$y_\sigma = x_\sigma - F'(x_\sigma)^{-1} F(x_\sigma)$$

$$z_\sigma = y_\sigma + \frac{1}{3}\left[F'(x_\sigma)^{-1} + 2\left(F'(x_\sigma) - 3F'(y_\sigma)\right)^{-1}\right] F(x_\sigma) \qquad (9.3)$$

$$x_{\sigma+1} = z_\sigma + \frac{1}{3}\left[4\left(F'(x_\sigma) - 3F'(y_\sigma)\right)^{-1} - F'(x_\sigma)^{-1}\right] F(z_\sigma).$$

DOI: 10.1201/9781003128915-9

The application of $F(x) = 0$ is mentioned in the standard books [13, 14, 18, 22]. The definition of the Fréchet derivative can be found for example in [17]. These methods use two operators, two Fréchet derivative evaluations and two linear operator inversions. The sixth convergence order of methods was given in Cordero et al. [9], Soleymani et al. [21], and Esmaeili and Ahmadi [11], respectively. The conclusions were obtained for the special case when $\mathbb{X} = \mathbb{Y} = \mathbb{R}^i$, using Taylor series with hypotheses up to the seventh derivative even though it does not appear in the methods. So, these hypotheses restrict the applicability of the methods. Let us consider a motivational example. We assume the following function F on $\mathbb{X} = \mathbb{Y} = \mathbb{R}$ and $\mathbb{D} = [-\frac{1}{2}, \frac{3}{2}]$ such as:

$$F(\kappa) = \begin{cases} \kappa^3 \ln \kappa^2 + \kappa^5 - \kappa^4, & \kappa \neq 0 \\ 0, & \kappa = 0 \end{cases}, \tag{9.4}$$

that leads to

$$F'(\kappa) = 3\kappa^2 \ln \kappa^2 + 5\kappa^4 - 4\kappa^3 + 2\kappa^2,$$

$$F''(\kappa) = 6\kappa \ln \kappa^2 + 20\kappa^3 - 12\kappa^2 + 10\kappa,$$

$$F'''(\kappa) = 6\ln \kappa^2 + 60\kappa^2 - 12\kappa + 22.$$

We note that $F'''(\kappa)$ is not bounded in $\mathbb{D}$. Therefore, results requiring the existence of $F''(\kappa)$ or higher cannot be applied for studying the convergence of (9.1)–(9.3). Moreover, no computable error bounds $\|x_\sigma - x_*\|$, where x_* solves the equation $F(x) = 0$, or any information regarding the uniqueness of the solution are provided using Lipschitz-type functions. Furthermore, the convergence criteria can not be compared, since they are based on different hypotheses. We address all these problems by using only the first derivative. Moreover, we rely on the computational order of convergence (COC) or approximated computational order of convergence $(ACOC)$ to determine the c-order (Computational order of convergence) not requiring derivatives of order higher than one. The new technique uses the same set of conditions for the three methods. Furthermore, it can also be used to extend the applicability of other methods along the same lines.

The rest of the chapter includes the following sections. Section 9.1 analyses the local convergence of the proposed technique. Section 9.2 discusses several applications experiments.

9.1 LOCAL CONVERGENCE

Let us introduce some real functions and parameters to be used later as follows in the local convergence analysis.

Suppose that equation

$$w_0(\zeta) = 1, \tag{9.5}$$

has a minimal positive solution ρ_0, where $w_0 : I \to I$ is continuous, increasing, with $w_0(0) = 0$, and $I = [0, \infty)$. Consider functions $w : I_0 \to I$, $v : I_0 \to I$ to be continuous, increasing, with $w(0) = 0$, and $I_0 = [0, \rho_0)$.

Suppose that

$$\frac{v(0)}{3} - 1 < 0. \tag{9.6}$$

Define functions g_1 and h_1 on I_0 as follows:

$$g_1(\zeta) = \frac{\int_0^1 w\left((1-\theta)\zeta\right)d\theta + \frac{1}{3}\int_0^1 v(\theta\zeta)d\theta}{1 - w_0(\zeta)},$$

$$h_1(\zeta) = g_1(\zeta) - 1.$$

By (9.6) and these definitions, we have $h_1(0) = \frac{v(0)}{3} - 1 < 0$ and $h_1(\zeta) \to \infty$ as $t \to \rho_0^-$. Denote by r_1 the minimal solution of equation $h_1(\zeta) = 0$ in the interval $(0, \rho_0)$ with assured existence by the intermediate value theorem.

Suppose that the equation

$$p(\zeta) = 1 \tag{9.7}$$

has a minimal positive solution ρ_p, where

$$p(\zeta) = \frac{1}{2}\left[3w_0\left(g_1(\zeta)\zeta\right) + w_0(\zeta)\right].$$

Set $I_1 = [0, \rho_1)$, where $\rho_1 := \min\{\rho_0, \rho_p\}$. Define functions g_2 and h_2 on the interval I_1 by

$$g_2(\zeta) = \frac{\int_0^1 w\left((1-\theta)\zeta\right)d\theta}{1 - w_0(\zeta)} + \frac{3}{4}\frac{\left[w_0\left(g_1(\zeta)\zeta\right) + w_0(\zeta)\right]\int_0^1 v(\theta\zeta)d\theta}{\left(1 - p(\zeta)\right)\left(1 - w_0(\zeta)\right)},$$

$$h_2(\zeta) = g_2(\zeta) - 1.$$

We get again $h_2(0) = -1$ and $h_2(\zeta) \to \infty$ as $\zeta \to \rho_1^-$. Denote by r_2 the smallest solution of equation $h_2(\zeta) = 0$ in the interval $(0, \rho_1)$.

Suppose that equation

$$w_0\left(g_2(\zeta)\zeta\right) = 1 \tag{9.8}$$

has a minimal positive solution ρ_2.

Set $I_2 := [0, \rho)$, where $\rho = \min\{\rho_1, \rho_2\}$. Next, define functions g_3 and h_3 on the interval I_2 by

$$g_3(\zeta) = \left[\frac{\int_0^1 w\left((1-\theta)\Lambda\right)d\theta}{1 - w_0(\Lambda)} + \frac{\left[3w_0\left(\Phi\right) + 2w_0\left(\Lambda\right) + w_0(\zeta)\right]\int_0^1 v\left(\theta\Lambda\right)d\theta}{2\left(1 - w_0(\Lambda)\right)\left(1 - p(\zeta)\right)}\right]g_2(\zeta),$$

$$h_3(\zeta) = g_3(\zeta) - 1.$$

where,

$$\Lambda = g_2(\zeta)\zeta \text{ and } \Phi = g_1(\zeta)\zeta$$

We obtain $h_3(0) = -1$ and $h_3(\zeta) \to \infty$ as $\zeta \to \rho^-$. Denote by r_3 the minimal solution of equation $h_3(\zeta) = 0$ in the interval $(0,\rho)$. Define a radius of convergence r by

$$r = \min\{r_j\}, \quad j = 1,2,3. \tag{9.9}$$

It follows that for all $\zeta \in I_3 := [0,r)$

$$0 \leq w_0(\zeta) < 1, \tag{9.10}$$

$$0 \leq w_0\left(g_2(\zeta)\zeta\right) < 1, \tag{9.11}$$

$$0 \leq p(\zeta) < 1, \tag{9.12}$$

$$0 \leq g_j(\zeta) < 1. \tag{9.13}$$

The hypotheses $(A_i, \ i = 1,2,\ldots5)$ used in the local convergence analysis of all three methods are:

(A_1) $F : \Omega \subset \mathbb{X} \to \mathbb{Y}$, is Fréchet differentiable and there exists $x_* \in \Omega$ with $F(x_*) = 0$ and $F'(x_*)^{-1}\ell(\mathbb{Y},\mathbb{X})$.

(A_2) There exists function $w_0 : I \to I$ continuous, increasing with $w_0(0) = 0$ such that for each $x \in \Omega$

$$\left\| F'(x_*)^{-1}\left(F'(x) - F'(x_*)\right) \right\| \leq w_0(\|x - x_*\|).$$

Set $\Omega_0 = \Omega \cap U(x_*,\rho_0)$.

(A_3) There exist functions $w : I_0 \to I$ and $v :: I_0 \to I$ continuous and increasing with $w(0) = 0$, such that for each $x,y \in \Omega_0$

$$\left\| F'(x_*)^{-1}\left(F'(x) - F'(y)\right) \right\| \leq w(\|x - x_*\|)\|x - y\|.$$

and

$$\left\| F'(x_*)^{-1}F'(x) \right\| \leq v(\|x - x_*\|).$$

(A_4) The ball $\bar{U}(x_*,r) \subset \Omega$, ρ_0, ρ_p and ρ_2 are defined in previous expressions.

(A_5) There exists $r_* \geq r$ such that

$$\int_0^1 w_0(\theta r_*)d\theta < 1.$$

Set $\Omega_1 = \Omega \cap U^*(x_*,r_*)$.

Next, we provide the local convergence analysis of method (9.1) using the hypotheses (A) and the aforementioned symbols.

Theorem 9.1. Suppose that the hypotheses (A) hold. Then, starting from any $x_0 \in U(x_*,r) - \{x_*\}$, the sequence $\{x_\sigma\}$ generated by method (9.1) is well defined, remains in $U(x_*, r)$ for each $\sigma = 0, 1, 2, 3, \ldots$ and $\lim_{\sigma \to \infty} x_\sigma = x_*$. Moreover, the following error estimates are available

$$\|y_\sigma - x_*\| \le g_1(\|x_\sigma - x_*\|)\|x_\sigma - x_*\| \le \|x_\sigma - x_*\| < r, \qquad (9.14)$$

$$\|z_\sigma - x_*\| \le g_2(\|x_\sigma - x_*\|)\|x_\sigma - x_*\| \le \|x_\sigma - x_*\|, \qquad (9.15)$$

$$\|x_{\sigma+1} - x_*\| \le g_3(\|x_\sigma - x_*\|)\|x_\sigma - x_*\| \le \|x_\sigma - x_*\|, \qquad (9.16)$$

where the functions g_j are given previously and the radius r is defined by (9.9). Furthermore, x_* is the only solution of equation $F(x) = 0$ in the set Ω_1 given below (A_5).

Proof. Inequations (9.14)–(9.16) are shown by using mathematical induction. Using (9.9) and (9.10), A_1 and A_2 we have for all $x \in U(x_*, r)$

$$\left\| F'(x)^{-1}\left(F'(x) - F'(x_*)\right) \right\| \le w_0(\|x - x_*\|) \le w_0(r) < 1. \qquad (9.17)$$

By the Banach lemma on invertible operators [4, 5, 17], expression (9.17), $F'(x)^{-1} \in \ell(\mathbb{Y}, \mathbb{X})$ and

$$\left\| F'(x)^{-1} F'(x) \right\| \le \frac{1}{w_0(\|x - x_*\|)}. \qquad (9.18)$$

Then, y_0 is well defined by the first substep of method (9.1). By A_1 and A_3, we can write

$$F(x) = F(x) - F(x_*) = \int_0^1 F'\left(x_* + \theta(x_0 - x_*)\right) d\theta(x_0 - x_*)$$

and so, by the second hypothesis in (A_3), we have

$$\left\| F'(x)^{-1} F'(x) \right\| = \left\| F'(x_*)^{-1} \int_0^1 F'\left(x_* + \theta(x_0 - x_*)\right) d\theta(x_0 - x_*) \right\|$$
$$\int_0^1 v\left(\theta\|x_0 - x_*\|\right) d\theta \|x_0 - x_*\|. \qquad (9.19)$$

In view of method (9.1) (for $\sigma = 0$), expressions (9.9), (9.13) (for $j = 1$), hypothesis (A_3), expression (9.18) (for $x = x_0$) and (9.19), we obtain

$$\|y_0 - x_*\| = \left\| \left(x_0 - x_* - F'(x_0)^{-1}F(x_0)\right) + \frac{1}{3}F'(x_0)^{-1}F(x_0) \right\|$$

$$\le \left\| \left(x_0 - x_* - F'(x_0)^{-1}F(x_0)\right) \right\| + \frac{1}{3}\left\| F'(x_0)^{-1}F(x_0) \right\|$$

$$\le \left\| F'(x_0)^{-1}F(x_*) \right\| \left\| \int_0^1 F'(x_*)^{-1}\left[F'\left(x_* + \theta(x_0 - x_*)\right) - F'(x_0)\right] d\theta(x_0 - x_*) \right\|$$

$$+ \frac{1}{3}\left\| F'(x_0)^{-1}F(x_*) \right\| \left\| F'(x_0)^{-1}F(x_0) \right\|$$

$$\le \frac{\left[\int_0^1 w\left((1-\theta)\|x_0 - x_*\|\right) d\theta + \frac{1}{3}\int_0^1 v\left(\theta\|x_0 - x_*\|\right) d\theta\right] \|x_0 - x_*\|}{w_0(\|x_0 - x_*\|)}$$

$$= g_1(\|x_0 - x_*\|)\|x_0 - x_*\| \le \|x_0 - x_*\| < r,$$

$$(9.20)$$

so that, $y_0 \in U(x_*, r)$ and (9.14) holds for $\sigma = 0$.
By expressions (9.9), (9.11), and (9.20), we have

$$\left\| \left(2F'(x_*) \right)^{-1} \left[3F'(y_0) - F'(x_0) - 3F'(x_*) + F'(x_*) \right] \right\|$$

$$\leq \frac{1}{2} \left[3 \left\| F'(x_*)^{-1} \left(F'(y_0) - F'(x_*) \right) \right\| + \left\| F'(x_*)^{-1} \left(F'(x_0) - F'(x_*) \right) \right\| \right]$$

$$\leq \frac{1}{2} \left[3w_0(\|y_0 - x_*\|) + w_0(\|x_0 - x_*\|) \right]$$

$$\leq p(\|x_0 - x_*\|) \leq p(r) < 1,$$

so that

$$\left(3F'(y_0) - F'(x_0) \right)^{-1} \in \ell(\mathbb{Y}, \mathbb{X}),$$

and

$$\left\| \left(3F'(y_0) - F'(x_0) \right)^{-1} F'(x_*) \right\| \leq \frac{1}{2(1 - p(\|x_0 - x_*\|))}.$$

Then, z_0 is well defined by the second substep of method (9.1) for $\sigma = 0$. Next, by the second substep of method (9.1) for $\sigma = 0$, we can write

$$z_0 - x_* = \left(x_0 - x_* - F'(x_0)^{-1}F(x_0) \right) + F'(x_0)^{-1}F(x_0) - \frac{1}{2}F'(x_0)^{-1}F(x_0)$$

$$- F'(x_0) \left(\Gamma \right)^{-1} F'(x_0)^{-1} F(x_0)$$

$$= \left(x_0 - x_* - F'(x_0)^{-1}F(x_0) \right) + \left[\frac{1}{2}I - F'(x_0) \left(\Gamma \right)^{-1} \right] F'(x_0)^{-1}F(x_0)$$

$$= \left(x_0 - x_* - F'(x_0)^{-1}F(x_0) \right) + \left[\frac{\Gamma}{2} - F'(x_0) \right] \left(\Gamma \right)^{-1} F'(x_0)^{-1}F(x_0)$$

$$= \left(x_0 - x_* - F'(x_0)^{-1}F(x_0) \right) + \frac{3}{2} \left(F'(y_0) - F'(x_0) \right) \left(\Gamma \right)^{-1} F'(x_0)^{-1}F(x_0).$$
$$(9.21)$$

where,

$$\Gamma = 3F'(y_0) - F'(x_0)$$

Hence, by expressions (9.9), (9.13) (for $j = 2$) and (9.19)–(9.21), we obtain

$$\|z_0 - x_*\| = \left[\frac{\int_0^1 w\left((1 - \theta)\|x_0 - x_*\| \right) d\theta}{1 - w_0(\|x_0 - x_*\|)} \right.$$

$$\left. + \frac{3\left(w_0(\|y_0 - x_*\|) + w_0(\|x_0 - x_*\|) \right) \int_0^1 v\left(\theta\|x_0 - x_*\| \right) d\theta}{4\left(1 - w_0(\|x_0 - x_*\|) \right) \left(1 - p(\|x_0 - x_*\|) \right)} \right] \|x_0 - x_*\|$$

$$= g_2(\|x_0 - x_*\|)\|x_0 - x_*\| \leq \|x_0 - x_*\| < r.$$
$$(9.22)$$

So, $z_0 \in U(x_*, r)$ and expression (9.15) holds for $\sigma = 0$.

In view of method (9.1) for $\sigma = 0$, x_1 is well defined $\left(F'(z_0)^{-1} \in \ell(\mathbb{Y}, \mathbb{X})\right.$ by (9.18) for $x = z_0\big)$. Then, we can write

$$x_1 - x_* = \left(z_0 - x_* - F'(z_0)^{-1} F(z_0)\right) + \left[F'(z_0)^{-1} - 2\left(3F'(y_0) - F'(x_0)\right)^{-1}\right] F(z_0),$$

which further yields

$$\|x_1 - x_*\| = \left[\frac{\int_0^1 w\left(\theta\lambda\right) d\theta}{1 - w_0(\lambda)} \right.$$
$$\left. + \frac{\left[2\left(w_0(\lambda_1) + w_0(\lambda)\right) + w_0(\lambda_1) + w_0(\|x_0 - x_*\|)\right] \int_0^1 v\left(\theta\lambda\right) d\theta}{2\left(1 - w_0(\lambda)\right)\left(1 - p(\|x_0 - x_*\|)\right)} \right] \lambda$$
$$= g_3(\|x_0 - x_*\|)\|x_0 - x_*\| \leq \|x_0 - x_*\| < r,$$

$$(9.23)$$

where,

$$\lambda = \|z_0 - x_*\| \quad \text{and} \quad \lambda_1 = \|y_0 - x_*\|$$

so that $x_1 \in U(x_*, r)$ and expression (9.16) holds for $\sigma = 0$. So far we have shown that estimates (9.14)–(9.16) holds for $\sigma = 0$. If we simply replace x_0, y_0, z_0 and x_1 by x_m, y_m, z_m and x_{m+1}, $(m = 1, 2, 3, ..., \sigma - 1)$ respectively, in the preceding computations, then we obtain

$$\|y_{m+1} - x_*\| \leq g_1(\|x_{m+1} - x_*\|)\|x_{m+1} - x_*\| \leq \|x_{m+1} - x_*\| < r$$
$$\|z_{m+1} - x_*\| \leq g_2(\|x_{m+1} - x_*\|)\|x_{m+1} - x_*\| \leq \|x_{m+1} - x_*\| < r$$

and

$$\|x_{m+2} - x_*\| \leq g_3(\|x_{m+1} - x_*\|)\|x_{m+1} - x_*\| < r.$$

By the above estimations

$$\|x_{m+1} - x_*\| \leq c\|x_m - x_*\| < r, \quad c = g_3(\|x_0 - x_*\|) \in [0, 1),$$

we deduce that $\lim_{m \to \infty} x_m = x_*$, with $x_{m+1} \in U(x_*, r)$. Consider $y_* \in \Omega_1$ with $F(y_*) = 0$ and set

$$S = \int_0^1 F'\left(x_* + \theta(y_* - x_*)\right) d\theta.$$

By (A_2) and (A_5), we obtain

$$\left\|F'(x_*)^{-1}\left(S - F'(x_*)\right)\right\| \leq \int_0^1 w_0(\theta\|y_* - x_*\|) d\theta$$
$$\leq \int_0^1 w_0(\theta r_*) d\theta < 1,$$

so that $S^{-1} \in \ell(\mathbb{Y}, \mathbb{X})$. Then, $x_* = y_*$ follows from the identity $0 = F(y_*) - F(x_*) = S(y_* - x_*)$. $\qquad \square$

Secondly for the method (9.2), the conclusion of the Theorem 9.1 holds but r is defined by

$$r^{(2)} = \min\{r_1, r_2, r_3^{(2)}\}, \tag{9.24}$$

so that $r_3^{(2)}$ is the minimal positive solution of equation $h_3^2(\zeta) = 0$, which $h_3^2(\zeta) = g_3^2(\zeta) - 1$ and

$$g_3^{(2)}(\zeta) = \left[1 + \frac{1}{4}q(\zeta)\left(\frac{\int_0^1 v\left(\theta g_2(\zeta)\zeta\right)d\theta}{1 - w_0(\zeta)}\right)\right]g_2(\zeta),$$

where

$$q(\zeta) = \frac{3w_0\left(g_1(\zeta)\zeta\right) + w_0(\zeta) + 4}{2\left(1 - p(\zeta)\right)}.$$

Notice also that g_1, h_1, g_2, h_2, r_1 and r_2 are the same as in Theorem 9.1. Functions $g_3^{(2)}$, $h_3^{(2)}$ and q appear due to the estimates

$$\left\|I + 2F'(x_\sigma)\left(3F'(y_\sigma) - F'(x_\sigma)\right)^{-1}\right\|$$

$$= \left\|\left[3\left(F'(y_\sigma) - F'(x_\sigma)\right) + 2F'(x_\sigma)\right]\left[3F'(y_\sigma) - F'(x_\sigma)\right]^{-1}\right\|$$

$$= \left\|\left[3\left(F'(y_\sigma) - F'(x_\sigma)\right) + \left(F'(x_\sigma) - F'(x_*)\right) + 4F'(x_*)\right]\left[3F'(y_\sigma) - F'(x_\sigma)\right]^{-1}\right\|$$

$$\leq \frac{3w_0(\|y_\sigma - x_*\| + w_0(\|x_\sigma - x_*\|) + 4}{2\left(1 - p(\|x_\sigma - x_*\|)\right)} \leq q(\|x_\sigma - x_*\|)$$

and

$$\|x_{\sigma+1} - x_*\| \leq \|z_\sigma - x_*\| + \frac{1}{4}\left\|\left[I + 2F'(x_\sigma)\left(3F'(y_\sigma) - F'(x_\sigma)\right)^{-1}\right]F'(x_\sigma)^{-1}F(z_\sigma)\right\|$$

$$\leq \left[1 + \frac{1}{4}q(\|x_\sigma - x_*\|)\frac{\int_0^1 v\left(\theta\|z_\sigma - x_*\|\right)d\theta}{1 - w_0(\|x_\sigma - x_*\|)}\right]\|z_\sigma - x_*\|$$

$$\leq g_3^{(2)}(\|x_\sigma - x_*\|)\|x_\sigma - x_*\| \leq \|x_\sigma - x_*\| < r^{(2)}.$$

Hence, we arrive at the following theorem.

Theorem 9.2. Suppose that the conditions (A) hold, but with r^2 and $g_3^{(2)}$ replaced by r and g_3, respectively. Then, the same conclusions hold for method (9.2), but with (9.16) replaced by

$$\|x_{\sigma+1} - x_*\| \leq g_3^{(2)}(\|x_\sigma - x_*\|)\|x_\sigma - x_*\| \leq \|x_\sigma - x_*\|. \tag{9.25}$$

Finally, for the local convergence of method (9.3), we introduce the functions

$$g_2^{(3)}(\zeta) = g_1(\zeta) + \frac{\left(w_0(\zeta) + w_0\left(g_1(\zeta)\zeta\right)\right)\int_0^1 v(\theta\zeta)d\theta}{2\left(2 - w_0(\zeta)\right)\left(1 - p(\zeta)\right)},$$

$$h_2^{(3)}(\zeta) = g_2^{(3)}(\zeta) - 1,$$

$$g_3^{(3)}(\zeta) = \left[1 + \frac{3\left(w_0(\zeta) + w_0\left(g_1(\zeta)\zeta\right)\right)\int_0^1 v\left(\theta g_2^{(3)}(\zeta)\zeta\right)d\theta}{2\left(2 - w_0(\zeta)\right)\left(1 - p(\zeta)\right)}\right]g_2^{(3)}(\zeta),$$

$$h_3^{(3)}(\zeta) = g_3^{(3)}(\zeta) - 1.$$

Let us denote by $r_2^{(3)}$ and $r_3^{(3)}$ the minimal positive solutions of equations $h_2^{(3)}(\zeta) = 0$ and $h_3^{(3)}(\zeta) = 0$, respectively. Set

$$r^{(3)} = \min\{r_1, r_2^{(3)}, r_3^{(3)}\}. \tag{9.26}$$

These functions are defined due to the estimates

$$\|z_\sigma - x_*\|$$

$$= \left\|y_\sigma - x_* + \frac{1}{3}F'(x_\sigma)^{-1}\left[\left(F'(x_\sigma) - 3F'(y_\sigma) + 2F'(x_\sigma)\right)\left(F'(x_\sigma) - 3F'(y_\sigma)\right)^{-1}F(x_\sigma)\right]\right\|$$

$$= \left\|(y_\sigma - x_*) + F'(x_\sigma)^{-1}\left(F'(x_\sigma) - F'(y_\sigma)\right)\left(F'(x_\sigma) - 3F'(y_\sigma)\right)^{-1}F(x_\sigma)\right\|$$

$$\leq \left[g_1(\|x_\sigma - x_*\|) + \frac{\left(w_0(\|x_\sigma - x_*\|) + w_0(\|y_\sigma - x_*\|)\right)\int_0^1 v(\theta\|x_\sigma - x_*\|)d\theta}{2\left(1 - w_0(\|x_\sigma - x_*\|)\right)\left(1 - p(\|x_\sigma - x_*\|)\right)}\right]\|x_\sigma - x_*\|$$

$$\leq g_2^{(3)}(\|x_\sigma - x_*\|)\|x_\sigma - x_*\| \leq \|x_\sigma - x_*\|$$

and

$$\|x_{\sigma+1} - x_*\|$$

$$= \left\|z_\sigma - x_* + \left(F'(x_\sigma) - 3F'(y_\sigma)\right)^{-1}\left[4F'(x_\sigma) - \left(F'(x_\sigma) - 3F'(y_\sigma)\right)\right]F'(x_\sigma)^{-1}F(z_\sigma)\right\|$$

$$= \left\|(z_\sigma - x_*) + 3\left(F'(x_\sigma) - 3F'(y_\sigma)\right)^{-1}\left(F'(x_\sigma) - F'(y_\sigma)\right)F'(x_\sigma)^{-1}F(z_\sigma)\right\|$$

$$\leq \left[1 + \frac{3\left(w_0(\|x_\sigma - x_*\|) + w_0(\|y_\sigma - x_*\|)\right)\int_0^1 v(\theta\|z_\sigma - x_*\|)d\theta}{2\left(1 - w_0(\|x_\sigma - x_*\|)\right)\left(1 - p(\|x_\sigma - x_*\|)\right)}\right]\|x_\sigma - x_*\|$$

$$\leq g_3^{(3)}(\|x_\sigma - x_*\|)\|x_\sigma - x_*\| \leq \|x_\sigma - x_*\|.$$

Theorem 9.3. Let us consider hypotheses (A), but with $g_2^{(3)}, g_3^{(3)}$ and r^3 replacing by g_2, g_3 and r, respectively. Then, the conclusions of Theorem 9.1 hold for method (9.3), but with (9.15) and (9.16) replaced by

$$\|z_\sigma - x_*\| = g_2^{(3)}(\|x_\sigma - x_*\|)\|x_\sigma - x_*\| \le \|x_\sigma - x_*\|. \tag{9.27}$$

and

$$\|x_{\sigma+1} - x_*\| = g_3^{(3)}(\|x_\sigma - x_*\|)\|x_\sigma - x_*\| \le \|x_\sigma - x_*\|, \tag{9.28}$$

respectively.

9.2 APPLICATIONS

The theoretical results developed in the previous sections are illustrated numerically in this section. We denote the methods (9.1), (9.2) and (9.3) by (CM), (SM) and (EA), respectively. We consider two real life problems and two standard nonlinear problems that are illustrated in examples 9.1–9.4. The results are listed in Tables 9.1, 9.2, 9.4 and 9.5. Additionally, we obtain the COC approximated by means of

$$\xi = \frac{\ln \frac{\|x_{\sigma+1} - x_*\|}{\|x_\sigma - x_*\|}}{\ln \frac{\|x_\sigma - x_*\|}{\|x_{\sigma-1} - x_*\|}}, \quad \text{for } \sigma = 1, 2, \dots \tag{9.29}$$

or $ACOC$ [23] by:

$$\xi^* = \frac{\ln \frac{\|x_{\sigma+1} - x_\sigma\|}{\|x_\sigma - x_{\sigma-1}\|}}{\ln \frac{\|x_\sigma - x_{\sigma-1}\|}{\|x_{\sigma-1} - x_{\sigma-2}\|}}, \quad \text{for } \sigma = 2, 3, \dots \tag{9.30}$$

We adopt $\varepsilon = 10^{-100}$ as the error tolerance and the terminating criteria to solve nonlinear system or scalar equations are: (i) $\|x_{\sigma+1} - x_\sigma\| < \varepsilon$, and (ii) $\|F(x_\sigma)\| < \varepsilon$. The computations are performed with the package *Mathematica* 9 with multiple precision arithmetics.

Example 9.1. Following the example presented in introduction, for $x_* = 1$ we can set

$$w_0(t) = w(t) = 96.662907t \text{ and } v(t) = 2.$$

In Table 9.1, we present radii for example 9.1.

Cases	CM	SM	EA
r_1	0.0022989	0.0022989	0.0022989
r_2	0.0017993	0.0017993	-
r_3	0.0015625	-	-
$r_3^{(2)}$	-	0.001022	-
$r_2^{(3)}$	-	-	0.0013240
$r_3^{(3)}$	-	-	0.00081403
r	0.0015625	-	-
$r^{(2)}$	-	0.001022	-
$r^{(3)}$	-	-	0.00081403
x_0	1.001	1.0009	1.0008
σ	3	3	3
ξ	6.0000	6.0000	6.0000

(On the basis of obtained results, we conclude that method CM has a larger radius of convergence.)

Table 9.1: Radii for Example 9.1

Example 9.2. Let $\mathbb{X} = \mathbb{Y} = \mathbb{R}^3$ and $\Omega = \bar{\mathbb{S}}(0,\ 1)$. Assume F on Ω with $v = (x,\ y,\ z)^T$ as

$$F(u) = F(u_1, u_2, u_3) = \left(e^{u_1} - 1,\ \frac{e-1}{2} u_2^2 + u_2,\ u_3 \right)^T, \qquad (9.31)$$

where $u = (u_1, u_2, u_3)^T$. Define the Fréchet-derivative as

$$F'(u) = \begin{bmatrix} e^{u_1} & 0 & 0 \\ 0 & (e-1)u_2 + 1 & 0 \\ 0 & 0 & 1 \end{bmatrix}.$$

Then, for $x_* = (0,\ 0,\ 0)^T$ and $F'(x_*) = F'(x_*)^{-1} = diag\{1,\ 1,\ 1\}$, we have

$$w_0(t) = (e-1)t,\ \ w(t) = e^{\frac{1}{e-1}}t\ \text{ and } v(t) = e^{\frac{1}{e-1}}.$$

We obtain the convergence radii depicted in Table 9.2.

Cases	CM	SM	EA
r_1	0.15441	0.15441	0.15441
r_2	0.11011	0.11011	-
r_3	0.096467	-	-
$r_3^{(2)}$	-	0.065471	-
$r_2^{(3)}$	-	-	0.092584
$r_3^{(3)}$	-	-	0.059581
r	0.096467	-	-
$r^{(2)}$	-	0.065471	-
$r^{(3)}$	-	-	0.059581
x_0	(0.094,0.094,0.094)	(0.063,0.063,0.063)	(0.054,0.054,0.054)
σ	3	3	3
ξ	6.0000	6.0000	6.0000

(Among three methods, the larger radius of convergence belong to the method CM.)

Table 9.2: Radii for example 9.2

Example 9.3. The kinematic synthesis problem for steering [7], is given as

$$[E_i(x_2 \sin(\psi_i) - x_3) - F_i(x_2 \sin(\varphi_i) - x_3)]^2$$
$$+ [F_i(x_2 \cos(\varphi_i) + 1) - F_i(x_2 \cos(\psi_i) - 1)]^2$$
$$- [x_1(x_2 \sin(\psi_i) - x_3)(x_2 \cos(\varphi_i) + 1) - x_1(x_2 \cos(\psi_i) - x_3)(x_2 \sin(\varphi_i) - x_3)]^2$$
$$= 0,$$

where,

$$E_i = -x_3 x_2(\sin(\varphi_i) - \sin(\varphi_0)) - x_1(x_2 \sin(\varphi_i) - x_3) + x_2(\cos(\varphi_i) - \cos(\varphi_0)),$$

$$F_i = -x_3 x_2 \sin(\psi_i) + (-x_2)\cos(\psi_i) + (x_3 - x_1)x_2 \sin(\psi_0) + x_2 \cos(\psi_0) + x_1 x_3,$$

and,

$$i = 1, 2, 3.$$

In Table 9.3, we present the values of ψ_i and φ_i (in radians).

i	ψ_i	φ_i
0	1.3954170041747090114	1.7461756494150842271
1	1.7444828545735749268	2.0364691127919609051
2	2.0656234369405315689	2.2390977868265978920
3	2.4600678478912500533	2.4600678409809344550

Table 9.3: Values of ψ_i and φ_i (in radians) for Example 9.3.

The approximated solution is for $\Omega = \bar{\mathbb{S}}(x_*, 1)$

$$x_* = (0.9051567\ldots, 0.6977417\ldots, 0.6508335\ldots)^T.$$

Then, we get

$$w_0(t) = w(t) = 3t \text{ and } v(t) = 2.$$

We provide the radii of convergence for Example 9.3 in Table 9.4.

Cases	CM	SM	EA
r_1	0.074074	0.074074	0.074074
r_2	0.057977	0.057977	-
r_3	0.050345	-	-
$r_3^{(2)}$	-	0.032936	-
$r_2^{(3)}$	-	-	0.042662
$r_3^{(3)}$	-	-	0.026229
r	0.050345	-	-
$r^{(2)}$	-	0.032936	-
$r^{(3)}$	-	-	0.026229
x_0	(0.945,0.737,0.690)	(0.933,0.726,0.678)	(0.929,0.722,0.674)
σ	3	3	3
ξ	6.1328	6.1377	4.8142

Table 9.4: Radii for Example 9.3

Example 9.4. Let us consider that $\mathbb{X} = \mathbb{Y} = C[0, 1]$, $\Omega = \bar{\mathbb{S}}(0, 1)$ and introduce the space of maps continuous in $[0, 1]$ having the max norm. We consider the following function φ on $\mathbb{A}$:

$$\Psi(\phi)(x) = \Psi(x) - \int_0^1 x\tau\phi(\tau)^3 d\tau, \qquad (9.32)$$

which further yields:

$$\Psi'(\phi(\mu))(x) = \mu(x) - 3\int_0^1 x\tau\phi(\tau)^2\mu(\tau)d\tau, \text{ for } \mu \in \Omega.$$

We have $x_* = 0$ and

$$w_0(t) = \frac{3}{2}, \ w(t) = 3t \ \text{and} \ v(t) = 2.$$

We list the radii of convergence for example 9.4 in Table 9.5.

Cases	CM	SM	EA
r_1	0.111111	0.111111	0.111111
r_2	0.105542	0.105542	-
r_3	0.0922709	-	-
$r_3^{(2)}$	-	0.0594758	-
$r_2^{(3)}$	-	-	0.0718454
$r_3^{(3)}$	-	-	0.0465723
r	0.0922709	-	-
$r^{(2)}$	-	0.0594758	-
$r^{(3)}$	-	-	0.0465723

(CM has a larger radius of convergence as compared to other two methods.)

Table 9.5: Radii of convergence for example 9.4

9.3 EXERCISES

1. Solve the initial value problem

$$\dot{x}(t) = 1 + \cos(x(t)), x(0) = 0,$$

using the three developed methods.

2. The predator-prey population models describe the interaction of a prey population X and a predator population Y. Assume that their interaction is modeled by the system of ordinary differential equations

$$\dot{x} = x - \frac{1}{4}x^2 - \frac{1}{10}xy + 1$$
$$\dot{y} = -\frac{1}{4}y + \frac{1}{7}xy + \frac{2}{3}.$$

3. Solve the initial value problem

$$\dot{x} = (x(t) + t, x(0) = 1)$$

by the iteration methods developed in this chapter. Perform 3 steps.

4. Solve the Fredholm-type integral equation

$$x(t) = \sum_0^1 \frac{t \cdot s}{10} ds + 1$$

by the iteration methods developed in this chapter. Perform 3 steps.

5. Solve the Volterra-type integral equation

$$x(t) = \sum_0^t \frac{t \cdot s}{10} ds + 1$$

by the iteration methods developed in this chapter. Perform 3 steps.

6. Solve equation

$$x = \frac{\sin x}{2} + 1$$

by using algorithms developed in this chapter.

7. Repeat the previous problem for equation

$$x = \frac{\sin x + \cos 2x}{10}.$$

REFERENCES

1. Amat, S., Argyros, I.K., Busquier, S., and Herńandez-Veŕon, M.A. 2018. On two high-order families of frozen Newton-type methods. *Numerical Linear Algebra with Applications*, 25. 1:1–17.
2. Amat, S., Berḿudez, C., Herńandez-Veŕon, M.A., and Martínez, E. 2016. On an efficient k–step iterative method for nonlinear equations. *J. Comput. Appl. Math.*, 302:258–271.
3. Amat, S., Busquier, S., Berḿudez, C., and Plaza, S. 2012. On two families of high order Newton type methods. *Appl. Math. Lett.*, 25. 12:2209–2217.
4. Argyros, I.K., and George, S. 2020. On the complexity of extending the convergence region for Traub's method. *J. Complex.*, 56. $https://doi.org/10.1016/j.jco.2019.101423$.
5. Argyros, I.K., and Magreñán, Á.A. 2017. *Iterative Methods and Their Dynamics with Applications*. New York: CRC Press, Taylor & Francis
6. Artidiello, S., Cordero, A., Torregrosa, J.R., and Vassileva, M.P. 2015. Multidimensional generalization of iterative methods for solving nonlinear problems by means of weigh-function procedure. *Appl. Math. Comput.*, 268:1064–1071.
7. Awawdeh, F. 2010. On new iterative method for solving systems of nonlinear equations. *Numer. Algor.*, 54:395–409.
8. Burden, R.L., and Faires, J.D. 2001. *Numerical Analysis*. Boston: PWS Publishing Company.
9. Cordero, A., Hueso, J.L., Martínez, E., Torregrosa, J.R. A modified Newton-Jarratt's composition. *Numer. Algor.* **2010**, *55*, 87–99.
10. Cordero, A., and Torregrosa, J.R. 2007. Variants of Newton's method using fifth-order quadrature formulas. *Appl. Math. Comput.*, 190. 1:686–698.
11. Esmaeili, H., and Ahmadi, M. 2015. An efficient three step method to solve system of nonlinear equations. *Appl. Math. Comput.*, 266:1093–1101.
12. Ezquerro, J.A., and Hernández, M.A. 2009. New iterations of R-order four with reduced computational cost. *BIT Numer. Math.*, 49:325–342.
13. Iliev, A., and Kyurkchiev, N. 2010. *Nontrivial Methods in Numerical Analysis: Selected Topics in Numerical Analysis*. Saarbrucken: LAP LAMBERT Academic Publishing.
14. Magreñán, Á.A. 2014. Different anomalies in a Jarratt family of iterative root-finding methods. *Appl. Math. Comput.*, 233:29–38.
15. Magreñán, Á.A. 2014. A new tool to study real dynamics: The convergence plane. *Appl. Math. Comput.*, 248:215–224.
16. Magrenan, Á.A., and Argyros, I.K. 2017. *A Contemporary Study of Iterative Methods: Convergence, Dynamics and Applications*. Academic Press, Elsevier.
17. Ortega, J.M., and Rheinboldt, W.C. 1970. *Iterative solution of nonlinear equations in several variables*. New-York: Academic Press.
18. Ostrowski, A.M. 1973. *Solution of equations in Euclidean and Banach spaces, Pure and Applied Mathematics*. vol. 9. New York-London: Academic Press.
19. Petkovic, M.S., Neta, B., Petkovic, L., and Džunič, J. 2013. *Multipoint methods for solving nonlinear equations*. Elsevier.
20. Rheinboldt, W.C. 1978. An adaptive continuation process for solving systems of nonlinear equations. *Polish Academy of Science, Banach Ctr. Publ.*, 3:29–142.
21. Soleymani, F., Lotfi, T., and Bakhtiari, O. 2014. A multi-step class of iterative methods for nonlinear systems. *Ptim. Lett.*, 8:1001–1015.
22. Traub, J.F. 1964. *Iterative methods for the solution of equations*. Englewood Cliffs, N.J.: Prentice Hall Series in Automatic Computation
23. Weerakoon, S., and Fernando, T.G.I. 2000. A variant of Newton's method with accelerated third order convergence. *Appl. Math. Lett.*, 13:87–93.

10 Constrained Generalized Equations

In this chapter, we aim to address an extended version of Newton's method for solving constrained generalized equations. This method can be seen as a combination of the classical Newton's method applied to generalized equations with a procedure to obtain a feasible inexact projection. Using the contraction mapping principle, we establish a local analysis of the proposed method under appropriate assumptions, namely metric regularity or strong metric regularity and Lipschitz continuity. Metric regularity is assumed to guarantee that the method generates a sequence that converges to a solution. Under strong metric regularity, we show the uniqueness of the solution in a suitable neighborhood, and that all sequences starting in this neighborhood converge to this solution. We also require the assumption of Hölder continuity to establish the convergence rate for the method. The novelty of our work is that in particular, we present a finer convergence analysis than in earlier works using our new idea of the restricted convergence domain. A favorable to us comparison is also given using a numerical experiment.

We propose a Newton-type method for solving the following generalized equations subject to a set of constraints:

$$x \in C, \qquad 0 \in f(x) + F(x), \tag{10.1}$$

where $f : \Omega \to \mathbb{R}^n$ is a continuously differentiable function, $\Omega \subseteq \mathbb{R}^n$ is an open set, $C \subset \Omega$ is a closed convex set, and $F : \Omega \rightrightarrows \mathbb{R}^n$ is a multifunction with a closed nonempty graph. As is well known, if $C = \mathbb{R}^n$ thet the inclusion (10.1) becomes the so called generalized equation

$$0 \in f(x) + F(x), \tag{10.2}$$

which is an abstract model for various problems in classical analysis and its applications, such as system of equalities and inequalities and variational inequality problem, which covers a wide range of problems in nonlinear programming. Additional comments about (10.2) can be found in [2, 3, 11, 12, 16, 18].

Newton's method for solving the generalized equation (10.2), for an initial point x_0, is defined as follows

$$0 \in f(x_k) + f'(x_k)(x_{k+1} - x_k) + F(x_{k+1}), \qquad k = 0, 1, \ldots. \tag{10.3}$$

Studies dealing with this method include, but are not limited to, [2, 3, 16]; see also [16, Section 6C]. When $F \equiv \{0\}$, (10.2) becomes the standard Newton's method for solving $f(x) = 0$. Now, if $F = N_D$ is the normal cone mapping of a convex set $D \subset \mathbb{R}^n$, then (10.3) is the known version of Newton's method for solving variational inequality; see [11, 16]. In particular, if (10.2) represents the Karush-Kuhn-Tucker optimality conditions for a nonlinear programming problem, then (10.3) describes a well-known sequential quadratic programming method; see for example [16, pag. 334].

Constrained Variational Inequality Problem (CVIP for short), see [10], and in particular, Split Variational Inequality Problem (SVIP for short), see [10, 27], can be stated as special cases of constrained generalized equations (10.1). Indeed, if U and Ω are closed convex sets in $\mathbb{R}^n$, $h : \mathbb{R}^n \to \mathbb{R}^n$ is a continuous function, $C = U \cap \Omega$ and $F = N_U(.)$ then the constrained generalized equation (10.1) becomes the problem

$$\text{find} \quad x_* \in U \cap \Omega \quad \text{such that} \quad 0 \in h(x_*) + N_U(x_*), \tag{10.4}$$

which can be rewritten equivalently as the following constrained variational inequality problem:

$$\text{find} \quad x_* \in U \cap \Omega \quad \text{such that} \quad \langle h(x_*), x - x_* \rangle \geq 0, \qquad \forall\, x \in U. \tag{10.5}$$

Thus, (10.5) can be seen as a special instance of the constrained generalized equation (10.1). We note that the classical variational inequality problem it is not equivalent to the above CVIP, since in (10.5) the point x_* must belongs to $U \cap \Omega$.

Moreover, for functions $f, g : \mathbb{R}^n \to \mathbb{R}^n$, $A : \mathbb{R}^n \to \mathbb{R}^n$ a linear operator and $U, \Omega \subset \mathbb{R}^n$ nonempty, closed convex sets, the SVIP is formulated as follows:

$$\text{find} \quad x_* \in U \quad \text{such that} \quad \langle f(x_*), x - x_* \rangle \geq 0, \qquad \forall\, x \in U,$$

and such that the point $y_* = Ax_* \in \Omega$ satisfies

$$\langle g(y_*), y - y_* \rangle \geq 0, \qquad \forall\, y \in \Omega.$$

By taking $D := U \times \Omega$ and $V := \{w = (x, y) \in \mathbb{R}^n \times \mathbb{R}^n : Ax = y\}$ the SVIP is equivalent to the following CVIP:

$$\text{find} \quad w_* \in D \cap V \quad \text{such that} \quad \langle h(w_*), w - w_* \rangle \geq 0, \qquad \forall\, w \in D,$$

where $w = (x, y)$ and $h : \mathbb{R}^n \times \mathbb{R}^n \to \mathbb{R}^n \times \mathbb{R}^n$ is defined by $h(x, y) := (f(x), g(y))$. See [12, Lemma 5.1]. Therefore, the SVIP is equivalent to the following constrained generalized equation:

$$\text{find} \quad w_* \in D \cap V \quad \text{such that} \quad 0 \in h(w_*) + N_D(w_*),$$

where $D := U \times \Omega$ and $V := \{w = (x, y) \in \mathbb{R}^n \times \mathbb{R}^n : Ax = y\}$.

It is worth noting that SVIP is quite general and includes several problems as special cases. For instance, *Split Minimization Problem* and *Common Solutions to Variational Inequalities Problem*. See, for example, [1, 10, 11]. See [10, 27] for an extensive discussion on this problem.

It is known that if F is the zero mapping, i.e., $F \equiv \{0\}$, then the problem (10.1) reduces to a constrained system of nonlinear equations, i.e., that of solving $f(x) = 0$

such that $x \in C$. This class of problems has been addressed in several studies, and various methods have been proposed for solving them. See, for example, [8, 25, 35].

Inspired by the elegant works in [13, 24, 25], we introduce the center-Hölder condition to obtain a larger radius of convergence, a tigther error bounds on $\|x_k - x_*\|$ and a more precise information on the uniqueness of the solution. We emphasize that these advantages do not require additional hypotheses than those in [13, 24, 25]. Besides, we propose Newton's method with a feasible inexact projection for solving the inclusion (10.1). We also establish the local convergence of the proposed method under appropriate assumptions, namely metric regularity or strong metric regularity. Metric regularity is assumed to guarantee that the Newton-InexP method generates a sequence that converges to a solution. Under strong metric regularity, we demonstrate the uniqueness of a solution in a suitable neighborhood, and that every sequence starting in this neighborhood converges to that solution.

The remainder of this chapter is organized as follows. In Section 10.1, we present the notations and some results that are used throughout the paper. In Section 10.2, we describe the proposed method, and present its local convergence analysis. In Section 10.3, we present applications to deal with the idea of the restricted convergence domain.

10.1 NOTATION AND AUXILIARY RESULTS

The *open* and *closed balls* of radius $\mu > 0$, centered at x, will be respectively defined by

$$B_\mu(x) := \{y \in \mathbb{R}^n \; : \; \|x - y\| < \mu\}, \qquad B_\delta[x] := \{y \in \mathbb{R}^n \; : \; \|x - y\| \le \mu\}.$$

Let $\mathscr{L}(\mathbb{R}^n, \mathbb{R}^n)$ be the *vector space consisting of all continuous linear mappings* $T : \mathbb{R}^n \to \mathbb{R}^n$ and the *norm* of T is defined by $\|T\| := \sup\{\|Tx\| \; : \; \|x\| \le 1\}$. Let $\Omega \subseteq \mathbb{R}^n$ be an open set and $f : \Omega \to \mathbb{R}^n$ be differentiable at all $x \in \Omega$. The *graph* of the multifunction $F : \mathbb{R}^n \rightrightarrows \mathbb{R}^n$ is the set $\operatorname{gph} F := \{(x, u) \in \mathbb{R}^n \times \mathbb{R}^n \; : \; u \in F(x)\}$. The *domain* and *range* of the multifunction F, respectively, are the sets $\operatorname{dom} F := \{x \in \mathbb{R}^n \; : \; F(x) \ne \varnothing\}$ and $\operatorname{rge} F := \{u \in \mathbb{R}^n \; : \; u \in F(x) \text{ for some } x \in \mathbb{R}^n\}$. The *inverse* of F is the multifunction $F^{-1} : \mathbb{R}^n \rightrightarrows \mathbb{R}^n$ defined by $F^{-1}(u) := \{x \in \mathbb{R}^n \; : \; u \in F(x)\}$. The *partial linearization* of $f + F$ at $x \in \Omega$ is the multifuntion $L_{f+F}(x, \cdot) : \Omega \rightrightarrows \mathbb{R}^m$ defined by

$$L_{f+F}(x, y) := f(x) + f'(x)(y - x) + F(y). \tag{10.6}$$

For sets C and D in $\mathbb{R}^n$, the *distance from x to D* and the *excess of C beyond D* are respectively defined by

$$d(x, D) := \inf_{y \in D} \|x - y\|, \qquad e(C, D) := \sup_{x \in C} d(x, D), \tag{10.7}$$

where the convention is adopted that $d(x, D) = +\infty$ when $D = \varnothing$, $e(\varnothing, D) = 0$ when $D \ne \varnothing$, and $e(\varnothing, \varnothing) = +\infty$. In the following, we present the notion of metric regularity, which plays an important role in the subsequent analysis.

Definition 10.1. Consider a multifunction $G : \mathbb{R}^n \rightrightarrows \mathbb{R}^n$ and $(\bar{x}, \bar{y}) \in \text{gph}(G)$. We say that G is metrically regular at $\bar{x}$ for $\bar{y}$ with constant $\kappa > 0$, if there exist some neighborhoods U_a of $\bar{x}$ and V_b of $\bar{y}$ such that

$$d(x, G^{-1}(y)) \leq \kappa d(y, G(x)) \text{ for all } x \in U_a, \; y \in V_b.$$

Moreover, if the mapping $V_b \ni u \mapsto G^{-1}(u) \cap U_a$ is single-valued, then G is called strongly metrically regular at $\bar{x}$ for $\bar{y}$, with associated constants $\kappa > 0$, $a > 0$, and $b > 0$.

The last result of this section is the famous fixed-point theorem for multifunctions. It was proved in [24,26] in the mid-1990s and it is crucial to proving our main result.

Theorem 10.1. Consider a multifunction $\Phi : \mathbb{R}^n \rightrightarrows \mathbb{R}^n$, a point $\bar{x} \in \mathbb{R}^n$, and non-negative scalars α and θ such that $0 \leq \rho < 1$. If

1. the set gph $\Phi \cap (B_\rho(\bar{x}) \times B_\rho(\bar{x}))$ is closed,
2. $d(\bar{x}, \Phi(\bar{x})) < \rho(1 - \lambda)$,
3. $e(\Phi(u) \cap B_\rho(\bar{x}), \Phi(v)) \leq \lambda \|u - v\|$, for all $u, v \in B_\rho(\bar{x})$,

then Φ has a fixed point in $B_\rho(\bar{x})$; that is, there exists an $x \in B_\rho(\bar{x})$ such that $x \in \Phi(x)$.

10.2   THE LOCAL NEWTON METHOD

In this section, we present the Newton-InexP method for solving the problem (10.1). In our analysis we use the assumption of metric regularity and strong metric regularity. Besides, the center-Hölder is assumed and we use a procedure to obtain a feasible inexact projection onto the feasible set. Next, we present the concept of a feasible inexact projection.

Definition 10.2. Let $C \subset \mathbb{R}^n$ be a closed convex set, with $x \in C$ and $\theta \geq 0$. The *feasible inexact projection mapping* relative to x with error tolerance θ, denoted by $P_C(\cdot, x, \theta) : \mathbb{R}^n \rightrightarrows C$ is the set-valued mapping defined as follows:

$$P_C(y, x, \theta) := \left\{ w \in C : \langle y - w, z - w \rangle \leq \theta \|y - x\|^2, \quad \forall z \in C \right\}.$$

Each point $w \in P_C(y, x, \theta)$ is called a *feasible inexact projection of y onto C with respect to x and with error tolerance θ*.

Remark 10.1. Because $C \subset \mathbb{R}^n$ is a closed convex set, [9, Proposition 2.1.3, p. 201] implies that for each $y \in \mathbb{R}^n$ we have $P_C(y) \in P_C(y, x, \theta)$, where P_C denotes the exact projection mapping. Therefore, $P_C(y, x, \theta) \neq \varnothing$ for all $y \in \mathbb{R}^n$ and $x \in C$. If $\theta = 0$ in Definition 10.2, then $P_C(y, x, 0) = \{P_C(y)\}$ for all $y \in \mathbb{R}^n$ and $x \in C$. *We used $P_C(y, x, 0) = P_C(y)$ instead of $P_C(y, x, 0) = \{P_C(y)\}$.*

The next result presents a basic property of the feasible inexact projection and its proof can be find in [25, Lemma 4]. See also [13].

Lemma 10.1. Let $y, \tilde{y} \in \mathbb{R}^n$, $x, \tilde{x} \in C$, and $\theta \geq 0$. Then, for any $w \in P_C(y, x, \theta)$, we have

$$\|w - P_C(\tilde{y}, \tilde{x}, 0)\| \leq \|y - \tilde{y}\| + \sqrt{2\theta}\|y - x\|.$$

The conceptual Newton's method, named the Newton-InexP method, for solving (10.1), with a feasible inexact projection, and with $x_0 \in C$ and $\{\theta_k\} \subset [0, +\infty)$ as the input data, is formally described as follows.

Newton-InexP method

Step 0. Let $x_0 \in C$ and $\{\theta_j\} \subset [0, +\infty)$ be given, and set $k = 0$.

Step 1. If $f(x_k) + F(x_k) \ni 0$, then **stop**; otherwise, compute $y_k \in \mathbb{R}^n$ such that

$$f(x_k) + f'(x_k)(y_k - x_k) + F(y_k) \ni 0. \qquad (10.8)$$

Step 2. If $y_k \in C$, set $x_{k+1} = y_k$; otherwise take any $x_{k+1} \in C$ satisfying

$$x_{k+1} \in P_C(y_k, x_k, \theta_k). \qquad (10.9)$$

Step 3. Set $k \leftarrow k+1$, and go to **Step 1**.

Remark 10.2. In **Step 1**, we check if the current iterate x_k is a solution of the problem (10.1). Otherwise, we compute a point y_k satisfying the inclusion (10.8). Because the point y_k in **Step 1** may be infeasible for the constraint set C, the Newton-InexP method applies a procedure to obtain a feasible inexact projection, and consequently the new iterate x_{k+1} in C. In particular, the point x_{k+1} obtained in (10.9) is an approximate feasible solution for the projection subproblem $\min_{z \in C}\{\|z - y_k\|^2/2\}$, satisfying $\langle y_k - x_{k+1}, z - x_{k+1}\rangle \leq \theta_k\|y_k - x_k\|^2$ for any $z \in C$. As we will see, the specific choice of the tolerance θ_k is essential to establish the local convergence of the Newton-InexP method.

We introduce some different types of Lipschitz conditions, so we can compare them to each other.

Definition 10.3. Let $\Omega \subset \mathbb{R}^n$ be an open set, $\zeta \in (0, 1]$ and $f : \Omega \to \mathbb{R}^m$ be continuously differentiable in Ω. Assume that $C \subset \Omega$ is a closed convex set, $x_* \in C$, and there exists $L_0 > 0$ such that

$$\|f'(x) - f'(x_*)\| \leq L_0\|x - y\|^{\zeta}, \qquad \forall x \in \Omega. \qquad (10.10)$$

Define the set

$$S := B_r(x_*) \cap B_{\left(\frac{1}{L_0}\right)^{\frac{1}{\zeta}}}(x_*),$$

where $r := \sup\{t \geq 0 : B_t(x_*) \subset \Omega\}$.

Definition 10.4. Let Ω, f, C, x_*, ζ be as in Definition 10.3. Assume that there exists $L > 0$ such that

$$\|f'(x) - f'(y)\| \leq L\|x - y\|^{\zeta}, \qquad \forall\, x, y \in S. \tag{10.11}$$

Definition 10.5. Let Ω, f, C, x_*, ζ be as in Definition 10.3. Assume that there exists $L_1 > 0$ such that

$$\|f'(x) - f'(y)\| \leq L_1\|x - y\|^{\zeta}, \qquad \forall\, x, y \in \Omega. \tag{10.12}$$

Remark 10.3. It follows from the definition of S that

$$S \subset B_r(x_*), \tag{10.13}$$

so

$$L_0 \leq L_1 \tag{10.14}$$

and

$$L \leq L_1. \tag{10.15}$$

It is worth noting that $L = L(L_0)$.

Next, we first show that L_0 and L can be replace L_1 in the results obtained in [13, 24, 25]. Secondly, we show the following advantages:

1. Larger radius of convergence, so more initial points become available;
2. Tigther error bounds on $\|x_k - x_*\|$ leading to fewer iterates to obtain a desired predetermined error tolerance and
3. A more precise information on the uniqueness of the solution (in a larger ball).

It is worth noting that these advantages do not require additional hypotheses than those in [13, 24, 25], since in practice the computation of constant L_1 requires L_0 and L as special cases.

We take advantage of the observation that the iterates lie in S which is a more precise location that B_r used in [13, 24, 25] leading to the tighter than L_1 constants L_0 and L. In the following, we state our main result for the Newton-InexP method. The proof constitutes a combination of the results that will be studied in the sequel.

Theorem 10.2. Let $\Omega \subset \mathbb{R}^n$ be an open set, $f : \Omega \to \mathbb{R}^m$ be continuously differentiable in Ω, and $F : \Omega \rightrightarrows \mathbb{R}^m$ be a set-valued mapping with closed graph. Assume that $C \subset \Omega$ is a closed convex set, $x_* \in C$, $f(x_*) + F(x_*) \ni 0$, there exist $L_0, L > 0$ such that (10.10) and (10.11) hold. Assume that the set-valued mapping $\Omega \ni y \mapsto L_{f+F}(x_*, y)$ is metrically regular at x_* for 0, with constants $\kappa > 0$, $a > 0$ and $b > 0$. Let $\zeta \in (0, 1]$, $\{\theta_k\} \subset [0, 1/2)$, $\tilde{\theta} := \sup_k \theta_k < \frac{1}{2}$ and

$$r_* := \min\left\{ r, \left(\frac{(\zeta + 1)\left(1 - \sqrt{2\tilde{\theta}}\right)}{\kappa\left[\left(1 + \sqrt{2\tilde{\theta}}\right)L + (\zeta + 1)\left(1 - \sqrt{2\tilde{\theta}}\right)L_0\right]} \right)^{\frac{1}{\zeta}}, a, \left[\frac{b(\zeta + 1)}{(\zeta + 1)L_0 + L} \right]^{\frac{1}{\zeta + 1}} \right\}. \tag{10.16}$$

Then, for every $x_0 \in C \cap B_{r_*}(x_*) \setminus \{x_*\}$, there exists a sequence $\{x_k\}$ generated by the Newton-InexP method that solves (10.1), associated to $\{\theta_k\}$ and starting at x_0, which is contained in $B_{r_*}(x_*) \cap C$ and converges to x_* with the following rate of convergence:

$$\|x_* - x_{k+1}\| \le e_k \|x_* - x_k\|, \qquad k = 0, 1, \ldots, \tag{10.17}$$

where

$$e_k = \left[\left(1 + \sqrt{2\theta_k}\right) \frac{\kappa L \|x_* - x_k\|^{\zeta}}{(\zeta + 1)(1 - \kappa L_0 \|x_* - x_k\|^{\zeta})} + \sqrt{2\theta_k} \right] \|x_* - x_k\|.$$

As a consequence, if $\lim_{k \to +\infty} \theta_k = 0$ then $\{x_k\}$ converges to x_* superlinearly. Furthermore, if the mapping $L_{f+F}(x_*, \cdot)$ is strongly metrically regular at x_* for 0, then x_* is the unique solution of (10.1) in $B_{r_*}(x_*)$, and every sequence generated by the Newton-InexP method associated to $\{\theta_k\}$ and starting at $x_0 \in C \cap B_{r_*}(x_*) \setminus \{x_*\}$ satisfies (10.17) and converges to x_*.

From now on, we assume that all the assumptions of Theorem 10.2 hold except the strong metric regularity, which will be considered to hold only when explicitly stated. Next, we present some preliminary results.

Lemma 10.2. The following inequality holds for all $p, q \in B_r(x_*)$ and $\zeta \in (0, 1]$

$$\|f(q) - f(p) - f'(p)(q - p)\| \le \frac{L}{\zeta + 1} \|p - q\|^{\zeta + 1}.$$

Moreover, if $\|x_* - p\| < r_*$, then

$$\|f(x_*) - f(p) - f'(p)(z - p) + f'(x_*)(z - x_*)\| < b, \qquad \forall z \in B_{r_*}(x_*).$$

Proof. Because $q + (1 - \tau)(p - q) \in B_r(x_*)$ for all $\tau \in [0, 1]$ and f is continuously differentiable in Ω, we can use (10.11) to find

$$\|f(q) - f(p) - f'(p)(q - p)\| \le \int_0^1 \|f'(p) - f'(q + (\tau - 1)(q - p))\| \|q - p\| \, d\tau$$

$$\le L \|p - q\|^{\zeta + 1} \int_0^1 \tau^{\zeta} \, d\tau = \frac{L}{\zeta + 1} \|p - q\|^{\zeta + 1},$$

which is the first inequality of the lemma. We proceed to prove the second inequality. For this purpose, let $0 < \|x_* - p\| < r_*$ and $0 < \|x_* - z\| < r_*$. By applying the first inequality of this lemma together with the Lipschitz conditions in (10.10) and (10.11), we get

$$\|f(x_*) - f(p) - f'(p)(z - p) + f'(x_*)(z - x_*)\|$$
$$\le \|f(x_*) - f(p) - f'(p)(x_* - p)\| + \|f'(p) - f'(x_*)\| \|x_* - z\|$$
$$\le \frac{L}{\zeta + 1} \|p - x_*\|^{\zeta + 1} + L_0 \|x_* - p\| \|x_* - z\|^{\zeta}.$$

Taking into account that $\|x_* - p\| < r_*$, $\|x_* - z\| < r_*$ and the assumption on r_*, the desired inequality follows from the last inequality. $\qquad \square$

To state the next result, for each fixed $x \in \mathbb{R}^n$ we define the following auxiliary set-valued mapping $\Phi_x : \Omega \rightrightarrows \mathbb{R}^n$:

$$\Phi_x(y) := L_{f+F}\left(x_*, f(x_*) - f(x) - f'(x)(y-x) + f'(x_*)(y-x_*)\right)^{-1}, \qquad (10.18)$$

where $\mathbb{R}^m \ni u \mapsto L_{f+F}(x_*, u)^{-1} := \{z \in \mathbb{R}^n \ : \ u \in L_{f+F}(x_*, z)\}$ is the inverse of L_{f+F} defined in (10.6). Therefore, $y \in \Phi_x(y)$ if and only if x and y satisfy the following inclusion:

$$f(x) + f'(x)(y-x) + F(y) \ni 0,$$

i.e., y is the *next Newton's iterate* from x. In the next lemma, we establish existence of a fixed point of Φ_x for all x in a suitable neighborhood of x_*. Moreover, we present an important bound on the distance between x_* and this fixed point, and establish its uniqueness under strong metric regularity. The statement of this result is as follows.

Lemma 10.3. If $0 < \|x_* - x\| < r_*$, then there exists a fixed point $y \in \Phi_x(y)$ such that

$$\|x_* - y\| \le \frac{\kappa L \|x_* - x\|^{\zeta+1}}{(\zeta+1)(1 - \kappa L_0 \|x_* - x\|^{\zeta})}. \qquad (10.19)$$

In particular, $y \in B_{r_*}(x_*)$. In addition, if $L_{f+F}(x_*, \cdot)$ is strongly metrically regular at x_* for 0, then for all $x \in B_{r_*}(x_*)$ the mapping Φ_x has only one fixed point in $B_{r_*}(x_*)$ satisfying (10.19).

Proof. To prove the first part of the lemma, we will first prove the following two inequalities:

(i) $d\left(x_*, \Phi_x(x_*)\right) \le \rho\left(1 - \kappa L_0 \|x_* - x\|^{\zeta}\right)$;

(ii) $e\left(\Phi_x(p) \cap B_\rho[x_*], \Phi_x(q)\right) \le \kappa L_0 \|x_* - x\|^{\zeta} \|p - q\|, \qquad \forall\, p, q \in B_\rho[x_*],$

where the scalar $\rho > 0$ is defined by

$$\rho := \frac{\kappa L \|x_* - x\|^{\zeta+1}}{(\zeta+1)(1 - \kappa L_0 \|x_* - x\|^{\zeta})}. \qquad (10.20)$$

In order to prove (i), first note that the definition of the mapping Φ_x given in (10.18) implies that

$$d(x_*, \Phi_x(x_*)) = d\left(x_*, \ L_{f+F}(x_*, f(x_*) - f(x) - f'(x)(x_* - x))^{-1}\right).$$

Thus, taking into account that the second part of Lemma 10.2 with $p = x$ and $z = x_*$ implies that $\|f(x_*) - f(x) - f'(x)(x_* - x)\| < b$, and considering that $x_* \in B_a[x_*]$ and $0 \in L_{f+F}(x_*, x_*)$, we can apply Definition 10.1 to conclude that

$$d\left(x_*, \Phi_x(x_*)\right) \le \kappa \|f(x_*) - f(x) - f'(x)(x_* - x)\|.$$

Because Lemma 10.2 with $p = x$ and $q = x_*$ implies that

$$\|f(x_*) - f(x) - f'(x)(x_* - x)\| \le \frac{L}{\zeta+1}\|x_* - x\|^{\zeta+1},$$

combining the two last inequalities we obtain that

$$d(x_*, \Phi_x(x_*)) \le \frac{\kappa L}{\zeta+1} \|x_* - x\|^{\zeta+1},$$

which after some manipulation yields

$$d(x_*, \Phi_x(x_*)) \le \frac{\kappa L \|x_* - x\|^{\zeta+1}}{(\zeta+1)(1 - \kappa L_0 \|x_* - x\|^{\zeta})} \left(1 - \kappa L_0 \|x_* - x\|^{\zeta}\right).$$

This inequality, together with Definition (10.20), proves item (i). To prove item (ii), we take $p, q \in B_\rho[x_*]$. Owing to Definition (10.20), taking into account the assumption on r_* and $\|x_* - x\| < r_*$, we can verify that $\rho < r_*$. Thus, Lemma 10.2 implies that $\|f(x_*) - f(x) - f'(x)(p-x) + f'(x_*)(p-x_*)\| < b$ and $\|f(x_*) - f(x) - f'(x)(q-x) + f'(x_*)(q-x_*)\| < b$. Because $e(\varnothing, \Phi_x(q)) = 0$, we can assume without loss of generality that $\Phi_x(p) \cap B_a[x_*] \ne \varnothing$ for all $p \in B_\rho[x_*]$. Let $z \in \Phi_x(p) \cap B_a[x_*]$. Then, from Definition 10.1 with $\bar{x} = x_*$, $\bar{u} = 0$, $x = z$, $u = f(x_*) - f(x) - f'(x)(q-x) + f'(x_*)(q-x_*)$, and $G = L_{f+F}(x_*, \cdot)$, we have

$$d(z, \Phi_x(q)) \le \kappa d\left(f(x_*) - f(x) - f'(x)(q-x) + f'(x_*)(q-x_*), L_{f+F}(x_*, z)\right).$$

Because $z \in \Phi_x(p)$ implies that $f(x_*) - f(x) - f'(x)(p-x) + f'(x_*)(p-x_*) \in L_{f+F}(x_*, z)$, by using the definition of distance given in (10.7), we obtain

$$d\left(f(x_*) - f(x) - f'(x)(q-x) + f'(x_*)(q-x_*), L_{f+F}(x_*, z)\right)$$
$$\le \|[f'(x) - f'(x_*)](p-q)\|.$$

Hence, combining the two last inequalities we conclude that

$$d(z, \Phi_x(q)) \le \kappa \|[f'(x) - f'(x_*)](p-q)\|.$$

Taking the supremum with respect to $z \in \Phi_x(p) \cap B_a[x_*]$ in the last inequality and using the definition of excess given in (10.7), we have

$$e\left(\Phi_x(p) \cap B_a[x_*], \Phi_x(q)\right) \le \kappa \|[f'(x) - f'(x_*)](p-q)\|.$$

Because $\rho < r_* \le a$, we have $e(\Phi_x(p) \cap B_\rho[x_*], \Phi_x(q)) \le e(\Phi_x(p) \cap B_a[x_*], \Phi_x(q))$. Hence, from the last inequality and the properties of the norm, we obtain

$$e\left(\Phi_x(p) \cap B_\rho[x_*], \Phi_x(q)\right) \le \kappa \|f'(x) - f'(x_*)\| \|p-q\|.$$

By using the fact that f' is center Lipschitz continuous with constant $L_0 > 0$, the latter inequality becomes

$$e\left(\Phi_x(p) \cap B_\rho[x_*], \Phi_x(q)\right) \le \kappa L_0 \|x_* - x\|^{\zeta} \|p-q\|,$$

and thus item (ii) is proved. Taking into account the assumptiom on r_* we conclude that $\kappa L_0 \|x_* - x\|^{\zeta} < 1$. So, we can apply Theorem 10.1 with $\Phi = \Phi_x$, $\bar{x} = x_*$, and

$\lambda = \kappa L_0 \|x_* - x\|^\zeta$ to conclude that there exists $y \in B_\rho[x_*]$, i.e., the inequality (10.19) holds, with that $y \in \Phi_x(y)$. To prove that $y \in B_{r_*}(x_*)$, we using again the assumption on r_* and (10.19) to conclude that

$$\|x_* - y\| \le \frac{\kappa L \|x_* - x\|^\zeta}{(\zeta + 1)(1 - \kappa L_0 \|x_* - x\|^\zeta)} \|x_* - x\| \le \|x_* - x\| < r_*,$$

which implies the desired result. Therefore, the proof of the first part of the lemma is complete. Next, we assume that $L_{f+F}(x_*, \cdot)$ is strongly metrically regular at x_* for 0. Suppose that there exist $\hat{y}$ and $\tilde{y} \in B_\rho[x_*] \subset B_{r_*}(x_*)$ such that $\hat{y} \in \Phi_x(\hat{y})$ and $\tilde{y} \in \Phi_x(\tilde{y})$. We know that the mapping $z \mapsto L_{f+F}(x_*, z)^{-1} \cap B_a[x_*]$ is single-valued on $B_b[0]$, and thus the definition of Φ_x in (10.18) and the second part of Lemma 10.2 imply that $\hat{y} = \Phi_x(\hat{y})$ and $\tilde{y} = \Phi_x(\tilde{y})$. Using the definition of excess in (10.7), item (ii), and the assumption on r_*, we obtain

$$\|\hat{y} - \tilde{y}\| = e(\Phi_x(\hat{y}) \cap B_\rho[x_*], \Phi_x(\tilde{y})) \le \kappa L_0 \|x_* - x\|^\zeta \|\hat{y} - \tilde{y}\| < \|\hat{y} - \tilde{y}\|,$$

which is a contradiction. Hence, we conclude $\hat{y} = \tilde{y}$. $\qquad\square$

The next lemma plays an important role in the convergence analysis. In particular, it will be used to prove the well-definedness of a sequence $\{x_k\} \subset B_{r_*}(x_*) \cap C$ and its convergence to a solution of the problem (10.1).

Lemma 10.4. If $\theta \ge 0$, $x \in C \cap B_{r_*}(x_*) \backslash \{x_*\}$ and $y \in \Phi_x(y)$ satisfies (10.19), then it holds that

$$\|x_* - w\| \le \left[\left(1 + \sqrt{2\theta}\right) \frac{\kappa L \|x_* - x\|^\zeta}{(\zeta + 1)(1 - \kappa L_0 \|x_* - x\|^\zeta)} + \sqrt{2\theta}\right] \|x_* - x\|, \ \forall w \in P_C(y, x, \theta).$$

(10.21)

In addition, if $\theta < 1/2$, then $P_C(y, x, \theta) \subset B_{r_*}(x_*) \cap C$.

Proof. Take $w \in P_C(y, x, \theta)$. Then, applying Lemma 10.1 with $\tilde{y} = x_*$ and $\tilde{x} = x_*$, we have

$$\|P_C(x_*, x_*, 0) - w\| \le \|x_* - y\| + \sqrt{2\theta}(\|x_* - x\| + \|x_* - y\|).$$

(10.22)

On the other hand, because $\|x_* - x\| < r_*$, by applying Lemma 10.3 and some manipulations, we conclude that

$$\|P_C(x_*, x_*, 0) - w\| \le \left[\left(1 + \sqrt{2\theta}\right) \frac{\kappa L \|x_* - x\|^\zeta}{(\zeta + 1)(1 - \kappa L_0 \|x_* - x\|^\zeta)} + \sqrt{2\theta}\right] \|x_* - x\|.$$

Hence, owing to the fact that $P_C(x_*, x_*, 0) = x_*$, the last inequality and (10.22) yield (10.21). The conditions (10.16) together with $\theta < 1/2$ imply

$$\|x_* - w\| < \|x_* - x\|, \qquad \forall w \in P_C(y, x, \theta),$$

and because $\|x_* - x\| < r_*$ we obtain that $P_C(y, x, \theta) \subset B_{r_*}(x_*)$. Because $P_C(y, x, \theta) \subset C$, the last statement of the lemma follows. $\qquad\square$

Next, let us study the uniqueness of the solution for (10.1) in the neighborhood $B_{r_*}(x_*)$.

Lemma 10.5. If the mapping $L_{f+F}(x_*,\cdot)$ is strongly metrically regular at x_* for 0, then x_* is the unique solution of (10.1) in $B_{r_*}(x_*)$.

Proof. Let $\hat{x}$ be a solution of (10.1) in $B_{r_*}(x_*)$. Thus, $\|\hat{x}-x_*\|<r_*$, which together with the first part of Lemma 10.2 implies that

$$\|f(\hat{x})-f(x_*)-f'(x_*)(\hat{x}-x_*)\|\le\frac{L}{\zeta+1}\|\hat{x}-x_*\|^{\zeta+1}<b. \tag{10.23}$$

Moreover, considering that $x_*\in B_{r_*}[x_*]$ and $r_*\le a$, we can apply Definition 10.1 to conclude that

$$d(x_*,L_{f+F}(x_*,-f(\hat{x})+f(x_*)+f'(x_*)(\hat{x}-x_*))^{-1})\le$$
$$\kappa d(-f(\hat{x})+f(x_*)+f'(x_*)(\hat{x}-x_*),L_{f+F}(x_*,x_*)).$$

Thus, owing to the fact that $0\in L_{f+F}(x_*,x_*)$, we can apply the first inequality in (10.23) and the definition of distance in (10.7) to conclude that

$$d(x_*,L_{f+F}(x_*,-f(\hat{x})+f(x_*)+f'(x_*)(\hat{x}-x_*))^{-1})\le\frac{\kappa L}{\zeta+1}\|\hat{x}-x_*\|^{\zeta+1}.$$

On the other hand, because the mapping $L_{f+F}(x_*,\cdot)$ is strongly metrically regular at x_* for 0, the mapping $z\mapsto L_{f+F}(x_*,z)^{-1}\cap B_a[x_*]$ is single-valued on $B_b[0]$. Furthermore, we know that $0\in f(\hat{x})+F(\hat{x})=f(\hat{x})-f(x_*)-f'(x_*)(\hat{x}-x_*)+L_{f+F}(x_*,\hat{x})$. Hence, we conclude that $\hat{x}=L_{f+F}(x_*,-f(\hat{x})+f(x_*)+f'(x_*)(\hat{x}-x_*))^{-1}$, and we obtain from the last inequality that

$$\|\hat{x}-x_*\|\le\frac{\kappa L_0}{2}\|\hat{x}-x_*\|^2.$$

If $\|\hat{x}-x_*\|\ne 0$, then last inequality implies that $\|\hat{x}-x_*\|\ge 2/(\kappa L_0)>2/[\kappa(2L_0+L)]\ge r_*$, which is absurd, because $\|\hat{x}-x_*\|<r_*$. Therefore, $\|\hat{x}-x_*\|=0$, and thus x_* is the unique solution of (10.1) in $B_{r_*}(x_*)$. $\square$

Our final task in this section is to prove Theorem 10.2. The proof comprises a convenient combination of Lemmas 10.3, 10.4, and 10.5.

10.2.1 PROOF OF THEOREM

Proof. First, we will show by induction on k that there exists a sequence $\{x_k\}$ generated by the Newton-InexP method solving (10.1), associated to $\{\theta_k\}$ and starting in x_0, which satisfies the following two conditions:

$$x_{k+1}\in B_{r_*}(x_*)\cap C,$$

$$\|x_*-x_{k+1}\|\le\left[\left(1+\sqrt{2\theta_k}\right)\frac{\kappa L\|x_*-x_k\|^{\zeta}}{(\zeta+1)(1-\kappa L_0\|x_*-x_k\|^{\zeta})}+\sqrt{2\theta_k}\right]\|x_*-x_k\|, \tag{10.24}$$

for all $k = 0, 1, \ldots$. Take $x_0 \in C \cap B_{r_*}(x_*) \setminus \{x_*\}$ and $k = 0$. Because $\|x_* - x_0\| < r_*$, applying the first part of Lemma 10.3 with $x = x_0$, we obtain that there exists $y_0 \in \Phi_{x_0}(y_0)$ such that $y_0 \in B_{r_*}(x_*)$. If $y_0 \in C$, then $x_1 = y_0 \in B_{r_*}(x_*) \cap C$, and by using (10.19) we can conclude that (10.24) holds for $k = 0$. Otherwise if $y_0 \notin C$, then take $x_1 \in P_C(y_0, x_0, \theta_0)$. Moreover, by using the first part of Lemma 10.4 with $x = x_0$, we obtain that (10.24) holds for $k = 0$. Furthermore, the conditions (10.16) imply that

$$\left[\left(1 + \sqrt{2\theta_0} \right) \frac{\kappa L \|x_* - x\|^\zeta}{(\zeta + 1)(1 - \kappa L_0 \|x_* - x\|^\zeta)} + \sqrt{2\theta_0} \right] < 1,$$

and so the second part of Lemma 10.4 give us that $x_1 \in B_{r_*}(x_*) \cap C$. Therefore, there exits x_1 satisfying (10.24) for $k = 0$. Assume for induction that the two assertions in (10.24) hold for $k = 0, 1, \ldots, j - 1$. Because $x_j \in B_{r_*}(x_*) \cap C$, we can apply Lemma 10.3 with $x = x_j$ to conclude that there exists $y_j \in \Phi_{x_j}(y_j)$ such that $y_j \in B_{r_*}(x_*)$. If $y_j \in C$, then $x_{j+1} = y_j \in B_{r_*}(x_*) \cap C$, and (10.19) implies that (10.24) holds for $k = j$. Otherwise, if $y_j \notin C$ then take $x_{j+1} \in P_C(y_j, x_j, \theta_j)$. Hence, using first part of Lemma 10.4 we obtain that the inequality in (10.24) holds for $k = j$. Because (10.16) implies that $(1 + \sqrt{2\theta_j})[(\kappa L \|x_* - x_j\|^\zeta)/(2(1 - \kappa L_0 \|x_* - x_j\|^\zeta))] + \sqrt{2\theta_j} < 1$, the second part of Lemma 10.4 yields that $x_{j+1} \in B_{r_*}(x_*) \cap C$. Thus, there exists x_{j+1} satisfying (10.24) for $k = j$, and the induction step is complete. Therefore, there exists a sequence $\{x_k\}$ generated by the Newton-InexP method solving (10.1), associated to $\{\theta_k\}$ and starting in x_0, and it satisfies the two conditions in (10.24). Now, we proceed to prove that the sequence $\{x_k\}$ converges to x_*. Indeed, because $\|x_* - x_k\| < r_*$ for all $k = 0, 1, \ldots$, $\tilde{\theta} = \sup_k \theta_k < 1/2$ and

$$r_*^\zeta \leq \frac{(\zeta + 1)(1 - \sqrt{2\tilde{\theta}})}{\kappa \left[\left(1 + \sqrt{2\tilde{\theta}} \right) L + (\zeta + 1) \left(1 - \sqrt{2\tilde{\theta}} \right) L_0 \right]},$$

we conclude from (10.24) that $\|x_* - x_{k+1}\| < \|x_* - x_k\|$. This implies that the sequence $\{\|x_* - x_k\|\}$ converges. Let us say that $t_* = \lim_{k \to +\infty} \|x_* - x_k\| \leq \|x_* - x_0\| < r_*$. Because $\{x_k\} \subset B_{r_*}(x_*) \cap C$, we can conclude that $t_* < r_*$. On the other hand, by combining the inequality in (10.24) with the second condition in (10.16), we obtain

$$\|x_* - x_{k+1}\| \leq \left[\left(1 + \sqrt{2\tilde{\theta}} \right) \frac{\kappa L \|x_* - x_k\|^\zeta}{(\zeta + 1)(1 - \kappa L_0 \|x_* - x_k\|^\zeta)} + \sqrt{2\tilde{\theta}} \right] \|x_* - x_k\|,$$

for all $k = 0, 1, \ldots$. Thus, taking the limit in this inequality as k goes to $+\infty$, we have

$$t_* \leq \left[\left(1 + \sqrt{2\tilde{\theta}} \right) \frac{\kappa L t_*^\zeta}{(\zeta + 1)(1 - \kappa L_0 t_*^\zeta)} + \sqrt{2\tilde{\theta}} \right] t_*.$$

If $t_* \neq 0$, we obtain from the last inequality that

$$\frac{(\zeta + 1)(1 - \sqrt{2\tilde{\theta}})}{\kappa \left[\left(1 + \sqrt{2\tilde{\theta}} \right) L + (\zeta + 1) \left(1 - \sqrt{2\tilde{\theta}} \right) L_0 \right]} \leq t_*,$$

which contradicts the first assertion in (10.16), because $t_* < r_*$. Hence, $t_* = 0$, and consequently the sequence $\{x_k\}$ converges to x_*. In particular, if $\lim_{k\to+\infty}\theta_k = 0$, then by taking the limit in (10.17) as k goes to $+\infty$ we obtain $\limsup_{k\to+\infty}[\|x_* - x_{k+1}\|/\|x_* - x_k\|] = 0$, i.e., the sequence $\{x_k\}$ converges to x_* superlinearly.

Furthermore, if the mapping $L_{f+F}(x_*, \cdot)$ is strongly metrically regular at x_* for 0, then Lemma 10.5 implies that x_* is the unique solution of (10.1) in $B_{r_*}(x_*)$. To prove the last statement of the theorem, take $x_0 \in C \cap B_{r_*}(x_*)\backslash\{x_*\}$. Then, the second part of Lemma 10.3 implies that there exist a *unique* $y_0 \in B_{\rho_0}(x_*)$ such that $y_0 \in \Phi_{x_0}(y_0)$, i.e., there exists a unique solution y_0 of (10.8) for $k = 0$, where

$$\rho_0 := \frac{\kappa L \|x_* - x_0\|^{\zeta+1}}{\zeta + 1(1 - \kappa L_0 \|x_* - x_0\|^{\zeta})}.$$

Furthermore, Lemma 10.4 implies that every $x_1 \in P_C(y_0, x_0, \theta_0)$ satisfies (10.17) for $k = 0$. Thus, proceeding by induction we can prove that the every sequence $\{x_k\}$ generated by the Newton-InexP method to solve (10.1), associated to $\{\theta_k\}$ and starting in x_0, satisfies (10.17), and by using similar argument as above we can prove that such a sequence converges to x_*. Therefore, the proof is complete.   □

Remark 10.4. If $L = L_0 = L_1$ and $\zeta = 1$ then our results reduce to the corresponding ones in [13]. But if $L_0 < L$, or $L < L_1$ then our results improve the ones in [13,25] with advantages as already stated in Section 3. For example, consider replace L_0 and L by L_1 in (10.10) and (10.11) and let r_*^1, e_k^1 stand for the corresponding radius in [13,25]. Then, we have

$$r_*^1 \le r_*, \tag{10.25}$$

and

$$e_k \le e_k^1. \tag{10.26}$$

Finally, our uniqueness ball $B_{r_*}(x_*)$ is larger than $B_{r_*^1}(x_*)$ given in [13,25]. Hence we have extend the aplicability of Newton-InexP method without additional hypotheses. Concrete examples where (10.25) and (10.26) are strict can be find in [4,5].

10.3   APPLICATIONS

Next, we also present such an example to deal with the restricted convergence domain.

Example 10.1. Let $F \equiv 0$, $C = \{0\}$, $n = m = 3$ and $\kappa = 1$. Define function f on $\Omega = B_1(0)$ by

$$f(x) = \left(e^u - 1, \frac{e-1}{2}v^2 + v, w\right)^T,$$

for $x = (u, v, w)^T$ and $x^* = (0, 0, 0)^T$. Then

$$f'(x) = \begin{bmatrix} e^u & 0 & 0 \\ 0 & (e-1)v + 1 & 0 \\ 0 & 0 & 1 \end{bmatrix}.$$

So for $f'(x_*) = I$,

$$L_0 = e - 1 < L = e^{\frac{1}{L_0}} < L_1 = e. \tag{10.27}$$

Then, using the old formula from [13, 25], $r_*^1 = \frac{2}{3L_1}$ and our new formula $r_* = \frac{2}{2L_0 + L}$, and by using values (10.27) on them we get

$$r_*^1 = 0.24\ldots < r_* = 0.38\ldots.$$

Then, (10.25) and (10.26) are strict. Moreover, the old uniqueness of the solution ball is $B_{\frac{2}{L_1}}(0)$ which is included in our ball $B_{\frac{2}{L_0}}(0)$. Hence, we validate the advantages as stated after Definition 10.5. Thus, the applicability of Newton's method (10.3) is extended. It is worth noting that the computation of L_1 requires that of L_0 and L as special cases. So, these advantages are obtained under the same computational cost as in [13, 25].

10.4   EXERCISES

1. Solve the initial value problem

$$\dot{x}(t) = 1 + \cos(x(t)), x(0) = 0.$$

2. The predator-prey population models describe the interaction of a prey population X and a predator population Y. Assume that their interaction is modeled by the system of ordinary differential equations

$$\dot{x} = x - \frac{1}{4}x^2 - \frac{1}{10}xy + 1$$
$$\dot{y} = -\frac{1}{4}y + \frac{1}{7}xy + \frac{2}{3}.$$

(Assume $x(0) = y(0) = 0$). Solve the system.

3. Solve the Fredholm-type integral equation

$$x(t) = \sum_0^1 \frac{t \cdot s}{10} ds + 1$$

by the iteration method developed in this chapter. Perform 3 steps.

4. Solve the Volterra-type integral equation

$$x(t) = \sum_0^t \frac{t \cdot s}{10} ds + 1$$

by the iteration method developed in this chapter. Perform 3 steps.

5. Solve equation

$$x = \frac{\sin x}{2} + 1$$

by using algorithm developed in this chapter.

6. Repeat the previous problem for equation

$$x = \frac{\sin x + \cos 2x}{10}.$$

REFERENCES

1. Abbas, M., AlShahrani, M., Ansari, Q.H., Iyiola, O.S., and Shehu, Y. 2018. Iterative methods for solving proximal split minimization problems. *Numer. Algorithms*, 78. 1:193–215.
2. Aragón Artacho, F.J., Belyakov, A., Dontchev, A.L., López, M. 2014. Local convergence of quasi-Newton methods under metric regularity. *Comput. Optim. Appl.*, 58. 1:225–247.
3. Aragón Artacho, F.J., Dontchev, A.L., Gaydu, M., Geoffroy, M.H., Veliov, V.M. 2011. Metric regularity of Newton's iteration. *SIAM J. Control Optim.*, 49. 2:339–362.
4. Argyros, I. K. 2008. *Convergence and Application of Newton Type Iterations*. Germany: Springer Berlin.
5. Argyros, I. K., and Magréñan, A. A. 2017. *A Contemporary Study of Iterative Methods*. San Diego, CA: Academic Press, Elsevier.
6. Auslender, A., and Teboulle, M. 2003. *Asymptotic Cones and Functions in Optimization and Variational Inequalities*. Springer Monographs in Mathematics. New York: Springer-Verlag.
7. Behling, R., Fischer, A., Herrich, M., Iusem, A., Ye, and Y. 2014. A Levenberg-Marquardt method with approximate projections. *Comput. Optim. Appl.*, 59. 1-2:5–26.
8. Bellavia, S., and Morini, B. 2006. Subspace trust-region methods for large bound-constrained nonlinear equations. *SIAM J. Numer. Anal.*, 44. 4:1535–1555.
9. Bertsekas, D.P. 1999. *Nonlinear programming, second edn*. Belmont, MA: Athena Scientific Optimization and Computation Series. Athena Scientific.
10. Censor, Y., Gibali, A., and Reich, S. 2012. Algorithms for the split variational inequality problem. *Numer. Algorithms*, 59. 2:301–323.
11. Censor, Y., Gibali, A., Reich, S., and Sabach, S. 2012. Common solutions to variational inequalities. *Set-Valued Var. Anal.*, 20. 2:229–247.
12. Daniel, J.W. 1973. Newton's method for nonlinear inequalities. *Numer. Math.*, 21:381–387.
13. De Oliveira, F. R., Ferreira, O. P., and Silva, G. N. 2019. Newton's method with feasible inexact projections for solving constrained generalized equations. *Comput. Optim. Appl.*, 72:159–177.
14. Dontchev, A.L. 1996. Local analysis of a Newton-type method based on partial linearization. In: *The mathematics of numerical analysis* (Park City, UT, 1995), *Lectures in Appl. Math.*, 32: 295–306.
15. Dontchev, A.L. 1996. Uniform convergence of the Newton method for Aubin continuous maps. *Serdica Math. J.*, 22. 3: 283–296.
16. Dontchev, A.L., and Rockafellar, R.T. 2013. Convergence of inexact Newton methods for generalized equations. *Math. Program.*, 139. 1-2. Ser. B:115–137.
17. Dontchev, A.L., and Rockafellar, R.T. 2014. *Implicit functions and solution mappings, second edn*. Springer Series in Operations Research and Financial Engineering. New York: Springer.
18. Ferreira, O.P. 2015. A robust semi-local convergence analysis of Newton's method for cone inclusion problems in Banach spaces under affine invariant majorant condition. *J. Comput. Appl. Math.*, 279:318–335.
19. Ferreira, O.P., and Silva, G.N. 2017. Kantorovich's theorem on Newton's method for solving strongly regular generalized equation. *SIAM J. Optim.*, 27. 2: 910–926.
20. Ferreira, O.P., and Silva, G.N. 2018. Local convergence analysis of Newton's method for solving strongly regular generalized equations. *J. Math. Anal. Appl.*, 458. 1:481–496.
21. Ferris, M.C., and Pang, J.S. 1997. Engineering and economic applications of complementarity problems. *SIAM Rev.*, 39. 4:669–713.

22. Frank, M., and Wolfe, P. 1956. An algorithm for quadratic programming. *Nav. Res. Log.* 95–110.
23. Fukushima, M., Luo, Z.Q., and Tseng, P. 2001. Smoothing functions for second-order-cone complementarity problems. *SIAM J. Optim.*, 12. 2:436–460 (electronic).
24. Gonçalves, M.L.N., and Oliveira, F.R. 2018. An inexact Newton-like conditional gradient method for constrained nonlinear systems. *Appl. Num. Math.*, 132: 22–34.
25. Gonçalves, M.L.N., and Melo, J.G. 2017. A Newton conditional gradient method for constrained nonlinear systems. *J. Comput. Appl. Math.*, 311: 473–483.
26. Gould, N.I.M., and Toint, P.L. 2002. Numerical methods for large-scale non-convex quadratic programming. In: *Trends in Industrial and Applied Mathematics* (Amritsar, 2001), *Appl. Optim.*, 72:149–179. Kluwer Acad. Publ., Dordrecht.
27. He, H., Ling, C., Xu, H.K. 2015. A relaxed projection method for split variational inequalities. *J. Optim. Theory Appl.*, 166. 1:213–233.
28. Izmailov, A.F., and Solodov, M.V. 2010. Inexact Josephy-Newton framework for generalized equations and its applications to local analysis of Newtonian methods for constrained optimization. *Comput. Optim. Appl.*, 46. 2: 347–368.
29. Josephy, N.H. 1979. Newton's method for generalized equations and the pies energy model. Ph.D. thesis, Department of Industrial Engineering, University of Wisconsin–Madison.
30. Kanzow, C. 2001. An active set-type Newton method for constrained nonlinear systems. In: *Complementarity: Applications, Algorithms and Extensions (Madison, WI, 1999), Appl. Optim.*, 50: 179–200. Kluwer Acad. Publ. Dordrecht.
31. Kimiaei, M. 2017. A new class of nonmonotone adaptive trust-region methods for nonlinear equations with box constraints. *Calcolo*, 54. 3: 769–812.
32. La Cruz, W. 2014. A projected derivative-free algorithm for nonlinear equations with convex constraints. *Optim. Methods Softw.*, 29. 1. 24–41.
33. Lan, G., Zhou, Y. 2016. Conditional gradient sliding for convex optimization. *SIAM J. Optim.*, 26. 2:1379–1409.
34. Marini, L., Morini, B., and Porcelli, M. 2018. Quasi-Newton methods for constrained nonlinear systems: complexity analysis and applications. *Comput. Optim. Appl.*, 71. 1:147–170.
35. Monteiro, R.D.C., and Pang, J.S.: 1999. A potential reduction Newton method for constrained equations. *SIAM J. Optim.*, 9. 3:729–754.
36. Moudafi, A. 2011. Split monotone variational inclusions. *J. Optim. Theory Appl.*, 150. 2:275–283.
37. Nocedal, J., and Wright, S.J. 2006. *Numerical optimization, second edn.* Springer Series in Operations Research and Financial Engineering. New York: Springer.
38. Robinson, S.M. 1972. Extension of Newton's method to nonlinear functions with values in a cone. *Numer. Math.*, 19:341–347.
39. Robinson, S.M. 1979. Generalized equations and their solutions, Part I: Basic theory. *Math. Program. Stud.*, 10:128–141.
40. Robinson, S.M. 1980. Strongly regular generalized equations. *Math. Oper. Res.*, 5. 1:43–62.
41. Robinson, S.M. 1982. Generalized equations and their solutions, Part II: Applications to nonlinear programming. *Math. Program. Stud.*, 19:200–221.
42. Robinson, S.M. 1983. Generalized equations. In: *Mathematical Programming: The State of the Art*, Springer, Berlin.

43. Uko, L.U. 1996. Generalized equations and the generalized Newton method. *Math. Program.*, 73. 3, Ser. A: 251–268.

44. Vanderbei, R.J. 1996. Linear programming: foundations and extensions, *International Series in Operations Research & Management Science*, vol. 4. Boston, MA: Kluwer Academic Publishers.

45. Zhang, Y., and Zhu, D.t. 2010. Inexact Newton method via Lanczos decomposed technique for solving box-constrained nonlinear systems. *Appl. Math. Mech. (English Ed.)*, 31. 12:1593–1602.

11 Inexact Gauss–Newton Method for Solving Least Squares Problems

The aim of this chapter is to extend the applicability of Gauss–Newton method for solving underdetermined nonlinear least squares problems in cases not covered before. The novelty of the chapter is the introduction of a restricted convergence domain. We find a more precise location where the Gauss–Newton iterates lie than in earlier studies. Consequently the Lipschitz constants are at least as small as the ones used before. This way and under the same computational cost, we extend the local as well the semilocal convergence of Gauss–Newton method. The new developments are obtained under the same computational cost as in earlier studies, since the new Lipschitz constants are special cases of the constants used before. Applications further justify the theoretical results.

We consider the *nonlinear least squares* problem

$$\min_{x \in \Omega} \ \zeta(x) := \frac{1}{2} F(x)^T F(x), \tag{11.1}$$

where $\Omega \subseteq \mathbb{R}^n$ is an open set, $F : \Omega \to \mathbb{R}^m$ is a continuously differentiable nonlinear function. A wide variety of applications can be found in mathematical programming literature, see for example [5–19].

It is not hard to see that finding the stationary points of ζ is equivalent to solving the following nonlinear equation

$$\nabla \zeta(x) = F'(x)^T F(x) = 0, \tag{11.2}$$

thus, the Newton's method for solving (11.2) can be used to solving (11.1). However, Newton's method for solving (11.2) requires the computation of the Hessian matrix of ζ at each iteration, and this may be difficult especially for large scale problems (see for instance [8]). A generalization of the Newton method called the Gauss–Newton method (GN), can be used to solve the problem (11.1). This iterative algorithm computes the sequence

$$x_{k+1} = x_k - F'(x_k)^\dagger F(x_k), \qquad k = 0, 1, \ldots,$$

where $F'(x_k)^\dagger$ denotes the Moore-Penrose inverse of the linear operator $F'(x_k)$. Many authors have studied the local as well as semi-local convergence of the Gauss–Newton method; see for instance [1–5].

The study of convergence of iterative algorithms is usually centered into two categories: semi-local and local convergence analysis. The semi-local convergence is

DOI: 10.1201/9781003128915-11

based on the information around an initial point, to obtain conditions ensuring the convergence of these algorithms, while the local convergence is based on the information around a solution to find estimates of the computed radii of the convergence balls. Local results are important since they provide the degree of difficulty in choosing initial points.

In [2] Bao *et al*, considered the case when $m \leq n$, i.e., the problem (11.1) is underdetermined, and they proposed some approximate Gauss–Newton methods for solving (11.1) and they studied the convergence of their method under the full row rank assumption. For that purpose, they noted that, in the case when $F'(x_k)$ is of full row rank, $F'(x_k)^\dagger = F'(x_k)^T (F'(x_k)F'(x_k)^T)^{-1}$ and thus, solving the Gauss–Newton step, $d_k := -F'(x_k)^\dagger F(x_k)$, is equivalent to solving the equation

$$F'(x_k)F'(x_k)^T s_k = -F(x_k), \tag{11.3}$$

and setting the step $d_k := F'(x_k)^T s_k$.

Bao *et al.*, [2], proposed the truncated Gauss–Newton method for solving (11.1) which solves (11.3) inexactly and compute d_k by $d_k := F'(x_k)^T s_k$ and, they considered the assumption that the Fréchet derivatives are Lipschitz continuous and of full row rank and Kantorovich convergence criteria of their algorithm are established.

An usual assumption used to obtain quadratic convergence of Newton's method, is the Lipschitz continuity of F' in a neighborhood of the solution, see [1, 2, 9, 11, 17]. Indeed, ensuring control of the derivative is an important consideration in the convergence analysis of Newton's method. On the other hand, a couple of studies have dealt with the issue of convergence analysis of Newton's method, by relaxing the assumption of Lipschitz continuity of F', see for example [1, 3, 5, 9].

Here, we extend the convergence domain of truncated Gauss–Newton method even further than [2, 5, 7, 12, 19, 21] using our new idea of restricted convergence domains. The idea introduced in this paper can be used on other iterative methods. Numerical examples are also provided to show that our results apply to solve equation but not earlier ones [1, 3]. To achieve this goal, we first introduce the center-Lipschitz condition which determines a subset of the original domain for the mapping containing the iterates. The classical Kantorovich's condition is then related to the subset instead of the original domain. This way, the center-Lipschitz condition is more precise than if they were depending on the original domain of the mapping as in earlier studies. The new techique leads to: weaker sufficient convergence conditions, tighter error bounds on the distance involved and an at least as precise information on the location of the solution. These advantages are obtained under the same computational cost as in earlier studies, since in practice the new conditions are special cases of the Kantorovich's condition.

The remainder of this paper is organized as follows. In Section 11.1, some notations and important results used throughout the paper are presented. In Sections 11.2 and 11.3, the convergence analysis for algorithm TGNU (I) and TGNU (II) are obtained. Section 11.4 contains Applications showing the superiority of the new results.

11.1 MAJORIZING SEQUENCES

The convergence analysis that follows in Section 3 is based on some scalar functions. Define function $f_\beta(.) : \mathbb{R} \cup \{0\} \to \mathbb{R}$ by

$$f_\beta(t) = \alpha t^2 - \beta t + \mu, \tag{11.4}$$

whrere $\alpha > 0$, $\beta > 0$ and $\mu \geq 0$. Suppose that

$$4\alpha\mu \leq \beta^2, \tag{11.5}$$

then f_β has two distinct positive zeros if (11.5) is a strict inequality and one positive zero if equality holds in (11.5).

The smaller zero denoted by t_* is given by

$$t_* = \frac{\beta - \sqrt{\beta^2 - 4\alpha\mu}}{2\alpha}. \tag{11.6}$$

Let $K > 0$, $L > 0$, $\delta \geq 0$ and $\omega \in [0,1)$ be parameters with $K \leq L$. Define

$$\alpha(M, \omega, \delta) = \frac{M(1+\omega)}{2[1 + M\omega\delta(1 + \omega\delta(1+\omega))]},$$

$$\beta(M, \omega, \delta) = 1 - \frac{\omega[1 - M\delta(1+\omega)]}{1 + M\omega\delta(1 + \omega\delta(1+\omega))},$$

$$\mu = \delta(1+\omega)$$

where $M = K$ or L. Set $\alpha := \alpha(K, \omega, \delta)$, $\beta := \beta(K, \omega, \delta)$, $\alpha_1 := \alpha(L, \omega, \delta)$ and $\beta_1 := \beta(L, \omega, \delta)$.

Similarly function $f_{\beta_1}(.) : \mathbb{R} \cup \{0\} \to \mathbb{R}$ is defined by

$$f_{\beta_1}(t) = \alpha_1 t^2 - \beta_1 t + \mu \tag{11.7}$$

and if

$$4\alpha_1\mu \leq \beta_1^2, \tag{11.8}$$

we can set

$$t^* = \frac{\beta_1 - \sqrt{\beta_1^2 - 4\alpha_1\mu}}{2\alpha_1} \tag{11.9}$$

to be the smallest zero of function f_{β_1}.

It was shown in [2, Lemma 3.1] that, if

$$L\delta \leq H \tag{11.10}$$

holds, then (11.8) holds. Clearly, by replacing α_1, β_1, L by α, β, K, respectively in the proof of the lemma, we have that, if

$$K\delta \leq H \tag{11.11}$$

holds, then (11.5) holds, where

$$H = \frac{(1-\omega)^2}{(1+\omega)(\sqrt{(2\omega^2-\omega+1)^2+2\omega(1-\omega)^3}+2\omega^2-\omega+1)}. \qquad (11.12)$$

Condition (11.10) is the sufficient convergence criterion for the semi-local convergence of method TGNU I, II. We shall show that (11.10) can be replaced by (11.11). Notice that

$$L\delta \leq H \Rightarrow K\delta \leq H. \qquad (11.13)$$

Implication (11.13) does not imply

$$K\delta \leq H \Rightarrow L\delta \leq H$$

unless, if $K = L$. Hence, the applicability of the methods TGNU I, II studied in Section 11.2 and Section 11.3 is extended. Moreover, the error bounds are also improved. Similarly, it is easy to see that

$$4\alpha_1\mu \leq \beta_1^2 \Rightarrow 4\alpha\mu \leq \beta^2, \qquad (11.14)$$

since $\alpha \leq \alpha_1$ and $\frac{\beta_1^2}{\alpha_1^2} \leq \frac{\beta^2}{\alpha^2}$.

The implication (11.14) does not imply

$$4\alpha\mu \leq \beta^2 \Rightarrow 4\alpha_1\mu \leq \beta_1^2, \qquad (11.15)$$

unless, if $K = L$.

We need to define some majorizing sequences for method TGNU I, II. First, we define sequences used in [2]:

$$t_0 = 0, \quad t_{k+1} = t_k - \frac{f_{\beta_1}(t_k)}{f'_{1_1}(t_k)} = t_k - \frac{\alpha_1(t_k-t_{k-1})^2+(1-\beta_1)(t_k-t_{k-1})}{2\alpha_1 t_{k-1}}, \qquad (11.16)$$

$$\hat{t}_0 = 0, \quad \hat{t}_{k+1} = \hat{t}_k - \frac{f_{\beta_1}(\hat{t}_k)}{f'_{\beta_1}(t_k)} = \hat{t}_k - \frac{\alpha_1(\hat{t}_k-\hat{t}_{k-1})^2+(1-\beta_1)(\hat{t}_k-\hat{t}_{k-1})}{2\alpha_1\hat{t}_k-\beta_1}, \qquad (11.17)$$

where $f_{1_1}(t) = \alpha_1 t^2 - t + \mu$.

The following result was given in [2].

Lemma 11.1. Let t_*, $\{t_k\}$, $\{\hat{t}_k\}$ be given, respectively by (11.9), (11.16) and (11.17). Suppose that (11.8) holds. Then, the following items hold for each $k = 0, 1, \ldots$

$$t_k < t_{k+1} < t_*, \quad \hat{t}_k < \hat{t}_{k+1} < t_*, \qquad (11.18)$$

$\{t_k\}$ and $\{\hat{t}_k\}$ increasingly converge to t_*. Moreover,

$$\hat{t}_{k+1} - \hat{t}_k \leq \frac{\mu}{\beta_1} \quad \text{for each} \quad k = 0, 1, \ldots \qquad (11.19)$$

and for

$$\gamma_1 = \frac{\beta_1 - \sqrt{\beta_1^2 - 4\alpha_1 \mu}}{\beta_1 + \sqrt{\beta_1^2 - 4\alpha_1 \mu}}$$

$$t_* - \hat{t}_k = \frac{\gamma_1^{2^k - 1}}{\sum_{i=0}^{2^k - 1} \gamma_1^i} t_* \quad \text{for each} \quad k = 0, 1, \dots \tag{11.20}$$

Next, we define the new sequences that shall be shown to be majorizing for method TGNU I, II for $0 \le K_0 \le K$:

$$\bar{r}_0 = 0, \quad \bar{r}_{k+1} = \bar{r}_k - \frac{(1 - K\bar{r}_k)[\alpha(\bar{r}_k - \bar{r}_{k-1})^2 + (1 - \beta)(\bar{r}_k - \bar{r}_{k-1})]}{(1 - K_0\bar{r}_k)(2\alpha_1 \bar{r}_k - 1)} \tag{11.21}$$

$$r_0 = 0, \quad r_{k+1} = r_k - \frac{\alpha(r_k - r_{k-1})^2 + (1 - \beta)(r_k - r_{k-1})}{2\alpha_1 r_k - 1)} = r_k - \frac{f_\beta(r_k)}{f_\beta'(r_k)}, \tag{11.22}$$

$$\bar{s}_0 = 0, \quad \bar{s}_{k+1} = \bar{s}_k - \frac{(1 - K\bar{s}_k)[\alpha(\bar{s}_k - \bar{s}_{k-1})^2 + (1 - \beta)(\bar{s}_k - \bar{s}_{k-1})]}{(1 - K_0\bar{s}_k)(2\alpha\bar{s}_k - \beta)} \tag{11.23}$$

and

$$\hat{s}_0 = 0, \quad \hat{s}_{k+1} = \hat{s}_k - \frac{\alpha(\hat{s}_k - \hat{s}_{k-1})^2 + (1 - \beta)(\hat{s}_k - \hat{s}_{k-1})}{2\alpha\hat{s}_k - \beta} = \hat{s}_k - \frac{f_\beta(\hat{s}_k)}{f_\beta'(\hat{s}_k)} \tag{11.24}$$

Remark 11.1. (a) It follows from the definition of sequences $\{r_k\}, \{\hat{s}_k\}, \{\bar{r}_k\}, \{\hat{r}_k\}$ and a simple inductive argument that for $k = 1, 2, \dots$ and

$$0 \le \bar{r}_k \le r_k, \quad 0 \le \bar{r}_{k+1} - \bar{r}_k \le r_{k+1} - r_k, \tag{11.25}$$

and

$$0 \le \bar{s}_{k+1} \le \bar{s}_k, \quad 0 \le \bar{s}_{k+1} - \bar{s}_k \le \hat{s}_{k+1} - \hat{s}_k, \tag{11.26}$$

so sequences $\{\bar{r}_k\}, \{\hat{r}_k\}$ under the hypotheses of next lemma increasingly converge to their unique least upper bounds $\bar{r}^* = \lim_{k \to +\infty} \bar{r}_k$ and $\bar{s}^* = \lim_{k \to +\infty} \bar{s}_k$ such that

$$\bar{r}^* \le t^* \quad \text{and} \quad \bar{s}^* \le t^*. \tag{11.27}$$

So far sequences $\{\bar{r}_k\}, \{\bar{s}_k\}$ were shown to converge under the same hypotheses (see (11.5) as $\{r_k\}, \{\hat{s}_k\}$. However, these sequences may converge under even weaker hypotheses than (11.5). Such results can be found in [1–5, 17].

(b) If we only suppose (11.8), then as noted above (11.5) also holds. A simple inductive argument again, the definition of the sequences lead to (11.25) and (11.26). Then, clearly we have the analog of Lemma 11.1 for sequence $\{r_k\}$.

Lemma 11.2. Let t_*, $\{r_k\}$, $\{\hat{r}_k\}$ be given, respectively by (11.6), (11.22) and (11.24). Suppose that (11.5) holds. Then, the following items hold

$$r_k < r_{k+1} < t_*, \qquad \hat{s}_k < \hat{s}_{k+1} < t_*, \tag{11.28}$$

$\{r_k\}$ and $\{\hat{s}_k\}$ increasingly converge to t_*. Moreover,

$$\hat{s}_{k+1} - \hat{s}_k \leq \frac{\mu}{\beta} \quad \text{for each} \quad k = 0, 1, \dots \tag{11.29}$$

and for $\gamma = \frac{\beta - \sqrt{\beta^2 - 4\alpha\mu}}{\beta + \sqrt{\beta^2 - 4\alpha\mu}}$

$$t_* - \hat{s}_k = \frac{\gamma^{2^k - 1}}{\sum_{i=0}^{2^k - 1} \gamma^i}. \tag{11.30}$$

Proof. Simply replace $\alpha_1, \beta_1, \gamma_1, t_k, \hat{t}_k, t^*, f_{\beta_1},$ (11.8) by $\alpha, \beta, \gamma, r_k, \hat{s}_k, t_*, f_\beta,$ (11.5), in the proof of Lemma 11.1, respectively. $\qquad\square$

Remark 11.2. In view of Remark 11.1, we have

$$0 \leq \bar{r}_k \leq r_k \leq t_k, \quad 0 \leq \bar{r}_{k+1} - \bar{r}_k \leq r_{k+1} - r_k \leq t_{k+1} - t_k \tag{11.31}$$

and

$$0 \leq \bar{\hat{s}}_{k+1} \leq \hat{s}_k \leq \hat{t}_k, \quad 0 \leq \bar{\hat{s}}_{k+1} - \bar{\hat{s}}_k \leq \hat{s}_{k+1} - \hat{s}_k \leq \hat{t}_{k+1} - \hat{t}_k. \tag{11.32}$$

Hence, in view of (11.30)–(11.32) not only the sufficient convergence criteria are weaker under our new approach (see (11.5), (11.11), replacing (11.8), (11.10), respectively) but also the error bounds on $\|x_{k+1} - x_k\|$, $\|x_k - x^*\|$ are also improved as well as the information on the location of the solution x^*. It is very important to notice that these advantages are obtained under the same computational cost as in [2]. Indeed in practice the computation of parameter L requires the computation of parameters K_0 and K as special cases.

11.2 CONVERGENCE ANALYSIS FOR ALGORITHM TGNU (I)

We present the semi-local and local convergence analysis of the truncated GN method denoted by Algorithm TGNU ($\{\varepsilon_k\}$), for solving problem (11.1) under condition (11.45) that follows.

Algorithm TGNU ($\{\varepsilon_k\}$)

Choose an initial point $x_0 \in \Omega \subseteq \mathbb{R}^n$. For each $k = 0, 1, \dots$, until convergence, do:
Step 1: Compute $F'(x_k)$.
Step 2: Solve (1.8) to find p_k such that the

$$\lambda_k := F'(x_k)F'(x_k)^T p_k + F(x_k) \tag{11.33}$$

satisfies

$$\|F'(x_0)^\dagger \lambda_k\| \leq \varepsilon_k \tag{11.34}$$

Step 3: Set $x_{k+1} = x_k + F(x_k)^T p_k$.

Let $Q : \mathbb{R}^n \to \mathbb{R}^m$ be a linear operator or an $m \times n$ matrix. We denote by Q^T the adjoint of Q. We say that a mapping $Q^\dagger : \mathbb{R}^m \to \mathbb{R}^n$ or an $n \times m$ matrix $Q^\dagger$ is the Moore-Penrose inverse of Q, if the following estimates are satisfied:

$$QQ^\dagger Q = Q, \quad Q^\dagger QQ^\dagger = Q^\dagger, \quad (QQ^\dagger)^T = QQ^T, \quad (Q^\dagger Q)^T = Q^\dagger Q.$$

Moreover, if Q is of full row rank, then $Q^\dagger = Q^T (QQ^T)^{-1}$ and $QQ^\dagger = I_{\mathbb{R}^m}$, where $I_{\mathbb{R}^m}$ stands for the $m \times m$ identity matrix. Furthermore, it follows easily from the definition of the Moore-Penrose inverse, that $(Q^\dagger P)^\dagger = P^\dagger Q$ if P and Q are full rank.

From now on we shall simply say full rank instead of simply full row rank. More details about the properties of Moore-Penrose inverse can be found in [7, 19, 21]. In the rest of the paper, we suppose that $m \leq n$, unless otherwise stated. Let $U(w, \delta)$ stands for the open ball in $\mathbb{R}^n$ with center w and of radius $\delta > 0$. Then, $\bar{U}(w, \delta)$ stands for its closure.

From now on, we shall often use the identity

$$x_{k+1} = x_k - F'(x_k)^\dagger F(x_k) + F'(x_k)^\dagger \lambda_k \tag{11.35}$$

implied by (11.33) and (11.34) provided that $F'(x_k)$ is of full rank.

Let $x_0 \in \mathbb{R}^n$. Define $R := \sup\{t \geq 0 \ : \ U(x_0, t) \subseteq \Omega\}$. Let also $\|.\|$ be the Euclidean vector norm or the induced matrix norm and set $\Omega_0 = U(x_0, R)$. We need the definition of the center Lipschitz condition.

Definition 11.1. A nonlinear operator $G(.) : \Omega_0 \to \mathbb{R}^m$ is said to satisfy the center Lipschitz condition at x_0 with center Lipschitz constant K_0 on Ω_0, if

$$\|G(x) - G(x_0)\| \leq K_0 \|x - x_0\| \quad \text{for each} \quad x, y \in \Omega_0. \tag{11.36}$$

Define $U_0 := \Omega_0 \cap U(x_0, \frac{1}{K_0})$.

Definition 11.2. A nonlinear operator $G(.) : \Omega_0 \to \mathbb{R}^m$ is said to satisfy the restricted Lipschitz condition with Lipschitz constant K on U_0, if

$$\|G(x) - G(y)\| \leq K \|x - y\| \quad \text{for each} \quad x, y \in U_0. \tag{11.37}$$

In earlier studies [2]– [5] the following condition was used instead of the combination of Definition 11.1 and Definition 11.2 that we shall use in the present study.

Definition 11.3. A nonlinear operator $G(.) : \Omega_0 \to \mathbb{R}^m$ is said to satisfy the Lipschitz condition with Lipschitz constant L on Ω_0, if

$$\|G(x) - G(y)\| \leq L \|x - y\| \quad \text{for each} \quad x, y \in \Omega_0. \tag{11.38}$$

Notice however that $U_0 \subseteq \Omega_0$. Hence, we have that

$$K \leq L. \tag{11.39}$$

We also have by Definition 11.1 and Definition 11.3 that

$$K_0 \leq L \tag{11.40}$$

and $\frac{L}{K_0}$ can be arbitrarily large [3–5].

In the rest of the paper, we shall assume that

$$K_0 \leq K. \tag{11.41}$$

Otherwise, i.e., if $K \leq K_0$ the results to follow will also hold with K_0 replacing K. Let us see why the new results improve the earlier results in [2]:

Let $x_0 \in \Omega_0$ be such that $F'(x_0)$ is of full row rank. Set $G(x) = F'(x_0)^\dagger F'(x)$. We need the auxiliary result.

Lemma 11.3. Let $x_0 \in \Omega_0$ be such that $F'(x_0)$ is of full row rank. Suppose that $F'(x_0)^\dagger F'(x)$ satisfies the Lipschitz condition with Lipschitz constant L on Ω_0. Then, for each $x \in U_0$, $F'(x)$ is of full row rank and

$$\|F'(x)^\dagger F'(x_0)\| \leq \frac{1}{1 - L\|x - x_0\|}. \tag{11.42}$$

If we simply use the more precise and needed K_0 instead of L used in the proof of the preceding lemma, we obtain:

Lemma 11.4. Let $x_0 \in \Omega_0$ be such that $F'(x_0)$ is of full row rank. Suppose that $F'(x_0)^\dagger F'(x)$ satisfies the center-Lipschitz condition with Lipschitz constant K_0 on Ω_0. Then, for each $x \in U_0$, $F'(x)$ is of full row rank and

$$\|F'(x)^\dagger F'(x_0)\| \leq \frac{1}{1 - K_0\|x - x_0\|}. \tag{11.43}$$

Notice that in view of (11.40), (11.43) is a more precise than (11.42) on the norm $\|F'(x)^\dagger F'(x_0)\|$. That exchange of upper bounds in the proofs of the results leads to a tighter convergence analysis. On the other hand, K can replace L on the upper bounds on $\|F'(x_0)^\dagger F'(x_k)\|$ leading again to more precise uppers bounds on this norm. Then, in view of (11.41), we can reproduce all the proofs of the semi-local results in [2] with simply K, K_0 replacing L, L_0, respectively.

The same technique can be used to improve the local results in [2]. It is convenient for the semi-local convergence analysis that follows to introduce some parameter, sequences and conditions.

Set

$$\rho = \|F'(x_0)^\dagger F(x_0)\|. \tag{11.44}$$

Let $\{\eta_k\}$ be a non-negative sequence satisfying $0 \leq \eta := \sup_{k \geq 0} \eta_k < 1$ for some $\eta \in [0, 1)$. We shall suppose that

$$\varepsilon_k \leq \|F'(x_0)^\dagger F(x_k)\| \tag{11.45}$$

and

$$\varepsilon_k \le \|F'(x_0)^\dagger F(x_k)\|^2. \tag{11.46}$$

Next, we present the semi-local convergence analysis of sequence $\{x_n\}$ generated by the Algorithm TGNU $(\{\varepsilon_k\})$.

Theorem 11.1. Suppose that there exists $x_0 \in \Omega_0$ be such that $F'(x_0)$ is of full row rank and $U(x_0, R) \subseteq \Omega$. Moreover, suppose that $F'(x_0)^\dagger F'(.)$ satisfies the center-Lipschitz condition at x_0 with Lipschitz constant K_0 on Ω_0 and the restricted Lipschitz condition with constant K on U_0, (11.11) holds and $r^* \le R$, where R, K_0, K, Ω_0, U_0 and R^* are defined previously. Then, sequence $\{x_k\}$ generated by (11.35) with starting point x_0 by the method with sequence $\{\varepsilon_k\}$ satisfying (11.45) is well defined, remains in $U(x_0, r^*)$ for each $k = 0, 1, ...$ and converges to a solution x_* of equation $F(x) = 0$ in $\bar{U}(x_0, r^*)$.

Moreover, the following error bound holds:

$$\|x_k - x_*\| \le r^* - r_k \quad \text{for each} \quad k = 0, 1, ... \tag{11.47}$$

Proof. By simply following the proof of Theorem 3.1 [2], replacing sequence $\{t_k\}$ by $\{r_k\}$ and the L_0, L by K_0, K, respectively, we obtain the estimates with the crucial differences (see also Remark 3.7).

The estimates for $j = 0, 1, ..., k$

$$\|F'(x_j)^\dagger F'(x_0)\| \le \frac{1}{1 - K_0\|x_j - x_0\|} \le \frac{1}{1 - K_0 r_j}, \tag{11.48}$$

$$\|F'(x_0)^\dagger F'(x_j)\| \le \frac{(1 - K r_j)(r_{j+1} - r_j)}{1 + \omega}, \tag{11.49}$$

and by (11.22), (11.35), (11.48), and (11.49), we get

$$\begin{aligned}
\|x_{j+1} - x_j\| &= \|F'(x_j)^\dagger F'(x_0)(-F'(x_0)^\dagger F(x_j) + F'(x_0)^\dagger \lambda_j\| \\
&\le (1 + K)\|F'(x_j)^\dagger F'(x_0)\|\|F'(x_0)^\dagger F(x_j)\| \le \bar{r}_{j+1} - \bar{r}_j \le r_{j+1} - r_j,
\end{aligned} \tag{11.50}$$

so

$$\|x_{k+1} - x_k\| \le \bar{r}_{j+1} - \bar{r}_j \le r_{j+1} - r_j. \tag{11.51}$$

The rest of the proof follows as in Theorem 3.1 with the noted modifications. $\quad\square$

Remark 11.3. (a) The corresponding estimates in [2] are less tight (see (11.5), (11.6), (11.8), (11.9), (11.10), (11.11), (11.13)–(11.15)),

$$\|F'(x_j)^\dagger F'(x_0)\| \le \frac{1}{1 - L\|x_j - x_0\|} \le \frac{1}{1 - L t_j},$$

$$\|F'(x_0)^\dagger F'(x_j)\| \le \frac{(1 - L t_j)(t_{j+1} - t_j)}{1 + \omega}$$

and

$$\|x_{j+1} - x_j\| \le t_{j+1} - t_j.$$

(b) If $m = n$, (11.45) and (11.46) reduce to

$$\|F'(x_0)^{-1}\lambda_k\| \le \varepsilon_k \|F'(x_0)^{-1}F(x_k)\| \quad \text{and} \quad \|F'(x_0)^{-1}\lambda_k\| \le \varepsilon_k \|F'(x_0)^{-1}F(x_k)\|^2$$
(11.52)

where,

$$k = 0, 1, \ldots$$

This way we obtain the next corollary that follows for inexact Newton methods [2, 4, 5, 14]. This corollary improves Corollary 3.1 (of Theorem 3.1 in [14]) which in turn improved the results in [4, 14].

Corollary 11.1. Suppose that $m = n$ and there exists $x_0 \in \Omega$ such that $F'(x_0)^{-1} \in L(\mathbb{R}^n, \mathbb{R}^n)$ and $U(x_0, R) \subseteq \Omega$. Moreover, suppose that $F'(x_0)^{-1}F'(.)$ satisfies the center-Lipschitz condition at x_0 with center Lipschitz constant K_0 on Ω_0 and the restricted Lipschitz condition with Lipschitz constant K on U_0, (11.11) holds and $r^* \le R_0$. Then, sequence $\{x_k\}$ generated by method (11.35) (with $F'(x)^{\dagger} = F'(x)^{-1}$) with sequence $\{\varepsilon_k\}$ satisfying (11.46) is well defined, remains in $U(x_0, r^*)$ and converges to a solution x_* of equation $F(x) = 0$ in $\bar{U}(x_0, r^*)$.

Moreover, the following error bound holds:

$$\|x_k - x_*\| \le r^* - r_k \quad \text{for each} \quad k = 0, 1, \ldots$$
(11.53)

Next, we present the local convergence analysis of method (11.35) with (11.11) satisfied using Theorem 11.1. To achieve this, we define

$$\hat{K} = \frac{K}{1 - K_0\rho} \quad \text{for some} \quad \rho > 0 \quad \text{to be determined later.}$$
(11.54)

Notice that

$$\hat{L} = \frac{L}{1 - \hat{L}\rho}$$
(11.55)

was used in [2]. We have that

$$\hat{K} \le \hat{L}.$$
(11.56)

This modification leads to advantages similar to the semi-local convergence casa and under the same computational cost (see also Remark 11.4 and the numerical section).

The proofs of the next three results are obtained from the corresponding ones in Lemma 3.2, Theorem 3.2 and Corollary 3.2 in [2], respectively by using the modification already suggested in Theorem 11.1. So these proofs are omitted (see also Remark 11.4).

Lemma 11.5. Suppose there exists $x^* \in \Omega$ solving equation $F(x) = 0$, such that $F'(x^*)$ is of full rank and $U(x^*, R) \subseteq \Omega$. Moreover, suppose that $F'(x^*)^{\dagger}F'(.)$ satisfies the center-Lipschitz condition at x^* with constant K_0 on Ω_0 (with $x_0 = x^*$) and the restricted Lipschitz condition with Lipschitz constant K on U_0 (with $x_0 = x^*$). Furthermore, suppose that $0 < \rho < \{R, \frac{1}{K_0}\}$. Then, the following items hold for each $x \in U(x^*, r)$:

1 $F'(x_0)$ is of full rank and $\|F'(x_0)^\dagger F'(x^*)\| \le \frac{1}{1-K_0\rho}$;

2 $F'(x_0)^\dagger F'(.)$ satisfies the Lipschitz condition with Lipschitz constant $\hat{K}$ given in (11.54 on $U(x_0, R-\rho)$;

3 $\delta \le \frac{K\rho^2}{2(1-K_0\rho)} + \rho$, where $\delta := \|F'(x_0)^\dagger F(x_0)\|$.

Theorem 11.2. Suppose that there exists $x^* \in \Omega$ solving equation $F(x) = 0$, so that $F'(x^*)$ is of full rank and $U(x^*, R) \subseteq D$. Moreover, suppose that $F'(x^*)^\dagger F'(.)$ satisfies the center-Lipschitz condition at x_0 with constant K_0 on D_0 and the restricted Lipschitz condition with Lipschitz constant K on U_0. Define

$$\rho := \min\left\{ \frac{(1-\omega)R}{1-\omega+4(1+\omega)^2}, \frac{1}{K}\left(1 - \frac{1}{\sqrt{1+2H}}\right)\right\}. \tag{11.57}$$

Then, sequence $\{x_k\}$ generated by method (3.3) for $x_0 \in U(x^*, \rho)$ and $\{\varepsilon_k\}$ satisfying (11.45 is well defined, remains in $U(x^*, \rho)$ for each $k = 0, 1, 2, \ldots$ and converges to a solution $\hat{x}_*$ of equation $F(x) = 0$.

We also have the following corollary of Theorem 11.2 for the GN method.

Corollary 11.2. Suppose that there exists $x^* \in \Omega$ solving equation $F(x) = 0$ such that $F'(x^*)$ is of full rank and $U(x^*, R) \subseteq D$. Suppose that $F'(x^*)^\dagger F'(.)$ satisfies the center-Lipschitz condition with constant K_0 on D_0 and the restricted Lipschitz condition with Lipschitz constant K on U_0. Define

$$\bar{\rho} := \min\left\{ \frac{R}{5}, \frac{2-\sqrt{2}}{2K}\right\}. \tag{11.58}$$

Then, sequence $\{x_k\}$ generated for $x_0 \in U(x^*, \bar{\rho})$ by GN method, is well defined, remains in $U(x^*, \bar{\rho})$ for each $k = 0, 1, 2, \ldots$ and converges to a solution $\hat{x}_*$ of equation $F(x) = 0$.

Remark 11.4. (a) The radii of convergence in [2] are given respectively by,

$$\rho_1 := \min\left\{ \frac{(1-\omega)R}{1-\omega+4(1+\omega)^2}, \frac{1}{L}\left(1 - \frac{1}{\sqrt{1+2H}}\right)\right\} \tag{11.59}$$

and

$$\bar{\rho}_1 := \min\left\{ \frac{R}{5}, \frac{2-\sqrt{2}}{2L}\right\}. \tag{11.60}$$

then, we have that

$$\rho_1 \le \rho \tag{11.61}$$

and

$$\bar{\rho}_1 \le \bar{\rho}. \tag{11.62}$$

That is we obtain an at least as large radius of convergence leading to a wider choice of initial points. Moreover, as already shown in Section 2 the error bounds on the distances $\|x_{n+1} - x_n\|$, $\|x_n - \hat{x}_*\|$ are tighter leading to fewer iterations to achieve a desired error tolerance. It is worth noticing that the preceding advantages are obtained under the same computational effort as in [2], since in practice the computation of L requires the computation of K_0 and K as special cases.

(b) The results obtained here can be improved even further, if we consider instead of the ball U_0 the ball U_1 defined by

$$U_1 := \Omega \cap U\left(x_1, \frac{1}{K_0} - \|F'(x_0)^\dagger F(x_0)\|\right).$$

Notice that $U_1 \subseteq U_0 \subseteq U_0 \subseteq \Omega$, so the Lipschitz constant K can be replaced by a constant $\hat{K}$ at least at small. Then, $\hat{K}$ can replace K in all preceding results. The iterates $\{x_k\}$ lie in U_1 according to the proofs (see also the Applications).

11.3 CONVERGENCE ANALYSIS FOR ALGORITHM TGNU (II)

We present the semi-local as well as the local convergence analysis of Algorithm TGNU (see method (11.35) for solving problem (1.1) under condition (11.46) along the same lines of Section 4 but some of the parameters are defined differently. We also use sequences $\{\bar{s}_k\}$, $\{\hat{s}_k\}$ instead of $\{\bar{r}_k\}$, $\{\hat{r}_k\}$, respectively (or $\{t_k\}$, $\{\hat{t}_k\}$, respectively in [2]).

The proofs are obtained as in Section 3 with the above modifications, therefore, these proof is omitted.

As in Section 3, it is convenient to introduce some parameters

$$\alpha := \frac{K(1+\omega\delta)}{2[1+K\delta^2\omega(1+\delta\omega)]} + \frac{\omega}{1+\omega(\delta-1)}, \quad \beta := 1 + \frac{\omega}{1+\omega(\delta-1)} \qquad (11.63)$$

and

$$\mu := 1 + \beta\delta(1+\omega\delta). \qquad (11.64)$$

If

$$\delta \leq \frac{1}{\sqrt{(K+2\omega)^2 + 2K\omega} + K + 2\omega} \qquad (11.65)$$

then (11.5) holds.

Theorem 11.3. Suppose that there exists $x_0 \in \Omega$ such that $F'(x_0)$ is of full rank and $U(x_0, R) \subseteq \Omega$. Moreover, suppose that $F'(x_0)^\dagger F'(.)$ satisfies the center-Lipschitz condition at x_0 with Lipschitz constant K_0 on Ω_0, the restricted Lipschitz condition with constant K on U_0, (11.65) holds and $\hat{s}^* \leq R$ amd $\{\varepsilon_k\}$ satisfies (11.46). Then, sequence $\{x_k\}$ generated by method (11.35) is well defined, remains in $U(x_0, \hat{s}^*)$ for each $k = 0, 1, \dots$ and converges quadratically to a solution x_* of equation $F(x) = 0$ in $\bar{U}(x_0, \bar{r}^*)$. Moreover, the following error bound holds:

$$\|x_k - x^*\| \leq \frac{\gamma^{2^k-1}}{\sum_{i=0}^{2^k-1}\gamma^i}\bar{r}^*, \qquad (11.66)$$

where $\bar{r}^*$, γ are given in Section 2 for α, β, μ given in (11.61) and (11.62).

Remark 11.5. (a) Let $H(\omega) := H$ be defined by (11.12) for each $\omega \in (0,1)$ and let $\hat{H}$ be defined by

$$\hat{H}(\omega,K) = \frac{K}{\sqrt{(K+2\omega)^2 + 2K\omega + K + 2\omega}} \tag{11.67}$$

for each $K \in [0,+\infty)$ and $\omega \in (0,1)$. Then, (11.11) and (11.65) are equivalent to

$$K\delta \leq \hat{H}(\omega,K) \tag{11.68}$$

and

$$K\delta \leq \hat{H}(\omega), \tag{11.69}$$

respectively.

(b) If $\varepsilon_k \equiv 0$, method (3.3) is reduced to the GN method. Choose $\omega_k \equiv 0$, then we have

$$H = \frac{1}{2}, \quad \bar{r}^* = \frac{1 - \sqrt{1 - 2K\delta}}{K}, \quad \text{and} \quad \gamma = \frac{1 - \sqrt{1 - 2K\delta}}{1 + \sqrt{1 - 2K\delta}}. \tag{11.70}$$

Then, we obtain from Theorem 11.3 the following improvement of the Kantorovich-like theorem for the GN method.

Corollary 11.3. Suppose that there exists $x_0 \in \Omega$ such that $F'(x_0)$ is of full rank and $U(x_0, R) \subseteq \Omega$. Moreover, suppose that $F'(x_0)^\dagger F'(.)$ satisfies the center-Lipschitz condition with Lipschitz constant K_0 on Ω_0, the restricted Lipschitz condition with constant K on U_0. Furthermore, suppose that

$$0 < K\delta \leq \frac{1}{2} \quad \text{and} \quad \bar{s}^* \leq R_0. \tag{11.71}$$

Then, sequence $\{x_k\}$ generated by method (11.35) is well defined, remains in $U(x_0, \bar{s}^*)$ for each $k = 0, 1, \ldots$ and converges quadratically to a solution x_* of equation $F(x) = 0$ in $\bar{U}(x_0, \bar{s}^*)$. Moreover, estimates (11.66) holds, where $\bar{s}^*$ and γ are given in (11.70).

Next, we present the local convergence analysis of method (11.35) under condition (11.46).

Theorem 11.4. Suppose there exists $x^* \in \Omega$ solving equation $F(x) = 0$ such that $F'(x^*)$ is of full rank and $U(x^*, R) \subseteq \Omega$. Moreover, suppose that $F'(x^*)^\dagger F'(.)$ satisfies the center-Lipschitz condition with Lipschitz constant K_0 on Ω_0 and the restricted Lipschitz condition with constant K on U_0. Define

$$\rho := \min\left\{ \frac{2}{11}R, \frac{1}{3(\sqrt{(K+\omega)^2 + K\omega + K + \omega})} \right\}. \tag{11.72}$$

Then, sequence $\{x_k\}$ generated for $x_0 \in U(x^*, \rho)$ by method (11.35) with sequence $\{\varepsilon_k\}$ satisfying (11.46) is well defined, remains in $U(x^*, \rho)$ for each $k = 0, 1, \ldots$ and converges quadratically to a solution $\hat{x}_*$ of equation $F(x) = 0$.

Remark 11.6. (a) The results of this section improve the correponding results Lemma 3.3, Theorem 3.3, Remark 3.4, Corollary 3.3 and Theorem 3.4 in [2] along the same lines of the arguments and comparison made in the Remarks of Section 2, Section 3 and this Section.
(b) The rest of the results in Section 4 in [2] are also immediately improved along the same lines. However, we leave the details to the motivated reader.

11.4 APPLICATIONS

We present numerical examples to show that the earlier results do not apply or if they apply, our results also apply and can do better. For simplicity we choose $m = n$ in the next two examples. The first example is given for the semi-local case.

Example 11.1. Let $m = n = 1$, $\Omega_0 = \Omega = U(0,1)$, $\omega = 0$, $\lambda_k = 0$, $x_0 = 1$, $R = 1$. Define function F on Ω_0 by

$$F(x) = x^3 - h, \quad \text{for some} \quad h \in (0, \frac{1}{2}). \tag{11.73}$$

Then, we have $\delta = \frac{1}{3}(1-h)$, $K_0 = 3-h$, $L = 2(2-h)$, $\alpha_1 = \frac{L}{2} = 2-h$, $\beta_1 = 1$ and $\mu = \delta$. Then, old condition (11.8) is not satisfied, since

$$4\alpha_1\mu > \beta_1^2 \quad \text{for each} \quad h \in (0, \frac{1}{2}). \tag{11.74}$$

Notice also that condition (11.8) in this special case is the famous for its simplicity and clarity Kantorovich sufficient convergence criterion for the convergence of Newton's method [2, 4, 11, 17] (see also (11.71) with $K = L$). Therefore, there is no guarantee under the old results that Newton's method converges to $x^* = \sqrt[3]{h}$ starting at $x_0 = 1$.

Using our results, we have
Case: $U_0 = \Omega_0 \cap U\left(x_0, \frac{1}{K_0}\right) = U\left(x_0, \frac{1}{K_0}\right)$.
Then, $F'(x_0)^{-1}F'(.)$ is restricted Lipschitz with $K = 2\left(\frac{4-h}{3-h}\right)$. We also have that $\alpha = \frac{K}{2}$, $\beta = 1$ and $\mu = \delta$.
Then, condition (11.5) becomes

$$\frac{4}{3}\frac{(4-h)(1-h)}{3-h} \leq 1 \tag{11.75}$$

which is satisfied for $h \in I_0 := [.46198316..., \frac{1}{2})$. The same range for h is obtained, if we use (11.71).
Case: $U_1 = \Omega_0 \cap U\left(x_1, \frac{1}{K_0} - \delta\right)$.
We need the computation

$$\|F'(x_0)^{-1}(F'(x) - F'(y))\| \leq (\|x - x_1\| + \|y - x_1\| + 2\|x_1\|)\|x - y\|$$
$$\leq 2\left[\left(\frac{1}{K_0} - \delta\right) + \frac{2+h}{3}\right]\|x - y\|,$$

so

$$\hat{K} = \frac{-2(2h^2 - 5h - 6)}{3 - h}.$$

Using these values, (11.5) is satisfied for all $h \in I := \left[.44137239, \frac{1}{2}\right]$ which extends the interval found in preceding case. Notice also that $\delta K_0 < 1$ for all $h \in I$. Therefore, if $h \in I_0$ or $h \in I$ our results guarantee the convergence of Newton's method to x^* starting at $x_0 = 1$.

Finally, as already noted in Section 2, if both (11.5) and (11.8) hold, then our results provide tighter error bounds on the distances $\|x_{n+1} - x_n\|$, $\|x_n - x^*\|$ and a more precise information on the location of the solution x^*. Notice also that $K_0 < L$ and $\hat{K} < L$ for all $h \in [0, \frac{1}{2})$ and $K_0 < \hat{K}$, if

$$h > 2 - \sqrt{3}, \quad K_0 > \hat{K}, \quad \text{if} \quad h < 2 - \sqrt{3}$$

and

$$K_0 = \hat{K}, \quad \text{if} \quad h = 2 - \sqrt{3}.$$

The second example concerns the local case.

Example 11.2. Let $m = n = 3$ and $\Omega = U(x^*, 1)$, so $R = 1$ and $\Omega_0 = \Omega$. Choose $\lambda_k = 0$, so $\omega = 0$. Define mapping F on Ω for $w = (x, y, z)^T$ by

$$F(w) = \left(e^x - 1, \frac{e - 1}{2}y^2 + y, z \right)^T. \tag{11.76}$$

Then, we have $x^* = (0, 0, 0)^T$ and the Fréchet-derivative is given by

$$F'(w) = \begin{bmatrix} e^x & 0 & 0 \\ 0 & (e-1)y + 1 & 0 \\ 0 & 0 & 1 \end{bmatrix}. \tag{11.77}$$

We obtain $L = e$, $K_0 = e - 1$, and $U_0 = \Omega_0 \cap U\left(x^*, \frac{1}{K_0}\right) = U\left(x^*, \frac{1}{K_0}\right)$, $K = e^{\frac{1}{K_0}}$. By using (11.57)–(11.60), we get that

$$\rho = \bar{\rho} = 0.16366659$$

and

$$\rho_1 = \bar{\rho}_1 = 0.10172259,$$

so $\rho_1 = \bar{\rho}_1 < \rho = \bar{\rho}$. Hence, our convergence radii are larger than the ones in [2]. It is worth noticing that larger radius of convergence implies a wider choice of initial points and fewer iterations to achieve a desired error tolerance. These improvements are important in computational mathematics.

11.5 EXERCISES

1. Solve the initial value problem
$$\dot{x}(t) = 1 + \cos(x(t)), x(0) = 0.$$

2. The predator-prey population models describe the interaction of a prey population X and a predator population Y. Assume that their interaction is modeled by the system of ordinary differential equations
$$\dot{x} = x - \frac{1}{4}x^2 - \frac{1}{10}xy + 1$$
$$\dot{y} = -\frac{1}{4}y + \frac{1}{7}xy + \frac{2}{3}.$$
   (Assume $x(0) = y(0) = 0$). Solve the system.

3. Consider the following equation in R^2:
$$x_1(t+1) = 0.8\sin\left(x_1(t) + \frac{\pi}{4}\right) + 0.2x_2(t)$$
$$x_2(t+1) = 0.8x_1(t) + 0.1x_2(t).$$

4. Solve the initial value problem
$$\dot{x} = (x(t) + t, x(0) = 1)$$
   by the iteration method developed in this chapter. Perform 3 steps.

5. Solve the Fredholm-type integral equation
$$x(t) = \sum_0^1 \frac{t \cdot s}{10} ds + 1$$
   by the iteration method developed in this chapter. Perform 3 steps.

6. Solve the Volterra-type integral equation
$$x(t) = \sum_0^t \frac{t \cdot s}{10} ds + 1$$
   by the iteration method developed in this chapter. Perform 3 steps.

7. Solve equation
$$x = \frac{\sin x}{2} + 1$$
   by using algorithm developed in this chapter.

8. Repeat the previous problem for equation
$$x = \frac{\sin x + \cos 2x}{10}.$$

REFERENCES

1. Argyros, I.K. 1998. On a new Newton-Mysovskii-type theorem with applications to inexact Newton-like methods and their discretizations. *IMA Journal of Numerical Analysis*, 18. 1:37–56.
2. Argyros, I.K. 1997. Inexact Newton methods and nondifferentiable operator equations on Banach spaces with a convergence structure. *Approximation Theory and its Applications*, 13. 3:91–103.
3. Argyros, I.K. and Hilout, S. 2012. Weaker conditions for the convergence of Newton's method. *J. Complexity*, 28:364–387.
4. Argyros, I.K., and Magreñán, Á.A. 2017. *Iterative Methods and Their Dynamics with Applications: A Contemporary Study*. New York: CRC Press.
5. Argyros, I.K., and Szidarovszky, F. 1993. *The Theory and Applications of Iteration Methods*. Boca Raton, Florida: CRC Press.
6. Bao, J.F., Li, C., Shen, W.P., Yao, J.C., and Guu, S.M. 2017. Approximate Gauss–Newton methods for solving underdetermined nonlinear least squares problems. *Applied Numerical Mathematics*, 111(Supplement C):92–110.
7. Ben-Israel, A., and Greville, T.N.E. 2003. *Generalized Inverses: Theory and Applications*.
8. Björck, A. 1996. *Numerical Methods for Least Squares Problems*. Society for Industrial and Applied Mathematics.
9. Brown, P.N. 1987. A local convergence theory for combined inexact-Newton/finite-difference projection methods. *SIAM Journal on Numerical Analysis*, 24(2):407–434.
10. Dembo, R.S., Eisenstat, S.C., and Steihaug, T. 1982. Inexact Newton methods. *SIAM J. Numer. Anal.*, 19(2):400–408.
11. Dennis, J.E., and Schnabel, R.B. 1996. *Numerical Methods for Unconstrained Optimization and Nonlinear Equations*, volume 16 of *Classics in Applied Mathematics*. Philadelphia: Society for Industrial and Applied Mathematics (SIAM).
12. Ezquerro, J.A., Hernández, M.A. 2017. *Newton's Method: An Update Approach of Kantorovich's Theory*. Cham Switzerland: Birkhauser, Springer.
13. Gutiérrez, J.M., and Hernández, M.A. 2000. Newton's method under weaker Kantorovich conditions. *IMA J. Numer. Anal.*, 20(4):521–532.
14. Guo, X. 2007. On semilocal convergence of inexact Newton methods. *Journal of Computational Mathematics*, 25(2):231–242.
15. Häußler, W.M. 1986. A Kantorovich-type convergence analysis for the Gauss–Newton method. *Numerische Mathematik*, 48(1):119–125.
16. Josephy, N. 1979. *Newton's Method for Generalized Equations and the PIES Energy Model*. University of Wisconsin–Madison.
17. Kantorovich, L.V., and Akilov, G.P. 1964. *Functional Analysis in Normed Spaces*. New York: The Macmillan Co.
18. Magreñán, Á.A. 2014. Different anomalies in a Jarrat family of iterative root finding methods. *Appl. Math. Comput.*, 233:29–38.
19. Magreñán, Á.A. 2014. A new tool to study real dynamics: The convergence plane. *Applied Mathematics and Computation*, 248(Supplement C):215–224.
20. Ortega, J, and Rheinboldt, W. 2000. *Iterative Solution of Nonlinear Equations in Several Variables*. Society for Industrial and Applied Mathematics.
21. Stewart, G.W., and Guang, J.S. 1990. *Matrix Perturbation Theory*. Academic Press.
22. Ypma, T.J. 1984. Local convergence of inexact Newton methods. *SIAM Journal on Numerical Analysis*, 21(3):583–590.

12 The Kantorovich's Theorem on Newton's Method for Solving Generalized Equation

The aim of this chapter is to extend the applicability of Newton's method for solving a generalized equation of the type $f(x) + F(x) \ni 0$ in Banach spaces, where f is a Fréchet differentiable function and F is a set-valued mapping. The novelty of the chapter is the introduction of a restricted convergence domain. Using the idea of a weaker majorant, the convergence of the method, the optimal convergence radius, and results of the convergence rate are established. That is we find a more precise location where the Newton iterates lie than in earlier studies. Consequently, the Lipschitz constants are at least as small as the ones used before. This way and under the same computational cost, we extend the semilocal convergence of the Newton iteration for solving $f(x) + F(x) \ni 0$. The strong regularity concept plays an important role in our analysis. We finally present numerical examples, where we can solve equations in cases not possible before without using additional hypotheses.

In the chapter, we deal with the problem of finding a point $x^* \in X$ satisfying the following generalized equation

$$f(x) + F(x) \ni 0, \tag{12.1}$$

where $f : \Omega \to Y$ is a Fréchet differentiable function, X and Y are Banach spaces, $\Omega \subseteq X$ is an open set and $F : X \rightrightarrows Y$ is a set-valued mapping. It is well-known that the generalized equation (12.1) is an abstract model for a wide range of problems in mathematical programming. See, for instance, [3, 5, 11, 12] as part of a whole. In the case $F \equiv \{0\}$, (12.1) becomes the nonlinear equation $f(x) = 0$. A particular case of problem (12.1) is when $F = -C$, where $C \subset X$ is a nonempty closed convex cone. Thus, problem (12.1) becomes

$$f(x) \in C. \tag{12.2}$$

If F is the normal cone mapping N_C of a convex set C in Y, then (12.1) is a variational inequality problem, which covers a wide range of problems in nonlinear programming, including linear and nonlinear complementary problems; additional comments about such problems can be found in [3, 5, 5, 11, 12, 18].

Newton's method for solving (12.1) utilizes the iteration

$$f(x_k) + f'(x_k)(x_{k+1} - x_k) + F(x_{k+1}) \ni 0, \qquad k = 0, 1, \dots \tag{12.3}$$

DOI: 10.1201/9781003128915-12

for x_0, a given initial point. As it is well-known, the generalized equation (12.1) has considerable scope in classical analysis and its applications. When $F \equiv 0$, the iteration (12.3) becomes the standard Newton method for solving the nonlinear equation $f(x) = 0$,

$$f(x_k) + f'(x_k)(x_{k+1} - x_k) = 0, \qquad k = 0, 1, \dots. \tag{12.4}$$

N. H. Josephy, in his Ph.D. thesis [11] studied Newton's method for solving $f(x) + N_C(x) \ni 0$, where $f : \Omega \to \mathbb{R}^m$ is continuously differentiable, $\Omega \subseteq \mathbb{R}^n$ is an open set, and $C \subset \mathbb{R}^m$ is a convex set. To validate the definition of the sequence generated by the method, the *strong regularity* property on $f + N_C$, a concept introduced by Robinson in [13], was used. If $X = Y$ and $N_C = \{0\}$, then strong regularity at x is equivalent to $f'(x)^{-1}$ being a continuous linear operator. An important case is when (12.1) represents Karush-Kuhn-Tucker's systems for the standard nonlinear programming problem with a strict local minimizer. In this case, the strong regularity of this system, along with the condition that the primal variable is an optimal solution, is equivalent to the linear independence of the gradients of the active constraints and a strong second-order sufficient optimality condition; see [4, Theorem 6].

A usual assumption used to obtain quadratic convergence of Newton's method (12.3), for solving equation (12.1), is the Lipschitz continuity of f' in a neighborhood of the solution. Indeed, ensuring control of the derivative is an important consideration in the convergence analysis of Newton's method. On the other hand, a couple of studies have dealt with the issue of convergence analysis of Newton's method, for solving the equation $f(x) = 0$, by relaxing the assumption of Lipschitz continuity of f', see for example [10, 16, 17].

The idea of the majorant function has been shown to be an appropriate and powerfull tool for the convergence of Newton-like methods. The convergence domain for such methods is small in general. In the present study, we extend the convergence domain for the Newton's method. To achieve this goal, we first introduce the center-Lipschitz condition which determines a subset of the original domain for the mapping containing the iterates. The majorant functions are then related to the subset instead of the original domain. This way, the majorant functions are more precise than if they were depending on the original domain of the mapping as in earlier studies. The new techique leads to: weaker sufficient convergence conditions, tighter error bounds on the distance involved and an at least as precise information on the location of the solution. These advantages are obtained under the same computational cost as in earlier studies, since in practice the new majorant functions are special cases of the old majorant functions. In [14, 15], (12.1) was considered for F maximal monotone acting between Hilbert spaces and a Kantorovich-type and local analysis were obtained under majorant condition. In [8], under strong regularity at the solution of (12.1) and a majorant condition, it was shown that (12.3) is locally convergent at a quadratic rate. Besides, another advantage of working with a majorant condition rests in the fact that it allows unifying of several convergence results pertaining to Newton's method; see [10, 16, 17]. The analysis presented provides a clear relationship between the majorant function and the function defining the generalized equation. In addition, it allows us to obtain the optimal convergence radius for the method with respect to the majorant condition and uniqueness of the solution. The analysis of this method, under Lipschitz's condition, is provided as a special case. In the concluding Section 12.3, we have presented examples, where earlier results [8–11,14,15] cannot be used to solve equations. However, our results can apply to solve these equations and under the same computational cost.

The remainder of this chapter is structured as follows. In Section 12.1, some notations and important results used throughout the paper are presented. In Section 12.2, the main result is stated and proved. In Section 12.3, some applications of this result are given.

12.1 PRELIMINARIES

The following notations and results are used throughout the paper. The *open* and *closed balls* at x with radius $\delta \geq 0$ are denoted, respectively, by $B(x,\delta) = \{y \in X : \|x-y\| < \delta\}$ and $B[x,\delta] = \{y \in X : \|x-y\| \leq \delta\}$. We denote by $L(X,Y)$ the *space consisting of all continuous linear mappings* $A : X \to Y$ and the *norm* of A by $\|A\| := \sup\{\|Ax\| : \|x\| \leq 1\}$. Let $\Omega \subseteq X$ be an open set and $f : \Omega \to Y$ be Fréchet differentiable at all $x \in \Omega$. The Fréchet derivative of f at x is the linear mapping $f'(x) : X \to Y$, which is continuous. The *graph* of the set-valued mapping $F : X \rightrightarrows Y$ is the set gph $F := \{(x,y) \in X \times Y : y \in F(x)\}$. The *domain* and the *range* of F are, respectively, the sets dom $F = \{x \in X : F(x) \neq \varnothing\}$ and rge $F = \{y \in Y : y \in F(x) \text{ for some } x\}$. The *inverse* of F is the set-valued mapping $F^{-1} : Y \rightrightarrows X$ defined by $F^{-1}(y) = \{x \in X : y \in F(x)\}$. The *partial linearization* of $f + F$ at $x \in \Omega$ is the set-valued mapping $L_f(x,\cdot) : \Omega \rightrightarrows Y$ defined by

$$L_f(x,y) := f(x) + f'(x)(y-x) + F(y). \tag{12.5}$$

Definition 12.1. Let $\Omega \subset X$ be open and nonempty. The mapping $T : \Omega \rightrightarrows Y$ is called strongly regular at x for y, when $y \in T(x)$ and there exist $r_x > 0$, $r_y > 0$, and $\lambda > 0$ such that $B(x,r_x) \subset \Omega$, the mapping $z \mapsto T^{-1}(z) \cap B(x,r_x)$ is single-valued from $B(y,r_y)$ to $B(x,r_x)$ and Lipschitzian on $B(y,r_y)$ with modulus λ, i.e., $\left\| T^{-1}(u) \cap B(x,r_x) - T^{-1}(v) \cap B(x,r_x) \right\| \leq \lambda \|u-v\|$, for all $u, v \in B(y,r_y)$.

Since $z \mapsto T^{-1}(z) \cap B(x,r_x)$ in Definition 12.1 is single-valued, for the sake of simplicity, we have used the notation $w = T^{-1}(z) \cap B(x,r_x)$ instead of $\{w\} := T^{-1}(z) \cap B(x,r_x)$. *Hereafter, we use this simplified notation.* For a detailed discussion on Definition 12.1; see [5, 6, 13]. The next lemma is a type of Banach Perturbation Lemma, its proof is similar to [13, Theorem 2.4] and is omitted here.

Lemma 12.1. Let X,Y be Banach spaces, a_0 be a point of Y, $F : X \rightrightarrows Y$ be a set-valued mapping and $A_0 : X \to Y$ be a bounded linear mapping. Suppose that $\bar{x} \in X$ and $0 \in A_0\bar{x} + a_0 + F(\bar{x})$. Assume that $A_0 + a_0 + F$ is strongly regular at $\bar{x}$ for 0 with modulus $\lambda > 0$. Then, there exist $r_{\bar{x}} > 0$, $r_{A_0} > 0$, $r_{a_0} > 0$, and $r_0 > 0$ such that, for any $A \in B(A_0, r_{A_0}) \subset L(X,Y)$ and $a \in B(a_0, r_{a_0}) \subset Y$ letting $T(A,a,\cdot) : B(\bar{x},r_{\bar{x}}) \rightrightarrows Y$ be defined as $T(A,a,x) := Ax + a + F(x)$, the mapping $y \mapsto T(A,a,y)^{-1} \cap B(\bar{x},r_{\bar{x}})$ is single-valued from $B(0,r_0) \subset Y$ to $B(\bar{x},r_{\bar{x}})$. Moreover, for each $A \in B(A_0, r_{A_0})$ and

$a \in B(a_0, r_{a_0})$ there holds $\lambda \|A - A_0\| < 1$ and the mapping $y \mapsto T(A,a,y)^{-1} \cap B(\bar{x}, r_{\bar{x}})$ is also Lipschitzian on $B(0, r_0)$ as follows

$$\left\| T(A,a,y_1)^{-1} \cap B(\bar{x}, r_{\bar{x}}) - T(A,a,y_2)^{-1} \cap B(\bar{x}, r_{\bar{x}}) \right\| \leq \frac{\lambda}{1 - \lambda \|A - A_0\|} \|y_1 - y_2\|,$$

for all
$$y_1, y_2 \in B(0, r_0).$$

Next, we establish a corollary to Lemma 12.1, which plays an important role in the sequel.

Corollary 12.1. Let X, Y be Banach spaces, $\Omega \subset X$ be open and nonempty, $f : \Omega \to Y$ be continuous with the Fréchet derivative f' continuous, and $F : X \rightrightarrows Y$ be a set-valued mapping. Suppose that $x_0 \in \Omega$ and $L_f(x_0, .) : \Omega \rightrightarrows Y$ is strongly regular at $x_1 \in \Omega$ for 0 with modulus $\lambda > 0$. Then, there exist $r_{x_1} > 0$, $r_0 > 0$, and $r_{x_0} > 0$ such that, for each $x \in B(x_0, r_{x_0})$, there holds $\lambda \|f'(x) - f'(x_0)\| < 1$, the mapping $z \mapsto L_f(x,z)^{-1} \cap B(x_1, r_{x_1})$ is single-valued from $B(0, r_0)$ to $B(x_1, r_{x_1})$ and Lipschitzian as follows

$$\left\| L_f(x,u)^{-1} \cap B(x_1, r_{x_1}) - L_f(x,v)^{-1} \cap B(x_1, r_{x_1}) \right\| \leq \frac{\lambda}{1 - \lambda \|f'(x) - f'(x_0)\|} \|u - v\|,$$

for all
$$u, v \in B(0, r_0).$$

Proof. See [9]. □

12.2   KANTOROVICH'S THEOREM FOR NEWTON'S METHOD

In this section, our objective is to state and prove Kantorovich's theorem for Newton's method for solving (12.1). To state the theorem, we need to set some important constants. We refer to the real numbers

$$r_{x_1} > 0, \qquad r_0 > 0, \qquad r_{x_0} > 0, \qquad (12.6)$$

as the three constants given by Corollary 12.1.

Further, we assume that Lipschitz continuity of f' is relaxed, i.e., we assume that f' satisfies the conditions stated in the next definitions.

Definition 12.2. Let X, Y be Banach spaces, $\Omega \subset X$ be open, $f : \Omega \to Y$ be continuous with Fréchet derivative f' continuous in Ω. Let $x_0 \in \Omega$, $R > 0$, and $\kappa := \sup\{t \in [0,R) : B(x_0, t) \subset \Omega\}$. A twice continuously differentiable function $\psi : [0, R) \to \mathbb{R}$ is a majorant function for f on $B(x_0, \kappa)$ with modulus $\lambda > 0$, if it satisfies the following inequality

$$\lambda \|f'(y) - f'(x)\| \leq \psi'(\|y - x\| + \|x - x_0\|) - \psi'(\|x - x_0\|), \qquad (12.7)$$

for all $x, y \in B[x_0, \kappa]$ and $\|y - x\| + \|x - x_0\| < R$. Moreover, there hold:

(C1) $\psi(0) > 0$, $\psi'(0) = -1$, for each $t \in [0, R)$;
(C2) ψ' is strictly increasing and convex;
(C3) $\psi(t) = 0$ for some $t \in (0, R)$.

Definition 12.3. Let X, Y be Banach spaces, $\Omega \subset X$ be open, $f : \Omega \to Y$ be continuous with Fréchet derivative f' continuous in Ω. Let $x_0 \in \Omega$, $R > 0$, and $\kappa := \sup\{t \in [0, R) : B(x_0, t) \subset \Omega\}$. A twice continuously differentiable function $\psi_0 : [0, R) \to \mathbb{R}$ is a center majorant function for f on $B(x_0, \kappa)$ with modulus $\lambda > 0$, if it satisfies the following inequality

$$\lambda \left\| f'(y) - f'(x_0) \right\| \leq \psi_0'(\|x - x_0\|) - \psi_0'(0), \tag{12.8}$$

for all $x, y \in B[x_0, \kappa]$. Moreover, suppose that

(C$_1^0$) $\psi_0(0) > 0$, $\psi_0'(0) = -1$, $\psi_0(t) \leq \psi(t)$, $\psi_0'(t) \leq \psi'(t)$;
(C$_2^0$) ψ_0' is strictly increasing and convex;
(C$_3^0$) $\psi_0(t) = 0$ for some $t \in (0, R)$.

Notice that there is a significant difference between the proof of our Theorem 12.1 and the corresponding one in [9], since we use a more flexible and accurate function ψ_0 instead of the less precise function ψ leading to the already aforementioned advantages and under the same computational cost. The statement of the main result is:

Theorem 12.1. Let $x_0 \in \Omega$, $R > 0$ and $\kappa := \sup\{t \in [0, R) : B(x_0, t) \subset \Omega\}$. Assume that $L_f(x_0, .) : \Omega \rightrightarrows Y$ is strongly regular at $x_1 \in \Omega$ for 0 with modulus $\lambda > 0$ and there exists a function $\psi_0 : [0, R) \to \mathbb{R}$ that satisfies the conditions in Definition 12.3. Then, the sequence $\{t_n^0\}$ generated by Newton's method for solving $\psi_0(t) = 0$

$$t_{n+1}^0 = t_n^0 - \frac{\psi_0(t_n^0)}{\psi_0'(t_n^0)}, \qquad k = 0, 1, \ldots, \tag{12.9}$$

is strictly increasing, $\{t_n^0\} \subset (0, t_0^*)$ and converges to t_0^*, where t_0^* is the smallest solution of $\psi_0(t) = 0$ in $[0, R)$. Furthermore, suppose that there exists a function $\psi : [0, R) \to \mathbb{R}$ that satisfies the conditions in Definition 12.2 for all $x, y \in B_0 := B[x_0, \kappa] \cap B[x_0, t_0^*]$ and

$$\|x_1 - x_0\| \leq \psi(0). \tag{12.10}$$

Let

$$\beta := \sup_{t \in [0, R)} -f(t), \qquad t_* := \min \psi^{-1}(\{0\}).$$

Additionally, for the constants r_0 and r_{x_0}, suppose that the following inequalities hold:

$$t_* \leq r_{x_0}, \qquad \frac{\psi''(t_*)}{2\lambda} \psi(0)^2 < r_0. \tag{12.11}$$

Then, the sequences generated by Newton's method for solving $0 \in f(x) + F(x)$ and $\psi(t) = 0$, with starting point x_0 and $t_0 = 0$, defined respectively by,

$$x_{k+1} := L_f(x_k, 0)^{-1} \cap B(x_1, r_{x_1}), \qquad t_{k+1} = t_k - \psi(t_k)/\psi'(t_k), \qquad k = 0, 1, \dots , \tag{12.12}$$

are well defined, $\{t_k\}$ is strictly increasing, $\{t_k\} \subset (0, t_*)$ and converges to t_*, and $\{x_k\} \subset B(x_0, t_*)$ and converges to $x_* \in B[x_0, t_*]$, which is the unique solution of $0 \in f(x) + F(x)$ in $B[x_0, t_*] \cap B[x_1, r_{x_1}]$. Moreover, $\{x_k\}$ and $\{t_k\}$ satisfies

$$\|x_* - x_k\| \le t_* - t_k, \qquad \|x_* - x_{k+1}\| \le \frac{t_* - t_{k+1}}{(t_* - t_k)^2} \|x_* - x_k\|^2, \tag{12.13}$$

for all k=0,1,..., and the sequences $\{x_k\}$ and $\{t_k\}$ converge Q-linearly as follows

$$\|x_* - x_{k+1}\| \le \frac{1}{2} \|x_* - x_k\|, \qquad t_* - t_{k+1} \le \frac{1}{2}(t_* - t_k), \qquad k = 0, 1, \dots . \tag{12.14}$$

Additionally, if the following condition holds

(C4) $\psi'(t_*) < 0$,

then the sequences, $\{x_k\}$ and $\{t_k\}$ converge Q-quadratically as follows

$$\|x_* - x_{k+1}\| \le \frac{\psi''(t_*)}{-2\psi'(t_*)} \|x_* - x_k\|^2, \qquad \|x_* - x_{k+1}\| \le \frac{\psi''(t_*)}{-2\psi_0'(t_*)} \|x_* - x_k\|^2 \tag{12.15}$$

$$t_* - t_{k+1} \le \frac{\psi''(t_*)}{-2\psi'(t_*)}(t_* - t_k)^2, \qquad t_* - t_{k+1} \le \frac{\psi''(t_*)}{-2\psi_0'(t_*)}(t_* - t_k)^2 \tag{12.16}$$

Remark 12.1. a)   In the earlier study [9] the following condition was studied instead of (12.8):

$$\lambda \|f'(y) - f'(x)\| \le \psi_1'(\|y - x\| + \|x - x_0\|) - \psi_1'(\|x - x_0\|), \tag{12.17}$$

for all $\tau \in [0, 1]$, $x \in B(x_0, \kappa)$ where $\psi_1 : [0, R) \to \mathbb{R}$ is twice continuously differentiable with $\psi_1(0) > 0$, $\psi_1'(0) = -1$ and ψ_1' convex and strictly increasing. The corresponding error estimates for

$$u_{k+1} = \left| \frac{u_k \psi_1'(u_k) - \psi_1(u_k)}{\psi_1'(u_k)} \right|, \qquad u_0 = \|x^* - x_0\| \tag{12.18}$$

are:

$$\|x^* - x_{k+1}\| \le \frac{u_{k+1}}{u_k} \|x^* - x_k\|^2, \qquad \frac{u_{k+1}}{u_k} \le \frac{\psi_1''(u_0)}{2|\psi_1'(0)|} \tag{12.19}$$

$$\|x^* - x_k\| \le u_0 \left(\frac{u_1}{u_0} \right)^{2^k - 1}, \qquad k = 0, 1, \dots, \text{ and} \tag{12.20}$$

the optimal convergence radius is $r = \rho_1$ if $\rho_1 < \kappa$ and solves

$$\frac{\psi_1(t) - t\psi_1'(t)}{t\psi_1'(t)} = 1. \tag{12.21}$$

By comparing (12.8) and (12.7) to (12.17), we see that

$$\psi'(t) \leq \psi_1'(t) \tag{12.22}$$

and

$$\psi_0'(t) \leq \psi_1'(t) \tag{12.23}$$

since, $B_0 \subseteq B(x^*, \kappa)$. Define functions φ and φ_1 on $B[0, R)$ by

$$\varphi(t) = \frac{\psi(t) - t\psi'(t)}{t\psi_0'(t)} - 1$$

and

$$\varphi_1(t) = \frac{\psi_1(t) - t\psi_1'(t)}{t\psi_1'(t)} - 1.$$

Then, we have in turn that

$$\begin{aligned}
\varphi(t) &= \frac{\int_0^1 [\psi'(\tau t) - \psi'(t)]d\tau}{\psi_0'(t)} - 1 \\
&\leq \frac{\int_0^1 [\psi_1'(\tau t) - \psi_1'(t)]d\tau}{\psi_1'(t)} - 1 = \varphi_1(t).
\end{aligned} \tag{12.24}$$

In particular, we have that

$$\varphi(\rho_1) \leq \varphi_1(\rho_1) = 0 \tag{12.25}$$

so $\rho_1 \leq \rho$. Moreover, in view of (12.22)–(12.24), the new error estimates are tigher than the corresponding ones given by (12.18)–(12.20) leading to more precise error bounds on the distances $\|x^* - x_{k+1}\|$ and $\|x^* - x_k\|$, i.e., at least as few iterates to obtain a desired error tolerance. Moreover, the information on the uniqueness of the solution is more precise, since

$$\bar{\bar{\sigma}} \leq \bar{\sigma},$$

where $\sigma_1 := \sup\{t \in (0, r) : \psi_1(t) < 0\}$, $\bar{\bar{\sigma}} = \min\{r_1, \rho_1\}$, $r_1 = \min\{\kappa, \rho_1\}$, if

$$\psi_0(t) \leq \psi_1(t), \quad t \in [0, R). \tag{12.26}$$

Condition (12.26) is assumed without loss of generality. It is also worth noticing that the preceding advantages are obtained under the same computational cost as in [9], since in practice the computation of function ψ_1 involves the computations of functions ψ_0 and ψ as special cases (see also the numerical example). Finally, notice that $\psi = \psi(\psi_0)$. That is, ψ depends on function ψ_0. The construction of ψ was not possible before without ψ_0, since only ψ_1 was used [9].

b)   (**C′1**) If $\psi'(t) \leq \psi'_0(t)$, $t \in [0, t_0^*)$ holds instead of (**C1**), then clearly, the
conclusions of Theorem 12.1 hold with (**C′1**) replacing (**C1**) and function
ψ_0 replacing ψ in the conclusions of this theorem. Moreover, the advantages
stated in (a) over the results in [9] hold with ψ_0 replacing ψ.

Remark 12.2. All the results about the sequences $\{t_n^0\}$ and $\{t_n\}$ are easily obtained
as in [9].

In Section 12.3, we present particular instances of Theorem 12.1 for some classes
of functions in the above examples. *Hereafter, we assume that all the assumptions in
Theorem 12.1 hold.*

12.2.1   SOME RESULTS

Now, we are going to establish some relationships between ψ_0 and $f + F$. The next
result is a consequence of Corollary 12.1.

Proposition 12.1. For any $x \in B(x_0, t_*)$, the mapping $z \mapsto L_f(x, z)^{-1} \cap B(x_1, r_{x_1})$ is
single-valued from $B(0, r_0)$ to $B(x_1, r_{x_1})$ and there holds

$$\left\| L_f(x, u)^{-1} \cap B(x_1, r_{x_1}) - L_f(x, v)^{-1} \cap B(x_1, r_{x_1}) \right\| \leq -\frac{\lambda}{\psi'_0(\|x - x_0\|)} \|u - v\|,$$

for all

$$u, v \in B(0, r_0).$$

Proof. Definitions of r_{x_1}, r_0, and r_{x_0} in (12.6) together with Corollary 12.1 imply
that, for any $x \in B(x_0, r_{x_0})$, the mapping $z \mapsto L_f(x, z)^{-1} \cap B(x_1, r_{x_1})$ is single-valued
from $B(0, r_0)$ to $B(x_1, r_{x_1})$ and there holds

$$\left\| L_f(x, u)^{-1} \cap B(x_1, r_{x_1}) - L_f(x, v)^{-1} \cap B(x_1, r_{x_1}) \right\| \leq \frac{\lambda}{1 - \lambda \|f'(x) - f'(x_0)\|} \|u - v\|,$$
$$\tag{12.27}$$

for all $u, v \in B(0, r_0)$. Since $\|x - x_0\| < t_*$ thus $\psi'_0(\|x - x_0\|) < 0$. Hence, (12.7) to-
gether with C_1^0 imply that

$$\lambda \|f'(x) - f'(x_0)\| \leq \psi'_0(\|x - x_0\|) - \psi'_0(0) < 1, \qquad \forall x \in B(x_0, t_*),$$

and then, using (12.11), i.e., $t_* \leq r_{x_0}$, (12.27) and C_1^0, the inequality of the proposition
follows. $\square$

The above proposition guarantees that, for $x \in B(x_0, t_*)$, the mapping $z \mapsto$
$L_f(x, z)^{-1} \cap B(x_1, r_{x_1})$ is single-valued from $B(0, r_0)$ to $B(x_1, r_{x_1})$. Thus, we define
the *Newton iteration mapping* $N_{f+F} : B(x_0, t_*) \to X$ by

$$N_{f+F}(x) := L_f(x, 0)^{-1} \cap B(x_1, r_{x_1}).   \tag{12.28}$$

It is easy to see that the definition of N_{f+F} can be rewritten as the following inclusions

$$0 \in f(x) + f'(x)(N_{f+F}(x) - x) + F(N_{f+F}(x)), \qquad N_{f+F}(x) \in B(x_1, r_{x_1}), x \in B(x_0, t_*).$$
$$(12.29)$$

Therefore, one can apply a *single* Newton iteration on any $x \in B(x_0, t_*)$ to obtain $N_{f+F}(x)$, which may not belong to $B(x_0, t_*)$. Thus, this is adequate to ensure the well-definedness of only one Newton iteration. To ensure that Newtonian iterations may be repeated indefinitely or, in particular, invariant on subsets of $B(x_0, t_*)$, we need some additional results. First, define some subsets of $B(x_0, t_*)$, in which, as we shall prove, Newton iteration mapping (12.28) are "well behaved". Define

$$K(t) := \left\{ x \in \Omega \; : \; \|x - x_0\| \le t, \quad \|L_f(x,0)^{-1} \cap B(x_1, r_{x_1}) - x\| \le -\frac{\psi(t)}{\psi_0'(t)} \right\}, \, t \in [0, t_*),$$
$$(12.30)$$

$$K := \bigcup_{t \in [0, t_*)} K(t).$$
$$(12.31)$$

Proposition 12.2. For each $0 \le t < t_*$ we have $K(t) \subset B(x_0, t_*)$ and $N_{f+F}(K(t)) \subset K(n_\psi(t))$. As a consequence, $K \subseteq B(x_0, t_*)$ and $N_{f+F}(K) \subset K$.

Proof. The proof is similar as in [9]. □

12.2.2 CONVERGENCE ANALYSIS

To prove the convergence results, which are consequences of the above results, first, we note that the definition (12.28) implies that the sequence $\{x_k\}$ defined in (12.12), can be formally stated as

$$x_{k+1} = N_{f+F}(x_k), \qquad k = 0, 1, \ldots,$$
$$(12.32)$$

or equivalently as,

$$0 \in f(x_k) + f'(x_k)(x_{k+1} - x_k) + F(x_{k+1}), \qquad x_{k+1} \in B(x_1, r_{x_1}), \, k = 0, 1, \ldots.$$
$$(12.33)$$

First, we show that the sequence $\{x_k\}$ generated by Newton's method converges to $x_* \in B[x_0, t_*]$, a solution of the generalized equation (12.1), and is well behaved with respect to the set defined in (12.30).

Proof of Theorem 12.1
Since the mapping $x \mapsto L_f(x_0, x)$ is strongly regular at x_1 for 0, thus Corollary 12.1 implies that

$$x_1 = L_f(x_0, 0)^{-1} \cap B(x_1, r_{x_1})$$

and the first Newton iterate is well defined. Using, $\mathbf{C_1^0}$, (12.10), (12.30) and (12.31) it is easy to see that

$$\{x_0\} = K(0) \subset K.$$
$$(12.34)$$

From Proposition 12.2 we have that $N_{f+F}(K) \subset K$. Thus, (12.34) and (12.32) imply that the sequence $\{x_k\}$ is well defined and $x_k \in K$. From the first inclusion in the second part of Proposition 12.2, we have that $\{x_k\} \subset B(x_0, t_*)$. By induction, we can prove that

$$x_k \in K(t_k), \qquad k = 0, 1 \dots . \tag{12.35}$$

Now, using (12.35) and (12.30), combined with (12.32), (12.28), and (12.12), we have

$$\|x_{k+1} - x_k\| = \|L_f(x_k, 0)^{-1} \cap B(x_1, r_{x_1}) - x_k\| \leq -\frac{\psi(t_k)}{\psi_0'(t_k)} \leq -\frac{\psi(t_k)}{\psi'(t_k)} = t_{k+1} - t_k, \tag{12.36}$$

where

$$k = 0, 1 \dots,$$

Taking into account that $\{t_k\}$ converges to t_*,, we can easily conclude from the above inequality that

$$\sum_{k=k_0}^{\infty} \|x_{k+1} - x_k\| \leq \sum_{k=k_0}^{\infty} t_{k+1} - t_k = t_* - t_{k_0} < +\infty,$$

for any $k_0 \in \mathbb{N}$. Hence, we conclude that $\{x_k\}$ is a Cauchy sequence in $B(x_0, t_*)$ and thus it converges to some $x_* \in B[x_0, t_*]$. Therefore, using (12.36) again, we conclude that $\|x_* - x_k\| \leq t_* - t_k$, for all $k = 0, 1 \dots$. Since f and f' are continuous in Ω, $B[x_0, t_*] \subset \Omega$ and F has a closed graph, it is easy to conclude that $f(x_*) + F(x_*) \ni 0$. The others statements of the theorem are similar to [9].

12.3 APPLICATIONS

In this section, we will present some special cases of Theorem 12.1. It is worth pointing out that there exist some classes of well known functions which a majorant function is available, below we will present two examples, namely, the classes of functions satisfying a Lipschitz-like and Smale's conditions, respectively. In this sense, the results obtained in Theorem 12.1 unify the convergence analysis of Newton's method for the classes of generalized equations involving these functions, for instance, Theorem 2 of [11] due to N. H. Josephy and, a particular instance of Theorem 2 of [3] due to A. L. Dontchev and a version of Smale's theorem on Newton's method for analytical functions, see [2].

Theorem 12.2. Let $\Omega \subset \mathbb{R}^n$, $x_0 \in \Omega$, $\lambda > 0$, and $f : \Omega \to \mathbb{R}^n$ be an analytic function. Let $F : \mathbb{R}^n \rightrightarrows \mathbb{R}^n$ be a set-valued mapping with a closed graph. Suppose that $L_f(x_0, .) : \Omega \rightrightarrows \mathbb{R}^n$ at x_0, is strongly regular at $x_1 \in \Omega$ for 0 with modulus $\lambda > 0$, $B(x_0, 1/\gamma) \subseteq \Omega$, where $\gamma = \sup_{n>1} \|[\lambda f^{(n)}(x_0)]/n!\|^{1/(n-1)} < +\infty$ and there exists $b > 0$ such that $\|x_1 - x_0\| \leq b$ and $b\gamma \leq 3 - 2\sqrt{2}$. Suppose that

$$t_* \leq r_{x_0}, \qquad \frac{4^3 \gamma b^2}{\lambda \left(3 - b\gamma + \sqrt{(b\gamma + 1)^2 - 8b\gamma}\right)^3} < r_0,$$

hold, where $t_* = (b\gamma + 1 - \sqrt{(b\gamma + 1)^2 - 8b\gamma})/4\gamma$. Then, the sequence generated by Newton's method for solving $f(x) + F(x) \ni 0$ with starting point x_0, $x_{k+1} := L_f(x_k, 0)^{-1} \cap B(x_1, r_{x_1})$, for all $k = 0, 1, \ldots$, is well defined, $\{x_k\}$ is contained in $B(x_0, t_*)$, and converges to the point x_*, which is the unique solution of $f(x) + F(x) \ni 0$ in $B[x_0, t_*] \cap B[x_1, r_{x_1}]$, where r_{x_1} is fixed in (12.6). Moreover, letting $\psi_0 : [0, R) \to \mathbb{R}$, $\psi : [0, 1/\gamma_0) \to \mathbb{R}$ be defined by $\psi_0(t) = t/(1 - \gamma_0 t) - 2t + b$, $\psi(t) = t/(1 - \gamma t) - 2t + b$, with $\gamma_0 \leq \gamma$, the sequence $\{x_k\}$ converges Q-linearly as follows $\|x_* - x_{k+1}\| \leq \|x_* - x_k\|/2$, for all $k = 0, 1, \ldots$. Additionally, if $b\gamma_0 < 3 - 2\sqrt{2}$, then $\{x_k\}$ converges Q-quadratically as follows

$$\|x_* - x_{k+1}\| \leq \frac{\gamma_0}{(1 - \gamma_0 t_*)[2(1 - \gamma t_*)^2 - 1]} \|x_* - x_k\|^2, \qquad k = 0, 1, \ldots.$$

Proof. Note that ψ_0 and ψ satisfy C_1^0, C_2^0, C_3^0, C_4^0, and (12.11). Therefore, the result follows from the Theorem 12.1. ☐

Remark 12.3. Similarly, we obtain the improvements of Wang's theory, if we choose for some $\gamma_0 \leq \gamma_1$ and $\gamma_0 \leq \gamma$ the functions

$$\psi_0(t) = \frac{t}{1 - \gamma_0 t} - 2t + b, \tag{12.37}$$

$$\psi(t) = \frac{t}{1 - \gamma t} - 2t + b \tag{12.38}$$

and

$$\psi_1(t) = \frac{t}{1 - \gamma_1 t} - 2t + b, \tag{12.39}$$

which can easily be seen to satisfy the conditions of Theorem 12.1 and those of Remark 12.1 (a) (or (b) if $\gamma \leq \gamma_0$).

The rest of the results in [9] can be improved as long as B_0 is a strict subset of $B[x_0, \kappa]$ along the same lines, since again the ψ function defined on B_0 is at least as tight as the ψ function defined on $B[x_0, \kappa]$ used in [9].

We present an academic and motivational numerical example, where the previously stated advantages in Remark 12.1 are obtained, when $F \equiv \{0\}$.

Example 12.1. Let $X = Y = \mathbb{R}$, $x_0 = 1$, $p \in [0, \frac{1}{2})$ and $\Omega = \bar{U}(x_0, 1 - p)$. Define function f on Ω by

$$f(x) = x^3 - p. \tag{12.40}$$

Then, we have by (12.17) that $\lambda = \frac{1}{3}$, $\kappa = 1 - p$ and $\psi_1(t) = (2 - p)t^2 - t + \frac{1}{3}(1 - p)$. The sufficient convergence Kantorovich condition is given by

$$h_1 = \frac{4}{3}(2 - p)(1 - p) \leq 1 \tag{12.41}$$

which however is not satisfied for all $p \in [0, \frac{1}{2})$. Hence, there is no guarantee that Newton's method starting at $x_0 = 1$ converges to $x^* = \sqrt[3]{p}$ under the earlier study [9].

Let us see that we can get under new approach. Condition (12.8) is satisfied, if we choose

$$\psi_0(t) = \frac{3-p}{2}t^2 - t + \frac{1}{3}(1-p).$$

The corresponding sufficient convergence condition is given by

$$h_0 = 4\left(\frac{3-p}{2}\right)\frac{1}{3}(1-p) \le 1 \tag{12.42}$$

which is satisfied provided that $p \in S := \left[\frac{4-\sqrt{10}}{2}, \frac{1}{2}\right)$. Notice that (12.42) is the sufficient condition for the modified Newton's method $f(x_k) + f'(x_0)(x_{k+1} - x_k) = 0$. In this case

$$t_0^* = \frac{1 - \sqrt{1 - h_0}}{3 - p} < \kappa, \tag{12.43}$$

so $B[x_0, \kappa] \cap B[x_0, t_0^*] = B[x_0, t_0^*]$. Therefore, condition (12.7) is satisfied, if we choose

$$\psi(t) = (1 + t_0^*)t^2 - t + \frac{1}{3}(1-p).$$

The sufficient convergence condition is given by

$$h = 4(1 + t_0^*)\frac{1}{3}(1-p) \le 1, \tag{12.44}$$

which is satisfied provided that $p \in \left[1 - \frac{3}{4(1+t_0^*)}, \frac{1}{2}\right)$ or if $p \in S$. Notice also that we have

$$\psi_0(t) < \psi(t) < \psi_1(t) \quad \text{for each} \quad t \in [0, \frac{1}{2}). \tag{12.45}$$

If we allow $p \in [0, 1)$, say e.g, $p = 0.6$, then conditions (12.41), (12.42) and (12.44) are satisfied. However, as already noted in Remark 12.1, in view of (12.45) the other advantages of our approach over the ones in [9] hold.

Hence, we have extended the applicability of the method for solving equation $f(x) = 0$.

12.4   EXERCISES

1. Solve the initial value problem

$$\dot{x}(t) = 1 + \cos(x(t)), \quad x(0) = 0.$$

2. The predator-prey population models describe the interaction of a prey population X and a predator population Y. Assume that their interaction is modeled by the system of ordinary differential equations

$$\dot{x} = x - \frac{1}{4}x^2 - \frac{1}{10}xy + 1$$

$$\dot{y} = -\frac{1}{4}y + \frac{1}{7}xy + \frac{2}{3}.$$

(Assume $x(0) = y(0) = 0$). Solve the system.

3. Solve the initial value problem

$$\dot{x} = (x(t) + t, x(0) = 1)$$

by the iteration method developed in this chapter. Perform 3 steps.

4. Solve the Fredholm-type integral equation

$$x(t) = \sum_0^1 \frac{t \cdot s}{10} ds + 1$$

by the iteration method developed in this chapter. Perform 3 steps.

5. Solve the Volterra-type integral equation

$$x(t) = \sum_0^t \frac{t \cdot s}{10} ds + 1$$

by the iteration method developed in this chapter. Perform 3 steps.

6. Solve equation

$$x = \frac{\sin x}{2} + 1$$

by using algorithm developed in this chapter.

7. Repeat the previous problem for equation

$$x = \frac{\sin x + \cos 2x}{10}.$$

REFERENCES

1. Argyros, I.K., and Magreñán, Á.A. 2017. *Iterative Methods and Their Dynamics with Applications: A Contemporary Study*. New York: CRC Press.
2. Blum, L., Cucker, F., Shub, M., and Smale, S. 1998. *Complexity and Real Computation*. New York: Springer-Verlag.
3. Dontchev, A.L. 1996. Local analysis of a Newton-type method based on partial linearization. In *The Mathematics of Numerical Analysis (Park City, UT, 1995)*, volume 32 of *Lectures in Appl. Math.* 295–306. Providence, RI.: Amer. Math. Soc.
4. Dontchev, A.L., and Rockafellar, R.T. 1996. Characterizations of strong regularity for variational inequalities over polyhedral convex sets. *SIAM J. Optim.*, 1087–1105.
5. Dontchev, A.L., and Rockafellar, R.T. 2009. *Implicit functions and solution mappings*. Dordrecht: Springer Monographs in Mathematics. Springer.
6. Dontchev, A.L., and Rockafellar, R.T. 2010. Newton's method for generalized equations: a sequential implicit function theorem. *Math. Program.*, 123(1, Ser. B):139–159.
7. Ferreira, O.P., and Silva, G.N. 2018. Inexact Newton's method for nonlinear functions with values in a cone. *Accepted for publication in Applicable Analysis*.
8. Ferreira, O.P., and Silva, G.N. 2018. Local convergence analysis of Newton's method for solving strongly regular generalized equations. *J. Math. Anal. Appl.*, 458. 1:481–496.
9. Ferreira, O.P., and Silva, G.N. 2017. Kantorovich's theorem on Newton's method for solving strongly regular generalized equation. *SIAM J. Optimization*, 27. 2:910–926.
10. Ferreira, O.P., and Svaiter, B.F. 2009. Kantorovich's majorants principle for Newton's method. *Comput. Optim. Appl.*, 42(2):213–229.
11. Josephy, N. 1979. *Newton's Method for Generalized Equations and the PIES Energy Model*. University of Wisconsin–Madison.
12. Robinson, S.M. 1972. Extension of Newton's method to nonlinear functions with values in a cone. *Numer. Math.*, 19:341–347.
13. Robinson, S.M. 1980. Strongly regular generalized equations. *Math. Oper. Res.*, 5(1):43–62.
14. Silva, G.N. 2016. Kantorovich's theorem on Newton's method for solving generalized equations under the majorant condition. *Applied Mathematics and Computation*, 286:178–188.
15. Silva, G.N. 2017. Local convergence of Newton's method for solving generalized equations with monotone operator. *Appl. Anal.*, http://dx.doi.org/10.1080/00036811.2017.1299860.
16. Wang, X. 1999. Convergence of Newton's method and inverse function theorem in Banach space. *Math. Comp.*, 68(225):169–186.
17. Zabrejko, P.P., and Nguen, D.F. 1987. The majorant method in the theory of Newton-Kantorovich approximations and the Pták error estimates. *Numer. Funct. Anal. Optim.*, 9(5-6):671–684.
18. Zhang, Y., Wang, J., and Guu, S. 2015. Convergence criteria of the generalized Newton method and uniqueness of solution for generalized equations. *J. Nonlinear Convex Anal.*, 16(7):1485–1499.

13 An Inverse Free Broyden's Method

Based on a center-Lipschitz-type condition and our idea of the restricted convergence domain, we present a new semi-local convergence analysis for an inverse free Broyden's method (BM) in order to approximate a locally unique solution of an equation in a Hilbert space setting. The operators involved have regularly continuous divided differences. This way we provide, weaker sufficient semi-local convergence conditions, tighter error bounds, and a more precise information on the location of the solution are provided in this schapter. Hence, our approach extends the applicability of BM under the same hypotheses as before. Finally, we consider some special cases.

In this chapter, we are concerned with the problem of approximating a locally unique solution $x^\star$ of equation

$$F(x) = 0, \tag{13.1}$$

where F is a continuous operator defined on a open convex subset Ω of a Hilbert space $\mathscr{B}_1$ with values in a Hilbert space $\mathscr{B}_2$.

Broyden's method BM

$$x_+ = x - AF(x), \quad y = F(x_+) - F(x), \quad A_+ = A - \frac{AF(x_+) < y, \cdot >}{< y, y >}, \tag{13.2}$$

where $\mathscr{L}(\mathscr{B}_2, \mathscr{B}_1) := \{A : \mathscr{B}_2 \longrightarrow \mathscr{B}_1,$ bounded and linear$\}$, and $< \cdot, \cdot >$ stands for the inner product.

Numerous convergence results for this type of methods have appeared in the literature [1, 3, 5, 8–11] (see also, e.g. [4], and the references there in). BM requires no inverse, so no linear subproblem needs to be solved at each iteration.

The convergence domain for such methods is small in general [12–15]. In the present study, we extend the convergence domain for BM. To achieve this goal, we first introduce the center-Lipschitz condition which determines a subset of the original domain for the operator containing the iterates. The scalar functions are then related to the subset instead of the original domain. This way, the scalar functions are more precise than if they were depending on the original domain. The new technique leads to : weaker sufficient convergence conditions, tighter error bounds on the distances involved, and an at least as precise information on the location of the solution. These advantages are obtained under the same computational cost as in earlier studies [8–11], since in practice the new functions are special cases of the old functions. This idea can be used to study other iterative methods requiring inverses of linear mappings [1–15].

The chapter is structured as follows. Section 13.1 contains some preliminary results for regularly continuous dd. In Section 13.2, we provide the semi-local

DOI: 10.1201/9781003128915-13

convergence analysis of BM. Finally, in Section 13.3, we provide special cases, as applications.

13.1   PRELIMINARIES: REGULARLY CONTINUOUS DD

In order to make the chapter as self–contained as possible, we reintroduce some definitions and some results on regularly continuous dd. The proofs are omitted, and can be found in [4,11]. In this section, $\mathscr{B}_1$ and $\mathscr{B}_2$ are Banach spaces, equipped with the norm $\| . \|$. We denote by $U(z,R) = \{x \in \mathscr{B}_1 : \| x - z \| < R, \}$ the open ball centered at z and of radius $R > 0$, whereas $\overline{U}(z,R)$ denotes its closure. For $x \in \mathscr{B}_1$, denote by $\mathscr{K}_x$ the subspace of operators vanishing at x $\mathscr{K}_x = \{A \in \mathscr{L}(\mathscr{B}_1,\mathscr{B}_2) : Ax = 0\}$. Let $\mathscr{N}$ be the class of increasing concave functions $v : \mathbb{R}^+ \longrightarrow \mathbb{R}^+$, with $v(0) = 0$. Note that $\mathscr{N}$ contains the functions in the form $\varphi(t) = ct^p$, $(c \geq 0$, and $p \in (0,1])$.

Definition 13.1. [11] An operator $[.,.;F]$ belonging in $\mathscr{L}(\mathscr{B}_1,\mathscr{B}_2)$ is called the first order divided difference (briefly dd) of F at the points x and y in $\mathscr{B}_1$ $(x \neq y)$, if the following secant equation holds $[x,y;F](y-x) = F(y) - F(x)$.

If F is Fréchet differentiable at x, then $[x,x;F] = F'(x)$. Otherwise, the following limit (if it exists) $\lim_{t \searrow 0}[x,x+th;F]h = \lim_{t \searrow 0} \dfrac{F(x+th) - F(x)}{t}$ vary according to h, with $\| h \| = 1$, and this limit is the Fréchet derivative (or the directional derivative) $F'(x)h$ of F in the direction h (i.e., if we suppose that F is Fréchet differentiable at x, then the Fréchet derivative is characterized as a limit of dd in the uniform topology of the space of continuous linear mappings of $\mathscr{B}_1$ into $\mathscr{B}_2$).

Remark 13.1.

(a)   Let $(x,y) \in \mathscr{B}_1 \times \mathscr{B}_2$, the set $\{A \in \mathscr{L}(\mathscr{B}_1,\mathscr{B}_2) : Ax = y\}$ constitue an affine manifold in $\mathscr{L}(\mathscr{B}_1,\mathscr{B}_2)$.

(b)   Let A and A_0 in $\mathscr{L}(\mathscr{B}_1,\mathscr{B}_2)$, and $(x,y) \in \mathscr{B}_1 \times \mathscr{B}_2$, such that $A_0 x = Ax = y$. Then $(A - A_0)x = 0$, and $A \in A_0 + \mathscr{K}_x$.

The following result gives some properties of set–valued mapping $\Upsilon_{x,y} : \mathscr{C}(\mathscr{B}_1,\mathscr{B}_2) \rightrightarrows \mathscr{L}(\mathscr{B}_1,\mathscr{B}_2)$ given by $\Upsilon_{x,y}(F) = [x,y;F]$ for the pair $(x,y) \in \mathscr{B}_1^2$.

Proposition 13.1.

(a) $\Upsilon_{x,y}(F) = F$ if and only if F is linear.

(b) $\Upsilon_{x,y}$ is linear, i.e., for F_1, F_2 in $\mathscr{C}(\mathscr{B}_1,\mathscr{B}_2)$, and $(\alpha,\beta) \in \mathbb{K}^2$ ($\mathbb{K} = \mathbb{R}$ or $\mathbb{C}$), we have
$$\Upsilon_{x,y}(\alpha F_1 + \beta F_2) = \alpha \Upsilon_{x,y}(F_1) + \beta \Upsilon_{x,y}(F_2).$$

(c)   If F is a composition of operators F_1 and F_2 (i.e., $F = F_1 \circ F_2$), then
$$\Upsilon_{x,y}(F) = \Upsilon_{F_2(x),F_2(y)}(F_1) \Upsilon_{x,y}(F_2).$$

Definition 13.2. [11] The dd $[x,y;F]$ is said to be w_1–regularly continuous on $\Omega \subseteq \mathscr{B}_1$ for $w_1 \in \mathscr{N}$ (call it regularity modulus), if the following inequality holds for each $x, y, u, v \in \Omega$

$$w_1^{-1} \left(\min\{\| [x,y;F] \|, \| [u,v;F] \|\} + \| [x,y;F] - [u,v;F] \| \right)$$

$$-w_1^{-1} \left(\min\{\| [x,y;F] \|, \| [u,v;F] \|\} \right) \leq \| x - u \| + \| y - v \| . \tag{13.3}$$

The dd $[x,y;F]$ is said to be regularly continuous on Ω, if it has there a regularity modulus.

We introduce a special notion (see also [5–7]).

Definition 13.3. The dd $[x,y;F]$ is said to be w_0- center regularly continuous on $\Omega \subset X$ for $w_0 \in \mathcal{N}$ (call it center regularity modulus), if for fixed $x_{-1}, x_0 \in \Omega$ the following inequality holds for each x, y in Ω

$$w_0^{-1} \left(\min\{\| [x,y;F] \|, \| [x_0,x_{-1};F] \|\} + \| [x,y;F] - [x_0,x_{-1};F] \| \right)$$

$$-w_0^{-1} \left(\min\{\| [x,y;F] \|, \| [x_0,x_{-1};F] \|\} \right) \leq \| x - x_0 \| + \| y - x_{-1} \| . \tag{13.4}$$

Clearly, we have that Definition 13.3 is a special case of Definition 13.2,

$$w_0(t) \leq w_1(t) \quad \text{for each} \quad t \in [0, \infty), \tag{13.5}$$

holds in general, and $\dfrac{w_1}{w_0}$ can be arbitrarily large [2,4]. If w_0, w_1 are linear functions ($w_1(t) = c_1 t$ and $w_0(t) = c_0 t$), then (13.4), and (13.5) become Lipschitz, and center–Lipschitz continuous conditions, respectively, i.e., the following hold respectively for each $(x,y,u,v) \in \Omega^4$:

$$\| [x,y;F] - [u,v;F] \| \leq c_1 (\| x - u \| + \| y - v \|) \tag{13.6}$$

and

$$\| [x,y;F] - [x_0,x_{-1};F] \| \leq c_0 (\| x - x_0 \| + \| y - x_{-1} \|). \tag{13.7}$$

Then, estimate (13.5) gives

$$c_0 \leq c_1. \tag{13.8}$$

We need the following auxiliary result.

Lemma 13.1. [9] If dd $[x,y;F]$ is w–regularly continuous on Ω, then we have

$$|w_1^{-1}(\| [x,y;F] \|) - w_1^{-1}(\| [u,v;F] \|)| \leq \|x - u\| + \|y - v\|, \quad \text{for each } (x,y,u,v) \in \Omega^4.$$

Then, the following holds for all $(x,y,u,v) \in \Omega^4$:

$$w_1^{-1}(\| [x,y;F] \|) \geq (w_1^{-1}(\| [u,v;F] \|) - \| x - u \| - \| y - v \|)^+, \tag{13.9}$$

where ρ^+ $(\rho \in \mathbb{R})$ denotes the nonnegative part of ρ: $\rho^+ = \max\{\rho, 0\}$.

In particular, if dd $[x,y;F]$ is w_0–regularly continuous on Ω (i.e., condition (13.4) holds), then, (13.9) holds, with w_0, x_0, and x_{-1} replacing w, u, and v, respectively.

Suppose that equation

$$w_0(t) = 1 \tag{13.10}$$

has at least one positive solution. Denote by r_0 the smallest such solution. Moreover, define

$$\Omega_0 = \Omega \cap U(x_0, r_0). \tag{13.11}$$

Notice also that we have a similar estimate for function w on Ω^4.

Definition 13.4. The dd $[x, y; F]$ is said to be restricted w−regularly continuous on $\Omega_0 \subset \Omega$ for $w \in \mathcal{N}$, if the following inequality holds for each $x, y, u, v \in \Omega_0$

$$w^{-1}\left(\min\{\| [x, y; F] \|, \| [u, v; F] \|\} + \| [x, y; F] - [u, v; F] \| \right)$$

$$\tag{13.12}$$

$$-w^{-1}\left(\min\{\| [x, y; F] \|, \| [u, v; F] \|\} \right) \leq \| x - u \| + \| y - v \|.$$

Notice that

$$w(t) \leq w_1(t) \text{ for each } t \in [0, r_0) \tag{13.13}$$

holds, since $\Omega_0 \subseteq \Omega$. Function w depends on function w_0. The construction of function w was not possible before in the earlier studies using only function w_1 [11]. Clearly, in those studies w can simply replace w_1, since the iterates lie in Ω_0 related to w, which is a more precise location than Ω used in [11] related to w_1. This modification leads to the already stated advantages, if strict inequality holds in (13.5) or (13.13). We suppose from now on until Remark 4.5 (b) that

$$w_0(t) \leq w(t) \text{ for each } t \in [0, r_0). \tag{13.14}$$

13.2   SEMI-LOCAL CONVERGENCE ANALYSIS OF BM

We present a semi-local convergence result for BM. The proofs are the proper modifications of the ones in [11], where, we use the more precise (13.4), (13.12) instead of (13.3). First, we denote

$$A_0 = [x_0, x_{-1}; F]^{-1}. \tag{13.15}$$

As in [11], for the selected dd $[x, y; F]$, such that (13.3) holds with w modulus, we associate the current iteration (x, A), and we consider $q = (\bar{t}, \bar{\gamma}, \bar{\delta})$, where

$$\bar{t} = \| x - x_0 \|, \quad \bar{\gamma} = \| x - x_- \|, \quad \bar{\delta} = \| x_+ - x \| = \| A F(x) \|.$$

Finally, denote q_+, and $\psi_w : \mathbb{R}^{+2} \longrightarrow \mathbb{R}^+$ by

$$q_+ = (\bar{t}_+, \bar{\gamma}_+, \bar{\delta}_+),$$

and

$$\psi_w(u, t) = w((u - t)^+ + t) - w((u - t)^+), \quad \text{for each } (u, t) \in \mathbb{R}^{+2},$$

respectively. Note that ψ_w is not increasing in the first argument, and not decreasing in the second, since w is concave, and increasing.

We provide now a result on q_+ using w and w_0−regularity.

Lemma 13.2. Under the hypotheses (13.4), and (13.12), the following estimates hold:

$$\bar{t}_+ := \| x_+ - x_0 \| \leq \bar{t} + \bar{\delta}, \tag{13.16}$$

$$\bar{\gamma}_+ := \| x_+ - x \| = \bar{\delta}, \tag{13.17}$$

and

$$\bar{\delta}_+ \leq \bar{\delta} \, e_{w_0,w}(q), \tag{13.18}$$

where,

$$e_{w_0,w}(q) = \frac{\psi_w(w^{-1}(\| A_0^{-1} \| - \bar{\gamma}_0 - 2\bar{t}_- - \bar{\gamma})^+, \bar{\gamma} + \bar{\delta})}{w_0(w_0^{-1}(\| A_0^{-1} \| - \bar{\gamma}_0 - 2\bar{t} - \bar{\delta}))}.$$

Proof. Estimates (13.16), and (13.17) follow from $\| x_+ - x_0 \| \leq \| x_+ - x \| + \| x - x_0 \| = \bar{t} + \bar{\delta}$, and the expression of $\bar{\delta}$, respectively.

We must show (13.18). We have

$$\bar{\delta}_+ \leq \| A_+ \| \, \| F(x_+) \|.$$

By the Banach lemma on invertible operators [4], and (13.15), we obtain

$$\| A_+^{-1} \| \geq \| A_0^{-1} \| - \| A_+^{-1} - A_0^{-1} \| = \| A_0^{-1} \|^{-1} - \| [x_+, x; F] - [x_0, x_{-1}; F] \|. \tag{13.19}$$

Using (13.3), we get in turn that

$$
\begin{aligned}
& \| [x,y;F] - [u,v;F] \| \\
& \leq w(w^{-1}(\min\{\| [x,y;F] \|, \| [u,v;F] \|\}) + \| [x,y;F] - [u,v;F] \|) \\
& \quad - \min\{\| [x,y;F] \|, \| [u,v;F] \|\} \\
& = w(\min\{w^{-1}(\| [x,y;F] \|), w^{-1}(\| [u,v;F] \|)\} + \| [x,y;F] - [u,v;F] \|) \\
& \quad - w(\min\{w^{-1}(\| [x,y;F] \|), w^{-1}(\| [u,v;F] \|)\}).
\end{aligned}
\tag{13.20}
$$

By Lemma 13.1, we have

$$w^{-1}(\| [u,v;F] \|) \geq (w^{-1}(\| [x,y;F] \|) - \| x - u \| - \| y - v \|)^+.$$

By (13.20), and the concavity of w, we get

$$
\begin{aligned}
& \| [x,y;F] - [u,v;F] \| \\
& \leq w(w^{-1}(\| [x,y;F] \| - \| x - u \| - \| y - v \|)^+ + \| x - u \| + \| y - v \|) \\
& \quad - w(w^{-1}(\| [x,y;F] \| - \| x - u \| - \| y - v \|)^+) \\
& = \psi_w(w^{-1}(\| [x,y;F] \|, \| x - u \| + \| y - v \|)).
\end{aligned}
\tag{13.21}
$$

Clearly, estimate (13.21) holds with w_0, x_+, x_0, and x_{-1} replacing w, y, u, and v, respectively. Consequently,

$$
\begin{aligned}
& \| [x_+, x; F] - [x_0, x_{-1}; F] \| \leq \psi_{w_0}(w_0^{-1}(\| [x_0, x_{-1}; F] \|), \| x_+ - x_0 \| + \| x - x_{-1} \|) \\
& \leq \psi_{w_0}(w_0^{-1}(\| A_0^{-1} \|, \| x_+ - x_0 \| + \| x - x_0 \| + \| x_0 - x_{-1} \|)) \\
& = \psi_{w_0}(w_0^{-1}(\| A_0^{-1} \|, \bar{t}_+ + \bar{t} + \bar{\gamma}_0)),
\end{aligned}
\tag{13.22}
$$

so,

$$\| A_+ \| \leq (\| A_0 \|^{-1} - \psi_{w_0}(w_0^{-1}(\| A_0^{-1} \|), \overline{\gamma}_0 + \overline{t}_+ + \overline{t}))^{-1} \implies \overline{\gamma}_0 + \overline{t}_+ + \overline{t} < w_0^{-1}(\| A_0^{-1} \|)$$
(13.23)

since, otherwise

$$\psi_{w_0}(w_0^{-1}(\| A_0^{-1} \|), \overline{t}_+ + \overline{t} + \overline{\gamma}_0) = w_0(\overline{\gamma}_0 + \overline{t}_+ + \overline{t}) \geq \| A_0^{-1} \| \geq \| A_0 \|^{-1}$$

and

$$\| A_0 \|^{-1} - \psi_{w_0}(w_0^{-1}(\| A_0^{-1} \|), \overline{t}_+ + \overline{t} + \overline{\gamma}_0) \leq 0.$$

We also have

$$\| A_+ \| \leq \frac{1}{w_0(w_0^{-1}(\| A_0 \|^{-1} - \overline{\delta} - 2\overline{t} - \overline{\gamma}_0))}.$$
(13.24)

Using (13.24), and since w_0 is concave and increasing, we deduce

$$\begin{aligned}
\psi_{w_0}(w_0^{-1}(\| A_0^{-1} \|, \overline{t}_+ + \overline{t} + \overline{\gamma}_0)) &= \| A_0 \|^{-1} - w_0(w_0^{-1}(\| A_0 \|^{-1} - \overline{t}_+ - \overline{t} - \overline{\gamma}_0)) \\
&\leq \| A_0 \|^{-1} - w_0(w_0^{-1}(\| A_0 \|^{-1} - \overline{\delta} - 2\overline{t} - \overline{\gamma}_0)).
\end{aligned}$$

By BM, we can have the identity

$$\begin{aligned}
F(x_+) &= F(x_+) - F(x) + F(x) \\
&= [x_+, x; F](x_+ - x) - A^{-1}(x_+ - x) \\
&= ([x_+, x; F] - [x, x_-; F])(x_+ - x).
\end{aligned}$$
(13.25)

Using (13.21), and (13.25), we obtain

$$\begin{aligned}
\| F(x_+) \| &\leq \overline{\delta} \| [x_+, x; F] - [x, x_-; F] \| \\
&\leq \psi_w(w^{-1}(\| A^{-1} \|), \overline{\gamma} + \overline{\delta}).
\end{aligned}$$

By (13.12), we get

$$\begin{aligned}
w^{-1}(\| A^{-1} \|) &= w^{-1}([x, x_-; F]) \\
&\geq (w^{-1}([x_0, x_{-1}; F] - \| x - x_0 \| - \| x_- - x_{-1} \|)^+) \\
&\geq (w^{-1}(\| A_0 \|^{-1} - \| x - x_0 \| - \| x_- - x_0 \| - \| x_0 - x_{-1} \|)^+) \\
&\geq (w^{-1}(\| A_0 \|^{-1} - \overline{t} - \overline{t}_- - \overline{\gamma}_0)^+) \\
&\geq (w^{-1}(\| A_0 \|^{-1} - \overline{\gamma} - 2\overline{t}_- - \overline{\gamma}_0)^+).
\end{aligned}$$

Consequently,

$$\psi_w(w^{-1}(\| A \|^{-1}), \overline{\gamma} + \overline{\delta}) \leq \psi_w(w^{-1}(\| A_0 \|^{-1} - \overline{\gamma} - 2\overline{t}_- - \overline{\gamma}_0)^+, \overline{\gamma} + \overline{\delta}),$$

$$\| F(x_+) \| \leq \overline{\delta} \, \psi_w(w^{-1}(\| A_0 \|^{-1} - \overline{\gamma} - 2\overline{t}_- - \overline{\gamma}_0)^+, \overline{\gamma} + \overline{\delta}),$$
(13.26)

and

$$\overline{\delta}_+ \leq \overline{\delta} \, e_{w,w_0}(q).$$

$\square$

We define the function χ_{w,w_0} for all $q = (t, \gamma, \delta)$ by $\chi_{w,w_0}(q) = q_+ = (t_+, \gamma_+, \delta_+)$,

$$t_+ = t + \delta, \quad \gamma_+ = \delta, \quad \delta_+ = \delta \frac{\psi_w((a - 2t + \gamma)^+, \gamma + \delta)}{w_0(a_0 - 2t - \delta)}, \tag{13.27}$$

where

$$a \leq w^{-1}(\|A_0\|^{-1}) - \|x_0 - x_{-1}\| \quad \text{and} \quad a_0 \leq w_0^{-1}(\|A_0\|^{-1}) - \|x_0 - x_{-1}\|.$$

In view of the definitions of a_0 and a, we can certainly assume $a \leq a_0$.

Remark 13.2. (a)   Since $2t - \gamma \leq 2t + \delta < a$, then

$$\psi_w((a - 2t + \gamma)^+, \gamma + \delta) = \psi_w(a - 2t + \gamma, \gamma + \delta) = w(a - 2t + \gamma) - w(a - 2t - \delta).$$

Consequently, we can simplify the third component δ_+ in the expression of χ_{w,w_0} by:

$$\delta_+ = \delta \frac{w(a - 2t + \gamma) - w(a - 2t - \delta)}{w_0(a_0 - 2t - \delta)}.$$

(b)   As $\bar{t}_0 = 0$, we can take $t_0 = 0$.

Consider the relation order "$\prec$" for $q = (t, \gamma, \delta)$ and $q' = (t', \gamma', \delta')$. We say that q' majorizes q, if

$$q \prec q' \iff t \leq t', \quad \gamma \leq \gamma' \text{ and } \delta \leq \delta'.$$

Lemma 13.3. Let $q = (t, \gamma, \delta)$ and $q' = (t', \gamma', \delta')$. Then

$$0 \prec q \prec q' \implies 0 \prec \chi_{w,w_0}(q) \prec \chi_{w,w_0}(q').$$

Proof. We suppose $q \prec q'$. Then, we obtain

$$t \leq t' \text{ and } \delta \leq \delta' \implies t_+ := t + \delta \leq t' + \delta' =: t'_+ \text{ and } \gamma_+ := \delta \leq \delta' =: \gamma'_+.$$

We show now $\delta_+ \leq \delta'_+$. Functions w, w_0 are concave and increasing, and by using Remark 13.2, we have

$$\begin{aligned}
\delta_+ &= \delta \frac{w(a - 2t + \gamma) - w(a - 2t - \delta)}{w_0(a_0 - 2t - \delta)} \\
&\leq \delta' \frac{w(a - 2t - \delta + (\delta + \gamma)) - w(a - 2t - \delta)}{w_0(a_0 - 2t' - \delta')} \\
&\leq \delta' \frac{w(a - 2t' - \delta' + (\delta + \gamma)) - w(a - 2t' - \delta')}{w_0(a_0 - 2t' - \delta')} \\
&\leq \delta' \frac{w(a - 2t' - \delta' + (\delta' + \gamma')) - w(a - 2t' - \delta')}{w_0(a_0 - 2t' - \delta')} = \delta'_+.
\end{aligned}$$

$\square$

Consider the sequence q_n with the initial iterate $q_0 = (t_0, \gamma_0, \delta_0)$ by

$$q_{n+1} = \chi_{w,w_0}(q_n). \tag{13.28}$$

Then, q_n is a majorizing sequence, if $\bar{q}_n \prec q_n$ for each $n \geq 0$, where $\bar{q}_n$ is produced by the n-th iteration (x_n, A_n) of BM.

Lemma 13.4. Let q_0 be an initial iterate for sequence $q_n = (t_n, \gamma_n, \delta_n)$ given by (13.28), such that $\bar{q}_0 \prec q_0$, and $2t_n + \delta_n < a$ for each $n \geq 0$. Then, the following hold for each $n \geq 0$:

(a)
$$\bar{q}_n \prec q_n.$$

(b)
$$\gamma_\infty = \delta_\infty = 0 \quad \text{and} \quad t_n = \sum_{k=0}^{k=n-1} \delta_k \leq 0.5\,(a - \delta_n),$$

where
$$\gamma_\infty = \lim_{n \to \infty} \gamma_n, \quad \delta_\infty = \lim_{n \to \infty} \delta_n \text{ and } t_\infty = \lim_{n \to \infty} t_n;$$

(c) The sequence (x_n, A_n) generated by BM from the initial iterate (x_0, A_0) converges to a solution (x^*, A_∞) of the system:

$$F(x) = 0 \quad \text{and} \quad A[x, x; F] = I;$$

(d) The solution x^* is unique in $U(x_0, a_0 - t_\infty)$;

(e)
$$\| F(x_{n+1}) \| \leq \delta_n\,(w(a - 2t_n + \gamma_n) - w(a - 2t_n - \delta_n)),$$

$$\| x_n - x_0 \| \leq t_n < t_\infty \leq 0.5\,a,$$

$$\Delta_n := \| x^* - x_n \| \leq t_\infty - t_n,$$

$$\| I - A_n[x^*, x^*; F] \| \leq \frac{w(a - 2t_n + \gamma_n) - w(a - 2t_\infty)}{w_0(a_0 - 2t_n + \gamma_n)},$$

and
$$\frac{\Delta_{n+1}}{\Delta_n} \leq \frac{w(a - 2t_n + \gamma_n) - w(a - t_n - t_\infty)}{w_0(a_0 - 2t_n + \gamma_n)}.$$

Proof. (a)   We show (a) using induction on n. We suppose that $\bar{q} \prec q$. By Lemma 13.2, we have

$$q_+ = (\bar{t}_+, \bar{\gamma}_+, \bar{\delta}_+) \prec (\bar{t} + \bar{\delta}, \bar{\delta}, \bar{\delta}\, e_{w_0, w}(q)), \tag{13.29}$$

where
$$w(w^{-1}(\| A_0 \|^{-1}) - \bar{\gamma}_0 - 2\bar{t} - \bar{\delta}) \geq w(a - 2t - \delta),$$

and

$$
\begin{aligned}
\psi_w((w^{-1}(\| A_0^{-1} \| -\overline{\gamma}_0 - 2\overline{t}_- - \overline{\gamma})^+, \overline{\gamma} + \overline{\delta})) \\
\le \psi_w((a - 2\overline{t}_- - \overline{\gamma})^+, \overline{\gamma} + \overline{\delta}) \\
\le \psi_w((a - 2t_- - \gamma)^+, \gamma + \delta) \\
= \psi_w(a - 2t + \gamma, \gamma + \delta) \\
= w(a - 2t + \gamma) - w(a - 2t - \delta).
\end{aligned}
$$

Then, we get the estimate

$$
\overline{\delta} \, e_{w_0, w}(q) \le \delta \, \frac{w(a - 2t + \gamma) - w(a - 2t - \delta)}{w_0(a_0 - 2t - \delta)}. \tag{13.30}
$$

By (13.29), and (13.30), we deduce

$$
\overline{q}_+ \prec \chi_{w, w_0}(q) = q_+.
$$

The induction is completed.

(b) By hypotheses of Lemma 2, $2t_n + \delta_n < a$ for all $n \ge 0$, so we deduce that $t_n < 0.5(a - \delta_n)$ for each $n \ge 0$. Then, $\{t_n\}$ is increasing and bounded. Thus, $\{t_n\}$ converges to a finite limit t_∞. Since $t_{n+1} = t_n + \delta_n$ and $\gamma_{n+1} = \delta_n$, for each $n \ge 0$, we obtain $\gamma_\infty = \delta_\infty = 0$, and $t_\infty \le 0.5\, a$.

(c) to (e) First, by (a), we have for $n, m \ge 0$

$$
\begin{aligned}
\| x_{n+m} - x_n \| & \le \sum_{k=n}^{k=n+m-1} \| x_{k+1} - x_k \| \\
& = \sum_{k=n}^{k=n+m-1} \overline{\delta}_k \le \sum_{k=n}^{k=n+m-1} \delta_k < \sum_{k=n}^{\infty} \delta_k = t_\infty - t_n.
\end{aligned} \tag{13.31}
$$

Hence, $\{x_n\}$ is a complete sequence in a Banach space, and as such it converges to $x^\star$. By letting $m \longrightarrow \infty$ in (13.31), we deduce for $n \ge 0$ the following estimate $\Delta_n = \| x^\star - x_n \| \le t_\infty - t_n$. Moreover, by (13.26), we get

$$
\begin{aligned}
\| F(x_{n+1}) \| & \le \overline{\delta}_n \, \psi_w((w^{-1}(\| A_0 \|^{-1} - \overline{\gamma}_0 - \overline{t}_n - \overline{t}_{n-1})^+, \overline{\gamma}_n + \overline{\delta}_n) \\
& \le \overline{\delta}_n \, \psi_w(a - 2t_n + \gamma_n, \gamma_n + \delta_n) \\
& = \overline{\delta}_n \, (w(a - 2t_n + \gamma_n) - w(a - 2t_n - \delta_n)),
\end{aligned}
$$

where $\delta_n \longrightarrow \infty$, so, $F(x^\star) = 0$. By letting $n \longrightarrow \infty$ in equality $A_n [x_n, x_{n-1}; F] = I$, we get $A_\infty [x^\star, x^\star; F] = 0$, and (c) is completed. Substituting A_+ by A_n in (13.24), we have

$$
\begin{aligned}
\| A_n \| & \le \frac{1}{w_0(w_0^{-1}(\| A_0 \|^{-1} - \overline{\gamma}_0 - \overline{t}_n - \overline{t}_{n-1})^+)} \\
& \le \frac{1}{w_0(a_0 - t_n - t_{n-1})} = \frac{1}{w_0(a_0 - 2t_n + \gamma_n)}.
\end{aligned} \tag{13.32}
$$

Using (13.21), we have

$$
\begin{aligned}
\| I - A_n [x^\star, x^\star; F] \| \\
\le \| A_n \| \, \| [x_n, x_{n-1}; F] - [x^\star, x^\star; F] \| \\
\le \| A_n \| \, \psi_w(w^{-1}(\| [x_n, x_{n-1}; F] \|), \| x_n - x^\star \| + \| x_{n-1} - x^\star \|).
\end{aligned} \tag{13.33}
$$

By Lemma 13.1, we get

$$
\begin{aligned}
w^{-1}&(\|\,[x_n,x_{n-1};F]\,\|) \\
&\geq (w^{-1}(\|\,[x_0,x_{-1};F]\,\|) - \|\,x_n - x_0\,\| - \|\,x_{n-1} - x_{-1}\,\|)^+) \\
&\geq (w^{-1}(\|\,A_0\,\| - \bar{t}_n - \bar{t}_{n-1} - \overline{\gamma}_0)^+) \\
&\geq a - 2t_n + \gamma_n.
\end{aligned}
\tag{13.34}
$$

Using (13.32)–(13.34), we obtain

$$
\|\,I - A_n\,[x^\star,x^\star;F]\,\| \;\leq\; \frac{w(a - 2t_n + \gamma_n) - w(a - 2t_\infty)}{w_0(a_0 - 2t_n + \gamma_n)}.
\tag{13.35}
$$

By the identity

$$
\begin{aligned}
x_{n+1} - x^\star &= x_n - x^\star - A_n F(x_n) \\
&= A_n([x_n,x_{n-1};F] - [x_n,x^\star;F])(x_n - x^\star),
\end{aligned}
$$

we get similarly as in (13.35)

$$
\begin{aligned}
\Delta_{n+1} &\leq \Delta_n \,\|\,A_n\,\|\,\|\,[x_n,x_{n-1};F] - [x_n,x^\star;F]\,\| \\
&\leq \Delta_n \,\frac{\psi_w(a - 2t_n + \gamma_n, \Delta_{n-1})}{w_0(a_0 - 2t_n + \gamma_n)} \\
&\leq \Delta_n \,\frac{\psi_w(a - 2t_n + \gamma_n, t_\infty - t_n + \gamma_n)}{w_0(a_0 - 2t_n + \gamma_n)} \\
&= \Delta_n \,\frac{w(a - 2t_n + \gamma_n) - w(a - t_n - t_\infty)}{w_0(a_0 - 2t_n + \gamma_n)}.
\end{aligned}
\tag{13.36}
$$

The proof of (e) is completed.

We prove now (d). Let $y^\star$ be a solution of $F(x) = 0$. Then, $F_0(y^\star) = 0$, where $F_0 = A_0 F$, and $F_0(y^\star) - F_0(x^\star) = 0 = [y^\star,x^\star;F_0](y^\star - x^\star)$. Then, we deduce by the Banach lemma on invertible operators [2,4], and Proposition 13.1 that $[y^\star,x^\star;F_0] = A_0\,[y^\star,x^\star;F]$ is not invertible, and $\|\,I - A_0\,[y^\star,x^\star;F]\,\| \geq 1$. Using (13.33), we have

$$
\begin{aligned}
1 \;&\leq\; \|\,I - A_0\,[x^\star,x^\star;F]\,\| \\
&\leq\; \|\,A_0\,\|\,\|\,[x_0,x_{-1};F] - [y^\star,x^\star;F]\,\| \\
&\leq\; \|\,A_0\,\|\,\psi_{w_0}(w_0^{-1}(\|\,[x_0,x_{-1};F]\,\|)), \|\,y^\star - x_0\,\| + \|\,x_{-1} - x^\star\,\|) \\
&\leq\; \|\,A_0\,\|\,\psi_{w_0}(w_0^{-1}(\|\,A_0^{-1}\,\|)), \|\,y^\star - x_0\,\| + \|\,x^\star - x_0\,\| + \|\,x_0 - x_{-1}\,\|),
\end{aligned}
$$

and

$$
\begin{aligned}
\|\,A_0^{-1}\,\| \\
\leq\; \psi_{w_0}&(w_0^{-1}(\|\,A_0\,\|^{-1}, \Lambda)) \\
=\; \mathscr{L}ft&\{\begin{array}{ll} \|\,A_0\,\|^{-1} - w_0(w_0^{-1}(\|\,A_0\,\|^{-1}, \Lambda)), & \text{if } \Lambda \leq w_0^{-1}(\|\,A_0\,\|^{-1}) \\ w_0(\Lambda), & \text{if } \Lambda \geq w_0^{-1}(\|\,A_0\,\|^{-1}), \end{array}\end{aligned}
$$

where

$$
\Lambda = \overline{\gamma}_0 + t_\infty + \|\,y^\star - x_0\,\|.
$$

It follows that

$$
\|\,y^\star - x_0\,\| \geq w_0^{-1}(\|\,A_0\,\|^{-1} - \overline{\gamma}_0 - t_\infty) \geq a_0 - t_\infty \geq 0.5\,a.
$$

$\square$

13.3   APPLICATIONS

In the special cases 1 and 2 that follow, we provide more precise estimates than [11, Section 4].

CASE 1: SEMI-LOCAL CONVERGENCE

Let x_0, x_{-1}, A_0, a, a_0, and q_0 given, such that

$$\bar{q}_0 \prec q_0 \quad \text{and} \quad 2t_n + \delta_n < a \quad \text{for each } n \geq 0. \tag{13.37}$$

Condition (13.37) guarantee the convergence of sequence (x_n, A_n). We denote by $\mathcal{Q}_c$ the set

$$\mathcal{Q}_c = \{(q_0, t_n, \delta_n) : (13.37) \quad \text{holds}\}.$$

In this subsection $w(t) = ct$, $w_0(t) = c_0 t$ with $c_0 \leq c$ and $w_1(t) = c_1 t$. The function χ_{w,w_0} defined in (13.27) by $\chi_{w,w_0}(q) = q_+ = (t_+, \gamma_+, \delta_+)$ for all $q = (t, \gamma, \delta)$, is simplified in the following form

$$t_+ = t + \delta, \quad \gamma_+ = \delta, \quad \delta_+ = \delta \, \frac{\bar{c}(\gamma + \delta)}{1 - \bar{c}_0 \left(\| x_0 - x_{-1} \| - 2t - \delta \right)}, \tag{13.38}$$

where

$$\bar{c} = c \, \| A_0 \|, \quad \bar{c}_0 = c_0 \, \| A_0 \| \quad \text{and } \bar{c}_1 = c_1 \, \| A_0 \| \, .$$

If in the denominator of δ_+, function w, and a replace w_0, and a_0, respectively, then, (13.38) becomes

$$t_+ = t + \delta, \quad \gamma_+ = \delta, \quad \delta_+ = \delta \, \frac{\gamma + \delta}{a - 2t - \delta}. \tag{13.39}$$

Define function Γ for $q = (t, \gamma, \delta)$ by

$$\Gamma(q) = (0.5 a - t)^2 - \delta (a - 2t + \gamma). \tag{13.40}$$

We present now two results on simplified generator χ_{w,w_0} given by (13.39).

Lemma 13.5. [3, 9, 11]

(a)   The function Γ given by (13.40) is an invariant of the generator (13.39):

$$\Gamma(q) = \Gamma(q_+).$$

(b)   For all $n \geq 0$,

$$2t_n + \delta_n < a \quad \Longleftrightarrow \quad \Gamma(0, \gamma_0, \delta_0) \geq 0 \Longleftrightarrow 4 \delta_0 (a + \gamma_0) \leq a^2$$
$$\Longleftrightarrow \quad t_n = 0.5 a - \delta_n - \sqrt{\delta_n (\gamma_n + \delta_n) + \Gamma(0, \gamma_0, \delta_0)}.$$

Theorem 13.1. Suppose that (13.6) and (13.7) hold. Let $x_{-1}, x_0, A_0, a_0, a, \gamma_0, \delta_0$, such that

$$\frac{\|A_0\|^{-1}}{c_0} - \|x_0 - x_{-1}\| \geq a_0, \qquad \frac{\|A_0\|^{-1}}{c} - \|x_0 - x_{-1}\| \geq a,$$

$$\|x_0 - x_{-1}\| \leq \gamma_0, \qquad \|A_0 F(x_0)\| \leq \delta_0,$$

and

$$4\,\delta_0\,(a + \gamma_0) \leq a^2. \tag{13.41}$$

Then, the following hold

(a) $(t_n, \gamma_n, \delta_n)$ generated by (13.39) started at $(0, \gamma_0, \delta_0)$ is well defined, and converges to $(t_\infty, 0, 0)$, where,

$$t_\infty = 0.5\,(a - \sqrt{a^2 - 4\,\delta_0\,(a + \gamma_0)});$$

(b) The sequence (x_n, A_n) generated by (BM) started at (x_{-1}, x_0, A_0) converges to a solution (x^*, A_∞) of the system

$$F(x) = 0 \quad \text{and} \quad A\,[x, x; F] = I;$$

(c) x^* is the unique solution of (13.1) in $U(x_0, r)$, where,

$$r = 0.5\,(a + \sqrt{a^2 - 4\,\delta_0\,(a + \gamma_0)});$$

(d) For each $n \geq 1$,

$$\|F(x_{n+1})\| \leq c\,\delta_n\,(\gamma_n + \delta_n), \qquad \Delta_n \leq \tilde{t}_\infty - \tilde{t}_n,$$

$$\|x_n - x_0\| \leq \tilde{t}_n - \tilde{t}_0,$$

$$\|I - A_n\,[x^*, x^*; F]\| \leq p_n,$$

and

$$\frac{\Delta_{n+1}}{\Delta_n} \leq q_n,$$

where

$$\tilde{t}_{-1} = 0, \quad \tilde{t}_0 = \|x_0 - x_{-1}\|, \quad \tilde{t}_1 = \tilde{t}_0 + \|A_0 F(x_0)\|,$$

$$\tilde{t}_n = \tilde{t}_{n-1} + \frac{\overline{c}\,(\tilde{t}_{n-1} - \tilde{t}_{n-3})\,(\tilde{t}_{n-1} - \tilde{t}_{n-2})}{1 - \overline{c}_0\,(-\tilde{t}_{n-1} - \tilde{t}_{n-2} + \tilde{t}_0)}, \quad (n \geq 2),$$

$$p_n = \frac{\overline{c}\,(\gamma_n + 2\,(\tilde{t}_\infty - \tilde{t}_n))}{\overline{c}_0\,(a_0 - 2\tilde{t}_n + \gamma_n)},$$

and

$$q_n = \frac{\overline{c}\,(\gamma_n + \tilde{t}_\infty - \tilde{t}_n)}{\overline{c}_0\,(a_0 - 2\tilde{t}_n + \gamma_n)},$$

with

$$\tilde{t}_\infty = \lim_{n \to \infty} \tilde{t}_n \leq t_\infty.$$

The estimates in Theorem 13.1 reduce to the corresponding ones in [11] for $c_0 = c = c_1$. Otherwise, the new estimates are more precise.

CASE 2: SEMI-LOCAL CONVERGENCE UNDER REGULAR CONTINUITY OF DD

Suppose that w is nonlinear, and let $q_0 = (t_0, \gamma_0, \delta_0)$. Define the scalar function $F(\cdot/q_0)$ on sequence $\{t_n\}$ $(n \geq 0)$ as follows

$$F(t_n/q_0) = \delta_n w(a - 2t_n + \gamma_n). \tag{13.42}$$

Sequence $F(t_n)$ is decreasing, and consequently, function F is invertible on $\{t_n\}$, i.e., for each $n \geq 0$:

$$t_n = F^{-1}(\delta_n w(a - 2t_n + \gamma_n)/q_0) \qquad [11]. \tag{13.43}$$

We present now two results on generator χ_{w,w_0} given by (13.27).

Lemma 13.6. [11]

(a) The function $F^{-1}(0/q)$ with initial iterate $q = (t, \gamma, \delta)$ is an invariant of the generator (13.27).

(b) For all $n \geq 0$, and $q_0 = (t_0, \gamma_0, \delta_0)$, we have the following equivalence

$$2t_n + \delta_n < a \Longleftrightarrow F^{-1}(0/q_0) \leq 0.5\,a.$$

Theorem 13.2. Suppose that (13.3), and (13.4) hold. Let $x_{-1}, x_0, A_0, a_0, a, \gamma_0, \delta_0$, such that

$$w_0^{-1}(\| A_0 \|^{-1}) - \| x_0 - x_{-1} \| \geq a_0, \quad w^{-1}(\| A_0 \|^{-1}) - \| x_0 - x_{-1} \| \geq a,$$

$$\| x_0 - x_{-1} \| \leq \gamma_0, \quad \| A_0 F(x_0) \| \leq \delta_0,$$

and

$$F^{-1}(0/(0, \gamma_0, \delta_0)) \leq 0.5\,a.$$

Then, the following hold

(a) $(t_n, \gamma_n, \delta_n)$ generated by (13.27) started at $(0, \gamma_0, \delta_0)$ is well defined, and converges to $(t_\infty, 0, 0)$, where

$$t_\infty = F^{-1}(0/(0, \gamma_0, \delta_0));$$

(b) The sequence (x_n, A_n) generated by BM started at (x_{-1}, x_0, A_0) converges to a solution (x^*, A_∞) of the system

$$F(x) = 0 \quad \text{and} \quad A[x, x; F] = I;$$

(c) x^* is the unique solution of (13.1) in $U(x_0, a_0 - t_\infty)$;

(d) For each $n \geq 1$,

$$\| F(x_{n+1}) \| \leq \delta_n (w(a - 2t_n + \gamma_n) - w(a - 2t_n - \delta_n)),$$

$$\| I - A_n [x^*, x^*; F] \| \leq \frac{w(a - 2t_n + \gamma_n) - w(a - 2t_\infty)}{w_0(a_0 - 2t_n + \gamma_n)},$$

and

$$\frac{\Delta_{n+1}}{\Delta_n} \leq \frac{w(a - 2t_n + \gamma_n) - w(a - t_n - t_\infty)}{w_0(a_0 - 2t_n + \gamma_n)}.$$

Remark 13.3.

(a) The results obtained in this study reduce to the corresponding ones in [11], if equality holds in (13.5) and (13.14). Otherwise, our results provide weaker sufficient convergence conditions, error bounds than in [11] (see also, the definition of a and a_0). Moreover, the information on the uniqueness of the solution $x^\star$ is more precise, since $a_0 - t_\infty > a - t_\infty$ (see also, Lemma 13.4–(b)).

As an example instead of studying the iteration in [11] corresponding to (13.39) and defined by

$$s_{-1} = 0, \quad s_0 = \| x_0 - x_{-1} \|, \quad s_1 = s_0 + \| A_0 F(x_0) \|$$
$$s_{n+1} = s_n + \frac{\overline{c}_1 (s_n - s_{n-1})(s_n - s_{n-2})}{1 - \overline{c}_1 (s_0 - s_n - s_{n-1})},$$

we study the more precise sequence defined by

$$\alpha_{-1} = 0, \quad \alpha_0 = \| x_0 - x_{-1} \|, \quad \alpha_1 = \alpha_0 + \| A_0 F(x_0) \|$$
$$\alpha_{n+1} = \alpha_n + \frac{\overline{c}(\alpha_n - \alpha_{n-1})(\alpha_n - \alpha_{n-2})}{1 - c_2 (\alpha_0 - \alpha_n - \alpha_{n-1})},$$

where $c_2 = \overline{c}$ or $c_2 = \overline{c}_0$. Using (13.3), (13.4) and our idea of recurrent functions but not (13.12), we have already obtained weaker sufficient convergence conditions for many iterative methods such as Newton's, Secant, and Newton–type methods (under very general conditions [1–7]). In particular, our work using regularly continuous divided differences can be found in [5].

(b) If $w(t) \le w_0(t)$ for all $t \in [0, r_0)$ holds instead of (13.14), then clearly function w_0 (still at least as small as function w_1) can replace w in the preceding results.

(c) If Ω_0 is replaced by $\Omega_0^* = \bigcup (x_1, r - \| A_0 F(x_0) \|)$ then in Definition 2.7 a function even tighter than w can be used, so, the results can be weakened even further, since $\Omega_0^* \subseteq \Omega_0$, and x_1 still depends on the initial data.

13.4   EXERCISES

1. Solve the initial value problem

$$\dot{x}(t) = 1 + \cos(x(t)), x(0) = 0,$$

2. The predator-prey population models describe the interaction of a prey population X and a predator population Y. Assume that their interaction is modeled by the system of ordinary differential equations

$$\dot{x} = x - \frac{1}{4}x^2 - \frac{1}{10}xy + 1$$
$$\dot{y} = -\frac{1}{4}y + \frac{1}{7}xy + \frac{2}{3}.$$

(Assume $x(0) = y(0) = 0$). Solve the system.

3. Solve the initial value problem

$$\dot{x} = (x(t) + t, x(0) = 1)$$

by the iteration method developed in this chapter. Perform 3 steps.

4. Solve the Fredholm-type integral equation

$$x(t) = \sum_0^1 \frac{t \cdot s}{10} ds + 1$$

by the iteration method developed in this chapter. Perform 3 steps.

5. Solve the Volterra-type integral equation

$$x(t) = \sum_0^t \frac{t \cdot s}{10} ds + 1$$

by the iteration method developed in this chapter. Perform 3 steps.

6. Solve equation

$$x = \frac{\sin x}{2} + 1$$

by using algorithm developed in this chapter.

7. Repeat the previous problem for equation

$$x = \frac{\sin x + \cos 2x}{10}.$$

REFERENCES

1. Argyros, I.K., 2004. On the convergence of Broyden's method. *Comm. Appl. Nonlinear Anal.*, 11:77–86.
2. Argyros, I.K. 2004. A unifying local–semilocal convergence analysis and applications for two–point Newton–like methods in Banach space. *J. Math. Anal. Appl.*, 298:374–397.
3. Argyros, I.K. 2007. On the convergence of Broyden–like methods. *Acta Math. Sin. (Engl. Ser.)*, 23:2087–2096.
4. Argyros, I.K. 2008. *Convergence and applications of Newton–type iterations*, New York: Springer–Verlag.
5. Argyros, I.K., and Hilout, S. 2009. On the convergence of some iterative procedures under regular smoothness. *PanAmerican Math. J.*, 19:17–34.
6. Argyros, I.K., and Hilout, S. 2010. A Kantorovich–type analysis of Broyden's method using recurrent functions. *J. Appl. Math. Comput.*, 32:353–368.
7. Argyros, I.K., and Hilout, S. 2011. A unifying theorem for Newton's method on spaces with a convergence structure, *J. Complexity*, 27:39–54.
8. Broyden, C.G. 1965. A class of methods for solving nonlinear simultaneous equations. *Math. Comp.*, 19:577–593.
9. Galperin, A. 2006. Secant method with regularity continuous divided differences. *J. Comput. Appl. Math.*, 193:574–575.
10. Galperin, A. 2009. Ulm's method without derivatives. *Nonlinear Anal.*, 71:2094–2113.
11. Galperin, A. 2015. Broyden's method for operators with regularly continuous divided differences. *J. Korean Math. Soc.*, 52. 1:43–65.
12. Hernández, M.A., Rubio, M.J., and Ezquerro, J.A. 2000. Secant–like methods for slving nonlinear integral equations of the Hammerstein type. *J. Comput. Appl. Math.*, 115:245–254.
13. Kantorovich, L.V., and Akilov, G.P. 1982. *Functional Analysis in normed spaces*, Oxford: Pergamon Press.
14. Kelley, C.T. 1995. *Iterative methods for linear and nonlinear equations*. Frontiers in Applied Mathematics, Philadelphia, PA: Society for Industrial and Applied Mathematics (SIAM).
15. Magaril–Il'yaev, G.G., and Tikhomirov, V.M. 2003. *Convex analysis: theory and applications*. Providence, RI: American Mathematical Society.

14 Complexity of a Homotopy Method for Locating an Approximate Zero

The goal of this chapter is to extend the applicability of a homotopy method for locating an approximate zero using Newton's method. The improvements are obtained using more precise Lipschitz-type functions than in earlier works and our new idea of restricted convergence regions. Moreover, these improvements are found under the same computational effort.

The convergence region and error analysis of iterative methods are very pessimistic in general for both the semi-local and local case [1–4,10–15]. The aim of the chapter is to extend the convergence region using the homotopy method. This goal is achieved using the same Lipschitz-type functions as before [4–9,12]. We achieve this goal, since we find a more precise location for the Newton iterates leading to at least as tight Lipschitz-type functions [4–6]. Let $F : D \subset \mathscr{B}_1 \longrightarrow \mathscr{B}_2$ be differentiable in the sense of Fréchet, D be a convex and open subset of $\mathscr{B}_1$ and $\mathscr{B}_1, \mathscr{B}_2$ be Banach spaces.

Let F' be one-to-one and onto, we introduce the Newton operator

$$N_F(x) := x - F'(x)^{-1}F(x) \tag{14.1}$$

and the corresponding Newton iteration

$$x_{n+1} = N_F(x_n) \text{ for all } n = 0, 1, 2, \ldots \tag{14.2}$$

where $x_0 \in D$ is an initial point. We are concerned with the problem of approximating a regular (to be precised in Section 2) solution w of

$$F(x) = 0 \tag{14.3}$$

utilizing a homotopy method of the form

$$\mathscr{H}(x,t) := F(x) - tF(x_0) \tag{14.4}$$

where $x_0 \in D$ is a given initial point and $t \in [0,1]$. Clearly this is a geometrical way of solving equation (14.3). Consider the line segment $M = \{tF(x_0) : t \in [0,1]\}$ and the set $F^{-1}(M)$. Suppose that $F'(x_0)$ is one-to-one and onto. Then, it follows by the implicit function theorem applied in a neighbourhood of x_0 that there exists a curve $x(t)$ solving the equation $F(x(t)) = tF(x_0)$ for $t \in [1 - \varepsilon, 1]$ and $\varepsilon > 0$. This curve solves the initial value problem (IVP)

$$\dot{x}(t) = -DF(x(t))^{-1}F(x_0), \quad x(1) = x_0. \tag{14.5}$$

DOI: 10.1201/9781003128915-14

It is well known that (14.5) has no solution on $[0,1]$, in general. But if it has a solution, one must follow $x(t)$ (numerically), which is given by $\mathcal{H}(x(t),t) = 0$ using the operator related to $\mathcal{H}(.,t)$. That is consider the sequence $\{s_n\}$ given by $s_0 = 1 > s_1 > \ldots > s_n > \ldots > 0$ such that

$$x_{n+1} = N_{\mathcal{H}(.,s_{n+1})}(x_k)$$

is an approximate zero of $x(s_{n+1})$, with

$$\mathcal{H}(x(s_{n+1}),s_{n+1}) = 0.$$

A convergence analysis of Newton sequence $\{x_n\}$ was given in the elegant work by Guttierrez et al. [9]. Here, we improve their results as already mentioned previously.

The study is structured as: The convergence of Newton's method is presented in Section 14.1 whereas Section 14.2 contains the special cases. Finally, in Section 14.3, we present some applications.

14.1 CONVERGENCE ANALYSIS

We need the Definition of an approximate zero.

Definition 14.1. [13] A $G-$regular ball is open so that $G'(x)$ is one-to-one and onto. A point x_0 is a regular approximate zero of G, provided there exists a ball $G-$regular containing a zero w of G and a sequence $\{x_n\}$ converging to w.

Let $L_0, \bar{L}, L : [0,+\infty) \longrightarrow [0,+\infty)$ be continuous and non-decreasing functions. These functions are needed for the introduction of the Lipschitz conditions that follow (see (14.7), (14.9) and (14.11)). We shall also suppose that there exists $z \in D$ so $G'(z)$ is continuous, one-to-one, onto and $G'(z)^{-1}$ exists. We need to introduce the following two Lipschitz conditions that follow.

Definition 14.2. The function $G'(z)^{-1}G'$ is L_0- center Lipschitz at z if there exist positive quantities v_0 and

$$\gamma_0 := \gamma_0(G,z) \tag{14.6}$$

satisfying for $a \in D, \gamma_0(\|a-z\|) \leq v_0$

$$\|G'(z)^{-1}(G'(a) - G'(z))\| \leq \int_0^{\gamma_0\|a-z\|} L_0(\tau)d\tau. \tag{14.7}$$

Definition 14.3. The function $G'(z)^{-1}G'$ is $\bar{L}-$center Lipschitz restricted at z, if there exist positive quantities $\bar{v}$ and

$$\bar{\gamma} := \bar{\gamma}(G,z) \tag{14.8}$$

satisfying for $a,b \in D_0 := D \cap \bar{U}(z,\frac{\bar{v}}{\bar{\gamma}})$

$$\bar{\gamma}(\|a-z\| + \tau\|a-b\|) \leq \bar{v}$$

and

$$\|G'(z)^{-1}(G'((1-\tau)a+\tau b)-G'(a))\| \leq \int_{\bar{\gamma}\|a-z\|}^{\bar{\gamma}(\|a-z\|+\tau\|b-a\|)} \bar{L}(\tau)d\tau \qquad (14.9)$$

for all $\tau \in [0,1]$.

Definition 14.4. [9] The function $G'(y_0)^{-1}G'$ is $L-$center Lipschitz at z if there exist positive quantities v and

$$\gamma := \gamma(G,z) \qquad (14.10)$$

satisfying for $a,b \in D$

$$\gamma(\|a-z\|+\tau\|a-b\|) \leq v$$

and

$$\|G'(z)^{-1}(G'((1-\tau)a+\tau b)-G'(a))\| \leq \int_{\gamma\|a-z\|}^{\gamma(\|a-z\|+\tau\|b-a\|)} L(\tau)d\tau \qquad (14.11)$$

for each $\tau \in [0,1]$.

Remark 14.1. Notice that (14.11) implies (14.7) and (14.9). We can certainly take $v_0 = v = \bar{v}$, $L_0(\tau) = L(\tau) = \bar{L}(\tau)$ for each $\tau \geq 0$, so for all $\tau \in [0,v]$

$$\gamma_0(\tau) \leq \gamma(\tau) \qquad (14.12)$$

and

$$\bar{\gamma}(\tau) \leq \gamma(\tau), \qquad (14.13)$$

since $D_0 \subset D$.

In what follows we shall assume that

$$\gamma_0(\tau) \leq \bar{\gamma}(\tau). \qquad (14.14)$$

If instead of (14.14)

$$\bar{\gamma}(\tau) \leq \gamma_0(\tau), \qquad (14.15)$$

holds then the following results are true with $\bar{L}$ replacing L_0 in all of them.

Lemma 14.1. Suppose that v_0 is the least positive number such that

$$\int_0^{v_0} L_0(\tau)d\tau = 1. \qquad (14.16)$$

Then $F'(x)$ is one-to-one, onto and

$$\|F'(x)^{-1}F'(z)\| \leq \left(1 - \int_0^{\gamma_0\|x-z\|} L_0(\tau)d\tau\right)^{-1} \text{ for all } x \in U(z, \frac{v_0}{\gamma_0}). \qquad (14.17)$$

The set $U(z, \frac{v_0}{\gamma_0})$ is called the γ_0−ball of z. We define similarly, the $\bar{\gamma}$ and γ−balls. As in [9], we assume the existence of $\bar{\varphi} : [0, \bar{v}) \longrightarrow [0, +\infty)$ satisfying $\bar{\varphi}(0) = 1$, where

$$\bar{\gamma}(F, x) = \bar{\varphi}(\bar{\gamma}(F, z)\|x - z\|)\bar{\gamma} \text{ for each } x \text{ in } U(z, \frac{\bar{v}}{\bar{\gamma}}). \tag{14.18}$$

Moreover, for $b = b(F, z) := \|F'(z)^{-1}F(z)\|$ we set

$$\bar{\alpha} := \bar{\alpha}(F, z) := \bar{\gamma}\bar{b}. \tag{14.19}$$

By simply using (14.17) instead of the less precise estimate (since $\gamma_0(\tau) \leq \gamma(\tau)$)

$$\|F'(x)^{-1}F'(x_0)\| \leq \left(1 - \int_0^{\gamma_0\|x - x_0\|} L(\tau)d\tau\right)^{-1} \text{ for all } x \in U(x_0, \frac{v}{\gamma}). \tag{14.20}$$

as well as $\bar{\gamma}, \bar{v}$ instead of γ, v, respectively, we can reproduce the proofs of the results of [9] in this setting.

The following result improves Theorem 1 in [9] which in turn generalizes the corresponding result by Meyer [12].

Theorem 14.1. Suppose: $F'(x_0)^{-1}F$ is $\bar{L}-$ and L_0- Lipschitz restricted at $x_0 \in D$;

$$\bar{\alpha}(F, x_0) \leq \int_0^{\bar{v}} \bar{L}(\tau)\tau d\tau \tag{14.21}$$

and

$$\bar{U}(x_0, \bar{v}) \subseteq D, \tag{14.22}$$

where $\bar{\alpha}$ is given by (14.19) and $\bar{v}$ is the smallest positive number such that

$$\int_0^{\bar{v}} \bar{L}(\tau)d\tau = 1. \tag{14.23}$$

Then, the solution of the IVP (14.5) exists in $U(x_0, \frac{v_1}{\bar{\gamma}})$ for each $t \in [0, 1]$, where $\bar{v}_1$ is the first positive root of $g_{\bar{a}}(t)$ less than or equal to $u_{\bar{L}/\bar{c}}$ where $g_{\bar{a}}(t) = \bar{a} - t + \int_0^t \bar{L}(\tau)(t - \tau)d\tau$. Therefore, $x(0)$ is a solution of equation (14.3).

Condition (14.21) is the usual Newton-Kantorovich type criterion [2, 3, 14].

Remark 14.2. If $L_0(s) \geq \bar{L}(s)$ for all $s \in [0, \bar{v}]$, then the results of Theorem 14.1 hold with $\bar{L}$ replacing L.

The Theorem 14.1 does not apply, if $\bar{\alpha} > \int_0^{\bar{v}} \bar{L}(s)ds$. That is why as in [9], we suppose that the solution of the IVP (14.5) is inside the $\bar{\gamma}$−ball of x_0. Then, we ask: How many $k-$steps are needed to approximate the zero x_k of $F = h(., 0)$?

Theorem 14.2. Let x_0 be an element of the $\bar{\gamma}$-ball of z. Set $v^* = \bar{\gamma}\|x_0 - z\|$ for $0 \le u < \bar{v}$, where $\bar{v}$ satisfies (14.23). Define function $\bar{q}$ on $[0, \bar{v}]$ by

$$\bar{q}(t) = \frac{\int_0^t \bar{L}(\tau)d\tau}{t(1 - \int_0^t L_0(\tau)d\tau)}. \tag{14.24}$$

Let $u_{\bar{L}}$ be such that

$$\bar{q}(u_{\bar{L}}) = 1. \tag{14.25}$$

Let $\bar{c} \ge 1$ and define function $g_{\bar{a}}$ on $[0, \bar{v}]$ by

$$g_{\bar{a}}(t) = \bar{a} - t + \int_0^t \bar{L}(\tau)(t - \tau)d\tau, \tag{14.26}$$

so that

$$\min\{u_{\bar{L}/\bar{c}} - \int_0^{u_{\bar{L}/\bar{c}}} \bar{L}(\tau)(u_{\bar{L}/\bar{c}} - \tau)d\tau, \int_0^{\bar{v}} \bar{L}(\tau)\tau d\tau\} \ge \bar{a} \tag{14.27}$$

with the smallest positive solution of equation $g_{\bar{a}}(t) = 0$ is not exceeding $u_{\bar{L}/\bar{c}}$. Set

$$p = \frac{\bar{\varphi}(u)(\bar{\alpha} + \int_0^{v^*} \bar{L}(\tau)(v^* - \tau)d\tau + v^*)}{(1 - \int_0^{u_{\bar{L}/\bar{c}}} L_0(\tau)d\tau)(1 - \int_0^u L_0(\tau)d\tau)}$$

$$q = \frac{\int_0^{u_{\bar{L}/\bar{c}}} \bar{L}(\tau)(u_{\bar{L}/\bar{c}} - \tau)d\tau + u_{\bar{L}/\bar{c}}}{1 - \int_0^{u_{\bar{L}/\bar{c}}} L_0(\tau)d\tau},$$

where $\bar{\varphi}$ is given in (14.18). Moreover, suppose $x(t)$ is the solution of the IVP is inside the $\bar{\gamma}$-ball of z. Let us also define sequence $\{s_n\}$ by

$$s_0 = 1, s_n > 0, s_{n-1} - s_n > s_n - s_{n+1} > 0, n \ge 0, \lim_{n \to +\infty} s_n = 0, \tag{14.28}$$

where

$$s_1 = 1 - \frac{1 - \int_0^u L_0(\tau)d\tau}{\bar{\varphi}(u)(\bar{\alpha} + \int_0^{v^*} \bar{L}(\tau)(v^* - \tau)d\tau + v^*)}.$$

Set w_n such that $F(w_n) = s_n F(x_0)$. Then, the following assertions hold:

(i)   Points w_n and w_{n+1}, are such that

$$\bar{\gamma}\varphi(u)\|w_{n+1} - w_n\| \le \bar{a}.$$

(ii)   Newton sequence $\{x_n\}$ generated by (14.28) and w_n are such that

$$\bar{\gamma}\varphi(v^*)\|x_n - w_n\| \le u_{\bar{L}/\bar{c}}.$$

(iii)   Set $\bar{N} = \frac{\int_0^{\bar{v}} \bar{L}(\tau)\tau d\tau - q}{p}$. The steps n required for x_n to be an approximate zero of w_n exceeds or is equal to

$$\left\lceil \frac{1 - \bar{N}}{1 - s_1} \right\rceil, \text{ if } s_n := \max\{0, 1 - n(1 - s_1)\},$$

$$\left[\frac{log\bar{N}}{log s_1}\right], \text{ if } s_n := s_1^n$$

$$\left[log_2\left(\frac{log\bar{N}}{log s_1} + 1\right)\right], \text{ if } s_n := s_1^{2^k-1}.$$

$$\|x_n - \bar{w}\| \le \bar{q}(\bar{u})^{2^n-1}\|x_0 - \bar{w}\|,$$

where $\bar{u} = \gamma(F, \bar{w})\|x_0 - \bar{w}\| < u_{\bar{L}}$ and $\bar{q}$ is given in (14.24).

Remark 14.3. If $L = L_0 = \bar{L}$, $\gamma_0 = \gamma = \bar{\gamma}$, then the preceding items coincide with the ones in [9]. But , if (14.12) or (14.2) hold as strict inequalities, then the new results constitute an improvement over the ones in [9]. These improvements are deduced using the same effort as in [9], because finding function L requires finding functions L_0 and $\bar{L}$. If $L_0 > \bar{L}$, then, the preceding results hold with $\bar{L}$ replacing L_0.

14.2   SPECIAL CASES

We consider specializations of the preceding results in the general (Kantorovich) case $\bar{L}(s) = 1$ and the analytic case $\bar{L}(s) = \frac{2}{(1-s)^3}$, respectively. Examples, where (14.14) and (14.15) hold as strict inequalities in the Kantorovich case can be found in [2, 3] whereas the examples in the analytic case can be found in [4]. To avoid repetitions, we refer the reader to [9], where $\alpha(F, x_0), \varphi, v, N, L$ are replaced by $\bar{\alpha}(F, x_0), \bar{\varphi}, u, \bar{N}$ $\bar{L}$, respectively.

Next, we present the α and γ Theorems improving the works in [9] which in turn improved the works by X. Wang [15] and Traub and Wozniakowski [14], respectively.

Theorem 14.3. Suppose: $F'(x_0)^{-1}F$ is $\bar{L}$ and L_0- center-Lipschitz restricted at x_0;

$$\bar{\alpha}(F, x_0) \le \int_0^{\bar{v}} \bar{L}(\tau)\tau d\tau,$$

where $\bar{v}$ is given in (14.23). Specialize function $\bar{g}_{\bar{\alpha}(F,x_0)}$ by

$$\bar{g}_{\bar{\alpha}(F,x_0)}(t) := \bar{g}(t) = \bar{\alpha}(F, x_0) - t + \int_0^t \bar{L}(\tau)(t - \tau)d\tau. \tag{14.29}$$

Then, the following items hold

(i)   There exist $\rho_1, \rho_2 \in \mathbb{R}$ with $\rho_1 \ne \rho_2$ such that $\bar{g}(\rho_1) = \bar{g}(\rho_2) = 0$ with $\bar{g}$ strictly convex and

$$\bar{g}(t) = (t - \rho_1)(t - \rho_2)\psi(t),$$

where

$$\psi(t) = \int_0^1 \int_0^1 \theta(\bar{L}(1 - \theta) + \theta s \rho_2 + \theta \tau t)d\tau d\theta.$$

and for $r_0 = 0$, $\lim_{n \longrightarrow +\infty} r_n = \lim_{n \longrightarrow +\infty} N_{\bar{g}}(r_{n-1}) = \bar{v}_1$.

(ii)   Equation $F(x) = 0$ has a solution $\bar{w}$ which is unique in $U(x_0, \frac{\bar{v}}{\bar{\gamma}(F,x_0)})$.

(iii)   Newton sequence $\{x_n\}$ defined by $x_{n+1} = N_F(x_n)$ exists, stays in $\bar{U}(x_0, \frac{\rho_1}{\bar{\gamma}(F,x_0)})$ and converges to $\bar{w}$, so that

$$\|x_n - \bar{w}\| \leq \|r_n - \rho_1\|$$

(iv)   If $\bar{g}(t) \geq \frac{\bar{\alpha}(F,x_0)}{\rho_1 \rho_2}$, then

$$\|x_n - \bar{w}\| \leq \frac{1}{\bar{\gamma}(F,x_0)z^n} \left(\frac{\rho_1}{\rho_2}\right)^{2^n - 1} \rho_1.$$

Theorem 14.4. Suppose:

(i) $\bar{w}$ solves $F(x) = 0$ and is a regular solution:

(ii) $F'(\bar{w})^{-1} F'(\bar{w})$ is $\bar{L}$ and L_0 center Lipschitz restricted for all $x \in U(\bar{w}, \frac{\bar{v}}{\bar{\gamma}(F,\bar{w})})$.
Then, Newton sequence $\{x_n\}$ generated by $x_0 = x, x_{n+1} = N_F(x_n)$ converges to $\bar{w}$ for all $x \in U(\bar{w}, \frac{u_{\bar{L}}}{\bar{\gamma}(F,\bar{w})})$, where $u_{\bar{L}}$ is given in (14.25). Moreover, we have the following:

$$\|x_n - \bar{w}\| \leq \bar{q}(\bar{u})^{2^n - 1} \|x_0 - \bar{w}\|.$$

Remark 14.4. If $L_0 = L = \bar{L}, \gamma_0 = \gamma = \bar{\gamma}$, the two preceding results reduce to Theorem 3 and Theorem 4 in [9], respectively, i.e., if (14.14) or (14.15) hold as strict inequalities, then the earlier results are improved (see also the numerical examples).

14.3   APPLICATIONS

We provide two examples for the Kantorovich case, where function has no positive roots. Hence the older results can not apply, but function $\bar{g}$ has roots, so the new results apply to solve equations.

Example 14.1. Let $\mathcal{B}_1 = \mathcal{B}_2 = \mathbb{R}$, $x_0 = 1$, $D = \{x : |x - x_0| \leq \lambda\}, \lambda \in [0, 1/2)$ and F defined by

$$F(x) = x^3 - \lambda. \tag{14.30}$$

Then, for $L_0(\tau) = L(\tau) = \bar{L}(\tau) = 1, v_0 = \bar{v} = v = 1$, we have

$$\bar{b} = b = \frac{1 - \lambda}{3}, \gamma_0(\tau) = 3 - \lambda, \gamma(\tau) = 2(2 - \lambda) \text{ and } \bar{\gamma}(\tau) = 2\left(1 + \frac{1}{3 - \lambda}\right).$$

Notice that

$$\gamma_0 < \gamma < \bar{\gamma}.$$

The functions g and $\bar{g}$ are then given, respectively by

$$g(t) = \frac{t^2}{2} - t + \frac{2}{3}(1 - \lambda)(2 - \lambda)$$

and

$$\bar{g}(t) = \frac{t^2}{2} - t + \frac{2}{3}(1-\lambda)(1+\frac{1}{3-\lambda}).$$

The Newton-Kantorovich condition (i.e., the discriminant d_g of g) is given by

$$d_g = 1 - \frac{4}{3}(1-\lambda)(2-\lambda) < 0 \text{ for each } \lambda \in [0,1/2) \qquad (14.31)$$

so function g has not positive roots. However, function $\bar{g}$ has positive roots, since the discriminant

$$d_{\bar{g}} = 1 - \frac{4}{3}(1-\lambda)(1+\frac{1}{3-\lambda}) > 0 \text{ for each } \lambda \in I = [0.4619832, 1/2). \quad (14.32)$$

Therefore, our Theorem 14.3 can be used to solve equation $F(x) = 0$ for all $\lambda \in I$.

Example 14.2. Let $\mathcal{B}_1 = \mathcal{B}_2 = \mathcal{C}[0,1]$. Let $D = \{x \in \mathcal{B}_1 : \|x\| \leq R\}$ for $R > 0$. Define F on D by

$$F(x)(s) = x(s) - f(s) - \delta \int_0^1 K(s,t)x(t)^3 dt, x \in \mathcal{B}_1, s \in [0,1], \qquad (14.33)$$

where $f \in \mathcal{B}_1$ is a fixed function and λ is given by

$$K(s,t) = \begin{cases} (1-s)t, & \text{if } t \leq s, \\ s(1-t), & \text{if } s \leq t. \end{cases}$$

Then, for each $x \in D$, $F'(x)$ is given by

$$[F'(x)(v)](s) = v(s) - 3\delta \int_0^1 K(s,t)x(t)^2 v(t) dt, v \in X, s \in [0,1].$$

Set $x_0(s) = f(s) = 1$. Then, we have $\|I - F'(x_0)\| \leq 3|\delta|/8$ if $|\delta| < 8/3$, then $F'(x_0)^{-1}$ exists and

$$\|F'(x_0)^{-1}\| \leq \frac{8}{8-3|\delta|}.$$

Moreover,

$$\|F(x_0)\| \leq \frac{|\delta|}{8},$$

so

$$b = \|F'(x_0)^{-1}F(x_0)\| \leq \frac{|\delta|}{8-3|\delta|}.$$

Furthermore, for $x, y \in D$, we have

$$\|F'(x) - F'(y)\| \leq \frac{1+3|\delta|\|x+y\|}{8}\|x-y\| \leq \frac{1+6R|\delta|}{8}\|x-y\|$$

and

$$\|F'(x) - F'(1)\| \leq \frac{1+3|\delta|(\|x\|+1)}{8}\|x-1\| \leq \frac{1+3|\delta|(1+R)}{8}\|x-1\|.$$

Let $\delta = 1.175$ and $R = 2$, we have $b = 0.26257...$, $\bar{\gamma}(\tau) = 2.76875...$, $\gamma_0(\tau) = 1.8875...$ and $\gamma(\tau) = 1.47314...$, $v_0 = \bar{v} = v = 1$. Using these values, we get that the discriminant d_g of g is

$$d_g = 1 - 1.02688 < 0,$$

but the discriminant $d_{\bar{g}}$ of $\bar{g}$ is

$$d_{\bar{g}} = 1 - 0.986217 > 0.$$

Hence, $\lim_{n \to \infty} x_n = x_*$ by Theorem 14.3, where x_* is a solution of equation $F(x)(s) = 0$, where F is given by (14.33).

14.4   EXERCISES

1. Solve the initial value problem

$$\dot{x}(t) = 1 + \cos(x(t)), x(0) = 0,$$

2. The predator-prey population models describe the interaction of a prey population X and a predator population Y. Assume that their interaction is modeled by the system of ordinary differential equations

$$\dot{x} = x - \frac{1}{4}x^2 - \frac{1}{10}xy + 1$$

$$\dot{y} = -\frac{1}{4}y + \frac{1}{7}xy + \frac{2}{3}.$$

(Assume $x(0) = y(0) = 0$). Solve the system.

3. Solve the initial value problem

$$\dot{x} = (x(t) + t, x(0) = 1)$$

by the iteration method developed in this chapter. Perform 3 steps.

4. Solve the Fredholm-type integral equation

$$x(t) = \sum_0^1 \frac{t \cdot s}{10} ds + 1$$

by the iteration method developed in this chapter. Perform 3 steps.

5. Solve the Volterra-type integral equation

$$x(t) = \sum_0^t \frac{t \cdot s}{10} ds + 1$$

by the iteration method developed in this chapter. Perform 3 steps.

6. Solve equation

$$x = \frac{\sin x}{2} + 1$$

by using algorithm developed in this chapter.

7. Repeat the previous problem for equation

$$x = \frac{\sin x + \cos 2x}{10}.$$

REFERENCES

1. Allgower, E. L., and George, K. 1990. *Numerical Continuation Methods*, Vol. 33. Berlin: Springer.
2. Argyros, I.K. 2007. *Computational Theory of Iterative Methods. Series: Studies in Computational Mathematics*. New York: Elsevier Publ. Co.
3. Argyros, I.K., and Hilout, S. 2012. Weaker conditions for the convergence of Newton's method. *J. Complexity*, 28:364–387.
4. Argyros, I.K., and George, S. 2017. Extending the applicability of Newton's method using Wang's– Smale's α-theory. *Carpathian Journal of Mathematics*, 33. 1:27 – 33.
5. Beltran, C., and Leykin, A. 2012. Certified numerical homotopy tracking. *Exp. Math.*, 21(1):69–83.
6. Cauchy, A. 1829. Sur la determination approximative des racines d'une equation algebrique on transcendante. In: *Lecons sur le Calcul Differentiel*. Bure freres. Paris. 573–609.
7. Dedieu, J. P., Malajovich, G., and Shub, M. 2013. Adaptive step-size selection for homotopy methods to solve polynomial equations. *IMA J. Numer. Anal.*, 33(1):1–29.
8. Fourier, J.B.J, 1831. *Analyse des equations determincees*, Vol.1. Paris: Firmin Didot.
9. Gutiérrez, J.M., Magrenãn, A. A., and Yakoubsohn, J. C. 2016. Complexity of a Homotopy method at the neighborhood of a zero. In. *Advances in Iterative methods for nonlinear equations*, ed. S. Amat, S. Busquier, Switzerland: Springer International Publishing.
10. Magrénãn, A. A. 2014. Different anomalies in a Jarratt family of iterative root finding methods. *Appl. Math. Comput.*, 233:29–38.
11. Magrénãn, A.A. 2014. A new tool to study real dynamics: The convergence plane. *Appl. Math. Comput.*, 248:29–38.
12. Meyer, G.H. 1968. On solving nonlinear equations with a one-parameter operator imbedding. *SIAM J. Numer. Anal.*, 5(4):739–752.
13. Smale, S. 1986. Newton's method estimates from data at one point. In: *The merging of Discipline New Directions in Pure, Applied, and Computational Mathematics*.185–196.
14. Traub, J.F., and Wozniakowski, H. 1979. Convergence and Complexity of Newton iteration for operator equations. *J. ACM*. 26(2): 250–258.
15. Xinghua, W. 2000. Convergence of Newton's method and uniqueness of the solution of equations in Banach space, *IMA J. Numer. Anal.*, 20(1):123–134.

15 Inexact Methods for Finding Zeros with Multiplicity

In this chapter, we present an extended ball convergence of inexact methods for approximating a zero of a nonlinear equation with multiplicity m, where m is a natural number. Many popular methods are special cases of the inexact method.

There is a plethora of problems in applied sciences and also in engineering can be written in a form like

$$F(x) = 0, \tag{15.1}$$

using mathematical modeling, where function $F : \Omega \subseteq \mathscr{B}_1 \longrightarrow \mathscr{B}_2$ is sufficiently many times differentiable, and $\Omega, \mathscr{B}_1, \mathscr{B}_2$ are convex subsets in $\mathbb{R}$. In this chapter, we pay attention to the case of a solution p with multiplicity $m > 1$, namely, $F(p) = 0, F^{(i)}(p) = 0$ for $i = 1, 2, \ldots m-1$, and $F^{(m)}(p) \neq 0$. The determination of solutions of multiplicity m is of great interest. As an example, in the study of electron trajectories, when the electron reaches a plate of zero speed, the function distance from the electron to the plate has a solution of multiplicity two. Moreover, the multiplicity of solutions appears in connection to Van Der Waals equation of state and other phenomena. The convergence order of iterative methods decreases, if the equation has solutions of multiplicity m. Modifications in the iterative function are needed to improve the order of convergence.

We present the ball convergence of the inexact method (IM) defined for each $n = 0, 1, 2, \ldots$ by

$$x_{n+1} = x_n - \varepsilon \frac{F(x_n)}{F'(x_n)} - \xi_n, \tag{15.2}$$

where x_0 is an initial point, $\varepsilon \in \mathbb{R}$ a parameter, and $\{\xi_n\} \in \mathbb{R}$ a sequence chosen in such a way as to force convergence for the sequence $\{x_n\}$ to a zero p of multiplicity m for function F. It is important to study the convergence of IM, since many popular methods are special cases of it.

Newton's method ($\varepsilon = 1, \xi_n = 0$):

$$x_{n+1} = x_n - \frac{F(x_n)}{F'(x_n)}. \tag{15.3}$$

Modified Newton's method ($\varepsilon = m, \xi_n = 0$):

$$x_{n+1} = x_n - m\frac{F(x_n)}{F'(x_n)}. \tag{15.4}$$

DOI: 10.1201/9781003128915-15

Laguerre third order method ($\varepsilon = m$) and also choose:

$$\xi_n = m\frac{F(x_n)}{F'(x_n)} - \frac{\lambda F(x_n)}{F'(x_n) \pm \sqrt{\frac{\lambda - m}{m}[(\lambda - 1)F'(x_n)^2 - \lambda F(x_n)F''(x_n)]}}.$$

Laguerre method further specializes to Euler-Chebyshev, Halley, Ostrowski and Hansen-Patrick method for $\lambda = 2$, $\lambda = 0$, $\lambda \longrightarrow \infty$ and $\lambda = \frac{1}{\mu} + 1$, respectively for $\mu \neq 0$.

Traub method ($\varepsilon = m$) and choose:

$$\xi_n = m\frac{F(x_n)}{F'(x_n)} - \frac{m(3-m)}{2}\frac{F(x_n)}{F'(x_n)} - \frac{m^2}{2}\frac{F(x_n)^2 F''(x_n)}{F'(x_n)^3}.$$

Osada method ($\varepsilon = m$) **and choose:**

$$\xi_n = m\frac{F(x_n)}{F'(x_n)} - \frac{1}{2}m(m+1)\frac{F(x_n)}{F'(x_n)} + \frac{1}{2}(m-1)^2\frac{F'(x_n)}{F''(x_n)}.$$

Neta et al fourth order method ($\varepsilon = m$) **and choose:**

$$\xi_n = m\frac{F(x_n)}{F'(x_n)} - s(t_n)\frac{F(y_n)}{F'(x_n)},$$

where $t_n = \frac{F(y_n)}{F'(x_n)}$ and s a real function satisfying some initial conditions. Other iterative methods of high convergence order can be found in [5, 6, 9, 12, 15] and the references therein.

Let $B(p, \lambda) := \{x \in B_1 : |x - p| < \lambda\}$ denote an open ball and let $\bar{B}(p, \lambda)$ denote its closure. It is said that $B(p, \lambda) \subseteq \Omega$ is a convergence ball for an iterative method, if the sequence generated by this iterative method converges to p, provided that the initial point $x_0 \in B(p, \lambda)$. But how close x_0 should be to p so that convergence can take place. Extending the ball of convergence is very important, since it shows the difficulty, we confront to pick initial points. It is desirable to be able to compute the largest convergence ball. This is usually depending on the iterative method and the conditions imposed on the function F and its derivatives. We can unify these conditions by expressing them as:

$$\|(F^{(m)}(p))^{-1}(F^{(m)}(x) - F^{(m)}(y))\| \leq \psi(\|x - y\|) \tag{15.5}$$

for all $x, y \in \Omega$, where $\psi : \mathbb{R}_+ \cup \{0\} \longrightarrow \mathbb{R}_+ \cup \{0\}$ is a continuous and nondecreasing function satisfying $\psi(0) = 0$. If we specialize function ψ, for $m \geq 1$ and

$$\psi(t) = \mu t^q, \mu > 0, \quad q \in (0, 1), \tag{15.6}$$

then, we obtain the conditions under which the preceding methods were studied in [4, 5, 12, 13, 16, 17]. However, there are cases where even (15.6) does not hold (see Example 15.1). Moreover, the smaller function ψ is chosen, the larger the radius of convergence becomes. The technique, we present next can be used for all preceding methods as well as for methods where $m = 1$. However, in this chapter, we only use it for IM. This way, we extend the results in [4, 5, 12, 13, 16, 17]. In view of (15.5) there always exists a function $\varphi_0 : \mathbb{R}_+ \cup \{0\} \longrightarrow \mathbb{R}_+ \cup \{0\}$ continuous, nondecreasing, and satisfying

$$\|(F^{(m)}(p))^{-1}(F^{(m)}(x) - F^{(m)}(p))\| \le \varphi_0(\|x - y\|) \qquad (15.7)$$

for all $x \in \Omega$. We can always choose $\varphi_0(t) = \psi(t)$ for all $t \ge 0$. However, in general

$$\varphi_0(t) \le \psi(t), t \ge 0 \qquad (15.8)$$

holds and $\frac{\psi}{\varphi_0}$ can be arbitrarily large [2]. Denote by r_0 the smallest positive solution of equation $\varphi_0(t) = 1$. Set $\Omega_0 := \Omega \cup B(p, r_0)$. We have again by (15.5) that there exists function $\varphi : [0, r_0) \longrightarrow \mathbb{R}_+ \cup \{0\}$ continuous, nondecreasing and satisfying $\varphi(0) = 0$ such that for each $x, y \in \Omega_0$

$$\|(F^{(m)}(p))^{-1}(F^{(m)}(x) - F^{(m)}(y))\| \le \varphi(\|x - y\|). \qquad (15.9)$$

Clearly, we have

$$\varphi(t) \le \psi(t), \text{ for all } t \in [0, r_0), \qquad (15.10)$$

since $\Omega_0 \subseteq \Omega$. It turns out that more precise (15.7)(see (15.8)) can be used than (15.5) to estimate upper bounds on the inverses of the functions involved (see (15.32) or (15.41)). Moreover, for the upper bounds on the numerators (see (15.33) or (15.42)) we can use (15.9) tighter than (15.5) (see (15.10)). This way we obtain (15.34) or (15.36) which are tighter than the corresponding ones using only ψ (or its special case (15.6)). This way we obtain a larger radius of convergence leading to a wider choice of initial guesses and at least as tight error bounds on the distances $|x_n - p|$ resulting to the computation of at least as few iterates to obtain a desired error tolerance (see also the applications). It is worth noticing that these advantages are obtained under the same computational cost as in earlier studies, since in practice the computation of function ψ (or (15.6)) requires the computation of functions φ_0 and ψ as special cases.

The rest of the chapter is structured as follows. Section 15.1 contains some auxiliary results on divided differences and derivatives. The ball convergence of IM is given in Section 15.2. The applications appear in the Section 15.3.

15.1 AUXILIARY RESULTS

We need the definition of divided differences, and their properties which can be found in [4, 13, 16, 17].

Definition 15.1. The divided differences $F[y_0, y_1, \ldots, y_k]$, on $k + 1$ distinct points $y_0, y_1, \ldots y_k$ of a function $F(x)$ are defined by

$$
\begin{aligned}
F[y_0] &= F(y_0) \\
F[y_0, y_1] &= \frac{F[y_0] - F[y_1]}{y_0 - y_1}, \\
&\vdots \\
F[y_0, y_1, \ldots, y_k] &= \frac{F[y_0, y_1, \ldots, y_{k-1}] - F[y_0, y_1, \ldots, y_k]}{y_0 - y_k}.
\end{aligned}
\qquad (15.11)
$$

If the function F is sufficiently differentiable, then its divided differences $F[y_0,y_1,\ldots,y_k]$ can be defined even if some of the arguments y_i coincide. For instance, if $F(x)$ has $k-$th derivative at y_0, then it makes sense to define

$$F[\underbrace{y_0,y_1,\ldots,y_k}_{k+1}] = \frac{F^{(k)}(y_0)}{k!}. \tag{15.12}$$

Lemma 15.1. The divided differences $F[y_0,y_1,\ldots,y_k]$ are symmetric functions of their arguments,i.e., they are invariant to permutations of the points $y_0,y_1,\ldots,y_k$.

Lemma 15.2. If the function F has $(k+1)-$th derivative, and p is a zero of multiplicity m, then for every argument x, the following formula holds

$$F(x) = F[y_0] + \sum_{i=1}^{k} F[y_0,y_1,\ldots,y_k] \prod_{j=0}^{i-1}(x-y_j) + F[y_0,y_1,\ldots,y_k,x] \prod_{i=0}^{k}(x-y_i). \tag{15.13}$$

Lemma 15.3. If the function F has $(m+1)-$th derivative, and p is a zero of multiplicity m, then for every argument x, the following formula holds

$$F(x) = F[\underbrace{p,p,\ldots,p}_{m},x](x-p)^m, \tag{15.14}$$

$$F'(x) = F[\underbrace{p,p,\ldots,p}_{m},x,x](x-p)^m + mF[\underbrace{p,p,\ldots,p}_{m},x](x-p)^{m-1}. \tag{15.15}$$

We need the following lemma on Genocchi's integral expression formula for divided differences:

Lemma 15.4. If the function F has continuous $k-$th derivative, then the following formula holds for any points $y_0,y_1,\ldots,y_k$

$$F[y_0,y_1,\ldots,y_k] = \int_0^1 \cdots \int_0^1 F^{)k)}(y_0 + \sum_{i=1}^{k}(y_i - y_{i-1}) \prod_{j=1}^{i} \theta_j) \prod_{i=1}^{k}(\theta_i^{k-i} d\theta_i). \tag{15.16}$$

We shall also use the following Taylor expansion with integral form reminder.

Lemma 15.5. Suppose that $F(x)$ is differentiable $n-$times in the ball $B(x_0,r), r>0$, and $F^{(n)}(x)$ is integrable from a to $x \in B(a,r)$. Then,

$$\begin{aligned} F(x) &= F(a) + F'(a)(x-a) + \frac{1}{2}F''(a)(x-a)^2 + \ldots + \frac{1}{n!}F^{(n)}(a)(x-a)^n \\ &\quad + \frac{1}{(n-1)!} \int_0^1 [F^{(n)}(a+t(x-a)) - F^{(n)}(a)](x-a)^n(1-t)^{n-1} dt, \end{aligned} \tag{15.17}$$

and

$$
\begin{aligned}
F'(x) &= F'(a) + F''(a)(x-a) + \frac{1}{2}F'''(a)(x-a)^2 \\
&\quad + \ldots + \frac{1}{(n-1)!}F^{(n)}(a)(x-a)^{n-1} \\
&\quad + \frac{1}{(n-2)!}\int_0^1 [F^{(n)}(a+t(x-a)) - F^{(n)}(a)](x-a)^{n-1}(1-t)^{n-2}dt.
\end{aligned}
$$

(15.18)

15.2 BALL CONVERGENCE

Let $\varphi_0 : \mathbb{R}_+ \cup \{0\} \longrightarrow \mathbb{R}_+ \cup \{0\}$ be a continuous and nondecreasing function with $\varphi_0(0) = 0$. Moreover, define functions $\beta_0, \beta : \mathbb{R}_+ \cup \{0\} \longrightarrow \mathbb{R}_+ \cup \{0\}$ by

$$
\beta_0(t) = (m-1)!(m-1)\int_0^1 \cdots \int_0^1 \varphi_0(t\prod_{i=1}^m \theta_i)\prod_{i=1}^m \theta_i^{m-i}d\theta_i
$$

and

$$
\beta(t) = (m-1)!\int_0^1 \cdots \int_0^1 \varphi_0(t\prod_{i=1}^{m-1} \theta_m)\prod_{i=1}^{m-1} \theta_i^{m-i}d\theta_i + \beta_0(t).
$$

We have that functions β_0, β are continuous and nondecreasing with $\beta_0(0) = \beta(0) = 0$. Suppose

$$
\beta(t) \longrightarrow 1 \text{ as } t \longrightarrow \text{ a positive number or } +\infty \tag{15.19}
$$

It follows from the intermediate value theorem that equation $\beta(t) = 1$ has solutions in $(0, +\infty)$. Denote by r_0 the smallest positive solution of equation $\beta(t) = 1$. Set $h_1(t) = 1 - \beta(t)$. Let $\varphi : [0, r_0) \longrightarrow \mathbb{R}_+ \cup \{0\}$ be a continuous and nondecreasing function satisfying $\varphi(0) = 0$. Furthermore, define functions α, h_0 and h on $[0, r_0)$ by

$$
\alpha(t) = (m-1)!\int_0^1 \cdots \int_0^1 \varphi(t\prod_{i=1}^{m-1} \theta_i(1-\theta_m))\prod_{i=1}^m \theta_i^{m-i}d\theta_i d\theta_m
$$

$$
h_0(t) = m^{-1}\alpha(t)t + m^{-1}|m - \varepsilon|\beta_0(t) + m^{-1}\alpha(t)at^b + \beta_0(t)at^{b-1},
$$

and

$$
h(t) = \frac{h_0(t)}{h_1(t)} - 1,
$$

where $a \geq 0$ and $b \geq 1$. We get that $h_0(0) = -1 < 0$ and $h(t) \longrightarrow +\infty$ as $t \longrightarrow r_0^-$. Denote by r the smallest solution of equation $h(t) = 0$ in $(0, r_0)$. Then, we have that for each $t \in [0, r)$

$$
0 \leq \beta(t) < 1 \tag{15.20}
$$

and

$$
0 \leq h(t) < 1. \tag{15.21}
$$

First, we show the ball convergence of method IM under conditions $(\mathscr{A})$:

$(\mathscr{A}_1)$ $F : \Omega \subseteq \mathscr{B}_1 \longrightarrow \mathscr{B}_2$ is continuously m−times Fréchet-differentiable
$(\mathscr{A}_2)$ Function F has a zero p of multiplicity $m, m = 1, 2, \ldots$.
$(\mathscr{A}_3)$ There exists function $\varphi_0 : \mathbb{R}_+ \cup \{0\} \longrightarrow \mathbb{R}_+ \cup \{0\}$ continuous and nondecreasing satisfying $\varphi_0(0) = 0$ such that for each $x \in \Omega$

$$\|F^{(m)}(p)^{-1}(F^{(m)}(x) - F^{(m)}(p))\| \leq \varphi_0(\|x - p\|)$$

Let $\Omega_0 = \Omega \cup B(p, r_0)$, where r_0 is defined previously.
$(\mathscr{A}_4)$ There exists $\varphi : [0, r) \longrightarrow \mathbb{R}_+ \cup \{0\}$ continuous and nondecreasing satisfying $\varphi(0) = 0$ such that for each $x, y \in \Omega_0$

$$\|F^{(m)}(p)^{-1}(F^{(m)}(x) - F^{(m)}(y))\| \leq \varphi(\|x - p\|)$$

$(\mathscr{A}_5)$ Condition (15.19) holds.
$(\mathscr{A}_6)$ $\bar{B}(p, r) \subseteq \Omega$.
$(\mathscr{A}_7)$ $|\xi_n| \leq a|x_n - p|^b$ for all $n = 0, 1, 2, \ldots$ and some $a \geq 0, b \geq 1$.

Theorem 15.1. Suppose that the $(\mathscr{A})$ conditions hold. Then, for starting point $x_0 \in B(p, r) - \{p\}$, the sequence $\{x_n\}$ generated by IM is well defined in $B(p, r)$, remains in $B(p, r)$ for all $n = 0, 1, 2, \ldots$ and converges to p.

Proof. We shall show that sequence

$$\delta_n = x_n - p \tag{15.22}$$

is non-increasing and converges to zero. Using $\delta_n = x_n - p$, method (15.2) for $n = 0$, Lemma 15.3 and the following formulae:

$$g(x) = F[\underbrace{p, p, \ldots, p}_{m}, x], \quad g_0(x) = F[\underbrace{p, p, \ldots, p}_{m}, x, x]. \tag{15.23}$$

$$F(x_0) = g(x_0)\delta_0^m, \tag{15.24}$$

and

$$F'(x_0) = [g_0(x_0)\delta_0 + mg(x_0)]\delta_0^{m-1}. \tag{15.25}$$

We can write

$$\delta_1 = \frac{g(p)^{-1}N}{g(p)^{-1}D}, \tag{15.26}$$

where

$$N = g_0(x_0)\delta_0^2 + [(m - \varepsilon)g(x_0) - g_0(x_0)\xi_0]\delta_0 - mg(x_0)\xi_0 \tag{15.27}$$

and

$$D = g_0(x_0)\delta_0 + mg(x_0). \tag{15.28}$$

In view of the definition of divided differences, we have

$$g_0(x_0)\delta_0 = F[\underbrace{p,p,\ldots,p}_{m-1},x_0,x_0] - g(x_0). \qquad (15.29)$$

Then, we obtain from (15.12) and (15.29) that

$$
\begin{aligned}
&|1 - (mg(p))^{-1}[h_0(x_0)\delta_0 + mg(x_0)]| \\
={}& |(mg(p))^{-1}[g_0(x_0)\delta_0 + mg(x_0) - mg(p)]| \\
={}& (m-1)!F^{(m)}(p)^{-1}(F[\underbrace{p,p,\ldots,p}_{m-1},x_0,x_0] - g(p) + (m-1)[g(x_0) - g(p)])|.
\end{aligned}
$$
$$(15.30)$$

By Lemma 15.4, we get

$$F[\underbrace{p,p,\ldots,p}_{m-1},x_0,x_0] \qquad (15.31)$$

$$
\begin{aligned}
={}& \int_0^1 \cdots \int_0^1 F^{(m)}\Big(p + \delta_0 \prod_{i=1}^{m-1}\theta_i\Big)\prod_{i=1}^{m}(\theta_i^{m-1}d\theta_i), \\
g(x_0) ={}& \int_0^1 \cdots \int_0^1 F^{(m)}\Big(p + \delta_0 \prod_{i=1}^{m}\theta_i\Big)\prod_{i=1}^{m}(\theta_i^{m-1}d\theta_i), \qquad (15.32) \\
g(p) ={}& \int_0^1 \cdots \int_0^1 F^{(m)}(p)\prod_{i=1}^{m}(\theta_i^{m-1}d\theta_i). \qquad (15.33)
\end{aligned}
$$

Substituting (15.30)–(15.33) into (15.29) using condition $(\mathscr{A}_3)$, $x_0 \in B(p,r)$, and the definition of r, we get

$$
\begin{aligned}
&|1 - (mg(p))^{-1}]g_0(x_0)\delta_0 + mg(x_0)]| \\
={}& (m-1)!\Big|\int_0^1 \cdots \int_0^1 F^{(m)}(p)^{-1}\Big(F^{(m)}\Big(p + \delta_0 \prod_{i=1}^{m-1}\theta_i\Big) - F^{(m)}(p)\Big)\prod_{i=1}^{m}(\theta_i^{m-i}d\theta_i) \\
&+ (m-1)F^{(m)}(p)^{-1}\Big(F^{(m)}\Big(p + \delta_0 \prod_{i=1}^{m-1}\theta_i\Big) - F^{(m)}(p)\Big)\prod_{i=1}^{m}(\theta_i^{m-i}d\theta_i)\Big| \\
\le{}& (m-1)!\Big(\int_0^1 \cdots \int_0^1 |F^{(m)}(p)^{-1}\Big(F^{(m)}\Big(p + \delta_0 \prod_{i=1}^{m-1}\theta_i\Big) - F^{(m)}(p)\Big)|\prod_{i=1}^{m}(\theta_i^{m-i}d\theta_i) \\
&+ (m-1)\int_0^1 \cdots \int_0^1 |F^{(m)}(p)^{-1}\Big(F^{(m)}\Big(p + \delta_0 \prod_{i=1}^{m-1}\theta_i\Big) - F^{(m)}(p)\Big)|\prod_{i=1}^{m}(\theta_i^{m-i}d\theta_i)|
\end{aligned}
$$

$$\leq (m-1)! [\int_0^1 \cdots \int_0^1 \varphi_0(|\delta_0| \prod_{i=1}^{m-1} \theta_i) \prod_{i=1}^{m} \theta_i^{m-i} d\theta_i$$

$$+ (m-1) \int_0^1 \cdots \int_0^1 \varphi_0(|\delta_0| \prod_{i=1}^{m} \theta_i) \prod_{i=1}^{m} \theta_i^{m-i} d\theta_i]$$

$$\leq \beta(|\delta_0|) < \beta(r) < 1.$$

(15.34)

It follows from the Banach perturbation lemma [1, 3] and (15.34) that, $g_0(x_0)\delta_0 + mg(x_0) \neq 0$, and

$$|(mg(p))^{-1} g_0(x_0)\delta_0 + mg(x_0))^{-1}| \leq \frac{1}{1-\beta(|\delta_0|)} < \frac{1}{1-\beta(r)}.$$

(15.35)

Moreover, using (15.29), (15.31), (15.32) and $(\mathscr{A}_4)$, we have in turn that

$$|(mg(p))^{-1} g_0(x_0)\delta_0|$$

$$= (m-1)! |\int_0^1 \cdots \int_0^1 F^{(m)}(p)^{-1}(F^{(m)}(p + \delta_0 \prod_{i=1}^{m-1} \theta_i)$$

$$- F^{(m)}(p + \delta_0 \prod_{i=1}^{m} \theta_i)) \prod_{i=1}^{m}(\theta_i^{m-i} d\theta_i)|$$

$$= (m-1)! |\int_0^1 \cdots \int_0^1 |F^{(m)}(p)^{-1}(F^{(m)}(p + \delta_0 \prod_{i=1}^{m-1} \theta_i)$$

$$- F^{(m)}(p + \delta_0 \prod_{i=1}^{m} \theta_i))| \prod_{i=1}^{m}(\theta_i^{m-i} d\theta_i)|$$

$$\leq (m-1)! \int_0^1 \cdots \int_0^1 \varphi_0(|\delta_0| \prod_{i=1}^{m-1} \theta_i(1-\theta_m)) \prod_{i=1}^{m} \theta_i^{m-i} d\theta_i d\theta_m$$

$$= \alpha(|\delta_0|) < \alpha(r) < 1.$$

(15.36)

Furthermore, we have

$$|g(p)^{-1} g(x_0)|$$

$$= |g(p)^{-1}(g(x_0) - g(p))|$$

$$= |(m-1)! F^{(m)}(p)^{-1}(m-1)(g(x_0) - g(p))|$$

$$= (m-1)(m-1)! \int_0^1 \cdots \int_0^1 |F^{(m)}(p)^{-1}(F^{(m)}(p + \delta_0 \prod_{i=1}^{m} \theta_i)$$

$$- F^{(m)}(p)| \prod_{i=1}^{m} \theta_i d\theta_1$$

$$\leq (m-1)(m-1)! \int_0^1 \cdots \int_0^1 \varphi_0(|\delta_0|) \prod_{i=1}^{m} \theta_i) \prod_{i=1}^{m} \theta_i^{m-i} d\theta_i.$$

(15.37)

Using (15.34)–(15.37), we obtain that

$$|\delta_1| \le c|\delta_0| < |\delta_0| < r, \tag{15.38}$$

where $c = h(|\delta_0|) \in [0,1)$, so $x_1 \in B(p,r)$. By simply replacing x_0, x_1, by x_k, x_{k+1}, we arrive at

$$|x_{k+1} - p| \le c|x_k - p| < r, \tag{15.39}$$

which shows $\lim_{k \longrightarrow +\infty} x_k = p$, and $x_{k+1} \in B(p,r)$.

□

Concerning the uniqueness of the solution p, we have:

Proposition 15.1. Suppose that conditions $(\mathscr{A})$ and

$$\frac{m}{(s_2 - s_1)^m} \int_{s_1}^{s_2} \varphi_0(|t - s_1|)|s_2 - t|^{m-1} dt < 1 \tag{15.40}$$

for all s_1, t, s_2 with $0 \le s_1 \le t \le s_2 \le \bar{r}$ for some $\bar{r} \ge r$ hold. Then, the solution p of equation $F(x) = 0$ is unique in $\Omega_0 = \Omega \cup \bar{B}(p, \bar{r})$.

Proof. Suppose that $p^* \in \Omega_0$ is a solution of equation $F(x) = 0$ with $p \ne p^*$. Without loss of generality suppose $p < p^*$. We can write

$$F(p^*) - F(p) = \frac{1}{(m-1)!} \int_p^{p^*} F^{(m)}(t)(p^* - t)^{m-1} dt. \tag{15.41}$$

Using $(\mathscr{A}_3)$ and (15.39), we obtain in turn that

$$\left| 1 - \left(\frac{(p^* - p)^m}{m} F^{(m)}(p) \right)^{-1} \int_p^{p^*} F^{(m)}(t)(p^* - t)^{m-1} dt \right|$$

$$= \left| \left(\frac{(p^* - p)^m}{m} F^{(m)}(p) \right)^{-1} \int_p^{p^*} [F^{(m)}(t) - F^{(m)}(p)](p^* - t)^{m-1} dt \right|$$

$$\le \frac{m}{(p^* - p)^m} \int_p^{p^*} \varphi_0(|t - p|)|p^* - t|^{m-1} dt < 1, \tag{15.42}$$

so $\left(\frac{(p^* - p)^m}{m} F^{(m)}(p) \right)^{-1} \int_p^{p^*} F^{(m)}(t)(p^* - t)^{m-1} dt$ is invertible, i.e., $\int_p^{p^*} F^{(m)}(t)(p^* - t)^{m-1} dt$ is invertible.

□

15.3 APPLICATIONS

We apply Theorem 15.1 to the case of the modified Newton method, so we choose $\xi_n = 0$. In the first example condition (15.5) is not satisfied. Hence, earlier results based on it [4,5,12,13,16,17] cannot guarantee the convergence of method (15.2), but our conditions hold, so convergence is assured. Hence, we extended the applicability of method (15.2). The second example is used to show how we choose functions φ_0 and φ appearing in Theorem 15.1.

Example 15.1. Let $\mathscr{B}_1 = \mathscr{B}_2 = \mathbb{R}, \Omega = [0,1], m = 2, p = 0$. Define function F on Ω by

$$F(x) = \frac{4}{35}x^{\frac{7}{2}} + \frac{1}{6}x^3 + \frac{1}{2}x^2. \tag{15.43}$$

We have by (15.43) $F'(x) = \frac{2}{5}x^{\frac{5}{2}} + \frac{x^2}{2} + x$, $F''(x) = x^{\frac{3}{2}} + x + 1$, $F'''(x) = x^{\frac{1}{2}} + 1$ and $F''(0) = 1$. Function F'' cannot satisfy (15.5) with ψ given by (15.6). Hence, the results in [4,5,12,13,16,17] cannot apply. However, the new results apply, since (15.7) and (15.9) are satisfied for $\varphi_0(t) = \varphi(t) = t^{\frac{3}{2}} + t$, respectively. The convergence radius is $r = 0.6511$ obtained by solving the equation $h(t) = 0$.

Example 15.2. Let $\mathscr{B}_1 = \mathscr{B}_2 = \mathbb{R}, \Omega = [-1,1], m = 2, p = 0$. Define function F on Ω by

$$F(x) = e^x - x - 1.$$

We get by (15.7) and (15.9), respectively that $\varphi_0(t) = (e-1)t$ and $\varphi(t) = et$. Then, the convergence radius is $r = 0.7163$ obtained by solving the equation $h(t) = 0$.

15.4 EXERCISES

1. Apply the methods defined in the introduction of this chapter to find the solution of the equation in the Example 15.1. Make a comparison between the results of these methods.

2. Do as in the previous exercise for Example 15.2.

REFERENCES

1. Amat, S., Hernández, M.A., and Romero, N. 2012. Semilocal convergence of a sixth order iterative method for quadratic equations. *Applied Numerical Mathematics*, 62:833–841.
2. Argyros, I.K. 2007. *Computational Theory of Iterative Methods*. Series: Studies in Computational Mathematics. 15. eds. C.K.Chui and L. Wuytack. New York: Elsevier Publ. Co.
3. Argyros, I.K. 2003. On the convergence and application of Newtons method under Hölder continuity assumptions. *Int. J. Comput. Math.*, 80:767–780.
4. Bi, W., Ren, H.M., and Wu, Q.B. 2011. Convergence of the modified Halley's method for multiple zeros under Höder continous derivatives. *Numer. Algor.*, 58:497–512.
5. Chun, C., and Neta, B. 2009. A third order modification of Newton's method for multiple roots. *Appl. Math. Comput.*, 211:474–479.
6. Hansen, E., and Patrick, M. 1977. A family of root finding methods. *Numer. Math.*, 27:257–269.
7. Magreñán, A.A. 2014. Different anomalies in a Jarratt family of iterative root finding methods, *Appl. Math. Comput.*, 233:29–38.
8. Magreñán, A.A., A new tool to study real dynamics: The convergence plane. *Appl. Math. Comput.*, 248:29–38.
9. Neta, B. 2008. New third order nonlinear solvers for multiple roots. *Appl. Math. Comput.*, 202:162–170.
10. Obreshkov, N. 1963. On the numerical solution of equations (Bulgarian). *Annuaire Univ. Sofia. Fac. Sci. Phy. Math.*, 56:73–83.
11. Osada, N. 1994. An optimal multiple root-finding method of order three. *J. Comput. Appl. Math.*, 52:131–133.
12. Petkovic, M.S., Neta, B., Petkovic, L., and Džunič, J. 2013. *Multipoint methods for solving nonlinear equations*. Elsevier.
13. Ren, H.M., and Argyros, I.K. 2010. Convergence radius of the modified Newton method for multiple zeros under Hölder continuity derivative, *Appl. Math. Comput.*, 217:612–621.
14. Schröder, E. 1870. Uber unendlich viele Algorithmen zur Auflosung der Gieichunger. *Math. Ann.*, 2:317–365.
15. Traub, J.F. 1982. *Iterative methods for the solution of equations*. AMS Chelsea Publishing.
16. Zhou, X., and Song, Y. 2011. Convergence radius of Osada's method for multiple roots under Hölder and Center-Hölder continuous conditions, *ICNAAM 2011(Greece)*. 1836–1839. AIP Conf. Proc.
17. Zhou, X., Chen, X., and Song, Y. 2014. On the convergence radius of the modified Newton method for multiple roots under the center- Hölder condition. *Numer. Algor.*, 65(2):221–232.

16 Multi-Step High Convergence Order Methods

The local convergence analysis of iterative methods is important since it indicates the degree of difficulty for choosing initial points. In this chapter, we introduce generalized multi-step high order methods for solving nonlinear equations. The local convergence analysis is given using hypotheses only on the first derivative which actually appears in the methods in contrast to earlier works using hypotheses on higher derivatives. This way we extend the applicability of these methods. The analysis includes computable radius of convergence as well as error bounds based on Lipschitz-type conditions not given in earlier studies. Numerical examples conclude this chapter.

Iterative regularization models are changing the face of the world by offering the scientists and mathematicians the opportunity to examine many real life problems, with a far greater generality and precision. To make use of the full power of the iterative methods, they must have a firm grip on numerical techniques developed for various mathematical models and their analysis. Application of the iterative schemes is found in any scientific field where real world problems are modeled into mathematical equations.

Iterative schemes/methods are generally used for certain class of numerical schemes, where the solution procedure start with an approximate value/function and then apply the method repeatedly to obtain a better approximation. Many mathematical equations are in the form (or get reduced to),

$$F(x) = 0, \tag{16.1}$$

where $F : D \subseteq \mathscr{B}_1 \longrightarrow \mathscr{B}_2$ is a Fréchet-differentiable operator, $\mathscr{B}_1$ and $\mathscr{B}_2$ are Banach spaces and D is a nonempty open convex subset of $\mathscr{B}_1$. Such equations can be linear or nonlinear in nature and there are various iterative schemes used to obtain the solution. Also, these iterative schemes are useful in solving many optimization problems from different disciplines. Many of these methods are firmly based on various calculus and functional analysis concepts and they can be effectively implemented by taking the advantage of the speed and power of modern computer technologies. In particular three step methods have been introduced in the special case when $\mathscr{B}_1 = \mathscr{B}_2 = \mathbb{R}^i$ (i a natural number) to solve nonlinear systems [1–26]. We introduce in a Banach space setting multi-step method consisting of $q + 2$ steps,

DOI: 10.1201/9781003128915-16

$q \in \mathbb{N}$ defined for each $n = 0, , 2, \ldots$ by

$$
\begin{aligned}
y_n &= \varphi_0(x_n) \\
z_n &= \varphi(x_n, y_n) \\
z_n^{(0)} &= z_n \\
z_n^{(1)} &= \varphi_1(x_n, y_n, z_n) \\
z_n(2) &= \varphi_2(x_n, y_n, z_n, z_n^{(1)}) \\
&\;\;\vdots \\
z_n^{(q)} &= x_{n+1} = \varphi_q(x_n, y_n, z_n, z_n^{(1)}, z_n^{(2)}, \ldots, z_n^{(q-1)}),
\end{aligned}
\tag{16.2}
$$

where x_0 is an initial point, $\varphi_0 : D \longrightarrow \mathcal{B}_1, \psi : D^2 \longrightarrow \mathcal{B}_1, \varphi_j : D^{j+2} \longrightarrow \mathcal{B}_1, j = 1, 2, \ldots, q$ are iteration operators. Usually φ is an iteration operator of convergence order $p \geq 2$. Numerous popular iterative methods are special cases of method (16.2) [1]- [27] (see Section 3).

The local convergence analysis usually involves Taylor expansions and conditions on higher order derivatives not appearing in these methods. Moreover, these approaches do not provide computable radius of convergence and error estimates on the distances $\|x_n - x^*\|$. Therefore the initial point is a shot in the dark. These problems limit the usage of these methods. That is why in the present chapter using only conditions on the first derivative, we address the preceding problems in the more general setting of methods (16.2) and Banach space.

We find computable radii of convergence as well as error bounds on the distances based on Lipschitz-type conditions. The order of convergence is found using computable order of convergence (COC) or approximate computational order of convergence (ACOC) [25] (see Remark 16.1) that do not require usage of higher order derivatives. This way we expand the applicability of three step method (16.2) under weak conditions.

The rest of the chapter is organized as follows: Section 16.1 contains the local convergence of method (16.2), where in the Section 16.2 applications can be found.

16.1 LOCAL CONVERGENCE ANALYSIS

The local convergence analysis of method (16.2) is based on some parameters and scalar functions that appear in the proof. Let $\bar{\rho}_0 > 0$. Suppose:

$(\mathscr{A}_0)$ There exists function $\psi_0 : [0, \infty) \longrightarrow [0, \infty)$ continuous and nondecreasing such that $\psi_0(0) = 0$ and for $\bar{\psi}_0(t) = \psi_0(t) - 1$

$$
\psi_0(0) < 1 \tag{16.3}
$$

and

$$
\bar{\psi}_0(t) \longrightarrow \text{a positive constant or } +\infty \tag{16.4}
$$

as $t \longrightarrow \lambda_0^- \leq \bar{\rho}_0 (\lambda_0 > 0)$. We have by (16.3) that $\bar{\psi}_0(0) = -1 < 0$. Then, by (16.4) and the intermediate value theorem equation $\bar{\psi}_0(t) = 0$ has solutions in $(0, \lambda_0)$. Denote by ρ_0 the smallest such solution.

($\mathscr{A}$) There exists function $\psi : [0, \lambda_0)^2 \longrightarrow [0, +\infty)$ continuous and nondecreasing such that $\psi(0,0) = 0$ and for $\bar{\psi}(t) = \psi(t,t) - 1$, $\psi(0,0) < 1$, $\bar{\psi}(t) \longrightarrow$ a positive constant or $+\infty$ as $t \longrightarrow \lambda^- \leq \lambda_0 (\lambda > 0)$. Denote by ρ the smallest solution of equation $\bar{\psi}(t) = 0$ in $(0, \lambda)$.

($\mathscr{A}_1$) There exists function $\psi_1 : [0, \lambda)^3 \longrightarrow [0, +\infty)$ continuous and nondecreasing such that $\psi_1(0,0,0) = 0$ and for $\bar{\psi}_1(t) = \psi_1(t,t,t) - 1$, $\psi_1(0,0,0) < 1$ $\bar{\psi}_1(t) \longrightarrow$ a positive constant or $+\infty$ as $t \longrightarrow \lambda_1^- \leq \lambda (\lambda_1 > 0)$. Denote by ρ_1 the smallest solution of equation $\bar{\psi}_1(t) = 0$ in $(0, \lambda_1)$.

$$\vdots$$

($\mathscr{A}_q$) There exists function $\psi_q : [0, \lambda_q)^{q+2} \longrightarrow [0, +\infty)$ continuous and nondecreasing such that $\psi_q(0, \underset{q\,times}{\overset{\cdots}{}}, 0) = 0$ and for $\bar{\psi}_q(t) = \psi_q(t, \underset{q\,times}{\overset{\cdots}{}}, t) - 1$, $\psi_q(0, \underset{q\,times}{\overset{\cdots}{}}, 0) < 1$ $\bar{\psi}_q(t) \longrightarrow$ a positive constant or $+\infty$ as $t \longrightarrow \lambda_q^- \leq \lambda (\lambda_{q-1} > 0)$. Denote by ρ_q the smallest solution of equation $\bar{\psi}_q(t) = 0$ in $(0, \lambda_q)$.

Define the radius of convergence r by

$$r = \min\{\rho_0, \rho, \rho_j\} \quad j = 1, 2, \ldots, q. \tag{16.5}$$

Then, for each $t \in [0, r)$

$$\psi_0(t) \quad < \quad 1 \tag{16.6}$$

$$\psi(t,t) \quad < \quad 1 \tag{16.7}$$

$$\psi_j(t, \underset{q+2\,times}{\overset{\cdots}{}}, t) \quad < \quad 1 \tag{16.8}$$

We shall adopt the notations for $x \in \mathscr{B}_1$ and $\mu > 0$ $U(x, \mu) = \{y \in \mathscr{B}_1 : \|x - y\| < \mu\}$ and $\bar{U}(x, \mu)\{y \in \mathscr{B}_1 : \|x - y\| \leq \mu\}$. The local convergence analysis of method (16.2) is based on conditions ($\mathscr{A}_0$), ($\mathscr{A}$), ($\mathscr{A}_j$), $j = 1, 2, \ldots, q$ and

($\mathscr{A}_{q+1}$) $F; D \subseteq \mathscr{B}_1 \longrightarrow \mathscr{B}_2$ is a continuously Fréchet-differentiable operator.
($\mathscr{A}_{q+2}$) There exists $x^* \in D$ such that $F(x^*) = 0$ and $F'(x^*)^{-1} \in L(\mathscr{B}_2, \mathscr{B}_1)$.
($\mathscr{A}_{q+3}$) There exists a continuous iteration operator $\varphi_0 : D \longrightarrow \mathscr{B}_1$ such that for each $x \in D$

$$\|\varphi_0(x) - x^*\| \leq \psi(\|x - x^*\|)\|x - x^*\|.$$

Set $D_0 = D \cap \bar{U}(x^*, \bar{\rho}_0)$.
($\mathscr{A}_{q+4}$) There exists continuous iteration operators $\varphi : D^2 \longrightarrow \mathscr{B}_1, \varphi_j : D^{j+2} \longrightarrow \mathscr{B}_1, j = 1, 2, \ldots, q$ such that for each $u_1, u_2, \ldots, u_{q+2} \in D_0$,

$$\|\varphi(u_1, u_2) - x^*\| \leq \psi(\|u_1 - x^*\|, \|u_2 - x^*\|)\|u_1 - x^*\|,$$

$$\|\varphi_j(u_1, u_2, \ldots, u_{j+2}) - x^*\| \leq \psi_j(\|u_1 - x^*\|, \|u_2 - x^*\|, \ldots, \|u_{j+2} - x^*\|)\|u_1 - x^*\|$$

and

$(\mathscr{A}_{q+5})$ $\bar{U}(x^*, r) \subseteq D$, where r is given in (16.5).

Next, we present the local convergence analysis of method (16.2) under the conditions "A" and the preceding notation.

Theorem 16.1. Suppose that the "A" conditions hold. Then sequence $\{x_n\}$ generated for $x_0 \in U(x^*, r) - \{x^*\}$ by method (16.2) is well defined in $U(x^*, r)$, remains in $U(x^*, r)$ for each $n = 0, 1, 2, \ldots$ and converges to x^*. Moreover, the following error bounds hold

$$\|y_n - x^*\| \leq \psi(\|x_n - x^*\|)\|x_n - x^*\| \leq \|x_n - x^*\| < r, \tag{16.9}$$

$$\|z_n - x^*\| \leq \psi(\|x_n - x^*\|, \|y_n - x^*\|)\|x_n - x^*\| \leq \|x_n - x^*\| \tag{16.10}$$

$$\begin{aligned}\|z_n^{(j)} - x^*\| &\leq \psi_j(\|x_n - x^*\|, \|y_n - x^*\|, \|z_n - x^*\|, \ldots, \|z_n^{(j-1)} - x^*\|)\|x_n - x^*\|, \\ &\leq \|x_n - x^*\| < r\end{aligned} \tag{16.11}$$

where the iteration functions are defined previously.

Proof. Using induction, condition $x_0 \in U(x^*, r) - \{x^*\}$,(16.5)-(16.8) and conditions $(\mathscr{A}_{q+1})$-$(\mathscr{A}_{q+5})$, we obtain that estimates (16.9)-(16.11) hold for $n = 0$ and $y_0, z_0, z_0^{(j)} \in U(x^*, r)$. That is we have

$$\|y_0 - x^*\| \leq \psi(\|x_0 - x^*\|)\|x_0 - x^*\| \leq \|x_0 - x^*\| < r, \tag{16.12}$$

$$\|z_0 - x^*\| \leq \psi(\|x_0 - x^*\|, \|y_0 - x^*\|)\|x_0 - x^*\| \leq \|x_0 - x^*\|, \tag{16.13}$$

$$\|x_1 - x^*\| \leq \psi(\|x_0 - x^*\|, \|x_0 - x^*\|) \leq \|x_0 - x^*\|,$$

$$\begin{aligned}\|z_0^{(j)} - x^*\| &\leq \psi_j(\|x_0 - x^*\|, \|y_0 - x^*\|, \|z_0 - x^*\|, \ldots, \|z_0^{(j-1)} - x^*\|) \\ &\quad \times \|x_0 - x^*\| \leq \|x_0 - x^*\| \\ &\leq \psi_j(\|x_0 - x^*\|, \|x_0 - x^*\|, \|x_0 - x^*\|, \ldots, \|x_0 - x^*\|) \\ &\quad \times \|x_0 - x^*\| \leq \|x_0 - x^*\|.\end{aligned} \tag{16.14}$$

By simply replacing $x_0, y_0, z_0, z_0^{(j)}$ by $x_k, y_k, z_k, z_k^{(j)}$, respectively we obtain estimates (16.9)- (16.11). Then, from the estimate

$$\|x_{k+1} - x^*\| = \|z_n^{(q)} - x^*\| \leq c\|x_n - x^*\|, \tag{16.15}$$

where $c = \psi_q(\|x_0 - x^*\|, \|x_0 - x^*\|, \ldots \|x_0 - x^*\|) \in [0, 1)$, we obtain $\lim_{k \to \infty} x_k = x^*$ and $x_{k+1} \in U(x^*, r)$.

$\square$

Remark 16.1. We can compute the computational order of convergence (COC) [25] defined by

$$\xi = \ln\left(\frac{\|x_{n+1} - x^*\|}{\|x_n - x^*\|}\right) / \ln\left(\frac{\|x_n - x^*\|}{\|x_{n-1} - x^*\|}\right)$$

or the approximate computational order of convergence

$$\xi_1 = \ln\left(\frac{\|x_{n+1} - x_n\|}{\|x_n - x_{n-1}\|}\right) / \ln\left(\frac{\|x_n - x_{n-1}\|}{\|x_{n-1} - x_{n-2}\|}\right).$$

This way we obtain in practice the order of convergence without resorting to the computation of higher order derivatives appearing in the method or in the sufficient convergence criteria usually appearing in the Taylor expansions for the proofs of those results.

16.2 APPLICATIONS

Application 3.1 Let us specialize method (16.2) by setting $\mathscr{B}_1 = \mathscr{B}_2 = \mathbb{R}^i, q = 1$, $\varphi_0(x_k) = x_k - F'(x_k)^{-1}F(x_k)$,

$$\varphi(x_k, y_k) = y_k - \bar{\tau}_k F'(x_k)^{-1}F(y_k) \text{ and } \psi(x_k, y_k, z_k) = z_k - \alpha_k F'(z_k)^{-1}F(z_k). \quad (16.16)$$

Then, method (16.2) reduces to method (5.3) in [28] defined by

$$\begin{aligned}
y_k &= x_k - F'(x_k)^{-1}F(x_k) \\
z_k &= y_k - \bar{\tau}_k F'(x_k)^{-1}F(y_k) \\
x_{k+1} &= z_k - \alpha_k F'(z_k)^{-1}F(z_k).
\end{aligned} \qquad (16.17)$$

It was shown in [28, Theorem 1, Theorem 5] that if operator F is sufficiently many times differentiable, $F'(x)$ is continuous on $D, F'(x^*)^{-1} \in L(\mathscr{B}_2, \mathscr{B}_1)$ then for x_0 sufficiently close to x^* method (16.2) converges with order $p \geq 2$ if and only if $\bar{\tau}_k$ and α_k satisfy certain conditions involving hypotheses on higher derivatives for F. Further special choices of $\bar{\tau}_k$ and α_k are given in the following table leading to other pth order methods.

	Mathods	Order	$\bar{\tau}_k$	α_k
1	Cordero et al. [6]	4	$2I - F'(x_k)^{-1}F(y_k)$	
2	Sharma [20]	4	$3I - 2F'(x_k)^{-1}[y_k, x_k; F]$	
3	Grau-Senchez et al. [12]	4	$(2[y_k, x_k; F] - F'(x_k)^{-1})F'(x_k)$	
4	Sharma et al. [22]	4	$r_k = I - \frac{3}{4}(s_k - I) + \frac{9}{8}(s_k - I)^2$ $s_k = F'(x_k)^{-1}F'(y_k)$ $r_k = \frac{1}{2}(-I + \frac{9}{4}F'(y_k)^{-1}F'(x_k) + \frac{3}{4}F'(x_k)^{-1}F'(x_k)$	
5	Gran-Sanchez et al. [12] Xiao et al. [26]	5	$\bar{\tau}_k = \frac{1}{2}(I + F'(y_k)^{-1}F'(x_k))$	$F'(y_k)^{-1}F'(x_k)$
6	Cordero et al. [5]	5	$2(I - F'(x_k)^{-1}F'(y_k))^{-1}$	$F'(y_k)^{-1}F'(x_k)$
7	Xiao et al. [26] Sharma et al. [21]	5	$y_k = x_k - aF'(x_k)^{-1}F(x_k)$ $\bar{\tau}_k = ((1 - \frac{1}{2a})I + \frac{1}{2a}F'(x_k)^{-1}F'(y_k))^{-1}$ $a = \frac{1}{2}, (F'(y_k)^{-1}F'(x_k) - I)\theta_k$	$-I + 2(\frac{1}{2a}F'(y_k) + (1 - \frac{1}{2a})F'(x_k))^{-1}F'(x_k)$ $2F'(y_k)^{-1}F'(x_k) - I$
8	Sharma et al. [23]	6	$3I - 2F'(x_k)^{-1}[y_k, x_k; F]$	$3I - 2F'(x_k)^{-1}[y_k, x_k; F]$
9	Xiao et al. [26]	6	$y_k = x_k - aF'(x_k)^{-1}F(x_k)$ $r_k = \frac{1}{2}(-I + \frac{9}{4}F'(y_k)^{-1})F'(x_k) + \frac{3}{4}F'(x_k))^{-1}F'(x_k)$ $- F'(x_k)^{-1}F(x_k))^{-1}F'(x_k)$	$(1 - \frac{1}{a})I + \frac{1}{a}F'(y_k)^{-1}F'(x_k)$ $(2[y_k, x_k; F] - F'(x_k))^{-1}F'(x_k)$
10	Grau-Sanchez et al. [12]	6	$(2[y_k, x_k; F] - F'(x_k)^{-1})F'(x_k)$	$2[y_k, x_k; F]^{-1}F'(x_k) - I$
11	Cordero et al. [6]	6	$a = \frac{1}{a}, (F'(x_k) - 2F'(y_k))^{-1}(3F'(x_k)\theta_k^{-1} - 4F'(x_k)$	$(F'(x_k) - 2F'(y_k))^{-1}F'(x_k)$

Table 16.1: Record of Specialized Methods

Application 3.2 Let $\mathscr{B}_1 = \mathscr{B}_2 = \mathbb{R}^i$, $\varphi(x_k, y_k) = \varphi_p(x_k, y_k)$ and ψ as in (16.16), (16.23) where φ_p denotes the iteration function of pth order. Then, again according to Theorem 6 in [28] the method (16.2) has order of convergence $p + 2$ under certain conditions of α_k. As an example, we present the choices given by

$$\alpha_k = \frac{1}{2}(5I - 3F'(x_k)^{-1}F'(y_k)), \tag{16.18}$$

$$\alpha_k = 3I - 2F'(x_k)^{-1}F'(y_k) \tag{16.19}$$

$$\alpha_k = F'(x_k)^{-1}F'(y_k) \tag{16.20}$$

$$\alpha_k = ((1 - \frac{1}{\alpha})F'(x_k) + \frac{1}{\alpha}F'(y_k))^{-1}F'(x_k) \tag{16.21}$$

$$\alpha_k = ((1 + \frac{1}{\alpha})F'(y_k) - \frac{1}{a}F'(x_k))^{-1}F'(y_k) \tag{16.22}$$

(see [3, 4, 28]).

Let us consider the special case of method (16.2) given by

$$\begin{aligned} y_k = \varphi_0(x_k) &= x_k - F'(x_k)^{-1}F(x_k) \\ x_{k+1} = \varphi(x_k, y_k) &= y_k - F'(x_k)^{-1}F(y_k). \end{aligned}$$

Suppose that

$$\|F'(x^*)^{-1}(F'(x) - F'(x^*))\| \leq w_0(\|x - x^*\|) \text{ for each } x \in D$$

$$\|F'(x^*)^{-1}(F'(x) - F'(y))\| \leq w(\|x - y\|)$$

and

$$\|F'(x^*)^{-1}F'(x)\| \leq v(\|x - x^*\|)$$

for each $x, y \in D_0$, where $w_0 : [0, +\infty) \longrightarrow [0, +\infty), w : [0, \bar{\rho}_0) \longrightarrow [0, +\infty), v : [0, \bar{\rho}_0) \longrightarrow [0, +\infty)$ are continuous nondecreasing functions with $w_0(0) = w(0) = 0$ and

$$\bar{\rho}_0 = \sup\{t \geq 0 : w_0(t) < 1\}.$$

Then, we have in turn

$$\begin{aligned} \|x_{k+1} - x^*\| &\leq \|F'(y_k)^{-1} \int_0^1 [F'(x^* + \theta(y_k - x^*)) - F'(x^*))(y_k - x^*)d\theta\| \\ &\quad + \|F'(y_k)^{-1}(F'(x_k) - F'(y_k))F'(x_k)^{-1}F(y_k)\| \\ &\leq \frac{\int_0^1 w(\theta\|y_k - x^*\|)d\theta\|y_k - x^*\|}{1 - w_0(\|y_k - x^*\|)} \\ &\quad + \frac{(w_0(\|x_k - x^*\|) + w_0(\|y_k - x^*\|))\int_0^1 v(\theta\|y_k - x^*\|)d\theta\|y_k - x^*\|}{(1 - w_0(\|y_k - x^*\|))(1 - w_0(\|x_k - x^*\|))} \\ &\leq \frac{\int_0^1 w(\theta g_1(\|x_k - x^*\|)\|x_k - x^*\|)d\theta g_1(\|x_k - x^*\|)\|x_k - x^*\|}{1 - w_0(\|x_k - x^*\|)} \end{aligned}$$

so we can choose $\psi_0(t) = \frac{\int_0^1 w((1-\theta)t)d\theta}{1-w_0(t)}$, $\psi(s,t) = \psi(t) = [\frac{\int_0^1 w((1-\theta)\psi_0(t)t)d\theta}{1-w_0(\psi_0(t)t)} +$

$\frac{(w_0(t)+w_0(\psi_0(t)t))\int_0^1 v(\theta\varphi_0(t)t)d\theta}{(1-w_0(t))(1-w_0(\psi_0(t)t))}]\psi_0(t) = 0$, $r = \min\{r_1, r_2\}$ respectively, where r_1, r_2 are the smallest positive solutions of equations $\bar{\psi}_0(t) = 0$ and $\bar{\psi}(t) = 0$. Using the above choice we present the following examples:

Example 16.1. Let us consider a system of differential equations governing the motion of an object and given by

$$F_1'(x) = e^x, \; F_2'(y) = (e-1)y + 1, \; f_3(z) = 1$$

with initial conditions $F_1(0) = F_2(0) = F_3(0) = 0$. Let $F = (F_1, F, F_3)$. Let $\mathscr{B}_1 = \mathscr{B}_2 = \mathbb{R}^3, D = \bar{U}(0,1), p = (0,0,0)^T$. Define function F on D for $w = (x,y,z)^T$ by

$$F(w) = (e^x - 1, \frac{e-1}{2}y^2 + y, z)^T.$$

The Fréchet-derivative is defined by

$$F'(v) = \begin{bmatrix} e^x & 0 & 0 \\ 0 & (e-1)y + 1 & 0 \\ 0 & 0 & 1 \end{bmatrix}.$$

Notice that using the $(\mathscr{A})$ conditions, we get for $\alpha = 1$, $w_0(t) = (e-1)t$, $w(t) = e^{\frac{1}{e-1}}t$, $v(t) = e^{\frac{1}{e-1}}$. The radii are

$$r_1 = 0.3827, \; r_2 = 0.2523 = r.$$

Example 16.2. Let $\mathscr{B}_1 = \mathscr{B}_2 = C[0,1]$, the space of continuous functions defined on $[0,1]$ be equipped with the max norm. Let $D = \bar{U}(0,1)$. Define function F on D by

$$F(\varphi)(x) = \varphi(x) - 5 \int_0^1 x\theta\varphi(\theta)^3 d\theta. \tag{16.23}$$

We have that

$$F'(\varphi(\xi))(x) = \xi(x) - 15 \int_0^1 x\theta\varphi(\theta)^2\xi(\theta)d\theta, \text{ for each } \xi \in D.$$

Then, we get that $x^* = 0$, so $w_0(t) = 7.5t$, $w(t) = 15t$ and $v(t) = 2$. Then the radii are

$$r_1 = 0.0667, \; r = r_2 = 0.0439.$$

Example 16.3. Let $\mathscr{B}_1 = \mathscr{B}_2 = C[0,1]$, and $D = \bar{U}(0,1)$. Consider the equation

$$x(s) = \int_0^1 T(s,t)(\frac{1}{2}x(t)^{\frac{3}{2}} + \frac{x(t)^2}{8})dt, \tag{16.24}$$

where the kernel T is the Green's function defined on the interval $[0, 1] \times [0, 1]$ by

$$T(s,t) = \begin{cases} (1-s)t, & t \leq s \\ s(1-t), & s \leq t, \end{cases} \tag{16.25}$$

Define operator $F : C[0, 1] \longrightarrow [0, 1]$ by

$$F(x(s)) = \int_0^1 T(s,t)(\frac{1}{2}x(t)^{\frac{3}{2}} + \frac{x(t)^2}{8})dt - x(s). \tag{16.26}$$

Then, we have

$$F'(x)\mu(s) = \mu(s) - \int_0^1 T(s,t)(\frac{3}{4}x(t)^{\frac{1}{2}} + \frac{x(t)}{4})\mu(t)dt. \tag{16.27}$$

Notice that $x^*(s) = 0$ is a solution of $F(x(s)) = 0$. Using (16.25), we obtain

$$\| \int_0^1 T(s,t)dt\| \leq \frac{1}{8}. \tag{16.28}$$

Then, by (16.27) and (16.28), we have that

$$\|F'(x) - F'(y)\| \leq \frac{1}{32}(3\|x-y\|^{\frac{1}{2}} + \|x-y\|). \tag{16.29}$$

We have $w_0(t) = w(t) = \frac{1}{32}(3t^{1/2}+t)$ and $v(t) = 1 + w_0(t)$. Then the radii are

$$r_1 = 19.4772, r_2 = 0.3889,$$

so we choose $r = 1$ since $\bar{U}(x^*, r) \subset D$ [5] - [28].

In view of (16.29) earlier results requiring hypotheses on the second Fréchet derivative or higher cannot be used to solve equation $F(x(s)) = 0$.

16.3   EXERCISES

1. Solve the initial value problem

   $$\dot{x}(t) = 1 + \cos(x(t)), x(0) = 0,$$

   using method (16.2).

2. The predator-prey population models describe the interaction of a prey population X and a predator population Y. Assume that their interaction is modeled by the system of ordinary differential equations

   $$\dot{x} = x - \frac{1}{4}x^2 - \frac{1}{10}xy + 1$$

   $$\dot{y} = -\frac{1}{4}y + \frac{1}{7}xy + \frac{2}{3}.$$

   (Assume $x(0) = y(0) = 0$). Solve the system.

3. Solve the initial value problem

$$\dot{x} = (x(t) + t, x(0) = 1)$$

by the iteration method (16.2). Perform 3 steps.

4. Solve the Fredholm-type integral equation

$$x(t) = \sum_0^1 \frac{t \cdot s}{10} ds + 1$$

by the iteration method (16.2). Perform 3 steps.

5. Solve the Volterra-type integral equation

$$x(t) = \sum_0^t \frac{t \cdot s}{10} ds + 1$$

by the iteration method (16.2). Perform 3 steps.

6. Solve equation

$$x = \frac{\sin x}{2} + 1$$

by using algorithm (16.2).

7. Repeat the previous problem for equation

$$x = \frac{\sin x + \cos 2x}{10}.$$

REFERENCES

1. Amat, S., Argyros, I.K., Busquier, S., and Hernandez, M.A. 2018. On two high-order families of frozen Newton-type methods. *Numerical Lin. Algebra Appl.*, 25:1–13.
2. Argyros, I.K. 2007. *Computational Theory of Iterative Methods. Series: Studies in Computational Mathematics*, New York: Elsevier Publ. Co.
3. Argyros, I.K. 2008. *Convergence and applications of Newton-type iterations*, New York: Springer Verlag.
4. Argyros, I.K., and Magreñañ, A.A. 2017. *Iterative Methods and Their Dynamics with Applications: A Contemporary Study*. CRC Press.
5. Cordero, A., Martinez, E., and Torregrosa, J.R. 2009. Iterative methods of order four and five for systems of nonlinear equations. *Appl. Math. Comput.*, 231:541–551.
6. Cordero, A., Torregrosa, J.R., and Vassileva, M.P. 2012. Pseudocomposition: a technique to design predictor-corrector methods for systems of nonlinear equations. *Appl. Math. Comput.*, 218:11496–11504.
7. Madru, K., and Jayaraman, J. 2017. Some higher order Newton-like methods for solving system of nonlinear equations and its applications. *Int. J. Appl. Comput. Math.*, 3:2213–2230.
8. Ezquerro, J.A., and Hernández, M.A. 2000. Recurrence relations for Chebyshev-type methods. *Appl. Math. Optim.*, 41(2):227–236 .
9. Ezquerro, J.A., Gutiérrez, J.M, and Hernández, M.A. 1998. Avoiding the computation of the second-Fréchet derivative in the convex acceleration of Newton's method. *J. Comput. Appl. Math.*, 96:1–12.
10. Ezquerro, J.A., and Hernández, M.A. 2000. Multipoint super-Halley type approximation algorithms in Banach spaces. *Numer. Funct. Anal. Optimiz.*, 21(7&8):845–858.
11. Ezquerro, J.A., and Hernández, M.A. 2000. A modification of the super-Halley method under mild differentiability condition. *J. Comput. Appl. Math.*, 114:405–409.
12. Grau-Sanchez, M., Grau, A., and Noguera, M. 2011. Ostrowski type methods for solving systems of nonlinear equations. *Appl. Math. Comput.*, 218:2377–2385.
13. Gutiérrez, J.M., Magreñán, A.A., and Romero, N. 2013. On the semi-local convergence of Newton-Kantorovich method under center-Lipschitz conditions. *Applied Mathematics and Computation*, 221:79–88.
14. Kantorovich, L.V., and Akilov, G.P. 1982. *Functional Analysis*, Oxford: Pergamon Press.
15. Magreñán, A.A. 2014. Different anomalies in a Jarratt family of iterative root finding methods. *Appl. Math. Comput.*, 233:29–38.
16. Magreñán, A.A. 2014. A new tool to study real dynamics: The convergence plane. *Appl. Math. Comput.*, 248:29–38.
17. Petkovic, M.S., Neta, B., Petkovic, L., and Džunič, J. 2013. *Multipoint methods for solving nonlinear equations*. Elsevier.
18. Potra, F.A., and Pták, V. 1984. Nondiscrete Induction and Iterative Processes. in: *Research Notes in Mathematics*, Vol. 103. Boston: Pitman
19. Rheinboldt, W.C. 1977. An adaptive continuation process for solving systems of nonlinear equations. In: *Mathematical Models and Numerical Methods (A.N.Tikhonov et al. eds.)*. pub.3. 129–142. Warsaw Poland: Banach Center.
20. Sharma, J.R., and Arora, H. 2013. On efficient weighted-Newton methods for solving system of nonlinear equations. *Appl. Math. Comput.*, 222:497–506.
21. Sharma, J.R., and Arora, H. 2014. Efficient Jarratt-like methods for solving systems of nonlinear equations. *Calcolo*, 51:193–210.

22. Sharma, J.R., and Gaupta, P. 2014. An efficient fifth order method for solving systems of nonlinear equations. *Comput. Math. Appl.*, 67:591–601.
23. Sharma, J.R., Guha, R.K., and Sharma, R. 2013. An efficient fourth-order weighted Newton method for systems of nonlinear equations. *Numer. Algorithms*, 62:307–323.
24. Traub, J.F. 1982. *Iterative Methods for the Solution of Equations*. AMS Chelsea Publishing.
25. Weerakoon, S., and Fernando, T.G.I. 2000. A variant of Newton's method with accelerated third-order convergence. *Appl. Math. Lett.*, 13:87–93.
26. Xiao, X.Y., and Yin, H.W. 2016. Increasing the order of convergence for iterative methods to solve nonlinear systems. *Calcolo*, 53:285–300.
27. Zhanlav, T., and Puzynin, I.V., The convergence of iteration based on a continuous analogue of Newton's method. *Comput. Math and Math. Phys.*, 32:729–737.
28. Zhanlav, T., Chun, C., Otgondoj, Kh., and Ulziibayar, V. 2019. High-order iterations for systems of nonlinear equations and their generalizations in Banach spaces. *International Journal of Computer Mathematics*, 1704–1724.

17 Two-Step Gauss-Newton Werner-Type Method for Least Squares Problems

The aim of this chapter is to extend the applicability of a two-step Gauss-Newton-Werner-type method (TGNWTM) for solving nonlinear least squares problems. The radius of convergence, error bounds and the information on the location of the solution are improved under the same information as in earlier studies. Numerical examples further validate the theoretical results.

Let i, j be natural numbers with $j \geq i$. Let also Ω be an open and convex subset of $\mathbb{R}^j$. We are concerned with the solution p of the least squares problem [4–9]:

$$\min_{x \in \Omega} f(x) := \frac{1}{2} F(x)^T F(x), \tag{17.1}$$

where $F : \Omega \longrightarrow \mathbb{R}^j$ is a Fréchet-differentiable mapping. Numerous problems can be brought in the form (17.1) using Mathematical Modeling [1–13]. The closed form solutions can only be found in special cases. That explains why most solution methods for problem (17.1) are iterative. Let $x_0, y_0 \in \Omega$ and set $z = \frac{x_0 + y_0}{2}$. In the present study, we provide the local convergence analysis of GNWTM defined for each $n = 0, 1, 2, \ldots$ by

$$\begin{aligned}
x_{n+1} &= x_n - A_n F(x_n) \\
y_{n+1} &= x_{n+1} - A_n F(x_{n+1}) \\
z_n &= \frac{x_n + y_n}{2},
\end{aligned} \tag{17.2}$$

where $A_n = [F'(z_n)^T F'(z_n)]^{-1} F'(z_n)^T$. If $i = j$, TGNWTM reduces to a Gauss-Newton-Werner-type method [3, 8, 9]. Notice that in each iteration the inversion of $[F'(z_n)^T F'(z_n)]^{-1}$ is required only once. Therefore, the computational cost is essentially the same as in the Gauss-Newton method. The LL^T decomposition of $[F'(z_n)^T F'(z_n)]^{-1}$ costs $O(n^3)$ floating-point operations (Flops) leading to the computation of x_{n+1}. It then follows from the second substep of method (17.2) that $O(n^2)$ Flops are needed for the computation of y_{n+1}.

The local convergence analysis of method (17.2) was given in the elegant paper by Shakhno et.al. in [9] (see also related work in [3, 8]). Their convergence analysis uses average Lipschitz continuity condition as well as Lipschitz conditions.

Using the concept of the average Lipschitz continuity [12] and our new idea of restricted convergence domains, we present a local convergence analysis with the

DOI: 10.1201/9781003128915-17

following advantages (**A**) over works using the similar information [3,4,8,9,11–13]:

(a)   Larger radius of convergence;
(b)   Tighter error bounds on the distances $\|x_n - p\|$;
(c)   An at least as precise information on the location of the solution p.

Achieving (a)–(c) is very important in computational sciences, since: (a)$'$ We obtain a wider choice of initial guesses; (b)$'$ Fewer iterates are required to obtain a desired error tolerance (c)$'$ Better information about the ball of convergence is obtained.

The rest of the chapter is structured as follows: Section 17.1 contains the local convergence analysis of method (17.2) whereas special cases and the applications are presented in Section 17.2.

17.1   LOCAL CONVERGENCE ANALYSIS

Set $U(w,\rho) = \{v \in \mathbb{R}^j : \|v - w\| < \rho\}$ to be the open ball in $\mathbb{R}^j$ and by $\bar{U}(w,\rho)$ to denote its closure. Let $R > 0$. Define parameter R_1 by $R_1 := \sup\{t \in [0,R] : \bar{U}(p,t) \subset \Omega\}$. The convergence analysis of numerous iterative methods has been given using the following concept due to Wang [12]:

Definition 17.1. Mapping $F : \bar{U}(p,R_1) \longrightarrow \mathbb{R}^i$ satisfies the Lipschitz condition with L_1 average on $U(p,R_1)$ if

$$\|F'(x) - F'(y)\| \leq \int_0^{\|x-y\|} L_1(u)du \text{ for each } x, y \in \bar{U}(p,R_1),$$

where L_1 is a positive non-decreasing function.

It turns out that the convergence analysis of iterative methods based on the preceding notion can be improved as follows:

Definition 17.2. Mapping $F : \bar{U}(p,R_1) \longrightarrow \mathbb{R}^i$ satisfies the center-Lipschitz condition with L_0 average on $U(p,R_1)$ if

$$\|F'(x) - F'(p)\| \leq \int_0^{\|x-p\|} L_0(u)du \text{ for each } x \in \bar{U}(p,R_1),$$

where L_0 is a positive non-decreasing function.

Clearly, we have that

$$L_0(u) \leq L_1(u) \text{ for each } u \in [0,R_1] \tag{17.3}$$

and $\frac{L_0}{L_1}$ can be arbitrary small [2–4]. Let $\beta > 0$ be a parameter. Suppose that equation

$$\beta \int_0^t L_0(u)du = 1 \tag{17.4}$$

has positive solutions. Denote by R_0 the smallest such solution. Notice for example that R_0 exists, if

$$\beta L_0(R_1)R_1 \geq 1. \tag{17.5}$$

Indeed, function $g(t) := \beta \int_0^t L(u)du - 1$ is such that $g(0) = -1 < 0$ and $g(R_1) = \beta L(R_1)R_1 - 1 > 0$. The existence of R_0 follows from the intermediate value theorem.

Definition 17.3. Mapping $F : \bar{U}(p,R_1) \longrightarrow \mathbb{R}^i$ satisfies the restricted Lipschitz condition with L average on $U(p,R_0)$ if

$$\|F'(x) - F'(y)\| \leq \int_0^{\|x-y\|} L(u)du \text{ for each } x,y \in \bar{U}(p,R_0),$$

where L is a positive non-decreasing function.

We have that

$$L(u) \leq L_1(u) \text{ for each } u \in [0,R_0], \tag{17.6}$$

since $R_0 \leq R_1$. Throughout this chapter, we suppose that

$$L_0(u) \leq L(u) \text{ for each } u \in [0,R_0] \tag{17.7}$$

unless, otherwise stated. Otherwise, i.e., if

$$L(u) \leq L_0(u) \text{ for each } u \in [0,R_0] \tag{17.8}$$

then the results that follow hold with L_0 replacing L. Moreover, we need the definitions:

Definition 17.4. [12] Let $F : \bar{U}(p,R_1) \longrightarrow \mathbb{R}^i$ be a twice Fréchet-differentiable mapping. We say that mapping F satisfies the Lipschitz condition with M_1 average on $U(p,R_1)$ if

$$\|F''(x) - F''(y)\| \leq \int_0^{\|x-y\|} M_1(u)du \text{ for each } x,y \in \bar{U}(p,R_1),$$

where M_1 is a positive nondecreasing function.

Definition 17.5. Let $F : \bar{U}(p,R_0) \longrightarrow \mathbb{R}^i$ be twice Fréchet-differentiable mapping. We say that mapping F satisfies the restricted Lipschitz condition with M average on $U(p,R_0)$ if

$$\|F''(x) - F''(y)\| \leq \int_0^{\|x-y\|} M(u)du \text{ for each } x,y \in \bar{U}(p,R_0),$$

where M is a positive nondecreasing function.

We have that

$$M(u) \leq M_1(u) \text{ for each } u \in [0, R_0]. \tag{17.9}$$

It is worth noticing that the definition of functions L and M (based on L_0 and R_0) was not possible in earlier studies using L_1 and M_1. That is $L = L(L_0, R_0, R_1), M = M(L_0, R_0, R_1)$, whereas $L_1 = L_1(R_1)$ and $M_1 = M_1(R_1)$. It turns out that L_0 can replace the less precise L in the computation of the upper bounds on the inverses of the operators involved and $\bar{U}(p, R_0), L, M$ can replace $\bar{U}(p, R_1), L_1, M_1$, respectively in the proofs of such results. Moreover, notice that the iterates x_n lie in $\bar{U}(p, R_0)$ which is a more precise location than $\bar{U}(p, R_1)$ used in earlier studies [2–4, 8, 11–13]. We shall make the chapter as self contained as possible by stating some standard auxiliary concepts and results.

Denote by $\mathbb{R}^{i \times j}$ the set of all $i \times j$ matrices. The Moore-Penrose pseudo-inverse is defined by $A^\dagger = (A^T A)^{-1} A^T$ for each full rank $A \in \mathbb{R}^{i \times j}$ [6].

Lemma 17.1. [2,6] Let $A, A_1 \in \mathbb{R}^{m \times n}$. Assume that $A_2 = A + A_1, \|A^\dagger\| \|A_1\| < 1$, and $rank(A) = rank(A_2)$. Then,

$$\|A_2^\dagger\| \leq \frac{\|A^\dagger\|}{1 - \|A^\dagger\| \|A_1\|}.$$

If $rank(A) = rank(A_2) = \min\{m, n\}$, the following holds

$$\|A_2 - A^\dagger\| \leq \frac{\sqrt{2} \|A^\dagger\|^2 \|A_1\|}{1 - \|A^\dagger\| \|A_1\|}.$$

Lemma 17.2. [5] Let $A, A_1 \in \mathbb{R}^{m \times n}$. Assume that $A_2 = A + A_1, \|A_1 A^\dagger\| < 1$, and $rank(A) = n$, then $rank(A_2) = n$.

Lemma 17.3. [12] Let $\varphi(t) = \frac{1}{t} \int_0^t P(u) du, 0 \leq t \leq \rho$, where $P(u)$ is a positive integrable function and monotonically non-decreasing on $[0, \rho]$. Then, $\varphi(t)$ is monotonically non-decreasing with respect to t.

Lemma 17.4. [12] Let $\psi(t) = \frac{1}{t^3} \int_0^t Q(u) du, 0 \leq t \leq \rho$, where $Q(u)$ is a positive integrable function and monotonically non-decreasing on $[0, \rho]$. Then, $\psi(t)$ is monotonically non-decreasing with respect to t.

As in [9] it is convenient for the local convergence analysis that follows to introduce some functions and parameters:

$$\alpha = \|F(p)\|, \beta = \|(F'(p)^T F'(p))^{-1} F'(p)^T\|$$

$$d(x) = \|x - p\|, s_0 = \max\{d(x_0), d(y_0)\},$$

$$\mu(t) = \mu(L_0, L, M)(t)$$

$$= \frac{\beta}{8} \int_0^t M(u)(t - u)^2 du$$

$$+ \beta t \left(\int_0^{\frac{3}{2}t} L(u) du + \int_0^t L_0(u) du \right) + \sqrt{2}\alpha\beta^2 \int_0^t L_0(u) du - t,$$

$$\gamma = \gamma(L_0, L, M)$$

$$= \frac{\beta \int_0^{d(x_0)} M(u)(d(x_0) - u)^2 du}{8 d(x_0)(1 - \beta \int_0^{d(x_0)} L_0(u) du)}$$

$$+ \frac{\beta d(x_0) \int_0^{\frac{d(x_0)+d(y_0)}{2}} L(u) du}{\frac{2d(x_0)+d(y_0)}{3}(1 - \beta \int_0^{d(z_0)} L_0(u) du)}$$

$$+ \frac{\sqrt{2}\alpha\beta^2 \int_0^{d(z_0)} L_0(u) du}{d(z_0)(1 - \beta \int_0^{d(z_0)} L_0(u) du)} < 1,$$

$$\delta = \frac{\beta \int_0^{d(x_0)} M(u)(d(x_0) - u)^2 du}{8 d(x_0)^3 (1 - \beta \int_0^{d(z_0)} L_0(u) du)},$$

$$\tau = \frac{\sqrt{2}\alpha\beta^2 \int_0^{d(z_0)} L_0(u) du}{d(z_0)(1 - \beta \int_0^{d(z_0)} L_0(u) du)},$$

$$\lambda = \frac{\beta \int_0^{\frac{d(x_0)+d(y_0)}{2}} L(u) du}{\frac{2d(x_0)+d(y_0)}{3}(1 - \beta \int_0^{d(z_0)} L_0(u) du)},$$

$$e_{n+1}^1 = \delta d(x_n)^3 + \lambda d(x_n) d(y_n) + \tau d(z_n),$$

$$e_{n+1}^2 = \delta d(x_{n+1})^3 + \frac{\lambda}{3}(d(x_n) + d(y_n) + d(x_{n+1})) d(x_{n+1}) + \tau d(z_n)$$

and

$$s_{n+1} = \max\{d(x_{n+1}), d(y_{n+1})\}.$$

Notice that if $L_0 = L = L_1$ and $M = M_1$, then the preceding definitions reduce to the corresponding ones in [9].

The local convergence analysis is based on the conditions ($\mathscr{C}$):

($\mathscr{C}_1$) Mapping $F : \bar{U}(p, R_1) \longrightarrow \mathbb{R}^i$ is twice Fréchet-differentiable, $F'(p)$ has full rank and p solves problem (17.1).

($\mathscr{C}_2$) $F'(x)$ satisfies: the center-Lipschitz condition with L_0 average on $\bar{U}(p, R_1)$ and the restricted Lipschitz condition with L average on $\bar{U}(p, R_0)$; $F''(x)$ satisfies the restricted Lipschitz condition with M average on $\bar{U}(p, R_0)$, where L_0, L and M are positive non-decreasing functions on $[0, \frac{3R_0}{2}]$.

($\mathscr{C}_3$)   Function μ has a minimal zero R^* in $[0, R_0]$, which also satisfies

$$\beta \int_0^{R^*} L_0(u)du < 1.$$

Then, we can show the following local convergence result for TGNWTM under the conditions ($\mathscr{C}$) and the preceding notation.

Theorem 17.1. Suppose that the conditions ($\mathscr{C}$) hold. Then, sequences $\{x_n\}$, $\{y_n\}, \{z_n\}$ generated for $x_0, y_0 \in \bar{U}(p, R^*) - \{p\}$ by TGNWTM are well defined in $\bar{U}(p, R^*)$, remain in $\bar{U}(p, R^*)$ for each $n = 0, 1, 2, \ldots$ and converge to p. Moreover, the following estimates hold

$$d(x_{n+1}) \le e_{n+1}^1, \tag{17.10}$$

$$d(y_{n+1}) \le e_{n+1}^2 \tag{17.11}$$

and

$$s_{n+1} \le \gamma s_k \le \ldots \le \gamma^{n+1} s_0, s_0. \tag{17.12}$$

Proof. The proof follows the corresponding one in [9] but there are differences where we use (L_0, L), M instead of L_1, M_1, respectively used in [9]. We shall use mathematical induction to show that iterates $\{x_k\}, \{y_k\}, \{z_k\}$ are well defined converge to p and the error estimates (17.10)–(17.12) are satisfied. Using TGNWTM for $n = 0$, we can write

$x_1 - p$
$= x_0 - p - [F'(z_0)^T F'(z_0)]^{-1} F'(z_0) F(x_0)$
$= [F'(z_0)^T F'(z_0)]^{-1} F'(z_0)^T [F'(z_0)(x_0 - p) - F(x_0) + F(p)]$
$\quad + [F'(p)^T F'(p)]^{-1} F'(p)^T F(p) - [F'(z_0)^T F'(z_0)]^{-1} F'(z_0)^T F(p)$
$= [F'(z_0)^T F'(z_0)]^{-1} F'(z_0)^T$
$\quad \times [(F'(\frac{x_0 + p}{2})(x_0 - p) - F(x_0) + F(p)) + (F'(z_0) - F'(\frac{x_0 + p}{2}))(x_0 - p)]$
$\quad + [F'(p)^T F'(p)]^{-1} F'(p)^T F(p) - [F'(z_0)^T F'(z_0)]^{-1} F'(z_0)^T F(p);$

and

$y_1 - p \ = \ x_1 - p - [F'(z_0)^T F'(z_0)]^{-1} F'(z_0)^T F(x_1)$
$\qquad = \ [F'(z_0)^T F'(z_0)]^{-1} F'(z_0)^T [F'(z_0)(x_1 - p) - F(x_1) + F(p)]$
$\qquad\quad + [F'(p)^T F'(p)]^{-1} F'(p)^T F(p) - [F'(z_0)^T F'(z_0)]^{-1} F'(z_0)^T F(p)$
$\qquad = \ [F'(z_0)^T F'(z_0)]^{-1} F'(z_0)^T$
$\qquad\quad \times [(F(\frac{x_1 + p}{2})(x_1 - p) - F(x_1) + F(p))$
$\qquad\quad + (F'(z_0) - F'(\frac{x_1 + p}{2}))(x_1 - p)]$
$\qquad\quad + [F'(p)^T F'(p)]^{-1} F'(p)^T F(p) - [F'(z_0)^T F'(z_0)]^{-1} F'(z_0) F(p).$

In view of the estimate

$$F(x) - F(y) - F'(\frac{x+y}{2})(x-y) = \frac{1}{4}\int_0^1 (1-t)[F''(\frac{x+y}{2}+\frac{t}{2}(x-y))$$
$$-F''(\frac{x+y}{2}+\frac{t}{2}(y-x))](x-y)^2 dt,$$

for $x = p$ and $y = x_0$, we obtain in turn

$$\|F(p) - F(x_0) - F'(\frac{x_0+p}{2})(p-x_0)\|$$
$$= \frac{1}{4}\|\int_0^1 (1-t)[F''(\frac{x_0+p}{2}+\frac{t}{2}(p-x_0))$$
$$-F''(\frac{x_0+p}{2}+\frac{t}{2}(x_0-p))](p-x_0)^2 dt\|$$
$$\leq \int_0^1 (1-t)\int_0^{t\|x_0-p\|} M(u)du\|x_0-p\|^2 dt$$
$$= \frac{1}{8}\int_0^{d(x_0)} M(u)(1-\frac{u}{d(x_0)})^2 dud(x_0)^2$$
$$= \frac{1}{8}\int_0^{d(x_0)} M(u)(d(x_0)-u)^2 du,$$

and

$$\|F'(\frac{x_0+y_0}{2}) - F'(\frac{x_0+p}{2})\| \leq \int_0^{d(y_0)/2} L(u)du.$$

By the central Lipschitz condition, we have that

$$\|([F'(p)^T F'(p)]^{-1}F'(p)^T\|\|F'(x)-F'(p)\| \leq \beta \int_0^{d(x)} L_0(u)du.$$

Moreover, by Lemma 17.1 and Lemma 17.2 and $(\mathscr{C}_1)$, for all $x \in U(p,R^*)$, we get

$$\|([F'(x)^T F'(x)]^{-1}F'(x)^T\| \leq \frac{\beta}{1-\beta\int_0^{d(x)} L_0(u)du}$$

and

$$\|([F'(x)^T F'(x)]^{-1}F'(x)^T - ([F'(p)^T F'(p)]^{-1}F'(p)^T\| \leq \frac{\sqrt{2}\beta^2 \int_0^{d(x)} L_0(u)du}{1-\beta\int_0^{d(x)} L_0(u)du}.$$

By the monotonicity of $L(u)$ and $M(u)$ with Lemmas 17.3 and 17.4, functions $\frac{1}{t}\int_0^t L(u)du$ and $\frac{1}{t^3}\int_0^t M(u)(t-u)^2 du$ are non-decreasing in t. That is, by $(\mathscr{C}_3)$

$$\gamma = \frac{1}{R_0}\left[\frac{\beta\int_0^{R_0} M(u)(R_0-u)^2 du}{8(1-\beta\int_0^{R_0} L(u)du)} + \frac{\beta R_0\int_0^{\frac{3}{2}R_0} L(u)du}{1-\beta\int_0^{R_0} L_0(u)du} + \frac{\sqrt{2}\alpha\beta^2\int_0^{R_0} L_0(u)du}{1-\beta\int_0^{R_0} L_0(u)du}\right]$$

$$< \frac{1}{R}\left[\frac{\beta\int_0^{R} M(u)(R-u)^2 du}{8(1-\beta\int_0^{R} L_0(u)du)} + \frac{\beta R\int_0^{\frac{3}{2}R} L(u)du}{1-\beta\int_0^{R} L_0(u)du} + \frac{\sqrt{2}\alpha\beta^2\int_0^{R} L_0(u)du}{1-\beta\int_0^{R} L_0(u)du}\right] \leq 1.$$

Thus, by Lemma 17.1–17.4 and condition $(\mathscr{C}_2)$, we have in turn

$$
\begin{aligned}
\|x_1 - p\| \ \leq\ & \ \|[F'(z_0)^T F'(z_0)]^{-1} F'(z_0)^T\| \\
& \times \|(F'(\tfrac{x_0 + p}{2})(x_0 - p) - F(x_0) + F(p)) \\
& + (F'(z_0) - F'(\tfrac{x_0 + p}{2}))(x_0 - p)\| \\
& + \|[F'(p)^T F'(p)]^{-1} F'(p)^T F(p) \\
& - [F'(z_0)^T F'(z_0)]^{-1} F'(z_0)^T F(p)\| \\
\leq\ & \ \frac{\beta d(x_0)^3 \int_0^{d(x_0)} M(u)(d(x_0) - u)^2 du}{8 d(x_0)^3 (1 - \beta \int_0^{d(x_0)} L_0(u) du)} \\
& + \frac{\beta d(x_0) d(y_0) \int_0^{d(y_0)/2} L(u) du}{d(y_0)(1 - \beta \int_0^{d(x_0)} L_0(u) du)} + \frac{\sqrt{2}\alpha\beta^2 \int_0^{d(x_0)} L_0(u) du}{d(z_0)(1 - \beta \int_0^{d(x_0)} L_0(u) du)} \\
<\ & \ \delta d(x_0)^3 + \lambda d(x_0) d(y_0) + \tau d(z_0) < \delta R_0 < R.
\end{aligned}
$$

In an analogous way, we get in turn

$$
\begin{aligned}
\|y_1 - p\| \ \leq\ & \ \|[F'(z_0)^T F'(z_0)]^{-1} F'(z_0)^T\| \\
& \times \|(F'(\tfrac{x_1 + p}{2})(x_1 - p) - F(x_1) + F(p)) \\
& + (F'(z_0) - F'(\tfrac{x_1 + p}{2}))(x_1 - p)\| \\
& + \|[F'(p)^T F'(p)]^{-1} F'(p)^T F(p) \\
& - [F'(z_0)^T F'(z_0)]^{-1} F'(z_0)^T F(p)\| \\
\leq\ & \ \frac{\beta d(x_1)^3 \int_0^{d(x_1)} M(u)(d(x_1) - u)^2 du}{8 d(x_1)^3 (1 - \beta \int_0^{d(x_1)} L_0(u) du)} \\
& + \frac{\beta d(x_1) d(z_0') \int_0^{d(z_0)/2} L(u) du}{d(z_0')(1 - \beta \int_0^{d(z_0)} L_0(u) du)} + \frac{\sqrt{2}\alpha\beta^2 \int_0^{d(z_0)} L_0(u) du}{d(z_0)(1 - \beta \int_0^{d(z_0)} L_0(u) du)} \\
\leq\ & \ \delta d(x_1)^3 + \frac{\lambda}{3} d(x_1)(d(x_0) + d(y_0) + d(x_1)) + \tau d(z_0) \\
<\ & \ \delta d(x_0)^3 + \frac{\lambda}{3} d(x_0)(d(x_0) + d(y_0)) + \tau d(z_0) < \delta R_0 < R,
\end{aligned}
$$

hold, where $d(z_0') = (d(x_0) + d(y_0) + d(x_1))/2$, so $x_1, y_1 \in U(p, R)$ and we also have that

$$
R_1 = \max\{\|x_1 - p\|, \|y_1 - p\|\} \leq \delta R_0,
$$

so (17.12) is satisfied. Suppose that $x_k, y_k \in U(p, R)$ and (17.12) hold for $k > 0$. By

TGNWTM for $k+1$ we get in turn that

$$\|x_{k+1} - p\| \leq \frac{\beta d(x_k)^3 \int_0^{d(x_k)} M(u)(d(x_k)-u)^2 du}{8d(x_k)^3(1-\beta \int_0^{d(x_k)} L_0(u)du)}$$

$$+\frac{\beta d(x_k)d(y_k)\int_0^{d(y_k)/2} L(u)du}{d(y_k)(1-\beta \int_0^{d(y_k)} L_0(u)du)} + \frac{\sqrt{2}\alpha\beta^2 d(z_k)\int_0^{d(z_k)} L_0(u)du}{d(z_k)(1-\beta \int_0^{d(z_k)} L_0(u)du)}$$

$$\leq \frac{\beta d(x_k)^3 \int_0^{d(x_0)} M(u)(d(x_0)-u)^2 du}{8d(x_0)^3(1-\beta \int_0^{d(x_0)} L_0(u)du)}$$

$$+\frac{\beta d(x_k)d(y_k)\int_0^{d(y_0)/2} L_0(u)du}{d(y_0)(1-\beta \int_0^{d(y_0)} L_0(u)du)} + \frac{\sqrt{2}\alpha\beta^2 d(z_k)\int_0^{d(z_0)} L_0(u)du}{d(z_0)(1-\beta \int_0^{d(z_0)} L_0(u)du)}$$

$$\leq \delta d(x_k)^3 + \lambda d(x_k)d(y_k) + \tau d(z_k) < \delta R_k < R$$

and

$$\|y_{k+1} - p\| \leq \frac{\beta d(x_{k+1})^3 \int_0^{d(x_{k+1})} M(u)(d(x_{k+1})-u)^2 du}{8d(x_{k+1})^3(1-\beta \int_0^{d(x_k)} L_0(u)du)}$$

$$+\frac{\beta d(x_{k+1})d(z_k')\int_0^{d(z_k')/2} L(u)du}{d(z_k')(1-\beta \int_0^{d(z_k)} L_0(u)du)} + \frac{\sqrt{2}\alpha\beta^2 d(z_k)\int_0^{d(z_k)} L(u)du}{d(z_k)(1-\beta \int_0^{d(z_k)} L_0(u)du)}$$

$$\leq \frac{\beta d(x_{k+1})^3 \int_0^{d(x_0)} M(u)(d(x_0)-u)^2 du}{8d(x_0)^3(1-\beta \int_0^{d(x_0)} L_0(u)du)}$$

$$+\frac{\beta d(x_{k+1})d(z_k')\int_0^{d(z_0')} L(u)du}{d(z_0')(1-\beta \int_0^{d(z_0)} L_0(u)du)} + \frac{\sqrt{2}\alpha\beta^2 d(z_k)\int_0^{d(z_0)} L_0(u)du}{d(z_0)(1-\beta \int_0^{d(z_0)} L_0(u)du)}$$

$$\leq \delta d(x_{k+1})^3 + \frac{\lambda}{3}(d(x_k)+d(y_k)+d(x_{k+1}))d(x_{k+1}) + \tau d(z_k)$$

$$< \delta R_k < R,$$

where $d(z_k') = (d(x_k)+d(y_k)+d(x_{k+1}))/2$. Furthermore, we obtain

$$R_{k+1} = \max\{\|x_{k+1} - p\|, \|y_{k+1} - p\|\} \leq \delta R_k \leq \delta^2 R_{k-1} \leq \ldots \delta^{k+1} R_0,$$

so $x_{k+1}, y_{k+1} \in U(p,R)$, (17.10), (17.12) hold, $\lim_{k\to\infty} x_k = p$ and $\lim_{k\to\infty} y_k = p$. □

Concerning the uniqueness of the solution p we have:

Proposition 17.1. Under the conditions ($\mathscr{C}$) further suppose that

$$\frac{\beta}{R^*} \int_0^{R^*} L_0(u)(R^* - u)du + \frac{\alpha\bar{\beta}}{R^*} \int_0^{R^*} L_0(u)du < 1, \qquad (17.13)$$

where $\bar{\beta} = \|[F'(p)^T F'(p)]^{-1}\|$. Then, limit point p is the only solution of problem (17.1) in $\bar{U}(p,R^*)$.

The proof follows from the corresponding one in [5] but we only use the center-Lipschitz condition.

17.2 APPLICATIONS

Remark 17.1.

(a) Set $\alpha = \|F(p)\| = 0$ in Theorem 17.1 and Proposition 17.1 to obtain the results in the case of zero residual.
(b) If L_0, L, M are constants, then we can obtain results of special cases.
(c) In the literature functions L_1 and M_1 are used instead of L and M, respectively [3, 5, 8, 9, 12]. Let us compare ratios and ball of convergence. Notice that in view of (17.3), (17.6), (17.7) and (17.9), we have

$$\mu(L_0, L, M)(t) \leq \mu(L_1, L_1, M_1)(t) \tag{17.14}$$

and

$$\gamma(L_0, L, M)(t) \leq \gamma(L_1, L_1, M_1)(t) \tag{17.15}$$

so

$$R^*(L_1, L_1, M_1)(t) \leq R^*(L_0, L, M)(t). \tag{17.16}$$

Therefore, our radius of convergence is larger and our ratio of convergence is smaller. Moreover the information on the location of the solution p is more precise, since only L is used in (17.13) [9]. Notice that these advantages are obtained under the same computational cost, since in practice the computation of L_1 and M_1 require the computation of the rest of the functions L_0, L and M as special cases.

In particular, we have by the error estimates that for $\alpha = 0, \tau = 0$ and

$$\begin{aligned}
d(x_{k+1}) &\leq d(x_k)(\delta d(x_k)^2 + \lambda d(y_k)) \\
d(y_{k+1}) &\leq d(x_{k+1})[\delta d(x_{k+1})^2 + \frac{\lambda}{3}(d(x_k) + d(x_{k+1}) + d(y_k))] \\
&\leq d(x_{k+1})[(\delta d(x_k) + \frac{2\lambda}{3})d(x_k) + \frac{\lambda d(y_k)}{3}] \\
&\leq d(x_{k+1})d(x_k)(\delta R^* + \lambda) \\
&= d(x_{k+1})d(x_k)\ell_1.
\end{aligned}$$

It then follows that for sufficiently large k

$$\begin{aligned}
d(x_{k+1}) &\leq d(x_k)(\delta d(x_k)^2 + \lambda d(y_k)) \\
&\leq d(x_k)(\delta d(x_k)^2 + \lambda \ell_1 d(x_k)d(x_{k-1})) \\
&\leq d(x_k)^2 d(x_{k-1})(\delta + \lambda \ell_1) \\
&= d(x_k)^2 d(x_{k-1})\ell_2
\end{aligned}$$

leading to the equation

$$x^2 - 2x - 1 = 0,$$

so the order of iterative method (17.2) is the positive root of the preceding equation which is $1 + \sqrt{2}$.

Next, we present an example to show that (17.14)–(17.16) hold as strict inequalities justifying the advantages as claimed at the introduction of this study.

Example 17.1. Let $X = \mathbb{R}^3, D = \bar{U}(0,1), p = (0,0,0)^T$. Define function F on D for $w = (x,y,z)^T$ by

$$F(w) = (e^x - 1, \frac{e-1}{2}y^2 + y, z)^T.$$

Then, the Fréchet-derivative is given by

$$F'(v) = \begin{bmatrix} e^x & 0 & 0 \\ 0 & (e-1)y+1 & 0 \\ 0 & 0 & 1 \end{bmatrix}.$$

Notice that using the (17.11) conditions, we get $L_0 = e - 1, L = M = e^{1/L_0}, L_1 = M_1 = e, \beta = 1, i = j = 3, \alpha = 0$. Then

$$R^*(L_1, L_1, M_1)(t) = 0.1468 < R^*(L_0, L, M)(t) = 0.2263,$$

which justify the improvements as stated in the introduction of this chapter.

17.3   EXERCISES

1. Solve the initial value problem

$$\dot{x}(t) = 1 + \cos(x(t)), x(0) = 0,$$

using method (17.2).

2. The predator-prey population models describe the interaction of a prey population X and a predator population Y. Assume that their interaction is modeled by the system of ordinary differential equations

$$\dot{x} = x - \frac{1}{4}x^2 - \frac{1}{10}xy + 1$$

$$\dot{y} = -\frac{1}{4}y + \frac{1}{7}xy + \frac{2}{3}.$$

(Assume $x(0) = y(0) = 0$). Solve the system.

3. Solve the initial value problem

$$\dot{x} = (x(t) + t, x(0) = 1)$$

by the iteration method (17.2). Perform 3 steps.

4. Solve the Fredholm-type integral equation

$$x(t) = \sum_0^1 \frac{t \cdot s}{10} ds + 1$$

by the iteration method (17.2). Perform 3 steps.

5. Solve the Volterra-type integral equation

$$x(t) = \sum_0^t \frac{t \cdot s}{10} ds + 1$$

by the iteration method (17.2). Perform 3 steps.

6. Solve equation
$$x = \frac{\sin x}{2} + 1$$
by using algorithm (17.2).

7. Repeat the previous problem for equation

$$x = \frac{\sin x + \cos 2x}{10}.$$

REFERENCES

1. Amat. S, Hernández. M.A , and Romero. N. 2012. Semilocal convergence of a sixth order iterative method for quadratic equations. *Applied Numerical Mathematics*, 62:833–841.
2. Argyros, I.K, and Hilout. S. 2011. Extending the applicability of the Gauss- Newton method under average Lipschitz-type conditions. *Numer. Algorithms*, 58. 1:23–52.
3. Argyros, I.K, and Hilout. S. 2011. On the semilocal convergence of Werner's method for solving equations using recurrent functions. *Punj. Univ. J. Math.*, 43:19–28.
4. Argyros, I.K, and Magreñán, A.A. 2017. *Iterative Methods and Their Dynamics with Applications*, New York: CRC Press.
5. Chen. J, and Li. W. 2005. Convergence of Gauss-Newton method's and uniqueness of the solution. *Appl. Math. Comput.*, 170:686-705.
6. Dennis. J.M, and Schnabel. R.B. 1983. *Numerical Methods for Unconstrained Optimization and Nonlinear Equations*, NY: Prentice-Hall.
7. Häubler., W.M. 1986. A Kantorovich-type convergence analysis for the Gauss-Newton method. *Numer. Math.*, 48:119–125.
8. Iakymchuk. R, and Shakhno. S.M. 2014. On the convergence analysis of a two-step modification of the Gauss-Newton method. *PAMM*, 14:813–814.
9. Iakymchuk, R.P., Shakhno, S.M., and Yarmola, H.P. 2017. Convergence analysis of a two-step modification of the Gauss-Newton method and its applications. *J. Numer. and Appl. Math.*, 3(126):61–74.
10. Magreñán, A.A. 2014. A new tool to study real dynamics: The convergence plane, *Appl. Math. Comput.*, 248:29–38.
11. Silva, G. N. 2018. Local convergence of Newton's method for solving generalized equations with monotone operator. *App. Anal.*, 7(97):1094–1105.
12. Wang., X. 2000. Convergence of Newton's method and uniqueness of the solution of equations in Banach space. *IMA J. Numer. Anal.*, 20:123–134.
13. Werner., W. 1979. Über ein Verfahren der Ordnung $1 = \sqrt{2}$ zur Nullstellenbestimmung, *Numer. Math.*, 32:333–342.

18 Convergence for $m-$Step Iterative Methods

In this chapter, we present a semi-local convergence analysis of $m-$step iterative methods in order to approximate a locally unique solution for Banach space valued equations. Our analysis extends the applicability of these methods. Using the center-Lipschitz condition, we determine a more precise domain containing the iterates leading to at least as tight Lipschitz constants as well as a finer semi-local convergence analysis than in earlier studies. Numerical examples are also presented, where the convergence criteria are tested and compared favorably to existing ones.

Let $\mathscr{B}_1, \mathscr{B}_2$ be Banach spaces, $D \subseteq \mathscr{B}_1$ be a nonempty convex set and consider $H : D \longrightarrow \mathscr{B}_2$ to be a continuously Fréchet differentiable operator. Numerous phenomena in applied mathematics, computational sciences and engineering can be reduced to a nonlinear equation like

$$H(x) = 0, \tag{18.1}$$

using mathematical modeling [5, 8, 9]. The analytical or closed form solution can be obtained in rare cases. That is why most researchers and practitioners resort to some type of numerical method which generates a sequence converging to a solution x_* of equation $H(x) = 0$.

Newton's method defined for each $n = 0, 1, 2, \ldots$ by

$$x_{n+1} = x_n - H'(x_n)^{-1} H(x_n), \tag{18.2}$$

where $x_0 \in D$ is the starting point is considered by most to be the most popular method for generating a sequence $\{x_n\}$ which converges quadratically to x_* under certain sufficient convergence criteria depending on x_0, H and some Lipschitz constants [1–9]. The convergence order can be increased using one-point methods without memory which however require higher than one Fréchet derivatives which are expensive to utilize. That is why several authors have studied the convergence of $m-$step methods ($m \in \mathscr{N}$) with frozen derivatives defined for each $n = 0, 1, 2, \ldots$ and $x_0 \in D$ by

$$
\begin{aligned}
x_0 &\in D, \\
x_n^{(1)} &= x_n^{(0)} - \Gamma_n H(x_n^{(0)}), \\
x_n^{(2)} &= x_n^{(1)} - \Gamma_n H(x_n^{(1)}), \\
&\ \ \vdots \\
x_n^{(k-1)} &= x_n^{(k-2)} - \Gamma_n H(x_n^{(k-2)}), \\
x_n^{(k)} &= x_n^{(k-1)} - \Gamma_n H(x_n^{(k-1)}), \ n \geq 0,
\end{aligned}
\tag{18.3}
$$

DOI: 10.1201/9781003128915-18

where $\Gamma_n = H'(x_n)^{-1}, x_n = x_n^{(0)}$ and $x_{n+1} = x_n^{(k)}$, with $k \geq 1$. Traub [9] showed that method (18.3) is of convergence order $m + 1$. We have three types of convergence, local, semi-local and global. The convergence domain for the semi-local convergence is usually very small in these studies limiting the applicability of the method. That is why in this chapter, we introduce the center Lipschitz condition through which an at least as small set as D is defined containing the iterates $\{x_n\}$. The Lipschitz constants in this smaller set are at least as small as the Lipschitz constants in D [7]. This crucial modification leads to a finer semi-local convergence analysis. The advantages ($\mathscr{A}$) are:

(i)   An at least as large convergence domain.
(ii)   Tighter error bounds on $\|x_{n+1} - x_n\|, \|x_n - x_*\|$ leading to the computation of at least as few iterates to obtain a predetermined error tolerance.
(iii) A more precise information on the location of the solution.

Notice that advantages ($\mathscr{A}$) are obtained under the same computational effort as in earlier works, since the new Lipschitz constants are special cases of the old Lipschitz constants (see also the numerical examples).

The rest of the chapter includes the semi-local convergence of method (18.3) in Section 18.1 and the applications in Section 18.2.

18.1   SEMI-LOCAL CONVERGENCE

We first need some auxiliary results in order to study the semi-local convergence of method (18.3).

Let $\eta > 0$ and $x_0 \in D$. Define parameter $\bar{\rho}$ by

$$\bar{\rho} = \sup\{t \geq 0 : U(x_0, t\eta) \subseteq D\}. \tag{18.4}$$

Definition 18.1. We say that H' is center-Lipschitz continuous on D, if there exists $L_0 > 0$ such that for each $x \in D$

$$\|H'(x) - H'(x_0)\| \leq L_0 \|x - x_0\|. \tag{18.5}$$

Lemma 18.1. Suppose that there exists $\rho \in (1, \bar{\rho}]$ such that $x_n^{(i)} \in U(x_0, \rho\eta)$ for each $i = 1, 2, \ldots, m, m = 1, 2, \ldots, n \in \mathbb{N}$ and (18.5) holds. Moreover, suppose that $H'(x_0)^{-1} \in \mathscr{L}(\mathscr{B}_2, \mathscr{B}_1)$ and there exists $\beta > 0$ such that

$$\|H'(x_0)^{-1}\| \leq \beta \tag{18.6}$$

and

$$\rho < \frac{1}{\beta L_0 \eta}. \tag{18.7}$$

Then, method (18.3) is well defined and

$$\|H'(x_n)^{-1}\| \leq \frac{\beta}{1 - \beta L_0 \eta \rho}. \tag{18.8}$$

Definition 18.2. We say that H' is Lipschitz continuous on D, if there exists $L_1 > 0$ such that for each $x, y \in D$

$$\|H'(x) - H'(y)\| \le L_1 \|x - y\|. \tag{18.9}$$

Remark 18.1.

(a) We have by (18.5) and (18.9) that

$$L_0 \le L_1, \tag{18.10}$$

where as $\frac{L_0}{L_1}$ can be arbitrarily small [4].

(b) Lemma 18.1 improves Lemma 1 in [7, p.313], where the less precise (18.9) was used to obtain instead of (18.8) that

$$\|H'(x_n)^{-1}\| \le \frac{\beta}{1 - \beta L_1 \eta \rho}. \tag{18.11}$$

Notice that (18.8) is a tighter, upper bounds on $\|H'(x_n)^{-1}\|$ than (18.11) which improves the semi-local convergence analysis of method (18.3) and leads together with (c), Definition 18.2 and Remark 18.1 to the aforementioned advantages ($\mathscr{A}$).

(c) Define the set $D_0 := \bar{U}(x_0, \bar{\rho}\eta) \cap U(x_0, \frac{1}{\beta L_0})$. Then, we have that

$$D_0 \subseteq \bar{U}(x_0, \bar{\rho}\eta) \tag{18.12}$$

and

$$D_0 \subseteq D. \tag{18.13}$$

Definition 18.3. We say that H' is Lipschitz continuous on D_0, if there exists $L > 0$ such that for each $x, y \in D_0$

$$\|H'(x) - H'(y)\| \le L \|x - y\|. \tag{18.14}$$

we assume from now on the hypotheses of Lemma 18.1 and introduce the notation:

$$\eta_0 = \eta$$
$$\beta_0 = \beta$$
$$h_0 = \beta_0 L \eta_0.$$

Remark 18.2. It follows from (18.4), (18.7)–(18.3) that

$$L \le L_1. \tag{18.15}$$

Notice also $L = L(x_0, L_0, D_0)$ and $L_1 = L_1(x_0, D)$. The constant L_0 helps us define D_0 and consequently L. Constants L_0 and L are special cases of L_1, so no additional (to L_1) computational cost is required to find these constants. By simply replacing L_1, D by L, D_0, respectively in the corresponding results in [7], we arrive at:

Lemma 18.2. The following items hold for $i = 2, 3, \ldots, m - 1$:

(a) $\|H(x_0^{(i)})\| \le L\rho\,\eta_0\|x_0^{(i)} - x_0^{(i-1)}\|,$

(b) $\|x_0^{(i)} - x_0^{(i-1)}\| \le h_0\rho\,\|x_0^{(i-1)} - x_0^{(i-2)}\|,$

(c) $\|x_0^{(i)} - x_0^{(0)}\| \le (1 + \frac{1}{2}h_0 + \frac{1}{2}h_0^2\rho + \ldots + \frac{1}{2}h_0^{i-1}\rho^{i-2})\eta_0,$

(d) $\|H(x_0^{(m)})\| \le L\rho\,\eta\,\|x_0^{(m)} - x_0^{(m-1)}\|,$

(e) $\|x_0^{(m)} - x_0^{(m-1)}\| \le h_0\rho\,\|x_0^{(m-1)} - x_0^{(m-2)}\|$
    and

(e) $\|x_0^{(m)} - x_0^{(0)}\| \le (1 + \frac{1}{2}h_0 + \frac{1}{2}h_0^2\rho + \ldots + \frac{1}{2}h_0^{m-1}\rho^{m-2})\eta_0.$

Moreover, we need the notation:

$$
\begin{aligned}
\beta_n &= \beta_1, \\
\eta_n &= \frac{1}{2}h_{n-1}^n\rho^{m-1}\eta_{n-1} \\
h_n &= \beta_n L\eta_n,
\end{aligned}
$$

$$
T_0(\rho) = \begin{cases} 1, & \text{if } m = 1 \\ 1 + \frac{1}{2}h_0 + \frac{1}{2}h_0^2\rho + \ldots + \frac{1}{2}h_0^{m-1}\rho^{m-2}), & \text{if } m = 2, 3, \ldots \end{cases} \tag{18.16}
$$

and

$$
T_n(\rho) = \begin{cases} 1, & \text{if } m = 1 \\ 1 + \frac{h_n}{2}(1 + h_n\rho + \cdots + (h_n\rho)^{m-2}), & \text{if } m > 1. \end{cases}
$$

Lemma 18.3. Suppose that the hypotheses of Lemma 18.1 hold. Then, the following items hold $x_n^{(i)}, x_{n+1} \in U(x_n, T_n(\rho)\eta_n)$, for each $i = 1, 2, \ldots, m - 1, n = 1, 2, \ldots$

$$
\|H(x_n)^{(1)})\| \le \frac{1}{2}L\eta_n\|x_n^{(1)} - x_n^{(0)}\|,
$$

$$
H(x_n^{(i)})\| \le L\rho\,\eta_n\|x_n^{(i)} - x_n^{(i-1)}\|
$$

for each $i = 2, 3, \ldots, m$ and for $i = 1, 2, \ldots, m-1$:

$$\begin{aligned}
\|x_n^{(i+1)} - x_n^{(i)}\| &\leq h_n \rho \|x_n^{(i)} - x_n^{(i-1)}\| \\
&\leq \frac{1}{2} h_n^i \rho^{i-1} \|x_n^{(i)} - x_n^{(i-1)}\|, \\
\|x_n^{(i+1)} - x_n^0\| &\leq (1 + h_n + \frac{1}{2} h_n^2 \rho + \cdots + \frac{1}{2} h_n^i \rho^{i-1}) \eta_n, \\
\|H(x_n^{(m-1)})\| &\leq L \rho \eta_n \|x_n^{(m-1)} - x_n^{(m-2)}\|, \\
\|x_n^{(m)} - x_n^{(m-1)}\| &= \|x_{m+1} - x_n^{(m-1)}\| \\
&\leq \rho h_n \|x_n^{(m-1)} - x_n^{(m-2)}\| \\
&\leq \frac{1}{2} h_n^{m-1} \rho^{m-2} \|x_n^{(1)} - x_n^{(0)}\|,
\end{aligned}$$

and

$$\begin{aligned}
\|x_n^{(m)} - x_n^{(0)}\| &= \|x_{n+1} - x_n^{(0)}\| \\
&\leq (1 + h_n + \frac{1}{2} h_n^2 \rho + \cdots + \frac{1}{2} h_n^{m-1} \rho^{m-2}) \eta_n.
\end{aligned}$$

Lemma 18.4. Suppose that

$$1 < \rho < \frac{2 - \sqrt{2h_0}}{2h_0}. \tag{18.17}$$

Then, sequences $\{h_n\}$ and $\{s_n(\rho)\}$ are decreasing for each $n \in \mathbb{N}$ and

$$h_0 < \frac{1}{2}. \tag{18.18}$$

Notice that (18.18) is implied for consistency by $1 < \frac{2 - \sqrt{2h_0}}{2h_0}$.

Next, we present the semi-local convergence of method (18.3):

Theorem 18.1. Suppose that conditions (18.5) and (18.3) hold and for a fixed number of steps m, equation

$$\varphi(t) = T_0(t) \left[1 + \frac{h_0^m t^{m-1}}{(1 - h_0 t)(2 - h_0^m t^{m-1})} \right] - t = 0 \tag{18.19}$$

has at least one positive solution. Denote by ρ the smallest such solution. Moreover, suppose that

$$h_0 < 2(1 - h_0 \rho)^2 \tag{18.20}$$

and

$$U(x_0, \rho \eta) \subset D. \tag{18.21}$$

Then, iteration $\{x_n\}$ generated by method (18.3) is well defined, remains in $\bar{U}(x_0, \rho \eta)$ for each $n = 0, 1, 2, \ldots$, and converges to a solution $x^* \in \bar{U}(x_0, \rho \eta)$ which is unique in $D_1 = U(x_0, \frac{2}{L\beta} - \rho\eta) \cap \bar{U}(x_0, \rho\eta)$ so that

$$\|x_n^{(i)} - x_0\| \leq T_0(\rho) \left(1 + \frac{h_0^m \rho^{m-1}}{(1 - h_0\rho)(2 - h_0^m \rho^{m-1})} \right) \eta \qquad (18.22)$$

and

$$\|x_n - x^*\| \leq \frac{2T_0(\rho)(\frac{h_0}{2})^n \eta}{2 - h_0}. \qquad (18.23)$$

Remark 18.3. The corresponding to (18.19)–(18.21) conditions in [1] are given, respectively by

$$\varphi(t) = S_0(t) \left[1 + \frac{g_0^m t^{m-1}}{(1 - g_0 t)(2 - g_0^m t^{m-1})} \right] - t = 0, \qquad (18.24)$$

$$g_0 < 2(1 - g_0 R)^2, \qquad (18.25)$$

$$U(x_0, R\eta) \subset D, \qquad (18.26)$$

where R solves (18.18)

$$g_0 = \beta L_1 \eta$$

and

$$S_0(R) = \begin{cases} 1, & \text{if } m = 1 \\ 1 = \frac{1}{2}g_0 + \frac{1}{2}g_0^2 R + \ldots + \frac{1}{2}g_0^{m-1} R^{m-2}, & \text{if } m = 2, 3, \ldots. \end{cases}$$

Notice that by (18.10)

$$h_0 \leq g_0 \quad T_0(t) \leq S_0(t), \quad \varphi(t) \leq \psi(t) \text{ and } \rho \leq R \qquad (18.27)$$

hold. Hence, the error bounds in [7] (using g_0) are improved as well as the uniquenes result (using L_1).

Authors usually prefer to leave equation (18.19) as unclustered as possible. This equation gives the radius of convergence ρ for method under some smallness conditions on η which is the accuracy of the initial point required to lead to convergence. It turns out that such conditions on ρ and η are not easy to find by just looking at equations such as (18.19). One can impose stronger convergence conditions that can lead to the computation of ρ and η. Let us provide such an example.
Suppose that

$$h_0 \leq \frac{1}{2}. \qquad (18.28)$$

Choose $\rho_0 = \rho_0(\eta)$ given by

$$\rho_0 = \frac{2 - \sqrt{2h_0}}{4h_0}. \qquad (18.29)$$

Then, it is easy to see that

$$1 < \rho_0 < \frac{2 - \sqrt{2h_0}}{2h_0}, \tag{18.30}$$

and

$$(1 - h_0\rho_0)(2 - h_0^m \rho_0^{m-1}) > 0. \tag{18.31}$$

There exist $\alpha_1, \alpha_2 \in (0,1)$ such that

$$h_0\rho_0 \leq \alpha_1 < 1 \tag{18.32}$$

and

$$h_0^m \rho_0^{m-1} \leq \alpha_2 < 1, \tag{18.33}$$

so

$$\frac{1}{1 - h_0\rho_0} \leq \frac{1}{1 - \alpha_1}, \tag{18.34}$$

and

$$\frac{1}{2 - h_0^m \rho_0^{m-1}} \leq \frac{1}{1 - \alpha_2}. \tag{18.35}$$

According to the proof of Theorem 18.1, we must have

$$T_0(t)\left[1 + \frac{h_0^m t^{m-1}}{(1 - h_0 t)(2 - h_0^m t^{m-1})}\right] \leq t. \tag{18.36}$$

In view of (18.34), (18.35) and (18.36), for

$$\rho_1 = \rho_1(\eta) := T_0(\rho_0)\left[1 + \frac{\alpha_2}{(1 - \alpha_1)(1 - \alpha_2)}\right] \tag{18.37}$$

we must have

$$\rho_1 \leq \rho \tag{18.38}$$

with equality in (18.38) being the best choice.

Remark 18.4. It follows from the proof of Theorem 18.1 that the set D_0 can be replaced by D_1 provided that (18.3) is replaced by

$$\|H'(x) - H'(y)\| \leq K\|x - y\| \tag{18.39}$$

for each $x, y \in D_1$, where $D_1 = D \cap U(x_1, \frac{1}{\beta L_0} - \|F'(x_0)^{-1}F(x_0)\|)$, provided that $\beta L_0 \eta \leq 1$. Then, condition (18.31) can also be replaced by

$$\bar{h}_0 = \beta K\eta \leq \frac{1}{2}. \tag{18.40}$$

Moreover, we have that

$$K \leq L \tag{18.41}$$

and

$$h_0 \leq \frac{1}{2} \Longrightarrow \bar{h}_0 \leq \frac{1}{2} \tag{18.42}$$

but not necessary vice versa unless if $K = L$.

18.2   APPLICATIONS

Example 18.1. Let $\mathscr{B}_1 = \mathscr{B}_2 = \mathbb{R}$, $x_0 = 1$ and $D = \bar{U}(x_0, 1 - p)$ for $p \in (0, 1)$. Define function H on D by $H(x) = x^3 - p$ and consider equation

$$H(x) = 0. \tag{18.43}$$

In view of the definitions on Lipschitz constants, we have

$$L_0 = 3(3 - p), \qquad L_1 = 6(2 - p), L = 6(1 + \frac{1}{\beta L_0}), \eta = \frac{1}{3}(1 - p), \tag{18.44}$$

$$K = \frac{2(-2p^2 + 5p + 6)}{3 - p} \text{ and } \beta = \frac{1}{3}.$$

The sufficient convergence for Newton's method and also needed in Theorem 2 [7] for the convergence of method (18.3) is given by the celebrated for its clarity and simplicity Kantorovich criterion given by

$$g_0 = \beta L_1 \eta \le \frac{1}{2}. \tag{18.45}$$

Notice that $L < L_1$. Hence, we have

$$g_0 \le \frac{1}{2} \implies h_0 \le \frac{1}{2}.$$

In view of (18.44), (18.45), (18.31) and (18.39), we have

$$g_0 \le \frac{1}{2} \text{ for each } p \in [\frac{1}{2}, 1).$$

$$h_0 = \beta L \eta \le \frac{1}{2} \text{ for each } p \in (0.461983163, 1).$$

Hence, there is no guarantee that Newton's method (18.2) or method (18.3) converges to $x_* = \sqrt[3]{p}$, for values $p \in (0.461983163, \frac{1}{2})$ under the old approach. However, under our approach Newton's method (18.2) converges in the interval $(0.461983163, 1)$. Hence, we extended the applicability of Newton's method (18.2). Concerning the convergence of method (18.3), ρ and $2(1 - h_0 \rho)^2$ are given in Table 18.1.

	$p = 0.93, h_0 = 0.0692$		$p = 0.83, h_0 = 0.1656$	
	ρ	$2(1 - h_0 \rho)^2$	ρ	$2(1 - h_0 \rho)^2$
1	1.0386	1.7044	1.1106	1.3321
2	1.0401	1.7224	1.1228	1.3705
3	1.1247	1.7008	1.4484	1.1558
4	0.0444	1.9877	0.1405	1.9080
5	0.3364	1.9080	0.0839	1.9448

Table 18.1: ρ and $2(1 - h_0 \rho)^2$ values

18.3   EXERCISES

1. Solve the initial value problem

$$\dot{x}(t) = 1 + \cos(x(t)), x(0) = 0,$$

using method (18.3).

2. The predator-prey population models describe the interaction of a prey population X and a predator population Y. Assume that their interaction is modeled by the system of ordinary differential equations

$$\dot{x} = x - \frac{1}{4}x^2 - \frac{1}{10}xy + 1$$

$$\dot{y} = -\frac{1}{4}y + \frac{1}{7}xy + \frac{2}{3}.$$

(Assume $x(0) = y(0) = 0$). Solve the system.

3. Solve the initial value problem

$$\dot{x} = (x(t) + t, x(0) = 1)$$

by the iteration method (18.3). Perform 3 steps.

4. Solve the Fredholm-type integral equation

$$x(t) = \sum_0^1 \frac{t \cdot s}{10} ds + 1$$

by the iteration method (18.3). Perform 3 steps.

5. Solve the Volterra-type integral equation

$$x(t) = \sum_0^t \frac{t \cdot s}{10} ds + 1$$

by the iteration method (18.3). Perform 3 steps.

6. Solve equation

$$x = \frac{\sin x}{2} + 1$$

by using algorithm (18.3).

7. Repeat the previous problem for equation

$$x = \frac{\sin x + \cos 2x}{10}.$$

REFERENCES

1. Amat, S., Argyros, I.K., Busquier, S., and Hernández, M.A. 2018. On two high order families of frozen Newton-type methods. *Numer. Linear. Algebra*, 25:21–26.
2. Argyros, I.K., Ezquerro, J.A, Gutierrez, J.M., Hernández, M.A, and Hilout, S. 2011. On the semi-local convergence of efficient Chebychev- Secant-type method. *J. Comput. Appl. Math.*, 235:3195–3206.
3. Argyros, I.K., and George, S. 2016. Expanding the applicability of Gauss-Newton method for convex composite optimization on Riemannian manifolds using restricted convergence domain. *Journal of Nonlinear Functional Analysis*, Article ID 27.
4. Argyros, I.K., and Magreñañ, A.A. 2017. *Iterative Methods and their dynamics with applications: A Contemporary Study*. CRC Press.
5. Argyros, I.K., George, S., and Thapa, N. 2018. *Mathematical Modeling For The Solution Of Equations And Systems Of Equations With Applications*. Volume-I, N.Y.: Nova Science Publishers.
6. Argyros, I.K., George, S., and Thapa, N. 2018. *Mathematical Modeling For The Solution Of Equations And Systems Of Equations With Applications*. Volume-II, N.Y.: Nova Science Publishers.
7. Hernández, M.A., Martinez, E., and Tervel, C. 2017. Semi-local convergence of K-step iterative process and its application for solving a special kind of conservative problems. *Numer. Algor.*, 76:309–331.
8. Kantorovich, L.V., and Akilov, G.P. 1964. *Functional Analysis in Normed Spaces*. International Series of Monographs in Pure and Applied Mathematics.
9. Traub, J.F. 1982. *Iterative Methods for the Solution of Equations*. American Mathematical Society.

19 Convergence of Newton's Method on Lie Groups

In this chapter, we extend the applicability of Newton's method used to approximate a solution of a mapping involving Lie valued operators. Using our idea of the restricted convergence region, we locate a more precise set containing the Newton iterates leading to tighter majorizing sequences than before. This way and under the same computational cost as before, we show the semi-local convergence of Newton's method with the following advantages over earlier works: weaker sufficient convergence criteria, tighter error bounds on the distances involved and at least as precise information on the location of the solution.

We are concerned with the problem of approximating a zero $x_\star$ of $\mathscr{C}^1$-mapping F : $G \longrightarrow Q$, where G is a Lie group and Q the Lie algebra of G that is the tangent space $T_e\, G$ of G at e, equipped with the Lie bracket $[.,.] : Q \times Q \longrightarrow Q$ [5–7, 17, 20, 22].

The study of numerical algorithms on manifolds for solving eigenvalue or optimization problems on Lie groups [1–25] is very important in Computational Mathematics. Newton-type methods are the most popular iterative procedures used to solve equations, when these equations contain differentiable operators. A local as well as a semilocal convergence of Newton-type methods has been given by several authors under various conditions [1–25]. There are two types of convergence results: the first uses information from the domain of an operator (see the well-known Kantorovich theorem [23]); where as the second uses information only at a point (see Smale's chapter [25]). In particular, a convergence analysis of Newton's method on manifolds can be found in [?, 1–10, 12, 14–18, 20–22, 25].

Newton's method (NM) with initial point $x_0 \in G$ was first introduced by Owren and Welfert [24] in the form

$$x_{n+1} = x_n \cdot \exp\left(-dF_{x_n}^{-1} F(x_n)\right) \quad (n \geq 0). \tag{19.1}$$

NM is undoubtedly the most popular method for generating a sequence $\{x_n\}$ approximating $x_\star$.

In this chapter, we are motivated by the work in [21] and optimization considerations. The following advantages are obtained in the semi-local case $(\mathscr{A}_1)$:

(i)   Weaker sufficient convergence criteria and a larger convergence region.
(ii) Tighter error estimates on the distances involved;

and

(iii) An at least as precise information on location of the solution $x_\star$.

That is, the applicability of NM is extended.

DOI: 10.1201/9781003128915-19   **349**

The rest of the chapter is structured as follows. Section contains the necessary background on Lie groups. In Section , we present the semi-local convergence of NM method.

19.1   BACKGROUND

In this section, we re-introduce standard concepts and notations from [5,6,17,20–22], to make the chapter as self contained as possible,.

"A Lie group $(G,\cdot)$ is a Hausdorff topological group with countable bases which also has the structure of a smooth manifold such that the group product and the inversion are smooth operations in the differentiable structure given on the manifold. The dimension of a Lie group is that of the underlying manifold and we shall always assume that it is finite. The symbol e designates the identity element of G. Let Q be the Lie algebra of the Lie group G which is the tangent space T_eG of G at e, equipped with Lie bracket $[\cdot,\cdot] : Q \times Q \to Q$. In the sequel we will make use of the left translation of the Lie group G. We define for each $y \in G$

$$\begin{aligned} L_y : G &\longrightarrow G \\ z &\longrightarrow y\cdot z \end{aligned} \qquad (19.2)$$

the left multiplication in the group. The differential of L_y at e denoted by $(dL_y)_e$ determines an isomorphism of $Q = T_eG$ with the tangent space T_yG via the relation

$$(dL_y)_e(Q) = T_yG,$$

or, equivalently,

$$Q = (dL_y)_e^{-1}(T_yG) = (dL_{y-1})_y(T_yG).$$

The exponential map is noted by exp and defined by

$$\begin{aligned} \exp : Q &\longrightarrow G \\ u &\longrightarrow \exp(u), \end{aligned}$$

which is certainly the most important construct associated to G and Q. Given $u \in Q$, the left invariant vector field $X_u : y \longrightarrow (dL_y)_e(u)$ determines an one-parameter subgroup of G: $\sigma_u : \mathbb{R} \longrightarrow G$ such that

$$\sigma_u(0) = e \quad and \quad \sigma_u'(t) = X_u(\sigma_u(t)) = (dL_{\sigma_u(t)})_e(u) \quad \forall t \in \mathbb{R}. \qquad (19.3)$$

Consequently, the exponential map is defined by the relation

$$\exp(u) = \sigma_u(1).$$

Consider $F : G \longrightarrow Q = T_eG$ be a $\mathscr{C}^1$-mapping. The differential of F at a point $x \in G$ is a linear map $F_x' : T_xG \longrightarrow Q$ defined by

$$F_x'(\triangle_x) = \frac{d}{dt}F(x\cdot\exp(t((dL_{x-1})_x)(\triangle_x)))|_{t=0} \text{ for any } \triangle_x \in T_xG. \qquad (19.4)$$

The differential F'_x can be expressed via a function $dF_x : Q \longrightarrow Q$ given by

$$dF_x = (F \circ L_x)'_e = F'_x \circ (dL_x)_e.$$

Thus, by (19.4), it follows that

$$dF_x(u) = F'_x((dL_x)_e(u)) = \frac{d}{dt}F(x \cdot \exp(tu))|_{t=0} \quad \text{for any} \quad u \in Q.$$

Therefore the following lemma is clear.

Lemma 19.1. Let $x \in G$, $u \in Q$ and $t \in \mathbb{R}$. Then

$$\frac{d}{dt}F(x \cdot \exp(-tu)) = -dF_{x \cdot \exp(-tu)}(u) \tag{19.5}$$

and

$$F(x \cdot \exp(tu)) - F(x) = \int_0^t dF_{x \cdot \exp(su)}(u)\,ds. \tag{19.6}$$

19.2 SEMI-LOCAL CONVERGENCE

We shall study the semi-local convergence of NM. In the rest of the chapter we assume $\langle \cdot, \cdot \rangle$ the inner product and $\| \cdot \|$ on Q. As in [5,6,21,22] we define a distance on G for $x, y \in G$ as follows:

$$m(x,y) = \text{dist inf} \quad \{ \quad \text{dist} \sum_{i=1}^{k} \| z_i \| : \text{there exist } k \geq 1 \text{ and } z_1, \cdots, z_k \in Q$$
$$\text{such that } y = x \cdot \exp z_1 \cdots \exp z_k \}. \tag{19.7}$$

By convention $\inf \emptyset = +\infty$. It is easy to see that $m(\cdot, \cdot)$ is a distance on G and the topology induced is equivalent to the original one on G. Let $w \in G$ and $r > 0$, we denote by $B(w,r) = \{y \in G : m(w,y) < r\}$ the open ball centered at w and of radius r. Moreover, we denote the closure of $B(w,r)$ by $\overline{B}(w,r)$. Let also $L(Q)$ denotes the set of all linear operators on Q.

Let L_0, L, L_1 be nondecreasing integrable functions on $[0, \rho)$, where $\rho > 0$ is such that $\int_0^\rho (\rho - t)L_0(t)dt \geq \rho$, $\int_0^\rho (\rho - t)L(t)dt \geq \rho$ and $\int_0^\rho (\rho - t)L_1(t)dt \geq \rho$. Define parameter r by $r = \sup\{t \geq 0 : B(x_0, r) \subseteq G\}$. We need the following definitions related to functions L_0, L and L_1.

Definition 19.1. Let $x_0 \in G$ and $M : G \longrightarrow L(Q)$. Operator M satisfies the center-L_0-Lipschitz condition on $B(x_0, r)$, if

$$\|M(x.\exp v) - M(x_0)\| \leq \int_0^{d(x_0,x)} L_0(t)dt \tag{19.8}$$

holds for each $v, v_i \in Q, i = 0, 1, \ldots m, x \in B(x_0, r)$ such that $x = x_0 \exp v_1 \exp v_2 \ldots \exp v_m$ and $d(x_0, x) < r$, where $d(x_0, x) = \sum_{i=0}^{m} \| v_i \|$.

Suppose that equation

$$\int_0^{\bar{\rho}} L_0(t)dt = 1 \tag{19.9}$$

for some $\bar{\rho} > 0$ has at least one positive solution. Denoted by ρ_0 the smallest such solution. Define the set B_0 by

$$B_0 = B(x_0, r_0), \quad r_0 = \min\{r, \rho_0\}. \tag{19.10}$$

Definition 19.2. Let $x_0 \in G$ and $M : G \longrightarrow L(Q)$. Operator M satisfies the restricted L–Lipschitz condition on B_0, if

$$\|M(x.\exp v) - M(x)\| \leq \int_{d(x_0,x)}^{d(x_0,x)+\|v\|} L(t)dt \tag{19.11}$$

holds for each $x \in B_0$.

Definition 19.3. [21] Let $x_0 \in G$ and $M : G \longrightarrow L(Q)$. Operator M satisfies the L_1–Lipschitz condition on $B(x_0, r)$, if

$$\|M(x.\exp v) - M(x)\| \leq \int_{d(x_0,x)}^{d(x_0,x)+\|v\|} L_1(t)dt \tag{19.12}$$

holds for each $x \in B(x_0, r)$.

It follows from (19.8)–(19.12) that

$$L_0(t) \leq L_1(t) \tag{19.13}$$

and

$$L(t) \leq L_1(t). \tag{19.14}$$

It is worth noticing that function L_1 was used in the semi-local convergence analysis in [21]. In the present chapter tighter function L shall replace L_1, since the iterates x_n lie in the more precise ball B_0 than $B(x_0, r)$ used in [21]. This way we obtain the advantages as already stated in the introduction. These advantages are obtained under the same cost, since the computation of function L_1 requires the computation of functions L_0 and L as special cases. Notice that function L_0 is used to define function L, i.e., $L = L(L_0)$.

We suppose from now on that

$$L_0(t) \leq L(t). \tag{19.15}$$

If

$$L(t) \leq L_0(t), \tag{19.16}$$

then function L_0 can replace L in the results that follow after Lemma 19.2.

We need the auxiliary Banach-type perturbation result.

Lemma 19.2. Let $\rho \in (0, \rho_0)$ and $x_0 \in G$ be such that $d_{F_{x_0}}$ is invertible. Suppose $d_{F_{x_0}}^{-1} d_F$ satisfies the center-L_0 Lipschitz condition on $B(x_0, \rho)$. Then, d_{F_x} is invertible and

$$\|d_{F_x}^{-1} d_{F_{x_0}} d_{F_{x_0}}^{-1} d_{F_{x_0}}\| \leq \frac{1}{1 - \int_0^{d(x_0, x)} L_0(t) dt}. \tag{19.17}$$

Proof. Set $z_0 = x_0$ and $z_{i+1} = z_i.exp\, v_i$ for each $i = 0, 1, 2, \ldots k$. Using (19.8) for $M = d_{F_{x_0}} d_F$, one has that

$$\|d_{F_x}^{-1} d_{F_{x_0}} d_{F_{x_0}}^{-1} (d_{F_{z_i.exp\, v_i}} - d_{F_{z_i}})\| \leq \int_{d(z_i, x_0)}^{d(z_{i+1}, x_0)} L(t) dt \text{ for each } 0 \leq i \leq k. \tag{19.18}$$

Since $z_{k+1} = x$, we have

$$\begin{aligned}
\|d_{F_x}^{-1} d_{F_{x_0}} d_{F_{x_0}} d_{F_x} - I_G\| &= \|d_{F_{x_0}}^{-1} (d_{F_{y_k.exp\, v_k}} - d_{F_{x_0}})\| \\
&\leq \sum_{i=0}^{k} \|d_{F_{x_0}}^{-1} (d_{F_{y_i.exp\, v_i}} - d_{F_{y_i}})\| \\
&= \int_0^{d(x, x_0)} L_0(t) dt < \int_0^{\rho_0} L_0(t) dt = 1. \tag{19.19}
\end{aligned}$$

The results follows from (19.19) and the Banach Lemma on invertible operators [23]. $\qquad\square$

Remark 19.1. The corresponding Lemma 2.1 in [21] arrived at the estimate

$$\|d_{F_x}^{-1} d_{F_{x_0}} d_{F_{x_0}}^{-1} d_{F_{x_0}}\| \leq \frac{1}{1 - \int_0^{d(x_0, x)} L_1(t) dt} \tag{19.20}$$

which is less tight than (19.17) (by (19.13)).

Let

$$\eta_0 = \int_0^\rho L_0(t) t\, dt. \tag{19.21}$$

The majorizing function φ shall be used. Define the majorizing function φ by

$$\varphi(t) = \eta - t + \int_0^t L(s)(t - s) ds \text{ for each } t \in [0, \rho]. \tag{19.22}$$

Some useful results are needed:

Proposition 19.1. [21] The function φ is monotonically decreasing on $[0, \rho_0]$ and monotonically increasing on $[\rho_0, \rho]$. Moreover, if $\eta \leq \eta_0, \varphi(t) = 0$ has a unique solution respectively in $[0, \rho_0]$ and $[\rho_0, \rho]$, which are denoted by r_1 and r_2.

Let $\{t_n\}$ denote the sequence generated by Newton's method with initial data $t_0 = 0$ for φ defined for each $n = 0, 1, 2, \ldots$ by

$$t_{n=1} = t_n - \varphi'(t_n)^{-1} \varphi(t_n) \text{ for each } n = 0, 1, 2, \ldots. \tag{19.23}$$

Proposition 19.2. [21] Suppose that $\eta \leq \eta_0$. Then the sequence $\{t_n\}$ generated by (19.23) is monotonically increasing and convergens to r_1.

Suppose that $x_0 \in G$ is such that $d_{F_{x_0}}^{-1}$ exists and set $\eta := \|d_{F_{x_0}}^{-1}\|$. Let η_0 given by (19.21) and r_1 be given by Proposition 19.1.

Theorem 19.1. Suppose that $d_{F_{x_0}}^{-1} d_F$ satisfies the center L_0−Lipschitz condition on B_0, L−Lipschitz condition on $B(x_0, r_1)$ and that

$$\eta = \|d_{F_{x_0}}^{-1}\| \leq \eta_0. \tag{19.24}$$

Then, the sequence $\{x_n\}$ generated by NM with initial point x_0 is well defined and converges to a zero x_* of F. Moreover, the following items hold for each $n = 0, 1, 2, \ldots$;

$$d(x_{n+1}, x_n) \leq \|d_{F_{x_n}}^{-1} F(x_n)\| \leq t_{n+1} - t_n; \tag{19.25}$$

$$d(x_n, x_*) \leq r_1 - t_n. \tag{19.26}$$

Proof. Set $v_n = -d_{F_{x_n}}^{-1} F(x_n)$ for each $n = 0, 1, 2, \ldots$. Using induction, we shall show that each v_n is well defined and

$$\rho(x_{k+1}, x_k) \leq \|v_n\| \leq t_{k+1} - t_k \tag{19.27}$$

holds for each $n = 0, 1, 2, \ldots$. Then, the sequence $\{x_k\}$ generated by NM starting at x_0 is well defined and converges to a zero x_* of F, since, by NM

$$x_{k+1} = x_k . exp \, v_k \text{ for each } n = 0, 1, 2, \ldots.$$

Moreover, items (19.25) and (19.26) hold for each n and the proof of the theorem is completed.

Note that v_0 is well defined by assumption and $x_1 = x_0 . exp \, v_0$. Hence, $\rho(x_1, x_0) \leq \|v_0\|$. Since $\|v_0\| = \| - d_{F_{x_0}}^{-1} F(x_0)\| = \eta = t_1 - t_0$, it follows that (19.23) is true for $k = 0$. Suppose that v_k is well defined and (19.27) holds for each $n \leq k - 1$. Then

$$\sum_{i=0}^{n-1} \|v_i\| \leq t_n - t_0 = t_n < r_1 \text{ and } x_n = x_0 . exp \, v_0 . \ldots . exp \, v_{n-1}. \tag{19.28}$$

By Lemma 19.2 $d_{F_{x_k}}^{-1}$ exists and

$$\|d_{F_{x_k}}^{-1} d_{F_{x_0}}\| \leq \frac{1}{1 - \int_0^1 t_k L_0(s) ds} = -\varphi'(t_k)^{-1}. \tag{19.29}$$

Hence, v_k is well defined. Using NM and (19.6), we can write

$$
\begin{aligned}
F(x_k) &= F(x_k) - F(x_{k-1}) - d_{F_{x_k}} v_{k-1} \\
&= \int_0^1 d_{F_{x_k}} . exp \, (tv_{k-1}) v_{k-1} dt - d_{F_{x_{k-1}}} v_{k-1} \\
&= \int_0^1 [d_{F_{x_{k-1}}} . exp \, (tv_{k-1}) - d_{F_{x_{k-1}}}] v_{k-1} dt. \tag{19.30}
\end{aligned}
$$

Then, using NM, (19.11), (19.29) and (19.30), we obtain in turn that

$$
\begin{aligned}
\|d_{F_{x_0}}^{-1} F(x_k)\| &\leq \int_0^1 \|d_{F_{x_0}}^{-1}[d_{F_{x_{k-1}}} \cdot exp\,(tv_{k-1}) - d_{F_{x_{k-1}}}]\| \|v_{k-1}\| dt \\
&\leq \int_0^1 \int_{\rho(x_{k-1},x_0)}^{\rho(x_{k-1},x_0)+t\|v_{k-1}\|} L(s)ds \|v_{k-1}\| dt \\
&\leq \int_0^1 \int_{t_{k-1}}^{t_{k-1}+t(t_k-t_{k-1})} L(s)ds(t_k-t_{k-1})dt \\
&= \int_{t_{k-1}}^{t_k} L(s)(t_k-s)ds \\
&= h(t_k) - h(t_{k-1}) - \varphi'(t_{k-1})(t_k-t_{k-1}) \\
&= \varphi(t_k).
\end{aligned}
\tag{19.31}
$$

By (19.29) and (19.31), we have in turn that

$$
\begin{aligned}
\|v_k\| &= \|-d_{F_{x_k}}^{-1} F(x_k)\| \\
&\leq \|d_{F_{x_k}}^{-1} d_{F_{x_0}}\| \|d_{F_{x_0}}^{-1} F(x_k)\| \\
&\leq -\varphi'(t_k)^{-1} \varphi(t_k) \\
&= t_{k+1} - t_k.
\end{aligned}
\tag{19.32}
$$

Then, by $x_{k+1} = x_k.exp\,v_k$, we get $d(x_{k+1},x_k) \leq \|v_k\|$ which together with (19.27) and (19.32) complete the induction.

$\square$

Remark 19.2. The majorizing sequence used in [21] is defined for each $n = 0,1,2,\ldots$ by

$$
u_{n+1} = u_n - \psi'(u_n)^{-1} \psi(u_n), \quad u_0 = 0,
\tag{19.33}
$$

where

$$
\psi(t) = \eta - t + \int_0^t L_1(s)(t-s)ds \text{ for each } t \in [0,\rho]
\tag{19.34}
$$

and the convergence criterion corresponding to (19.26) is

$$
\eta \leq \bar{\eta}_0,
\tag{19.35}
$$

where

$$
\int_0^{\bar{\rho}_0} L_1(t)dt = 1 \text{ and } \bar{\eta}_0 = \int_0^{\bar{\rho}_0} L_1(t)t\,dt.
\tag{19.36}
$$

Denote also by $\bar{r}_1$ and $\bar{r}_2$ the solutions of equation $\psi(t) = 0$ corresponding to r_1 and r_2, respectively. Notice that

$$
\varphi(t) \leq \psi(t),
\tag{19.37}
$$

$$
\bar{\rho}_0 \leq \rho_0
\tag{19.38}
$$

and
$$\bar{r}_1 \le r_1. \tag{19.39}$$

It turns out that (19.24) is weaker than (19.35). As an example, suppose functions L_0, L and L_1 are constants. Then, we have

$$\bar{\rho}_0 = \frac{1}{L_1}, \rho_0 = \frac{1}{L}, \bar{\eta}_0 = \frac{1}{2L_1} \text{ and } \eta_0 = \frac{1}{2L}. \tag{19.40}$$

Therefore (19.24) and (19.35) reduce respectively to the Kantorovich criteria for NM [23]
$$2L\eta \le 1 \tag{19.41}$$
and
$$2L_1\eta \le 1. \tag{19.42}$$

Then, in this case since $L \le L_1$, we have

$$2L_1\eta \le 1 \Longrightarrow 2L\eta \le 1 \tag{19.43}$$

but not necessarily vice versa unless, if $L = L_1$. It follows from (19.31) that sequence $\{s_n\}$ defined for each $n = 0,1,2,\dots$ by

$$\begin{aligned} s_0 &= 0, s_1 = \eta, \\ s_{n+2} &= s_{n=1} + \bar{\varphi}_0'(s_n)^{-1} \int_{s_{n-1}}^{s_n} L(t)(s_n - t)dt \end{aligned} \tag{19.44}$$

is also a majorizing sequence which converges to $s_* = r_1$ (under (19.24)), where

$$\bar{\varphi}_0(t) = \eta - t + \int_0^t L_0(s)(t-s)ds. \tag{19.45}$$

Moreover, we have
$$0 \le s_n \le t_n \text{ for each } n = 0,1,2,\dots, \tag{19.46}$$

$$0 \le s_{n+2} - s_{n+1} \le t_{n+2} - t_{n+1} \text{ for each } n = 0,1,2,\dots, \tag{19.47}$$
and
$$s_* = \lim_{n \to \infty} s_n \le r_1 = \lim_{n \to \infty}. \tag{19.48}$$

Hence, tighter sequence $\{s_n\}$ can replace $\{t_n\}$ in Theorem 19.1 (under (19.24)). It turns out that sequence $\{s_n\}$ can converge under weaker than (19.41) criteria. We refer the reader to our work in [4–15] for such criteria. The same advantages are obtained, if we specialize the "L" functions to "gamma" functions [21]. Our new technique of the restricted convergence region can be used to other iterative methods. Examples where (19.15) and (19.17) hold as strict inequalities can be found in [11–15]. The local convergence analysis can be improved along the same lines (see also [5, 12–15]).

19.3 EXERCISES

1. Provide the proofs of the results in this chapter.

2. Provide examples to test the convergence criteria.

3. Provide examples where (19.13) or (19.14) are strict.

REFERENCES

1. Absil, P.A., Baker, C.G., and Gallivan, K.A. 2007. Trust-region methods on Riemannian manifolds. *Found. Comput. Math.*, 7:303–330.
2. Adler, R.L., Dedieu, J.P., Margulies, J.Y., Martens, M., and Shub, M. 2002. Newton's method on Riemannian manifolds and a geometric model for the human spine. *IMA J. Numer. Anal.*, 22:359–390.
3. Alvarez, F., Bolte, J., and Munier, J. 2008. A unifying local convergence result for Newton's method in Riemannian manifolds. *Found. Comput. Math.*, 8:197–226.
4. Argyros, I.K., 2007. An improved unifying convergence analysis of Newton's method in Riemannian manifolds. *J. Appl. Math. Comput.*, 25:345–351.
5. Argyros, I.K. 2007. On the local convergence of Newton's method on Lie groups. *PanAmer. Math. J.*, 17:101–109.
6. Argyros, I.K. 2008. A Kantorovich analysis of Newton's method on Lie groups, *J. Concrete Appl. Anal.*, 6:21–32.
7. Argyros, I.K. 2008. *Convergence and Applications of Newton-type Iterations*, New York: Springer Verlag Publ.
8. Argyros, I.K. 2008. Newton's method in Riemannian manifolds. *Rev. Anal. Numér. Théor. Approx.*, 37:119–125.
9. Argyros, I.K. 2009. Newton's method on Lie groups. *J. Appl. Math. Comput.*, 31:217–228.
10. Argyros, I.K., and Hilout, S. 2009. Newton's method for approximating zeros of vector fields on Riemannian manifolds. *J. Appl. Math. Comput.*, 29:417–427.
11. Argyros, I.K., Hilout, S. 2012. Weaker conditions for the convergence of Newton's method. *J. Complexity*, 28:364–387.
12. Argyros, I.K., and Magreñán, A.A. 2017. *Iterative Methods and Their Dynamics with Applications*, New York: CRC Press.
13. Argyros, I. K, and Magréñan, A. A. 2018, *A Contemporary Study of Iterative Methods*, New York: Elsevier (Academic Press).
14. Argyros, I. K., George, S., and Thapa, N. 2018. *Mathematical Modeling For The Solution Of Equations And Systems Of Equations With Applications*, Volume I, NY: Nova Science Publishers.
15. Argyros, I. K., George, S., and Thapa, N. 2018. *Mathematical Modeling For The Solution Of Equations And Systems Of Equations With Applications*, Volume II, NY: Nova Science Publishers.
16. Dedieu, J.P., Priouret, P., and Malajovich, G. 2003. Newton's method on Riemannian mnifolds: Convariant α theory. *IMA J. Numer. Anal.*, 23:395–419.
17. Do Carmo, M.P. 1992. *Riemannian Geometry*, Boston: Birkhauser.
18. Ferreira, O.P., and Svaiter, B.F. 2002. Kantorovich's theorem on Newton's method in Reimannian manifolds. *J. Complexity*, 18:304–329.
19. Gutiérrez, J.M., and Hernández, M.A. 2000. Newton's method under weak Kantorovich conditions. *IMA J. Numer. Anal.*, 20:521–532.
20. Hall, B.C. 2003. *Lie Groups, Lie Algebras and Representations. An Elementary Introduction, Graduate Texts in Mathematics*, 222, NY: Springer-Verlag
21. He, J., Wang, J, and Yao, J. 2013. Convergence criteria of Newton's method on Lie groups. *Fixed point Theory and Applications*, 293:1–15.
22. Helgason, S. 1978. *Differential Geometry, Lie Groups and Symmetric Spaces*. Pure and Applied Mathematics, 80, New York-London: Academic Press, Inc. Harcourt Brace Jovanovich, Publishers.

23. Kantorovich, L.V., and Akilov, G.P. 1982. *Functional Analysis in Normed Spaces*, New York: Pergamon Press.
24. Owren, B., and Welfert, B. 2000. The Newton iteration on Lie groups. *BIT*, 40:121–145.
25. Smale, S. 1986. *Newton's Method Estimates from Data at a Point, The Merging of Disciplines: New Directions in Pure, Applied and Computational Mathematics*, eds. R. Ewing, K. Gross and C. Martin, 185–196. New-York: Springer Verlag Publ.

20 The Convergence Region of m-Step Iterative Procedures

The convergence region of iterative procedures is small in general, and it becomes smaller as m increases. This problem limits the choice of starting points, and consequently the applicability of these methods. The novelty of this chapter lies in the fact that, we extend the convergence region by using specializations of the Lipschitz constants used before. Further advantages include improved error estimations and uniqueness results. The results are tested favorably to us on examples.

Let $\mathscr{B}_1$ and $\mathscr{B}_2$ stand for Banach spaces, $\Omega \subset \mathscr{B}_1$ denote a convex and open set. Consider mapping $G : \Omega \longrightarrow \mathscr{B}_2$, to be nonlinear and Fréchet differentiable. One of the most challenging tasks is to find a solution x_* of

$$G(x) = 0. \tag{20.1}$$

Solving (20.1) is needed, since many problems from various disciplines can be made to look like equation (20.1) by resorting to Mathematical modeling. The solution x_* is rarely obtained in closed or analytical form. Hence, researchers develop iterative procedures that approximate x_* provided that the initial point $x_0 \in \Omega$ is close enough to x_*. A great effort is given recently to develop high convergence order iterative procedures [1–27].

Let m be a natural number. In this chapter, we give the semi-local convergence of the $m-$step Traub or Newton-type method defined as

$$
\begin{aligned}
x_n &= y_n^{(0)} \\
y_n^{(1)} &= y_n^{(0)} - G'(y_n^{(0)})^{-1} G(y_n^{(0)}) \\
y_n^{(2)} &= y_n^{(1)} - G'(y_n^{(0)})^{-1} G(y_n^{(1)}) \\
&\quad \cdots \\
x_{n+1} &= y_n^{(m)} = y_n^{(m-1)} - G'(y_n^{(0)})^{-1} G(y_n^{(m-1)}).
\end{aligned}
\tag{20.2}
$$

Notice that if $m = 1$, we obtain Newton's method [1–16].

Method (20.2) has convergence order $m + 1$ [1, 2, 21]. But it only uses the first derivative. This is in contrast to third order methods such as Euler, Chebyshev, Halley and other methods that utilize expensive higher order than one derivatives. That explains why many works have appeared dealing with the convergence and efficiency of method (20.2) [1, 2, 26]. A major drawback of these works is the convergence region which is small in general and becomes even smaller as m increases. This problem limits the availability of initial points. Motivated by this concern, we use an extension of our technique introduced in [10] for Newton's method combined with

DOI: 10.1201/9781003128915-20

our notion of the restricted convergence region to find a strict subset of Ω also containing the iterates. This techniques gives us at least as small constants as before leading to a larger convergence region as well as to the carrying out of less computations to obtain a predetermined error accuracy. The information on the uniqueness of x_* in a neighborhood of Ω is also improved. The novelty of this chapter lies in the observation that these developments involve not further conditions because the new parameters constitute specializations of the old ones.

The rest of the chapter lays out: In Section 20.1, the convergence of method (20.2) is studied using majorizing sequences. Some applications are given in Section 20.2.

20.1 SEMI-LOCAL CONVERGENCE

We base the semi-local convergence analysis of method (20.2) on some scalar majorizing sequences.

Lemma 20.1. Let ℓ_0, ℓ and $r_0^{(1)}$ be given positive parameters. Denote by γ the unique solution in the interval $(0,1)$ of the polynomial equation

$$p(t) = 0, \qquad (20.3)$$

where

$$p(t) = \ell_0 t^{m+1} - \frac{\ell}{2}(2 - t^m - t^{m-1}). \qquad (20.4)$$

Define the scalar sequence $\{t_n\}$ for each $n = 0, 1, 2, \ldots,$ and $i = 1, 2, \ldots, m-1$ by

$$t_0 = 0, \qquad (20.5)$$

$$r^{(0)} = t_n, r_{n+1}^{(1)} = t_{n+1} + \frac{\bar{k}(t_{n+1} - t_n + r_n^{(m-1)} - t_n)(t_{n+1} - r_n^{(m-1)})}{2(1 - \ell_0(t_{n+1} - t_0))}. \qquad (20.6)$$

$$r_n^{(m)} = t_{n+1}, r_n^{(i+1)} = r_n^{(i)} + \frac{\bar{k}(r_n^{(i)} - t_n + r_n^{(i-1)} - t_n)(r_n^{(i)} - r_n^{(i-1)})}{2(1 - \ell_0(t_n - t_0))} \qquad (20.7)$$

where $\bar{k} = \begin{cases} \ell_0, & n = 0 \\ \ell, & n = 1, 2, 3 \ldots \end{cases}$. Suppose that

$$\ell_0 t_2 < 1, 0 \leq \frac{\ell(t_2 - t_1 + r_1^{(m-1)} - t_1)}{2(1 - \ell_0 t_2)} \leq \gamma < 1 - \frac{\ell_0(r^{(2)} - r_0^{(1)})}{1 - \ell_0 r_0^1}. \qquad (20.8)$$

Then, sequence $\{t_n\}$ is bounded from above by $t_{**} = r_0^{(1)} + \frac{r_0^{(2)} - r_0^{(1)}}{1-\gamma}$, nondecreasingly convergent to its unique least upper bound denoted by t_* so that $0 \leq t_* \leq t_{**}$.

Moreover, the following items hold

$$r_n^{(i)} - r_n^{(i-1)} \leq \gamma(r_n^{(i-1)} - r_n^{(i-2)}) \leq \ldots \leq \gamma^{mn+i-2}(r_0^{(2)} - r_0^{(1)}) \qquad (20.9)$$

$$r_{n+1}^{(1)} - r_n^{(m)} \leq \gamma^{m(n+1)-1}(r_0^{(2)} - r_0^{(1)}), \qquad (20.10)$$

and

$$t_n = r_n^{(0)} \leq r_1^{(1)} \leq r_n^{(2)} \leq \ldots \leq r_n^{(m-1)} \leq r_n^{(m)} = t_{n+1}. \qquad (20.11)$$

The proof is a nontrivial extension of the one, we gave for Newton's method in [10] for $m = 1$.

Proof. The intermediate value theorem together with the estimations $p(0) = -\ell < 0$ and $p(1) = \ell_0$ assures that equation $p(t) = 0$ has at least one solution in the interval $(0,1)$. By $p'(t) = (m+1)\ell_0 t^k + \frac{\ell}{2}(tm + m - 1) > 0$ for each $t \in (0,1)$, function p increases on $(0,1)$, so it crosses the $x-$ axis only one time between 0 and 1. We name the unique solution of equation $p(t) = 0$ in $(0,1)$ by γ.

Clearly, if we show estimations (20.9)–(20.11) then sequence $\{t_n\}$ shall be non-decreasing. Evidently, (20.9)–(20.11) by (20.6)–(20.7) hold true, if

$$0 \leq \frac{\ell(r_j^{(i)} - t_j + r_j^{(i-1)} - t_j)}{2(1 - \ell_0 t_j)} \leq \gamma, \qquad (20.12)$$

$$0 \leq \frac{\ell(r_j^{(m)} - t_j + r_j^{(m-1)} - t_j)}{2(1 - \ell_0 t_{j+1})} \leq \gamma \qquad (20.13)$$

and

$$t_j = r_j^{(0)} \leq r_j^{(1)} \leq r_j^{(2)} \leq \ldots \leq r_j^{(m-1)} \leq r_j^{(m)} = t_{j+1}. \qquad (20.14)$$

If $r_0^{(i)} - r_0^{(i-1)} \geq 0$, then it follows from $r_0^{(i+1)} - r_0^{(i)} = \frac{\ell_0}{2}(r_0^{(i)} + r_0^{(i-1)})(r_0^{(i)} - r_0^{(i-1)})$ that $r_0^{(i+1)} - r_0^{(i)} \geq 0$. But $r_0^{(0)} = t_0^{(0)} = t_0 = 0$ and $r_0^{(1)} > 0$, so $r_0^{(i+1)} - r_0^{(i)} \geq 0$. Then (20.14) holds for $j = 0$. Then, by (20.8), estimations (20.12) and (20.13) hold for $m = 1$ by the left hand side inequality in (20.8). Next, assume (20.12)–(20.14) hold for $m = 2, 3, \ldots, n$. Notice that

$$0 \leq \frac{\ell(r_{j-1}^{(i)} - t_{j-1} + r_{j-1}^{(i-1)} - t_{j-1})}{2(1 - \ell_0 t_{j-1})}$$

$$\leq \frac{\ell(r_j^{(i)} - t_j + r_j^{i-1} - t_j)}{2(1 - \ell_0 t_j)}$$

$$\leq \frac{\ell(r_j^{(m)} - t_j + r_j^{(m-1)} - t_j)}{2(1 - \ell_0 t_{j+1})}. \qquad (20.15)$$

Hence, it suffices to show only (20.13). We need to use the induction hypotheses to obtain an upper bound on $r_n^{(i)}$:

$$
\begin{aligned}
r_n^{(i)} &\leq r_n^{(i-1)} + \gamma^{mn+i-2}(r_0^{(2)} - r_0^{(1)}) \\
&\leq r_n^{(i-2)} + \gamma^{mn+i-3}(r_0^{(2)} - r_0^{(1)}) + \gamma^{mn+i-2}(r_0^{(2)} - r_0^{(1)}) \\
&\vdots \\
&\leq r_n^{(1)} + (\gamma^{mn} + \ldots + \gamma^{mn+i-2})(r_0^{(2)} - r_0^{(1)}) \\
&\leq r_{n-1}^{(m)} + (\gamma^{mn-1} + \ldots + \gamma^{mn+i-2})(r_0^{(2)} - r_0^{(1)}) \\
&\vdots \\
&\leq r_0^{(2)} + \gamma(r_0^{(2)} - r_0^{(1)}) + \ldots + \gamma^{mn+i-2}(r_0^{(2)} - r_0^{(1)}) \\
&= r_0^{(2)} - (r_0^{(2)} - r_0^{(1)}) + (r_0^{(2)} - r_0^{(1)}) \\
&\quad + \gamma(r_0^{(2)} - r_0^{(1)}) + \ldots + \gamma^{mn+i-2}(r_0^{(2)} - r_0^{(1)}) \\
&= r_0^{(1)} + (1 + \gamma + \ldots + \gamma^{mn+i-2})(r_0^{(2)} - r_0^{(1)}) \\
&= r_0^{(1)} + \frac{1 - \gamma^{mn+i-1}}{1 - \gamma}(r_0^{(2)} - r_0^{(1)}) \\
&\leq r_0^{(1)} + \frac{r_0^{(2)} - r_0^{(1)}}{1 - \gamma} = t_{**}.
\end{aligned} \tag{20.16}
$$

Evidently (20.13) holds, if

$$
\frac{\ell(r_0^{(2)} - r_0^{(1)})[(\gamma^{mj-1} + \ldots + \gamma^{mj+m-2}) + (\gamma^{mj-1} + \ldots + \gamma^{mj+m-3})]}{2(1 - \ell_0(r_0^{(1)} + \frac{1 - \gamma^{mj+m-1}}{1 - \gamma}(r_0^{(2)} - r_0^{(1)}))} \leq \gamma \tag{20.17}
$$

or

$$
\frac{\ell(r_0^{(2)} - r_0^{(1)})[\frac{1 - \gamma^{mj+m-1}}{1 - \gamma} - \frac{1 - \gamma^{m(j-1)+m-1}}{1 - \gamma} + \frac{1 - \gamma^{mj+m-2}}{1 - \gamma} - \frac{1 - \gamma^{m(j-1)+m-1}}{1 - \gamma}]}{2(1 - \ell_0(r_0^{(1)} + \frac{1 - \gamma^{mj+m-1}}{1 - \gamma}(r_0^{(2)} - r_0^{(1)}))} \leq \gamma \tag{20.18}
$$

or

$$
\frac{\ell(r_0^{(2)} - r_0^{(1)})\gamma^{mj-2}(2 - \gamma^n - \gamma^{n-1})\frac{1}{1-\gamma}}{2(1 - \ell_0(r_0^{(1)} + \frac{1 - \gamma^{mj+m-1}}{1 - \gamma}(r_0^{(2)} - r_0^{(1)}))} \leq 1 \tag{20.19}
$$

or

$$
\frac{\ell(r_0^{(2)} - r_0^{(1)})\gamma^{mj-2}(2 - \gamma^n - \gamma^{n-1})}{2[(1 - \ell_0 r_0^{(1)})(1 - \gamma) - \ell_0(1 - \gamma^{mj+m-1})(r_0^{(2)} - r_0^{(1)})]} \leq 1. \tag{20.20}
$$

It is helpfull to define recurrent polynomials defined on $(0, 1)$ by

$$
\begin{aligned}
h_j(t) &= \frac{\ell}{2}(r_0^{(2)} - r_0^{(1)})t^{mj-2}(2 - t^m - t^{m-1}) \\
&\quad \ell_0(1 - t^{mj+m-1})(r_0^{(2)} - r_0^{(1)}) - (1 - \ell_0 r_0^{(1)})(1 - t). \tag{20.21}
\end{aligned}
$$

Then, instead of (20.20), we can prove

$$h_j(\gamma) \leq 0. \tag{20.22}$$

To achieve this we need a computation relating two consecutive polynomials:

$$
\begin{aligned}
h_{j+1}(t) &= \frac{\ell}{2}(r_0^{(2)} - r_0^{(1)})t^{m(j+1)-2}(2 - t^m - t^{m-1}) \\
&\quad -(1 - \ell_0 r_0^{(1)})(1 - t) + \ell_0(1 - t^{m(j+1)+m-1}) - h_j(t) + h_j(t) \\
&= \frac{\ell}{2}(r_0^{(2)} - r_0^{(1)})t^{mj-2}(2 - t^m - t^{m-1}) \\
&\quad -\frac{\ell}{2}(r_0^{(2)} - r_0^{(1)})t^{mj-2}(2 - t^m - t^{m-1}) \\
&\quad +\ell_0(1 - t^{m(j+1)+m-1})(r_0^{(2)} - r_0^{(1)}) \\
&\quad -\ell_0(1 - t^{mj+j-1})(r_0^{(2)} - r_0^{(1)}) + h_j(t) \\
&= h_j(t) + p(t)(1 - t^m)t^{mj-2}(r_0^{(2)} - r_0^{(1)}). \tag{20.23}
\end{aligned}
$$

Notice that by $p(\gamma) = 0$, and (20.21)

$$
\begin{aligned}
h_{j+1}(\gamma) &= h_j(\gamma) = h_\infty(\gamma) := \lim_{j \longrightarrow \infty} h_j(\gamma) \\
&= \ell_0(r_0^{(2)} - r_0^{(1)}) - (1 - \ell_0 r_0^{(1)})(1 - \gamma). \tag{20.24}
\end{aligned}
$$

Then, (20.22) holds, if

$$h_j(\gamma) \leq 0, \tag{20.25}$$

which is true by the left hand side inequality in condition (20.8). The induction for (20.12)-(20.14) is completed

□

Define $\mathscr{L}(\mathscr{B}_1, \mathscr{B}_2) =: \{T : \mathscr{B}_1 \longrightarrow \mathscr{B}_2 \text{ is a bounded linear operator}\}$, and $S(x, \rho) := \{y \in \mathscr{B}_1 : \|x - y\| < \rho, \rho > 0\}$, and $\bar{S}(x, \rho) = \{y \in \mathscr{B}_1 : \|x - y\| \leq \rho\}$.

Let us consider conditions (B) to be used together with the previous notations in the analysis of method (20.2).

(b1) $G : \Omega \subseteq \mathscr{B}_1 \longrightarrow \mathscr{B}_2$ is differentiable in the Fréchet sense. There exist $x_0 \in \Omega$, $r_0^{(1)} > 0$ such that $G'(x_0)^{-1} \in \mathscr{L}(\mathscr{B}_2, \mathscr{B}_1)$, and

$$\|G'(x_0)^{-1}G(x_0)\| \leq r_0^{(1)}.$$

(b2)   For all $x \in \Omega$, and some $\ell_0 > 0$

$$\|G'(x_0)^{-1}(G'(x_0) - G'(x))\| \leq \ell_0\|x_0 - x\|.$$

Let $S_0 = \Omega \cap S(x_0, \frac{1}{\ell_0})$.

(b3)   For all $x, y \in S_0$, and some $\ell > 0$

$$\|G'(x_0)^{-1}(G'(y) - G'(x))\| \le \ell \|y - x\|.$$

(b4)   Hypotheses of Lemma 20.1 hold.
(b5) $\bar{S}(x_0, t_*) \subseteq \Omega$, where t_* is given in Lemma 20.1.
(b6)   There exists $u \ge t_*$ such that $\ell_0(t_* + u) < 2$. Let $S_1 = \Omega \cap \bar{S}(x_0, u)$.

Theorem 20.1. Assume conditions (B) hold. Then, the following hold $\{x_n\} \subset \bar{S}(x_0, t_*)$, $\lim_{n \to \infty} x_n = x_* \in \bar{S}(x_0, t_*)$,

$$\|x_* - x_n\| \le t_* - t_n, \tag{20.26}$$

and x_* is the only solution of equation $F(x) = 0$, in the set S_1.

The proof is an extension of the corresponding one given by us in [10] for Newton's method ($m = 1$).

Proof. Mathematical induction is used to show items

(Q1) $\|y_n^{(1)} - x_n\| \le r_n^{(1)} - t_n$,
(Qi) $\|y_n^{(i)} - y_n^{(i-1)}\| \le r_n^{(i)} - r_n^{(i-1)}$ for all $i = 2, 3, \ldots, m - 1$.
(Qk) $\|x_{n+1} - y_n^{(i-1)}\| \le t_{n+1} - r_n^{(m-1)}$.

If $n = 0$, by (b1) $\|y_0^{(1)} - x_0\| = \|G'(x_0)^{-1}G(x_0)\| \le r_0^{(1)} \le t_*$, so $y_0^{(1)} \in \bar{S}(x_0, t_*)$, and (Q1) holds for $n = 0$. By the ith step of method (20.2), $i = 2, 3 \ldots, m - 1$, (b2), we have in turn that

$$
\begin{aligned}
\|y_1^{(1)} - x_1\| &= \|(G'(x_1)^{-1}G'(x_0)(G'(x_0)^{-1}G(x_1))\| \\
&\le \frac{\|G'(x_0)^{-1}(G(x_1) - G(y_0^{(m-1)})) - G'(x_0)(x_1 - y_0^{(m-1)})\|}{1 - \ell_0 \|x_1 - x_0\|} \\
&\le \frac{\ell_0(\|x_1 - x_0\| + \|y_0^{(m-1)} - x_0\|)\|x_1 - y_0^{(m-1)}\|}{2(1 - \ell_0 \|x_1 - x_0\|)} \\
&\le \frac{\ell_0(t_1 - t_0 + r_0^{(m-1)} - t_0)(t_1 - r_0^{(m-1)})}{2(1 - \ell_0(t_1 - t_0))} \\
&= r_1^{(1)} - t_1,
\end{aligned}
$$

and

$$\|y_1^{(1)} - x_0\| \le \|y_1^{(1)} - x_1\| + \|x_1 - x_-\| \le r_1^{(1)} - t_1 + t_1 - t_0 = r_1^{(1)} \le t_*,$$

so $y_0^{(1)} \in \bar{S}(x_0, t_*)$, and (Q1) holds. Next, using (b3), we obtain by method (20.2)

$$
\begin{aligned}
\|y_1^{(i)} - y_1^{(i-1)}\| &= \|(G'(x_1)^{-1}G'(x_0)(G'(x_0)^{-1}G(y_1^{(i-1)})))\| \\
&\leq \frac{\|G'(x_0)^{-1}(G(y_1^{(i-1)}) - G(y_1^{(i-2)})) - G'(x_1)(y_1^{(i-1)} - y_0^{(i-2)})\|}{2(1 - \ell_0\|x_1 - x_0\|)} \\
&\leq \frac{\ell(r_1^{(i-1)} - t_1 + r_1^{(i-2)} - t_1)(r_1^{(i-1)} - r_1^{(i-2)})}{2(1 - \ell_0(t_1 - t_0))} \\
&= r_1^{(i)} - r_1^{(i-1)},
\end{aligned}
$$

and

$$
\begin{aligned}
&\|y_0^{(i)} - y_0^{(i-1)}\| \\
&= \|G'(x_0)^{-1}(G(y_0^{(i-1)}) - G(y_0^{(i-2)}) + G(y_0^{(i-2)}))\| \\
&= \|G'(x_0)^{-1}[G(y_0^{(i-2)}) - G(y_0^{(i-2)}) - G'(x_0)(y_0^{(i-1)} - y_0^{(i-2)})]\| \\
&\leq \int_0^1 \|G'(x_0)^{-1}(G'(y_0^{(i-2)} + \tau(y_0^{(i-1)} - y_0^{(i-2)})) - G'(x_0)(y_0^{(i-1)} - y_0^{(i-2)}))d\tau\| \\
&\leq \frac{\ell_0}{2}(\|y_0^{(i-1)} - x_0\| + \|y_0^{(i-2)} - x_0\|)\|y_0^{(i-1)} - y_0^{(i-2)}\| \\
&\leq \frac{\ell_0}{2}(r_0^{(i-1)} - t_0 + r_0^{(i-2)} - t_0)(r_0^{(i-1)} - r_0^{(i-2)}) \\
&= r_0^{(i)} - r_0^{(i-1)}, \tag{20.27}
\end{aligned}
$$

where, we also used the estimate

$$
G'(x_n)(y_n^{(i)} - y_n^{(i-1)}) = -G(y_n^{(i-1)}). \tag{20.28}
$$

Then, we get

$$
\begin{aligned}
\|y_0^{(i)} - x_0\| &\leq \|y_0^{(i)} - y_0^{(i-1)}\| + \|y_0^{(i-1)} - y_0^{(i-2)}\| \\
&\quad + \ldots + \|y_0^{(1)} - x_0\| \\
&\leq r_0^{(i)} - r_0^{(i-1)} + r_0^{(i-1)} - r_0^{(i-2)} + \ldots + r_0^{(1)} - r_0^{(0)} \\
&= r_0^{(i)} \leq t_*, \tag{20.29}
\end{aligned}
$$

so $y_0^{(i)} \in \bar{S}(x_0, t_*)$. As in (20.27) but for $i = m$, we have that

$$
\begin{aligned}
&\|y_0^{(m)} - y_0^{(m-1)}\| \\
&= \|x_1 - y_0^{(m-1)}\| \\
&\leq \int_0^1 \|G'(x_0)^{-1}(G'(y_0^{(m-1)} + \tau(x_1 - y_0^{(m-1)})) - G'(x_0)(x_1 - y_0^{(m-1)}))d\tau\| \\
&\leq \frac{\ell_0}{2}(\|y_0^{(m-1)} - x_0\| + \|x_1 - x_0\|)\|x_1 - y_0^{(m-1)}\| \\
&\leq \frac{\ell_0}{2}(r_0^{(m-1)} - t_0 + t_1 - t_0)(t_1 - r_0^{(m-1)}) \\
&= t_1 - r_0^{(m-1)},
\end{aligned} \tag{20.30}
$$

and

$$
\begin{aligned}
\|x_1 - x_0\| &\leq \|x_1 - y_0^{(m-1)}\| + \|y_0^{(m-1)} - x_0\| \\
&\leq t_1 - r_0^{(m-1)} + r_0^{(m-1)} - t_0 = t_1 \leq t_*,
\end{aligned}
$$

so $x_1 \in \bar{S}(x_0, t_*)$ and Q_k holds for $n = 0$. Next, we show Q_i for $i = 1, 2, \ldots, m$ when, $n = 1$ by induction on i. Let $v \in S_0$. Then, by (b2), we get that

$$
\|G'(x_0)^{-1}(G'(v) - G'(x_0))\| \leq \ell_0 \|v - x_0\| < \ell_0 \frac{1}{\ell_0} = 1, \tag{20.31}
$$

so $G'(v)^{-1} \in \mathcal{L}(\mathcal{B}_2, \mathcal{B}_1)$ and

$$
\|G'(v)^{-1}G'(x_0)\| \leq \frac{1}{1 - \ell_0 \|v - x_)\|}, \tag{20.32}
$$

by the Banach lemma for invertible mappings [22, 27]. In particular, (20.32) holds for $v = x_1$. Then, by (b2), (20.6) and method (20.2) (first step of the kth step), we obtain in turn that

$$
\begin{aligned}
\|y_1^{(i)} - x_)\| &\leq \|y_1^{(i)} - y_1^{(i-1)}\| + \|y_1^{(i-1)} - x_0\| \\
&\leq r_1^{(i)} - r_1^{(i-1)} + r_1^{(i-1)} - t_0 = r_1^{(i)} \leq t_*,
\end{aligned}
$$

so $Q_1^{(i)} \in \bar{S}(x_0, t_*)$, and (Q_i) holds. By using $x_n, y_n^{(1)}, y_n^{(2)}, \ldots, y_n^{(m)}$ instead of $x_1, y_1^{(1)}, y_1^{(2)}, \ldots, y_1^{(m)}$, respectively items (Qi), $i = 1, 2, \ldots, m$ hold. In view of

$$
\begin{aligned}
\|x_{n+1} - x_n\| &= \|y_n^{(m)} - y_n^{(0)}\| \\
&\leq \sum_{q=0}^{m-1} \|y_n^{(m-q)} - y_n^{(m-q-1)}\| \\
&\leq \sum_{q=0}^{m-1} (r_n^{(m-q)} - r_n^{(m-q-1)}) \\
&= r_n^{(m)} - r_n^{(0)}.
\end{aligned} \tag{20.33}
$$

That is $\{x_n\}$ is a complete sequence in a Banach space $\mathscr{B}_1$. Hence, there exists $x_* \in \bar{S}(x_0, t_*)$ such that $\lim_{n \to \infty} x_n = x_*$. Then, we have by the estimation

$$\|G'(x_0)^{-1} G(y_n^{(i)})\| \le \ell(\|x_n^{(i)} - x_n\| + \|y_n^{(i-1)} - x_n\|) \|y_n^{(i)} - y_n^{(i-1)}\| \longrightarrow 0 \text{ as } n \longrightarrow \infty,$$

that $G(x_*) = 0$. The uniqueness proof is given in [10]. $\qquad \Box$

Remark 20.1.

(a) The point t_{**} can be used in Theorem 20.1 instead of t_*, if one wants a closed form point.

(b) As already noted in [10], we use initial data until the computation of t_2. This way, we relate the semi-local convergence condition (20.8) to $r_0^{(2)} - r_0^{(1)}$ (which is smaller) instead of $r_0^{(1)} - r_0^{(0)}$ which is traditionally done. It turns out that smaller Lipschitz constants can replace the ones in conditions (B). Indeed, define set $\Omega_2 = \Omega \cap S(x_1, \frac{1}{t_0} - r_0^{(1)})$ provided that $\ell_0 r_0^{(1)} < 1$. Clearly

$$\Omega_2 \subseteq \Omega_1, \tag{20.34}$$

and can replace it in Theorem 20.1. But then, by (20.34) the new Lipschitz constant corresponding to ℓ denoted by λ will be such that

$$\lambda \le \ell. \tag{20.35}$$

Then, λ can also replace ℓ in all estimations, and the resulting analysis will be even finer (see also the numerical example).

(c) The condition in [26] using $r_0^{(1)} - r_0^{(0)}$ in (20.8) instead of $r_0^{(2)} - r_0^{(1)}$ is given by

$$0 \le \frac{\bar{\ell}(t_1 + r_0^{(m-1)})}{2(1 - \ell_0 t_1)} \le \gamma < 1 - \ell_0 r_0^{(1)}, \tag{20.36}$$

where $\bar{\ell}$ is the Lipschitz constant on Ω not on S_0 (see (b3)).

20.2 APPLICATIONS

Example 20.1. Let $\mathscr{X} = \mathscr{Y} = \mathbb{R}$, $\Omega = \bar{U}(x_0, 1 - \xi)$, $x_0 = 1$ and $\xi \in [0, \frac{1}{2})$. Define function $\mathscr{H}$ on Ω by

$$\mathscr{H}(x) = x^3 - \xi.$$

Then, we get by (20.3)–(20.8) and (20.36) that for
(i) $m = 1$, $r_0^{(1)} = \frac{1}{3}(1 - \xi)$, $\ell_0 = 3 - \xi$, $\ell = 2(1 + \frac{1}{t_0})$, $\bar{\ell} = 2(2 - \xi)$. Then conditions of Lemma 20.1 are satisfied for $I_1 = [0.42289, 0.5)$, $I_2 = [0.43289, 0.5)$.
(ii) For $m = 2$, the conditions of Lemma 20.1 are satisfied for $\xi \in I_1 = [0.496599, 0.5)$ but the earlier conditions in [26] are satisfied for $\xi \in I_2 = [0.49669, 0.5)$, and $I_2 \subset I_1$.

Hence, in both cases $m = 1, 2$ we find infinite many values of ξ for which convergence is not guaranteed under the old condition (20.36) given in [26]. That is the usage of method (20.2) is extended.

20.3   EXERCISES

1. Solve the initial value problem

$$\dot{x}(t) = 1 + \cos(x(t)), x(0) = 0.$$

using method (20.2).

2. The predator-prey population models describe the interaction of a prey population X and a predator population Y. Assume that their interaction is modeled by the system of ordinary differential equations

$$\dot{x} = x - \frac{1}{4}x^2 - \frac{1}{10}xy + 1$$

$$\dot{y} = -\frac{1}{4}y + \frac{1}{7}xy + \frac{2}{3}.$$

(Assume $x(0) = y(0) = 0$). Solve the system.

3. Solve the initial value problem

$$\dot{x} = (x(t) + t, x(0) = 1)$$

by the iteration method (20.2). Perform 3 steps.

4. Solve the Fredholm-type integral equation

$$x(t) = \sum_0^1 \frac{t \cdot s}{10} ds + 1$$

by the iteration method (20.2). Perform 3 steps.

5. Solve the Volterra-type integral equation

$$x(t) = \sum_0^t \frac{t \cdot s}{10} ds + 1$$

by the iteration method (20.2). Perform 3 steps.

6. Solve equation

$$x = \frac{\sin x}{2} + 1$$

by using algorithm (20.2).

7. Repeat the previous problem for equation

$$x = \frac{\sin x + \cos 2x}{10}.$$

REFERENCES

1. Amat, S., Argyros, I.K., Busquier, S., and Herńandez-Veŕon, M.A. 2018. On two high-order families of frozen newton-type methods. *Numerical Linear Algebra with Applications*, 25(1).

2. Amat, S., Berḿudez, C., Herńandez-Veŕon, M.A., and Martinez., E. 2016. On an efficient k-step iterative method for nonlinear equations. *Journal of Computational and Applied Mathematics*, 302:258–271.

3. Amat, S., Busquier, S., Berḿudez, C., and Plaza, S., 2012. On two families of high order newton type methods. *Applied Mathematics Letters*, 25(12):2209–2217.

4. Amat, S., Busquier, S., and Plaza, S. 2004. Review of some iterative root-finding methods from a dynamical point of view. *Scientia*, 10(3):35.

5. Argyros, I.K. 2007. *Computational theory of iterative methods*, volume 15. Elsevier.

6. Argyros, I.K. 2008. On the semilocal convergence of a fast two-step newton method. *Revista Colombiana de Matematicas*, 42(1):15–24.

7. Argyros, I.K., George, S., and Thapa, N. 2018. *Mathematical Modeling For The Solution Of Equations And Systems Of Equations With Applications*, Volume-I, NY: Nova Science Publisher.

8. Argyros, I.K., George, S., and Thapa, N. 2018. *Mathematical Modeling For The Solution Of Equations And Systems Of Equations With Applications*, Volume-II, NY: Nova Science Publisher.

9. Argyros, I.K., Cordero, A., Magreñán, A.A., and Torregrosa, J.R. 2017. Third-degree anomalies of traubs method. *Journal of Computational and Applied Mathematics*, 309:511–521.

10. Argyros, I.K., and Hilout, S. 2012. Weaker conditions for the convergence of newtons method. *Journal of Complexity*, 28(3):364–387.

11. Argyros, I.K., and Hilout, S. 2013. *Computational methods in nonlinear analysis: efficient algorithms, fixed point theory and applications*. World Scientific.

12. Argyros, I.K., Magreñán, A.A., Orcos, L., and Sicilia, J.A. 2017. Local convergence of a relaxed two-step newton like method with applications. *Journal of Mathematical Chemistry*, 55(7):1427–1442.

13. Argyros, I.K., and Magréñan, A.A. 2018. *A contemporary study of iterative methods*, New York: Elsevier (Academic Press).

14. Argyros, I.K., and Magréñan, A.A. 2017. *Iterative methods and their dynamics with applications*, New York: CRC Press.

15. Cordero, A., Guasp, L., and Torregrosa, J.R. 2017. Choosing the most stable members of Kou's family of iterative methods. *Journal of Computational and Applied Mathematics*.

16. Cordero, A., and Torregrosa, J.R. 2015. Low-complexity root-finding iteration functions with no derivatives of any order of convergence. *Journal of Computational and Applied Mathematics*, 275:502–515.

17. Cordero, A., Torregrosa, J.R., and Vindel, P. 2012. Study of the dynamics of third-order iterative methods on quadratic polynomials. *International Journal of Computer Mathematics*, 89(13-14):1826–1836.

18. Ezquerro, J.A., and Herńandez-Veŕon, M.A. 2014. How to improve the domain of starting points for steffensen's method. *Studies in Applied Mathematics*, 132(4):354–380.

19. Ezquerro, J.A., and Herńandez-Veŕon, M.A., Majorizing sequences for nonlinear Fredholm Hammerstein integral equations. *Studies in Applied Mathematics*..

20. Ezquerro, J.A., Grau-Śanchez, M., Herńandez-Veŕon, M.A., and Noguera, M. 2015. A family of iterative methods that uses divided differences of first and second orders. *Numerical algorithms*, 70(3):571–589.

21. Herńandez-Veŕon, M.A., Martínez, E., and Teruel, C. 2017. Semilocal convergence of a k-step iterative process and its application for solving a special kind of conservative problems. *Numerical Algorithms*, 76(2):309–331.
22. Kantlorovich, L.V., and Akilov, G.P. 1982. *Functional analysis*. Pergamon press.
23. Magreñán, A.A., and Argyros, I.K. 2016. Improved convergence analysis for Newton-like methods. *Numerical Algorithms*, 71(4):811–826.
24. Magreñán, A.A., Cordero, A., Gutiérrez, J.M., and Torregrosa, J.R. 2014. Real qualitative behavior of a fourth-order family of iterative methods by using the convergence plane. *Mathematics and Computers in Simulation*, 105:49–61.
25. Magreñán, A.A., and Argyros, I.K., Two-step newton methods. *Journal of Complexity*, 30(4):533–553.
26. Moccari, M., and Lofti, T. 2020. Using majorizing sequences for the semi-local convergence of a high-order and multipoint iterative method along with stability analysis, Journal of Mathematical Extension.
27. Potra, F.A., and Pták, V. 1984. *Nondiscrete induction and iterative processes*, volume 103. Pitman Advanced Publishing Program.

21 Efficient Eighth Order Method in Banach Spaces Under Weak ω−Conditions

The aim of this chapter is to extend the applicability of a computationally efficient eighth order method for solving equations with Banach space valued operators. Using our idea of the restricted convergence region, the center-Lipschitz condition and hypotheses only on the first derivative, we present a finer local as well as a semi-local convergence analysis than in earlier similar works and methods. Applications are also provided in this chapter.

Let $\mathscr{E}_1, \mathscr{E}_2$ be Banach spaces and $\Omega \subset \mathscr{E}_1$ be a nonempty, convex and open set. Let also $\mathscr{L}(\mathscr{E}_1, \mathscr{E}_2)$ stand for the space of bounded linear operators from $\mathscr{E}_1$ into $\mathscr{E}_2$. By $U(w, \alpha)$, we denote the open ball in $\mathscr{E}_1$ with center $w \in \mathscr{E}_1$ and radius $\alpha > 0$. The ball $\bar{U}(w, \alpha)$ denote the closure of $U(w, \alpha)$. Mathematical modeling is used [1–7] to bring plethora of problems from diverse disciplines in a form like

$$F(x) = 0, \tag{21.1}$$

where $F : \Omega \longrightarrow \mathscr{E}_2$ is a continuously differentiable operator in the sense of Fréchet. Most solution methods for finding a locally unique solution x_* of equation (21.1) are iterative, where starting from an initial point x_0 a sequence is generated converging to x_* under some assumptions. Newton's method is the most popular method and converges quadratically. Chebyshev or Halley methods are cubically convergent but involve the computation of the seconf Fréchet derivative which is expensive in general. That is why numerous researchers have proposed fast second derivative free methods. However, as the convergence order increases the convergence region decreases in general and hypotheses on higher order derivatives (which do not appear in these methods) are required to show convergence, significantly limiting the applicability of these methods. In the present chapter, we address these problems.

The study of convergence of iterative methods is usually centered into two categories: semi-local and local convergence analysis. The semi-local convergence is based on the information around an initial point, to obtain conditions ensuring the convergence of these of these methods, while the local convergence is based on the information around a solution to find estimates of the computed radii of the convergence balls. Local results are important since they provide the degree of difficulty in choosing the initial points.

DOI: 10.1201/9781003128915-21

We study the local as well as the semi-local convergence analysis of eighth order method defined for each $n = 0, 1, 2, \ldots$ by

$$
\begin{aligned}
r_n &= x_n - F'(x_n)^{-1} F(x_n), \\
y_n &= \frac{1}{2}(r_n + x_n), \\
z_n &= \frac{1}{3}(4y_n - x_n), \\
u_n &= y_n + (F'(x_n) - 3F'(z_n))^{-1} F(x_n), \\
v_n &= u_n + 2(F'(x_n) - 3F'(z_n))^{-1} F(u_n) \\
x_{n+1} &= v_n + 2(F'(x_n) - 3F'(z_n))^{-1} F(v_n).
\end{aligned} \tag{21.2}
$$

The semi-local convergence analysis of method (21.2) was given in [5] using conditions up to the third derivative (H):

(h1) $\|F'(x_0)^{-1} F(x_0)\| \leq \eta$,
(h2) $\|F'(x_0)^{-1}\| \leq \beta$,
(h3) $\|F''(x)\| \leq \beta_1$ or $\|F'(y) - F'(x)\| \leq \beta_1 \|y - x\|$ for each $x, y \in \Omega$,
(h4) $\|F'''(x)\| \leq \beta_2$
(h5) $\|F'''(y) - F'''(x)\| \leq \lambda(\|y - x\|)$, where $\lambda : [0, +\infty) \longrightarrow [0, +\infty), \mu : [0, 1] \longrightarrow [0, +\infty)$ are continuous and non-decreasing functions satisfying

$$
\lambda(st) \leq \mu(s)\lambda(t), s \in [0, 1], t \in (-\infty, +\infty).
$$

Note that $F''(x)$ or $F'''(x)$ do not appear in method (21.2). Moreover, these conditions are not satisfied even for simple examples limiting the applicability of method (21.2) and other methods using similar conditions.
    Let $\mathscr{E}_1 = \mathscr{E}_2 = \mathbb{R}, \Omega = [-\frac{5}{2}, \frac{1}{2}]$. Define F on Ω by

$$
F(x) = x^3 \log x^2 + x^5 - x^4
$$

Then

$$
\begin{aligned}
F'(x) &= 3x^2 \log x^2 + 5x^4 - 4x^3 + 2x^2, \\
F''(x) &= 6x \log x^2 + 20x^3 - 12x^2 + 10x, \\
F'''(x) &= 6 \log x^2 + 60x^2 = 24x + 22.
\end{aligned}
$$

Obviously $F'''(x)$ is not bounded on Ω.
    We extend the local as well as the semi-local convergence of method (21.2) by determining a more precise than Ω region containing the iterates x_n. This way the Lipschitz constants and functions are tighter leading to:

(i)   Larger convergence region;
(ii) Tighter error bounds on the distances $\|x_{n+1} - x_n\|, \|x_n - x_*\|$;
(iii) At least as precise information on the location of the solution.
      Similarly, in the local case

(iv) Larger convergence ball leading to a larger choice of initial points, and also (ii) and (iii).

It is worth noticing that these advantages are obtained under the same computational effort, since in practice the computation of the old constants and functions require the computation of the new constants and functions as special cases.

The rest of the chapter contains the local and semi-local convergence analysis of method (21.2) in Sections 21.1 and 21.2, respectively. The applications appear in the concluding Section 21.3.

21.1   LOCAL CONVERGENCE

Let $\varphi_0 : [0, \infty) \longrightarrow [0, \infty)$ be a continuous and increasing function with $\varphi_0(0) = 0$. Suppose that equation

$$\varphi_0(t) = 1 \tag{21.3}$$

has at least one positive solution. Denote by ρ_1 the smallest such solution. Let $\varphi : [0, \rho_1) \longrightarrow [0, \infty)$ and $\psi : [0, \rho_1) \longrightarrow [0, \infty)$ be continuous and increasing with $\varphi(0) = 0$. Define functions g_1, h_1, g_2, h_2, g_3 on the interval $[0, \rho_1)$ by

$$
\begin{aligned}
g_1(t) &= \frac{\int_0^1 \varphi(\theta t)d\theta}{1 - \varphi_0(t)}, \\
h_1(t) &= g_1(t) - 1, \\
g_2(t) &= \frac{1}{2}(1 + g_1(t)), \\
h_2(t) &= g_2(t) - 1, \\
g_3(t) &= \frac{1}{3}(1 + 4g_2(t))t.
\end{aligned}
$$

We have $h_1(0) = -1$ and $h_1(t) \longrightarrow \infty$ as $t \longrightarrow \rho_1^-$. It follows by the intermediate value theorem that eqaution $h_1(t) = 0$ has at least one solution in $(0, \rho_1)$. Denote by r_1 the smallest such solution. We also get $h_2(0) = -\frac{1}{2}$ and $h_2(t) \longrightarrow \infty$ as $t \longrightarrow \rho_1^-$. Denote by r_2 the smallest solution of equation $h_2(t) = 0$ on the interval $(0, \rho_1)$. Set $r_3 = \frac{1}{3}(4g_1(r_1) + 1)r_1$.
Suppose that equation

$$p(t) = 1, \tag{21.4}$$

has at least one positive solution, where

$$p(t) = \frac{1}{2}(\varphi_0(t) + \varphi_0(g_3(t)t)). \tag{21.5}$$

Denote by ρ_2 the smallest such solution. Set $\rho = \min\{\rho_1, \rho_2\}$. Define functions g_4, h_4, g_5, h_5, g_6 and h_6 on the interval $[0, \rho)$ by

$$g_4(t) = \frac{\int_0^1 \varphi((1 - \theta)t)d\theta}{1 - \varphi_0(t)} + \frac{3(\varphi_0(t) + \varphi_0(g_1(t)t))\int_0^1 \psi(\theta t)d\theta}{4(1 - \varphi_0(t))(1 - p(t))},$$

$$h_4(t) = g_4(t) - 1,$$

$$g_5(t) = \left(1 + \frac{\int_0^1 \psi(\theta g_4(t)t)d\theta}{1 - p(t)}\right) g_5(t),$$

$$h_5(t) = g_5(t) - 1,$$

$$g_6(t) = \left(1 + \frac{\int_0^1 \psi(\theta g_5(t)t)d\theta}{1 - p(t)}\right) g_5(t),$$

and

$$h_6(t) = g_6(t) - 1.$$

We get $h_4(0) = h_5(0) = h_6(0) = -1$, $h_4(t) \longrightarrow \infty$, $h_5(t) \longrightarrow \infty$ and $h_6(t) \longrightarrow \infty$ as $t \longrightarrow \rho^-$. Denote by r_4, r_5 and r_6 the smallest solutions of equations $h_4(t) = 0, h_5(t) = 0$ and $h_6(t) = 0$ in $(0, \rho)$. Define parameter r by

$$r = \min\{r_i\}, \ i = 1, 2, 3, 4, 5, 6. \tag{21.6}$$

Then, for each $t \in [0, r)$

$$0 \le \varphi_0(t) < 1 \tag{21.7}$$

$$0 \le p(t) < 1 \tag{21.8}$$

and

$$0 \le g_i(t) < 1, \ i = 1, 2, \dots. \tag{21.9}$$

The local convergence analysis is based on the conditions (A):

(a1) $F : \Omega \longrightarrow \mathcal{E}_2$ is a continuously differentiable operator in the sense of Fréchet, and there exists $x_* \in \Omega$ such that $F(x_*) = 0$ and $F'(x_*)^{-1} \in \mathcal{L}(\mathcal{E}_2, \mathcal{E}_1)$.

(a2) There exists a continuous and increasing function $\varphi_0 : [0, \infty) \longrightarrow [0, \infty)$ with $\varphi_0(0) = 0$ such that for each $x \in \Omega$

$$\|F'(x_*)^{-1}(F'(x) - F'(x_*))\| \le \varphi_0(\|x - x_*\|).$$

Set $\Omega_0 = \Omega \cap U(x_*, \rho_1)$, where ρ_1 is given in (21.3).

(a3) There exist continuous and increasing functions $\varphi : [0, \rho_1) \longrightarrow [0, \infty)$, $\psi : [0, \rho_1) \longrightarrow [0, \infty)$ such that for each $x, y \in \Omega_0$

$$\|F'(x_*)^{-1}(F'(y) - F'(x))\| \le \varphi(\|y - x\|)$$

and

$$\|F'(x_*)^{-1}F'(x)\| \le \psi(\|x - x_*\|).$$

(a4) $\bar{U}(x_*, \bar{r})$, where $\bar{r} = \max\{r, g_3(r)r\}$ is given in (21.6).

(a5) There exists $r_* \ge r$ such that $\int_0^1 \varphi_0(\theta r_*)d\theta < 1$. Set $\Omega_1 = \Omega \cap U(x_*, r_*)$.

Next, we present the local convergence analysis of method (21.2) using the conditions (A) and the preceeding notation.

Theorem 21.1. Suppose that conditions (A) hold. Then, sequence $\{x_n\}$ generated by method (21.2) for $x_0 \in U(x_*, r) - \{x_*\}$ is well defined, remains in $U(x_*, r)$ for each $n = 0, 1, 2, \ldots$ and converges to x_*. Moreover, the following estimates hold

$$
\begin{align}
\|r_n - x_*\| &\leq g_1(\|x_n - x_*\|)\|x_n - x_*\| \leq \|x_n - x_*\| < r, & (21.10)\\
\|y_n - x_*\| &\leq g_2(\|x_n - x_*\|)\|x_n - x_*\| \leq \|x_n - x_*\|, & (21.11)\\
\|z_n - x_*\| &\leq g_3(\|x_n - x_*\|)\|x_n - x_*\| \leq r_3, & (21.12)\\
\|u_n - x_*\| &\leq g_4(\|x_n - x_*\|)\|x_n - x_*\| \leq \|x_n - x_*\| & (21.13)\\
\|v_n - x_*\| &\leq g_5(\|x_n - x_*\|)\|x_n - x_*\| \leq \|x_n - x_*\| & (21.14)
\end{align}
$$

and

$$
\|x_{n+1} - x_*\| \leq g_6(\|x_n - x_*\|)\|x_n - x_*\| \leq \|x_n - x_*\|. \tag{21.15}
$$

Furthermore, the limit point x_* is the only solution of equation $F(x) = 0$ in Ω_1, where Ω_1 is given in (a5).

Proof. Let $x \in U(x_*, r)$. Using (a1), (a2), (21.6) and (21.7), we have that

$$
\|F'(x_*)^{-1}(F'(x) - F'(x_*))\| \leq \varphi_0(\|x - x_*\|) \leq \varphi_0(r) < 1 \tag{21.16}
$$

which together with the Banach perturbation Lemma [6] shows that $F'(x_*)^{-1} \in \mathscr{L}(\mathscr{E}_2, \mathscr{E}_1)$ and

$$
\|F'(x)^{-1}F'(x_*)\| \leq \frac{1}{1 - \varphi_0(\|x - x_*\|)}. \tag{21.17}
$$

We also have that r_0, y_0 and z_0 are well defined by method (21.2) for $n = 0$. We can write

$$
r_0 - x_* = x_0 - x_* - F'(x_0)^{-1}F(x_0),
$$

so by (21.17) (for $x = x_0$), (a3), (21.6) and (21.9) (for $i = 1$), we get

$$
\begin{align}
\|r_0 - x_*\| &\leq \|F'(x_0)^{-1}F'(x_*)\| \\
&\quad \left\| \int_0^1 F'(x_*)^{-1}(F'(x_0 + \theta(x_0 - x_*)) - F'(x_0))(x_0 - x_*)d\theta \right\| \\
&\leq \frac{\int_0^1 \varphi((1 - \theta)\|x_0 - x_*\|)d\theta}{1 - \varphi_0(\|x_0 - x_*\|)}\|x_0 - x_*\| \\
&= g_1(\|x_0 - x_*\|)\|x_0 - x_*\| \\
&\leq \|x_0 - x_*\| < r, & (21.18)
\end{align}
$$

so (21.10) holds for $n = 0$ and $r_0 \in U(x_*, r)$. Then,

$$
\begin{aligned}
\|y_0 - x_*\| &= \|\frac{1}{2}(r_0 + x_0) - x_*\| \\
&= \frac{1}{2}\|r_0 - x_* + x_0 - x_*\| \\
&\leq \frac{1}{2}[\|r_0 - x_*\| + \|x_0 - x_*\|] \\
&\leq \frac{1}{2}(1 + g_1(\|x_0 - x_*\|))\|x_0 - x_*\| \\
&\leq g_2(\|x_0 - x_*\|)\|x_0 - x_*\|) \\
&\leq \|x_0 - x_*\| < r,
\end{aligned}
\tag{21.19}
$$

so (21.11) holds for $n = 0$ and $y_0 \in U(x_*, r)$. Next we get

$$
\begin{aligned}
\|z_0 - x_*\| &= \|\frac{1}{3}(4y_0 - x_0) - x_*\| \\
&\leq \frac{1}{3}[4\|y_0 - x_*\| + \|x_0 - x_*\|] \\
&\leq \frac{1}{3}[4g_2(\|x_0 - x_*\|) + 1)\|x_0 - x_*\| \\
&\leq g_3(\|x_0 - x_*\|)\|x_0 - x_*\| \leq r_3,
\end{aligned}
$$

so (21.12) holds for $n = 0$. Next, we show that $(F'(x_0) - 3F'(z_0))^{-1} \in \mathscr{L}(\mathscr{E}_2, \mathscr{E}_1)$. We have by (21.6), (21.8) and (a2) that

$$
\begin{aligned}
&\|(-2F'(x_*))^{-1}[(F'(x_0) - 3F'(z_0)) + 2F'(x_*)]\| \\
&\leq \frac{1}{2}[\|F'(x_*)^{-1}(F'(x_0) - F'(x_*))\| \\
&\quad + 3\|F'(x_*)^{-1}(F'(z_0) - F'(x_*))\|] \\
&\leq \frac{1}{2}(\varphi_0(\|x_0 - x_*\|) + 3\varphi_0(\|z_0 - x_*\|)) \\
&\leq \frac{1}{2}(\varphi_0(\|x_0 - x_*\|) + 3\varphi_0(g_3(\|x_0 - x_*\|)\|x_0 - x_*\|) \\
&= p(\|x_0 - x_*\|) < p(r) < 1,
\end{aligned}
\tag{21.20}
$$

so

$$
\|(F'(x_0) - 3F'(z_0))^{-1}F'(x_*)\| \leq \frac{1}{2(1 - p(\|x_0 - x_*\|))}
\tag{21.21}
$$

and u_0, v_0 and x_1 are well defined. Then, we can write by the fourth substep of method

(21.2) that

$$
\begin{aligned}
u_0 - x_* &= y_0 - x_* + (F'(x_0) - 3F'(z_0))^{-1}F(x_0) \\
&= \frac{1}{2}r_0 + \frac{1}{2}x_0 - x_* + (F'(x_0) - 3F'(z_0))^{-1}F(x_0) \\
&= \frac{1}{2}(x_0 - 4F'(x_0)^{-1}F(x_0)) - \frac{1}{2}x_0 - x_* \\
&\quad + (F'(x_0) - 3F'(z_0))^{-1}F(x_0) \\
&= x_0 - x_* - \frac{1}{2}F'(x_0)^{-1}F(x_0) \\
&\quad + (F'(x_0) - 3F'(z_0))^{-1}F(x_0) \\
&= x_0 - x_* - F'(x_0)^{-1}F(x_0) \\
&\quad + [\frac{1}{2}(F'(x_0))^{-1} + (F'(x_0) - 3F'(z_0))^{-1}]F(x_0) \\
&= x_0 - x_* - F'(x_0)^{-1}F(x_0) \\
&\quad + F'(x_0)^{-1}[\frac{1}{2}(F'(x_0) - 3F'(z_0)) + F'(x_0)](F'(x_0) - 3F'(z_0))^{-1}F(x_0) \\
&= x_0 - x_* - F'(x_0)^{-1}F(x_0) \\
&\quad + \frac{3}{2}F'(x_0)^{-1}[(F'(x_0) - F'(x_*)) + (F'(x_*) - F'(z_0))] \\
&\quad (F'(x_0) - 3F'(z_0))^{-1}F(x_0).
\end{aligned}
\tag{21.22}
$$

In view of (21.6), (21.9) (for $i = 2$), (21.16), (21.18), (21.21), (21.22) and the triangle inequality, we obtain in turn that

$$
\begin{aligned}
\|u_0 - x_*\| &\leq \left[\frac{\int_0^1 \varphi((1-\theta)\|x_0 - x_*\|)d\theta}{1 - \varphi_0(\|x_0 - x_*\|)} \right. \\
&\quad \left. \frac{3}{4} \frac{(\varphi_0(\|x_0 - x_*\|) + \varphi_0(\|z_0 - x_*\|))\int_0^1 \psi(\theta\|x_0 - x_*\|)d\theta}{(1 - \varphi_0(\|x_0 - x_*\|))(1 - p(\|x_0 - x_*\|))} \right] \|x_0 - x_*\| \\
&= g_4(\|x_0 - x_*\|)\|x_0 - x_*\| \leq \|x_0 - x_*\|,
\end{aligned}
\tag{21.23}
$$

which shows (21.13) for $n = 0$ and $u_0 \in U(x_*, r)$. By the fifth substep of method (21.2) for $n = 0$, (21.6), (21.9) (for $i = 3$), (21.21) and (21.23), we have in turn that

$$
\begin{aligned}
\|v_0 - x_*\| &\leq \|v_0 - x_*\| + 2\|(F'(x_0) - 3F'(z_0))^{-1}F(u_0)\| \\
&\leq \left(1 + \frac{\int_0^1 \psi(\theta\|u_0 - x_*\|)d\theta}{1 - p(\|x_0 - x_*\|)}\right) \|u_0 - x_*\| \\
&\leq \left(1 + \frac{\int_0^1 \psi(\theta g_4(\|x_0 - x_*\|))d\theta}{1 - p(\|x_0 - x_*\|)}\right) \|u_0 - x_*\| \\
&\leq g_5(\|x_0 - x_*\|)\|x_0 - x_*\| \leq \|x_0 - x_*\|,
\end{aligned}
\tag{21.24}
$$

which shows (21.14) for $n = 0$ and $v_0 \in U(x_*, r)$. Similarly, by the last substep of method (21.2) for $n = 0$, (21.6), (21.9) (for $i = 4$), (21.21) and (21.24), we get in turn that

$$
\begin{aligned}
\|x_1 - x_*\| &\leq \|v_0 - x_*\| + 2\|(F'(x_0) - 3F'(z_0))^{-1}F(v_0)\| \\
&\leq \left(1 + \frac{\int_0^1 \psi(\theta\|v_0 - x_*\|)d\theta}{1 - p(\|x_0 - x_*\|)}\right)\|v_0 - x_*\| \\
&\leq g_6(\|x_0 - x_*\|)\|x_0 - x_*\| \leq \|x_0 - x_*\|,
\end{aligned}
\tag{21.25}
$$

which shows (21.15) for $n = 0$ and $x_1 \in U(x_*, r)$. The induction for (21.12)-(21.15) is terminated, if we replace $x_0, r_0, y_0, z_0, u_0, v_0, x_1$ by $x_m, r_m, y_m, z_m, u_m, v_m, x_{m+1}$ in the preceding estimates. Then, from the inequation

$$
\|x_{m+1} - x_*\| \leq c\|x_m - x_*\| < r
\tag{21.26}
$$

where $c = g_6(\|x_0 - x_*\|) \in [0, 1)$, we conclude that $\lim_{m \to \infty} x_m = x_*$ and $x_{m+1} \in U(x_*, r)$. Let $y_* \in \Omega_1$ with $F(y_*) = 0$. Set $Q = \int_0^1 F'(x_* + \theta(y_* - x_*))d\theta$. Using (a5), we obtain in turn that

$$
\begin{aligned}
\|F'(x_*)^{-1}(Q - F'(x_*))\| &\leq \int_0^1 \varphi_0(\theta\|y_* - x_*\|)d\theta \\
&\leq \int_0^1 \varphi_0(\theta r_*)d\theta < 1,
\end{aligned}
\tag{21.27}
$$

so $Q^{-1} \in \mathscr{L}(\mathscr{E}_2, \mathscr{E}_1)$. Finally, from the identity

$$
0 = F(y_*) - F(x_*) = Q(y_* - x_*),
\tag{21.28}
$$

we conclude that $x_* = y_*$.

$\square$

21.2   SEMI-LOCAL CONVERGENCE

The semi-local convergence of method (21.2) given in [5] is extended as follows. We first modify conditions (H) to weaker conditions (C):

(c1)   (h1)

(c2)   (h2)

(c3) $\|F'(x) - F'(x_0)\| \leq \beta_0\|x - x_0\|$ for each $x \in \Omega$. Set $\Omega_2 = \Omega \cap U(x_0, \frac{1}{\beta_0})$

$\quad\quad\quad \|F''(x)\| \leq \bar{\beta}_1$ or $\|F'(y) - F'(x)\| \leq \bar{\beta}_1\|y - x\|$ for each $x, y \in \Omega_2$.

(c4) $\|F'''(x)\| \leq \bar{\beta}_2$ for each $x \in \Omega_2$

(c5) $\|F'''(y) - F'''(x)\| \leq \bar{\lambda}(\|y - x\|)$, for each $x, y \in \Omega_2$, where $\bar{\lambda} : [0, +\infty) \longrightarrow [0, +\infty), \bar{\mu} : [0, 1] \longrightarrow [0, +\infty)$ are continuous and non-decreasing functions satisfying

$$
\bar{\lambda}(st) \leq \bar{\mu}(s)\bar{\lambda}(t), s \in [0, 1], t \in (-\infty, +\infty).
$$

By comparing conditions (C) to conditions (H) and since

$$\Omega_2 \subseteq \Omega, \tag{21.29}$$

we see that

$$
\begin{align}
\bar{\beta}_0 &\leq \beta_1 \tag{21.30}\\
\bar{\beta}_1 &\leq \beta_1 \tag{21.31}\\
\bar{\beta}_2 &\leq \beta_2 \tag{21.32}\\
\bar{\lambda}(t) &\leq \lambda(t) \tag{21.33}
\end{align}
$$

and $\frac{\bar{\beta}_0}{\bar{\beta}_1}$ can be arbitrarily large [1]. Notice that the center Lipschitz condition is used to find $\bar{\beta}_0$ which helps with the definition of Ω_2. The iterates remain in Ω_2, which is more accurate location than Ω. Suppose that

$$\bar{\beta}_0 \leq \bar{\beta}_1, \tag{21.34}$$

otherwise, use $\bar{\beta}_0$ in place in the proofs in [5] instead of $\bar{\beta}_1$. To avoid repetitions replace $\beta_1, \beta_2, \lambda, \mu$ and rest of parameters and functions introduced in [5] by the bar functions $\bar{\beta}_1, \bar{\beta}_2, \bar{\lambda}, \bar{\mu}, \ldots$, respectively. Then, the conclusions in [5] hold in this new and extended setting leading to the improvements as already noted in the introduction of this chapter. In particular, the uniqueness ball in [5] is given by $B = U(x_0, \frac{2}{\beta_1 \beta} - K l) \cap \Omega$, whereas our uniqueness ball is given by $B_1 = U(x_0, \frac{2}{\bar{\beta}_0 \beta} - \bar{K} l) \cap \Omega$, since we use the more precise center-Lipschitz condition instead of (h3) (i.e. β_1) in the proof of the uniqueness part. But, we have

$$\frac{2}{\bar{\beta}_0 \beta} - \bar{K} l \geq \frac{2}{\beta_1 \beta} - K l,$$

so $B \subseteq B_1$. That is the uniqueness ball is extended. These improvements are obtained under the same computational effort, since in practice the computation of $\beta_1, \beta_2, \lambda, \mu, \ldots$ require the computation of $\bar{\beta}_0, \bar{\beta}_1, \bar{\beta}_2, \bar{\lambda}, \bar{\mu}, \ldots$, as special cases. Numerical examples, where Ω_2 is a strict subset of Ω or estimates (21.30)–(21.33) hold as strict inequalities can be found in [1–4].

21.3   APPLICATIONS

Example 21.1. Let $\mathscr{B}_1 = \mathscr{B}_2 = \mathbb{R}^3, \Omega = \bar{U}(0,1), x_* = (0,0,0)^T$. Define function F on Ω for $u = (x,y,z)^T$ by

$$F(u) = (e^x - 1, \frac{e-1}{2}y^2 + y, z)^T.$$

Then, the Fréchet-derivative is given by

$$
F'(v) = \begin{bmatrix} e^x & 0 & 0 \\ 0 & (e-1)y+1 & 0 \\ 0 & 0 & 1 \end{bmatrix}.
$$

Notice that using the (21.10)–(21.14), conditions, we get $\varphi_0(t) = (e-1)t, \varphi(t) = e^{\frac{1}{e-1}}t, \psi(t) = e^{\frac{1}{e-1}}$.

Then using the definition of r, we have that

$r_1 = r_2 = 0.38269191223238574472986783803208$
$r_3 = 0.45230295561228406331366613812861$
$r_4 = 0.11531385877864519517999042363954$
$r_5 = 0.04989652202561532801095012246150 9$
$r_6 = 0.02027978607366865207439765583785 6 = r$.

Example 21.2. Let $\mathscr{B}_1 = \mathscr{B}_2 = C[0,1]$, the space of continuous functions defined on $[0,1]$ and be equipped with the max norm. Let $\Omega = \overline{U}(0,1)$. Define function F on Ω by

$$F(\varphi)(x) = \varphi(x) - 5\int_0^1 x\theta\varphi(\theta)^3 d\theta. \tag{21.35}$$

We have that

$$F'(\varphi(\xi))(x) = \xi(x) - 15\int_0^1 x\theta\varphi(\theta)^2\xi(\theta)d\theta, \text{ for each } \xi \in \Omega.$$

Then, we get that $x_* = 0, \varphi_0(t) = 7.5t, \varphi(t) = 15t, \psi(t) = 2$. This way, we have that

$r_1 = r_2 = 0.06666666666666666666666666666667$
$r_3 = 0.12202661619405000270610628376744$
$r_4 = 0.02184981455588622806240906015773 4$
$r_5 = 0.00889637481134086284539908007218 4$
$r_60.00333324010218524309154264351207075 = r$.

Example 21.3. Let us return back to the motivational example. Then, we get that $\varphi_0(t) = \varphi(t) = 147t, \psi(t) = 2$. So, we obtain

$r_1 = r_2 = 0.04761904761904761904761904761904 8$
$r_3 = 0.06968847899876723372969422598544$
$r_4 = 0.01318711345538192171444880074204 8$
$r_5 = 0.00539179691393654007824931539971 66$
$r_6 = 0.00202656021499001814251128550381 51 = r$.

Example 21.4. Let $\mathscr{B}_1 = \mathscr{B}_2 = C[0,1], \Omega = \bar{U}(x_*,1)$ and consider the nonlinear integral equation of the mixed Hammerstein-type [1–4] defined by

$$x(s) = \int_0^1 G(s,t)(x(t)^{3/2} + \frac{x(t)^2}{2})dt,$$

where the kernel G is the Green's function defined on the interval $[0,1] \times [0,1]$ by

$$G(s,t) = \begin{cases} (1-s)t, & t \leq s \\ s(1-t), & s \leq t. \end{cases}$$

The solution $x_*(s) = 0$ is the same as the solution of equation (21.1), where $F : C[0,1] \longrightarrow C[0,1])$ is defined by

$$F(x)(s) = x(s) - \int_0^1 G(s,t)(x(t)^{3/2} + \frac{x(t)^2}{2})dt.$$

Notice that

$$\left\| \int_0^1 G(s,t)dt \right\| \leq \frac{1}{8}.$$

Then, we have that

$$F'(x)y(s) = y(s) - \int_0^1 G(s,t)\left(\frac{3}{2}x(t)^{1/2} + x(t)\right)dt,$$

so since $F'(x_*(s)) = I$,

$$\|F'(x_*)^{-1}(F'(x) - F'(y))\| \leq \frac{1}{8}\left(\frac{3}{2}\|x - y\|^{1/2} + \|x - y\|\right).$$

Then, we get that $\varphi_0(t) = \varphi(t) = \frac{1}{8}(\frac{3}{2}t^{1/2} + t)$, $\psi(t) = 1 + \lambda_0(t)$. So, we obtain
$r_1 = r_2 = 2.13333333333333333333333333333333333$
$r_3 = 0.886705940448744955872939499388612$
$r_4 = 0.893447716586591278442597285902597$
$r_5 = 0.389852038159823166552087059244513$
$r_6 = 0.1886265101922472708917410955105 = r.$

21.4   EXERCISES

1. Solve the initial value problem

$$\dot{x}(t) = 1 + \cos(x(t)), x(0) = 0,$$

2. The predator-prey population models describe the interaction of a prey population X and a predator population Y. Assume that their interaction is modeled by the system of ordinary differential equations

$$\dot{x} = x - \frac{1}{4}x^2 - \frac{1}{10}xy + 1$$

$$\dot{y} = -\frac{1}{4}y + \frac{1}{7}xy + \frac{2}{3}.$$

(Assume $x(0) = y(0) = 0$). Solve the system.

3. Solve the initial value problem

$$\dot{x} = (x(t) + t, x(0) = 1)$$

by the iteration method (21.2). Perform 3 steps.

4. Solve the Fredholm-type integral equation

$$x(t) = \sum_0^1 \frac{t \cdot s}{10}ds + 1$$

by the iteration method (21.2). Perform 3 steps.

5. Solve the Volterra-type integral equation

$$x(t) = \sum_0^t \frac{t \cdot s}{10} ds + 1$$

by the iteration method (21.2). Perform 3 steps.

6. Solve equation

$$x = \frac{\sin x}{2} + 1$$

by using algorithm (21.2).

7. Repeat the previous problem for equation

$$x = \frac{\sin x + \cos 2x}{10}.$$

REFERENCES

1. Argyros, I.K, Magréñan, A.A. 2018. *A Contemporary Study of Iterative Methods*. New York: Elsevier (Academic Press).
2. Argyros, I.K., and Magreñán, A.A., 2017. *Iterative Methods and Their Dynamics with Applications*. New Yor: CRC Press.
3. Argyros, I. K., George, S., and Thapa, N. 2018. *Mathematical Modeling For The Solution Of Equations And Systems Of Equations With Applications*, Volume-I, N.Y.: Nova Science Publisher.
4. Argyros, I. K., George, S., and Thapa, N. 2018. *Mathematical Modeling For The Solution Of Equations And Systems Of Equations With Applications*, Volume-II, N.Y.: Nova Science Publisher.
5. Jaiswal, J. P. 2018. Semi-local convergence of a computationally efficient eighth-order method in Banach spaces under $w-$continuity condition. *Iran. J. Sci. Technol Trans. Sci.*, 42:819–826..
6. Rheinboldt, W.C. 1978. An adaptive continuation process for solving systems of nonlinear equations. *Polish Academy of Science, Banach Ctr. Publ.*, 3(1):129–142.
7. Traub, J.F. 1982. *Iterative Methods for the Solution of Equations*, New York: Chelsea Publishing Company.

22 Schroder-Like Methods for Multiple Roots

A plethora of problems from diverse disciplines can be made to look like equation

$$F(x) = 0, \tag{22.1}$$

where $F : \Omega \subseteq S \longrightarrow S$ is differentiable function, Ω is an open and nonempty and $S = \mathbb{R}$ or $S = \mathbb{C}$. Motivated by the quadratically convergent Schröder's method [14] for multiple roots defined for all $n = 0, 1, 2, \ldots$ by

$$x_{n+1} = x_n - \frac{F'(x_n)F(x_n)}{(F'(x_n)^2 - F''(x_n)F(x))}, \tag{22.2}$$

where x_0 is an initial point, Behl et al. [4] introduced the 4-*th* order Schöder-like method of multiplicity $m > 1$ defined for all $n = 0, 1, 2, \ldots$, by

$$
\begin{aligned}
y_n &= x_n - \frac{2mF(x_n)}{m + 2F'(x_n)} \\
x_{n+1} &= x_n - \frac{2m\alpha F(x_n)}{(m-2)\beta F'(x_n) + (m+2)F'(y_n)} \sum_{i=0}^{2} \alpha_i \left(\frac{F'(y_n)}{F'(x_n)} \right)^i, \tag{22.3}
\end{aligned}
$$

where

$$
\begin{aligned}
\alpha &= -\frac{m(m^2 - 4)}{4(m\alpha_1 + 2u(m+3)\alpha_2)}, \\
\alpha_1 &\neq \frac{2q(m+3)\alpha_2}{m}, \\
\beta &= -\frac{q(m+2)[m^3\alpha_1 + 2q(m^3 + 3m^2 + 2m - 4)\alpha_2]}{m^2(m-2)(m\alpha_1 + 2q(m+3)\alpha_2)}, \\
\alpha_0 &= -\frac{q[m^4\alpha_1 + q(m^4 + 2m^3 + 4m^2 + 8m + 16)\alpha_2]}{m^2(m-2)(m\alpha_1 + 2q(m+3)\alpha_2)}, \\
q &= \left(\frac{m}{m+2} \right)^m
\end{aligned}
$$

and $\alpha_1, \alpha_2 \in S$ are free parameters.

The method was compared favorably to other fourth order methods using similar information. The defect of these methods are: no radius of convergence is known, so the initial point is a shot in the dark; no computable error bounds on the distances $\|x_{n+1} - p\|$ are given or information on the uniqueness of the root p about a certain ball of convergence.

DOI: 10.1201/9781003128915-22

In order to address all these problems, we introduce a new technique and show convergence under conditions that can easily be verified and also addressing the aforementioned concerns.

We first simplify method (22.3) and write it as an inexact single step for all $n = 0, 1, 2, \ldots$, as follows

$$x_{n+1} = x_n - \mu_n - \gamma \frac{F(x_n)}{F'(x_n)}, \tag{22.4}$$

where

$$
\mu_n = \frac{2m\alpha F(x_n)}{(m-2)\beta F'(x_n) + (m+2)F'(x_n - \frac{2mF(x_n)}{m+2F'(x_n)})}
$$

$$
\sum_{i=0}^{2} \alpha_i \left(\frac{F'(x_n - \frac{2mF(x_n)}{m+2F'(x_n)})}{F'(x_n)} \right)^i
$$

$$
-\gamma \frac{F(x_n)}{F'(x_n)}.
$$

22.1   BALL CONVERGENCE

We shall also use the following Taylor expansion with integral form reminder and other standard formuli [13, 16, 17].

Lemma 22.1. Suppose that $F(x)$ is differentiable n-times in the ball $B(x_0, r), r > 0$, and $F^{(n)}(x)$ is integrable from a to $x \in B(a, r)$. Then,

$$
\begin{aligned}
F(x) &= F(a) + F'(a)(x-a) + \frac{1}{2}F''(a)(x-a)^2 + \ldots + \frac{1}{n!}F^{(n)}(a)(x-a)^n \\
&\quad + \frac{1}{(n-1)!} \int_0^1 [F^{(n)}(a+t(x-a)) - F^{(n)}(a)](x-a)^n(1-t)^{n-1}dt,
\end{aligned}
$$

and

$$
\begin{aligned}
F'(x) &= F'(a) + F''(a)(x-a) + \frac{1}{2}F'''(a)(x-a)^2 \\
&\quad + \ldots + \frac{1}{(n-1)!}F^{(n)}(a)(x-a)^{n-1} \\
&\quad + \frac{1}{(n-2)!} \int_0^1 [F^{(n)}(a+t(x-a)) - F^{(n)}(a)](x-a)^{n-1}(1-t)^{n-2}dt.
\end{aligned}
$$

Let $\varphi_0 : \mathbb{R}_+ \cup \{0\} \longrightarrow \mathbb{R}_+ \cup \{0\}$ be a continuous and nondecreasing function with $\varphi_0(0) = 0$. Moreover, define functions $\beta_0, \beta : \mathbb{R}_+ \cup \{0\} \longrightarrow \mathbb{R}_+ \cup \{0\}$ by

$$
\beta_0(t) = (m-1)!(m-1) \int_0^1 \cdots \int_0^1 \varphi_0(t \prod_{i=1}^m \theta_i) \prod_{i=1}^m \theta_i^{m-i} d\theta_i
$$

and

$$\beta(t) = (m-1)! \int_0^1 \cdots \int_0^1 \varphi_0\left(t \prod_{i=1}^{m-1} \theta_m\right) \prod_{i=1}^{m-1} \theta_i^{m-i} d\theta_i + \beta_0(t).$$

We have that functions β_0, β are continuous and nondecreasing with $\beta_0(0) = \beta(0) = 0$. Suppose

$$\beta(t) \longrightarrow 1 \text{ as } t \longrightarrow \text{ a positive number or } +\infty \tag{22.5}$$

It follows from the intermediate value theorem that equation $\beta(t) = 1$ has solutions in $(0, +\infty)$. Denote by r_0 the smallest positive solution of equation $\beta(t) = 1$. Set $h_1(t) = 1 - \beta(t)$. Let $\varphi : [0, r_0) \longrightarrow \mathbb{R}_+ \cup \{0\}$ be a continuous and nondecreasing function satisfying $\varphi(0) = 0$. Furthermore, define functions α, h_0 and h on $[0, r_0)$ by

$$\alpha(t) = (m-1)! \int_0^1 \cdots \int_0^1 \varphi\left(t \prod_{i=1}^{m-1} \theta_i (1 - \theta_m)\right) \prod_{i=1}^{m} \theta_i^{m-i} d\theta_i d\theta_m$$

$$h_0(t) = m^{-1}\alpha(t)t + m^{-1}|m - \varepsilon|\beta_0(t) + m^{-1}\alpha(t)at^b + \beta_0(t)at^{b-1},$$

and

$$h(t) = \frac{h_0(t)}{h_1(t)} - 1,$$

where $a \geq 0$ and $b \geq 1$. We get that $h_0(0) = -1 < 0$ and $h(t) \longrightarrow +\infty$ as $t \longrightarrow r_0^-$. Denote by r the smallest solution of equation $h(t) = 0$ in $(0, r_0)$. Then, we have that for each $t \in [0, r)$

$$0 \leq \beta(t) < 1 \tag{22.6}$$

and

$$0 \leq h(t) < 1. \tag{22.7}$$

First, we show the ball convergence of method (22.4) under conditions $(\mathscr{A})$:

$(\mathscr{A}_1)$ $F : \Omega \subseteq \mathscr{B}_1 \longrightarrow \mathscr{B}_2$ is continuously m-times Fréchet-differentiable

$(\mathscr{A}_2)$ Function F has a zero p of multiplicity $m, m = 1, 2, \ldots$.

$(\mathscr{A}_3)$ There exists function $\varphi_0 : \mathbb{R}_+ \cup \{0\} \longrightarrow \mathbb{R}_+ \cup \{0\}$ continuous and nondecreasing satisfying $\varphi_0(0) = 0$ such that for each $x \in \Omega$

$$\|F^{(m)}(p)^{-1}(F^{(m)}(x) - F^{(m)}(p))\| \leq \varphi_0(\|x - y\|)$$

Let $\Omega_0 = \Omega \cup B(p, r_0)$, where r_0 is defined previously.

$(\mathscr{A}_4)$ There exists $\varphi : [0, r) \longrightarrow \mathbb{R}_+ \cup \{0\}$ continuous and nondecreasing satisfying $\varphi(0) = 0$ such that for each $x, y \in \Omega_0$

$$\|F^{(m)}(p)^{-1}(F^{(m)}(x) - F^{(m)}(y))\| \leq \varphi(\|x - y\|)$$

$(\mathscr{A}_5)$ Condition (22.5) holds.

$(\mathscr{A}_6)$ $\bar{B}(p, r) \subseteq \Omega$.

$(\mathscr{A}_7)$ $|\xi_n| \leq a|x_n - p|^b$ for all $n = 0, 1, 2, \ldots$ and some $a \geq 0, b \geq 1$.

Theorem 22.1. Suppose that the $(\mathscr{A})$ conditions hold. Then, for starting point $x_0 \in B(p,r) - \{p\}$, the sequence $\{x_n\}$ generated by IM is well defined in $B(p,r)$, remains in $B(p,r)$ for all $n = 0,1,2,\ldots$ and converges to p.

Proof. We shall show that sequence

$$\delta_n = x_n - p \tag{22.8}$$

is non-increasing and converges to zero. Using $\delta_n = x_n - p$, method (22.4) for $n = 0$, and the following formula:

$$g(x) = F[\underbrace{p,p,\ldots,p,x}_{m}], \; g_0(x) = F[\underbrace{p,p,\ldots,p,x,x}_{m}]. \tag{22.9}$$

$$F(x_0) = g(x_0)\delta_0^m, \tag{22.10}$$

and

$$F'(x_0) = [g_0(x_0)\delta_0 + mg(x_0)]\delta_0^{m-1}. \tag{22.11}$$

We can write

$$\delta_1 \; = \; \frac{g(p)^{-1}N}{g(p)^{-1}D}, \tag{22.12}$$

where

$$N = g_0(x_0)\delta_0^2 + [(m - \varepsilon)g(x_0) - g_0(x_0)\xi_0]\delta_0 - mg(x_0)\xi_0 \tag{22.13}$$

and

$$D = g_0(x_0)\delta_0 + mg(x_0). \tag{22.14}$$

In view of the definition of divided differences, we have

$$g_0(x_0)\delta_0 = F[\underbrace{p,p,\ldots,p}_{m-1},x_0,x_0] - g(x_0). \tag{22.15}$$

Then, we obtain from (22.15) that

$$
\begin{aligned}
& |1 - (mg(p))^{-1}[h_0(x_0)\delta_0 + mg(x_0)]| \\
= \; & |(mg(p))^{-1}[g_0(x_0)\delta_0 + mg(x_0) - mg(p)]| \\
= \; & (m-1)!F^{(m)}(p)^{-1}(F[\underbrace{p,p,\ldots,p}_{m-1},x_0,x_0] - g(p) + (m-1)[g(x_0) - g(p)])|.
\end{aligned}
$$

$$\tag{22.16}$$

Next, we get

$$F[\underbrace{p,p,\ldots,p}_{m-1},x_0,x_0] \tag{22.17}$$

$$= \int_0^1 \cdots \int_0^1 F^{(m)}\left(p+\delta_0\prod_{i=1}^{m-1}\theta_i\right)\prod_{i=1}^m(\theta_i^{m-1}d\theta_i),$$

$$g(x_0) = \int_0^1 \cdots \int_0^1 F^{(m)}\left(p+\delta_0\prod_{i=1}^{m}\theta_i\right)\prod_{i=1}^m(\theta_i^{m-1}d\theta_i), \tag{22.18}$$

$$g(p) = \int_0^1 \cdots \int_0^1 F^{(m)}(p)\prod_{i=1}^m(\theta_i^{m-1}d\theta_i). \tag{22.19}$$

Substituting (22.16)–(22.19) into (22.15) using condition $(\mathscr{A}_3)$, $x_0 \in B(p,r)$, and the definition of r, we get

$$|1-(mg(p))^{-1}]g_0(x_0)\delta_0 + mg(x_0)]|$$

$$= (m-1)!\Big|\int_0^1 \cdots \int_0^1 F^{(m)}(p)^{-1}\Big(F^{(m)}\big(p+\delta_0\prod_{i=1}^{m-1}\theta_i\big)-F^{(m)}(p)\Big)\prod_{i=1}^m(\theta_i^{m-i}d\theta_i)$$

$$+(m-1)F^{(m)}(p)^{-1}\Big(F^{(m)}\big(p+\delta_0\prod_{i=1}^{m-1}\theta_i\big)-F^{(m)}(p)\Big)\prod_{i=1}^m(\theta_i^{m-i}d\theta_i)\Big|$$

$$\leq (m-1)!\Big(\int_0^1 \cdots \int_0^1 \Big|F^{(m)}(p)^{-1}\big(F^{(m)}\big(p+\delta_0\prod_{i=1}^{m-1}\theta_i\big)-F^{(m)}(p)\big)\Big|\prod_{i=1}^m(\theta_i^{m-i}d\theta_i)$$

$$+(m-1)\int_0^1 \cdots \int_0^1 \Big|F^{(m)}(p)^{-1}\big(F^{(m)}\big(p+\delta_0\prod_{i=1}^{m-1}\theta_i\big)-F^{(m)}(p)\big)\Big|\prod_{i=1}^m(\theta_i^{m-i}d\theta_i)\Big|$$

$$\leq (m-1)!\Big[\int_0^1 \cdots \int_0^1 \varphi_0\big(|\delta_0|\prod_{i=1}^{m-1}\theta_i\big)\prod_{i=1}^m\theta_i^{m-i}d\theta_i$$

$$+(m-1)\int_0^1 \cdots \int_0^1 \varphi_0\big(|\delta_0|\prod_{i=1}^{m}\theta_i\big)\prod_{i=1}^m\theta_i^{m-i}d\theta_i\Big]$$

$$\leq \beta(|\delta_0|) < \beta(r) < 1.$$

$$\tag{22.20}$$

It follows from the Banach perturbation lemma [1, 3] and (22.20) that, $g_0(x_0)\delta_0 + mg(x_0) \neq 0$, and

$$|(mg(p))^{-1}g_0(x_0)\delta_0 + mg(x_0))^{-1}| \leq \frac{1}{1-\beta(|\delta_0|)} < \frac{1}{1-\beta(r)}. \tag{22.21}$$

Moreover, using (22.15), (22.17), (22.18) and ($\mathscr{A}_4$), we have in turn that

$$
\begin{aligned}
&|(mg(p))^{-1}g_0(x_0)\delta_0| \\
=\ &(m-1)!\left|\int_0^1 \cdots \int_0^1 F^{(m)}(p)^{-1}(F^{(m)}(p+\delta_0\prod_{i=1}^{m-1}\theta_i) \right. \\
&\left. -F^{(m)}(p+\delta_0\prod_{i=1}^{m}\theta_i))\prod_{i=1}^{m}(\theta_i^{m-i}d\theta_i)\right| \\
=\ &(m-1)!\left|\int_0^1 \cdots \int_0^1 |F^{(m)}(p)^{-1}(F^{(m)}(p+\delta_0\prod_{i=1}^{m-1}\theta_i) \right. \\
&\left. -F^{(m)}(p+\delta_0\prod_{i=1}^{m}\theta_i))|\prod_{i=1}^{m}(\theta_i^{m-i}d\theta_i)\right| \\
\le\ &(m-1)!\int_0^1 \cdots \int_0^1 \varphi_0(|\delta_0|\prod_{i=1}^{m-1}\theta_i(1-\theta_m))\prod_{i=1}^{m}\theta_i^{m-i}d\theta_i d\theta_m \\
=\ &\alpha(|\delta_0|) < \alpha(r) < 1. \qquad\qquad (22.22)
\end{aligned}
$$

Furthermore, we have

$$
\begin{aligned}
&|g(p)^{-1}g(x_0)| \\
=\ &|g(p)^{-1}(g(x_0)-g(p))| \\
=\ &|(m-1)!F^{(m)}(p)^{-1}(m-1)(g(x_0)-g(p))| \\
=\ &(m-1)(m-1)!\int_0^1 \cdots \int_0^1 |F^{(m)}(p)^{-1}(F^{(m)}(p+\delta_0\prod_{i=1}^{m}\theta_i) \\
&-F^{(m)}(p)|\prod_{i=1}^{m}\theta_i d\theta_1 \\
\le\ &(m-1)(m-1)!\int_0^1 \cdots \int_0^1 \varphi_0(|\delta_0|)\prod_{i=1}^{m}\theta_i\prod_{i=1}^{m}\theta_i^{m-i}d\theta_i. \qquad (22.23)
\end{aligned}
$$

Using (22.20) – (22.23), we obtain that

$$
|\delta_1| \ \le\ c|\delta_0| < |\delta_0| < r,
$$

where $c = h(|\delta_0|) \in [0,1)$, so $x_1 \in B(p,r)$. By simply replacing x_0, x_1, by x_k, x_{k+1}, we arrive at

$$
|x_{k+1} - p| \le c|x_k - p| < r, \qquad\qquad (22.24)
$$

which shows $\lim_{k\longrightarrow +\infty} x_k = p$, and $x_{k+1} \in B(p,r)$.   □

Concerning the uniqueness of the solution p, we have:

Proposition 22.1. Suppose that conditions $(\mathscr{A})$ and

$$\frac{m}{(s_2-s_1)^m}\int_{s_1}^{s_2}\varphi_0(|t-s_1|)|s_2-t|^{m-1}dt < 1 \qquad (22.25)$$

for all s_1,t,s_2 with $0 \le s_1 \le t \le s_2 \le \bar{r}$ for some $\bar{r} \ge r$ hold. Then, the solution p of equation $F(x)=0$ is unique in $\Omega_0 = \Omega \cup \bar{B}(p,\bar{r})$.

Proof. Suppose that $p^* \in \Omega_0$ is a solution of equation $F(x)=0$ with $p \ne p^*$. Without loss of generality suppose $p < p^*$. We can write

$$F(p^*)-F(p) = \frac{1}{(m-1)!}\int_p^{p^*}F^{(m)}(t)(p^*-t)^{m-1}dt. \qquad (22.26)$$

Using $(\mathscr{A}_3)$ and (22.24), we obtain in turn that

$$\left| 1 - \left(\frac{(p^*-p)^m}{m}F^{(m)}(p)\right)^{-1}\int_p^{p^*}F^{(m)}(t)(p^*-t)^{m-1}dt \right|$$

$$= \left| \left(\frac{(p^*-p)^m}{m}F^{(m)}(p)\right)^{-1}\int_p^{p^*}[F^{(m)}(t)-F^{(m)}(p)](p^*-t)^{m-1}dt \right|$$

$$\le \frac{m}{(p^*-p)^m}\int_p^{p^*}\varphi_0(|t-p|)|p^*-t|^{m-1}dt < 1, \qquad (22.27)$$

so $\left(\frac{(p^*-p)^m}{m}F^{(m)}(p)\right)^{-1}\int_p^{p^*}F^{(m)}(t)(p^*-t)^{m-1}dt$ is invertible, i.e., $\int_p^{p^*}F^{(m)}(t)(p^*-t)^{m-1}dt$ is invertible. $\qquad\square$

22.2 EXERCISES

1. Solve the exercises of Chapter 15 but use the methods developed in this chapter. Make comparisons.

REFERENCES

1. Argyros, I.K., and Magréñan, A.A. 2017. *Iterative Methods and Their Dynamics with Applications*, New York: CRC Press.
2. Argyros, I.K. 2007. *Computational Theory of Iterative Methods. Series: Studies in Computational Mathematics*, 15, eds. C.K. Chui and L. Wuytack, New York: Elsevier Publ. Co.
3. Argyros, I.K., and George, S. 2019. *Mathematical Modeling for the Solution of Equations and Systems of Equations with Applications*, Volume-III, N.Y.: Nova Science Publisher.
4. Behl, R., Al-Hamdan, W., Alsulami, A., Pansera, B., Salimi, S., Ferrara, M., 2020. New optimal families of Schröder's method suitable for multiple zeros, *Mathematics*.
5. Chun, C., and Neta, B. 2009. A third order modification of Newton's method for multiple roots. *Appl. Math. Comput.*, 211:474–479.
6. Hansen, E., and Patrick, M. 1977. A family of root finding methods. *Numer. Math.*, 27:257–269.
7. Magreñán, A.A. 2014. Different anomalies in a Jarratt family of iterative root finding methods. *Appl. Math. Comput.*, 233:29-38.
8. Magreñán, A.A. 2014. A new tool to study real dynamics: The convergence plane. *Appl. Math. Comput.*, 248:29-38.
9. Neta, B. 2008. New third order nonlinear solvers for multiple roots. *Appl. Math.Comput.*, 202:162–170.
10. Obreshkov, N. 1963. On the numerical solution of equations (Bulgarian). *Annuaire Univ. Sofia. Fac. Sci. Phy. Math.*, 56:73–83.
11. Osada, N. 1994. An optimal multiple root-finding method of order three. *J. Comput. Appl. Math.*, 52:131–133.
12. Petkovic, M.S., Neta, B., Petkovic, L., and Džunič, J. 2013. *Multipoint methods for solving nonlinear equations*. Elsevier.
13. Ren, H.M., and Argyros, I.K. 2010. Convergence radius of the modified Newton method for multiple zeros under Hölder continuity derivative. *Appl. Math. Comput.*, 217:612–621.
14. Schröder, E. 1870. Uber unendlich viele Algorithmen zur Auflosung der Gieichunger. *Math. Ann.*, 2:317–365.
15. Traub, J.F. 1982. *Iterative Methods for the Solution of Equations*. AMS Chelsea Publishing.
16. Zhou, X., and Song, Y. 2011. Convergence radius of Osada's method for multiple roots under Hölder and Center-Hölder continous conditions. *ICNAAM 2011(Greece), AIP Conf. Proc.*, 1389:1836-1839.
17. Zhou, X., Chen, X., and Song, Y. 2014. On the convergence radius of the modified Newton method for multiple roots under the center-Hölder condition. *Numer. Algor.*, 65(2):221–232.

23 Gauss-Newton Solver for Systems of Equations

Let E_1 and E_2 be real or complex Hilbert spaces, $D \subset E_1$ be an open and nonempty set, and $F : D \longrightarrow E_2$ be a continuously differentiable operator in the sense of Fréchet. We utilize the Gauss-Newton solver (GNS) defined for all $n = 0, 1, 2, \ldots$ by

$$x_{n+1} = x_n - F'(x_n)^+ F(x_n) \qquad (23.1)$$

to find a least squares solution x^* of equation

$$F(x) = 0, \qquad (23.2)$$

where $F'(x_n)^+$ stands for the Moore-Penrose inverse of the linear operator $F'(x_n)$ [14]. The ball of convergence is small and the error bounds are too pessimistic in general. Then, we do not have many initial points and may be too many iterations are needed to achieve a desired error tolerance. To address these problems we introduce a technique leading to a more precise region containing the iterate x_n. This way the majorant functions are tighter. The novelty of this approach is that the improved results are obtained under the same computational cost, since the new majorant functions are special cases of the old ones. In particular, we extend the results in [14] which in turn have extended earlier works [12–16]. Relevant works can be found in [1–26]. We refer to [14] for standard auxiliary results, formulas and notation.

23.1 BALL CONVERGENCE

We introduce Lipschitz-type conditions needed in the ball convergence that follows. Let $R > 0, \beta := \|F'(x^*)^+\|$ and $\kappa = \sup\{t \in [0,R) : U(x^*,t) \subset D\}$.

Definition 23.1. Operator F' is center-Lipschitz-like continuous if there exists a function $f_0 : [0,R) \longrightarrow \mathbb{R}$ continuously differentiable such that for all $x \in U(x^*,R)$

$$\beta\|F'(x) - F'(x^*)\| \leq f_0'(\|x - x^*\|) - f_0'(0).$$

Suppose that equation

$$f_0'(t) - f_0'(0) = 1 \qquad (23.3)$$

has a minimal positive solution ρ. Define the set D_0 by

$$D_0 = U(x^*,R) \cap U(x^*,\rho). \qquad (23.4)$$

DOI: 10.1201/9781003128915-23

Definition 23.2. Operator F' is restricted-Lipschitz-like continuous if there exists a function $f : [0,R) \longrightarrow \mathbb{R}$, $\rho_0 = \min\{\rho,R\}$ such that for all $x \in D_0$, $\theta \in [0,1]$

$$\beta \|F'(x) - F'(x^* + \theta(x - x^*))\| \leq f'(\|x - x^*\|) - f'(\theta \|x - x^*\|).$$

Definition 23.3. Operator F' is Lipschitz-like continuous if there exists a function $f_1 : [0,R) \longrightarrow \mathbb{R}$ such that for each $x \in U(x^*,R)$, $\theta \in [0,1]$

$$\beta \|F'(x) - F'(x^* + \theta(x - x^*))\| \leq f_1'(\|x - x^*\|) - f_1'(\theta \|x - x^*\|).$$

Remark 23.1. We have that

$$f_0'(t) \leq f_1'(t) \qquad (23.5)$$

and

$$f'(t) \leq f_1'(t) \qquad (23.6)$$

for all $t \in [0,\rho_0)$, since

$$D_0 \subset U(x^*,R). \qquad (23.7)$$

Notice that the center-Lipschitz condition is used to determine D_0 and $f = f(D_0,R)$. From now on we suppose

$$f_0'(t) \leq f'(t) \text{ for each } t \in [0,\rho). \qquad (23.8)$$

Otherwise we choose f' to be the largest of the two functions f_0 and f on subintervals of $[0,\rho_0)$.

In view of (23.5)–(23.8) the ball convergence analysis of GNS using f_0 and f is tighter than the ones using f_1 (given in [14].

Next, we state the ball convergence result for GNS under the preceding notation.

Theorem 23.1. Suppose that $F(x^*) = 0$, $F'(x^*)$ is injective, $F'(x)$ is center-Lipschitz-like, and restricted-Lipschitz-like.

(c1) $f(0) = 0$ and $f'(0) = -1$
(c2) f' is strictly increasing. Consider $v = \sup\{t \in [0,R) : f'(t) < 0\}$, $\gamma = \sup\{\delta \in (0,v) : \frac{\frac{f(t)}{f'(t)} - t}{t} < 1, t \in [0,\delta)\}$ and $r := \min\{\kappa,\gamma\}$.
       Then, sequences $\{x_n\}$ and $\{s_n\}$ with $x_0 \in U(x^*,r) - \{x^*\}$, $s_0 = \|x_0 - x^*\|$ defined by

$$x_{n+1} = x_n - F'(x_n)^+ F(x_n)$$

and

$$s_{n+1} = |s_n - \frac{f(s_n)}{f'(s_n)}|$$

are well defined, $\{s_n\}$ is strictly increasing, remains in $(0,r)$ and $\lim_{n \longrightarrow \infty} s_n = 0$. Moreover, $\{x_n\}$ is well defined, remains in $U(x^*,r)$ for all

$n = 0, 1, 2, \ldots$ and $\lim_{n \to \infty} x_n = x^*$, which is the unique solution in $U(x^*, \sigma)$, where $\sigma = \sup\{t \in [0, \bar{\kappa}) : f(t) < 0\}$, $\bar{\kappa} = \min\{\kappa, \rho\}$. Furthermore,

$$\lim_{n \to \infty} \frac{\|x_{n+1} - x^*\|}{\|x_n - x^*\|} = 0, \text{ and } \lim_{n \to \infty} \frac{s_{n+1}}{s_n} = 0.$$

Finally, if $\frac{f(\lambda)}{\lambda f'(\lambda)} - 1 = 1$ and $\lambda < \bar{\kappa}$, then $r = \lambda$ is the best possible radius of convergence. Additionally, if $0 \leq \mu \leq 1$

(c3) The function $(0, v) \longrightarrow \frac{\frac{f(t)}{f'(t)} - t}{t^{\mu + 1}}$ is strictly increasing, then $\{\frac{s_{n+1}}{s_n^{\mu+1}}\}$ is strictly increasing so that for each $n = 0, 1, 2, \ldots$

$$\|x_{n+1} - x^*\| \leq \frac{s_{n+1}}{s_n^{\mu+1}} \|x_n - x^*\|^{\mu+1} \leq \frac{s_1}{s_0^{\mu+1}} \|x_n - x^*\|^{\mu+1}.$$

Remark 23.2. In view of (23.5)–(23.8) we have

$$r_1 \leq r, \tag{23.9}$$

where r_1 is defined as r but f_1 (used in [14]) replaces f_0 and f.

As an example, let us specialize functions for $\mu = 1$ to $f_0(t) = \frac{K_0}{2} t^2 - t$, $f(t) = \frac{K}{2} t^2 - t$ and $f_1(t) = \frac{K_1}{2} t^2 - t$. Then, hypotheses of Theorem 23.1 are satisfied and

$$r_1 = \frac{2}{3L_1} \leq r. \tag{23.10}$$

Moreover, the error bounds are

$$\|x_{n+1} - x^*\| \leq \frac{L\|x_n - x^*\|^2}{2(1 - L\|x_n - x^*\|)}$$

which are better than the ones in [14] given by

$$\|x_{n+1} - x^*\| \leq \frac{L_1\|x_n - x^*\|^2}{2(1 - L_1\|x_n - x^*\|)}.$$

23.2 EXERCISES

1. Solve the initial value problem

$$\dot{x}(t) = 1 + \cos(x(t)), x(0) = 0,$$

using the method (23.1).

2. The predator-prey population models describe the interaction of a prey population X and a predator population Y. Assume that their interaction is modeled by the system of ordinary differential equations

$$\dot{x} = x - \frac{1}{4}x^2 - \frac{1}{10}xy + 1$$

$$\dot{y} = -\frac{1}{4}y + \frac{1}{7}xy + \frac{2}{3}.$$

(Assume $x(0) = y(0) = 0$). Solve the system.

3. Solve the initial value problem

$$\dot{x} = (x(t) + t, x(0) = 1)$$

by the iteration method (23.1). Perform 3 steps.

4. Solve the Fredholm-type integral equation

$$x(t) = \sum_0^1 \frac{t \cdot s}{10} ds + 1$$

by the iteration method (23.1). Perform 3 steps.

5. Solve the Volterra-type integral equation

$$x(t) = \sum_0^t \frac{t \cdot s}{10} ds + 1$$

by the iteration method (23.1). Perform 3 steps.

6. Solve equation

$$x = \frac{\sin x}{2} + 1$$

by using algorithm (23.1).

7. Repeat the previous problem for equation

$$x = \frac{\sin x + \cos 2x}{10}.$$

REFERENCES

1. Argyros, I.K., and Magréñan, A.A. 2017. *Iterative Methods and Their Dynamics with Applications*, New York: CRC Press.
2. Argyros, I.K., 2007. *Computational Theory of Iterative Methods. Series: Studies in Computational Mathematics*, New York: Elsevier Publ. Co.
3. Argyros, I.K., 1999, On Newton's method under mild differentiability conditions and applications. *Appl. Math. Comput.*, 102:177–183.
4. Argyros, I.K. 1999. A convergence analysis for Newton-like methods in Banach space under weak hypotheses and applications. *Tamkang J. Math.* 30:255–263.
5. Argyros, I.K., and George, S. 2019. *Mathematical Modeling for the Solution of Equations and Systems of Equations with Applications*, Volume-III, N.Y.: Nova SciencePublisher.
6. Chen, J. 2008. The convergence analysis of inexact Gauss-Newton methods for nonlinear problems. *Comput. Optim. Appl.*, 40(1):97–118.
7. Chen, J., and Li, W. 2005. Convergence of Gauss-Newton's method and uniqueness of the solution. *Appl. Math. Comput.*, 170(1):686–705.
8. Ferreira, O.P. 2009. Local convergence of Newton's method in Banach space from the view point of the majorant principle. *IMA J. Numer. Anal.*, 29(3):746–759.
9. Ferreira, O.P. 2011. Local convergence of Newton's method under majorant condition. *J. Comput. Appl. Math.*, 235(5):1515–1522.
10. Ferreira, O.P., and Goncalves, M.L.N. 2011. Local convergence analysis of inexact Newton-like methods under majorant condition. *Comput. Optim. Appl.*, 48(1):1–21.
11. Ferreira, O.P., and Svaiter, B.F. 2009. Kantorovich's majorants principle for Newton's method. *Comput. Optim. Appl.*, 42(2):213–29.
12. Ferreira, O.P., Goncalves, M.L.N., and Oliveira, P.R. 2011. Local convergence analysis of the Gauss-Newton method under a majorant condition. *J. Complexity*, 27(1):111–125.
13. Ferreira, O.P., Goncalves, M.L.N., and Oliveira, P.R. 2012. Local convergence analysis of inexact Gauss–Newton like methods under majorant condition. *J. Comput. Appl. Math.*, 236(9):2487–2498.
14. Goncalves, M.L.N. 2013. Local convergence of the Gauss-Newton method for injective-over determined systems of equations under a majorant condition. *Computers and Mathematics with Applications*, 66:490–499.
15. Huang, Z. 2004. The convergence ball of Newton's method and the uniqueness ball of equations under Hölder-type continuous derivatives. *Comput. Math. Appl.*, 47(2–3):247–251.
16. Li, C., Zhang, W.H., and Jin, X.Q. 2004. Convergence and uniqueness properties of Gauss-Newton's method. *Comput. Math. Appl.*, 47(6-7):1057–1067.
17. Smale, S. 1986. Newton's method estimates from data at one point, in: *The Merging of Disciplines: New Directions in Pure, Applied, and Computational Mathematics*,185–196 New York: Springer.
18. Wang, X. 2000. Convergence of Newton's method and uniqueness of the solution of equations in Banach space. *IMA J. Numer. Anal.*, 20(1):123–134.
19. Wang, X.H., and Li, C. 2003. Convergence of Newton's method and uniqueness of the solution of equations in Banach spaces.II. *Acta. Math. Sin. (Engl.Ser.)*, 19(2):405–412.
20. Xu, X., and Li, C. 2007. Convergence of Newton's method for systems of equations with constant rank derivatives. *J. Comput. Math.*, 25(6):705–718.
21. Xu, X., and Li, C. 2008. Convergence criterion of Newton's method for singular systems with constant rank derivatives. *J. Math. Anal. Appl.*, 345(2):689–701.

22. Stewart, G.W. 1969. On the continuity of the generalized inverse. *SIAM J. Appl. Math.*, 17:33–45.

23. Wedin, P.A. 1973. Perturbation theory for pseudo-inverses. *BIT, Nord. Tidskr. Inf. behandl.*, 13:217–232.

24. Dedieu, J.P., and Shub, M. 2000. Newton's method for overdetermined systems of equations. *Math. Comp.*, 69(231):1099–1115.

25. Okamoto, H., and Wunsch, M. 2007. A geometric construction of continuous, strictly increasing singular functions. *Proc. Japan. Acad. Ser. AMath. Sci.*, 83(7):114–118.

26. Takacs, L. 1978. An increasing continuous singular function. *Amer. Math. Monthly*, 85(1):35–37.

24 High-Order Iterative Schemes Under Kantorovich Hypotheses

In this chapter, we use restricted convergence regions to locate a more precise set than in earlier works containing the iterates of some high-order iterative schemes involving Banach space valued operators. This way the Lipschitz conditions involve tighter constants than before leading to weaker sufficient semilocal convergence criteria, tighter bounds on the error distances and an at least as precise information on the location of the solution. These improvements are obtained under the same computational effort since computing the old Lipschitz constants also requires the computation of the new constants as special cases. The same technique can be used to extend the applicability of other iterative schemes. Numerical examples further validate the new results.

Let E_1, E_2 be Banach spaces and $D \subset E_1$ be a non-empty convex set. Consider $F : D \to E_2$ to be a twice continuously differentiable operator in the sense of Fréchet. There ia a plethora of problems from many diverse disciplines such as mathematics, computational sciences, physics and also engineering that can be reduced to solving a non-linear equation of the form

$$F(x) = 0, \tag{24.1}$$

using mathematical modeling [1–17].

A locally unique solution x_* in closed form is desirable but this can rarely be achieved. That is why most researchers and practitioners develop iterative scheme which converges to x_*, if certain convergence criteria are satisfied. Newton's scheme defined for each $n = 0, 1, 2, \ldots$ by

$$x_{n+1} = x_n - F'(x_n)^{-1} F(x_n) \tag{24.2}$$

is the most popular iterative scheme which converges quadratically under certain conditions. To produce even higher convergence order the following scheme has been considered [1],

$$
\begin{aligned}
y_n &= x_n - F'(x_n)^{-1} F(x_n) \\
x_{n+1} &= y_n - (I + L_F(x_n) + L_F^2(x_n) G_F(x_n)) F'(x_n)^{-1} F(y_n),
\end{aligned} \tag{24.3}
$$

where $L_F(x) = F'(x)^{-1} F''(x) F'(x)^{-1} F(x)$ and $G_F : D \to Ł(E_1, E_1)$, where $Ł(E_1, E_1)$ stands for the space of bounded linear operators from E_1 into E_1. The operator G_F depends on F and its derivatives. The computation of F'' is expensive in general. As

DOI: 10.1201/9781003128915-24

an example for a system of i equations with i unknowns, the Fréchet derivative is a matrix with i^2 entries and the second Fréchet derivative has i^3 entries. However, if for example the second Fréchet derivative is diagonal by blocks or quadratic equations, then it is constant, so it is easy to evaluate.

Scheme (24.3) extends another one considered by Traub [17], when $E_1 = E_2 = R$. In this special case the convergence order was shown to be four. Numerous, well known high-order iterative schemes, such as Halley scheme, Super Halley, Chebyshev, Chebyshev-type scheme, Two step Newton scheme are obtained from (24.3). A major problem connected to these schemes is that the convergence region is very small in general.

In this chapter, we enlarge this region by using the same hypothesis as before [1]. This is done by considering an at least as tight region D_0 than D containing the iterates. It turns out that the new Lipschitz type constants are tighter than the old ones leading to weaker convergence criteria, tighter error bounds on the distances $\|x_{n+1} - x_n\|, \|x_n - x_*\|$ and an at least as precise information on the location of the solution x_*. It is worth noticing that these improvements are obtained under the same computational effort as before, since the computation of the old constants requires the computation of the new constants as special cases.

24.1 SEMI-LOCAL CONVERGENCE

We present the local convergence of scheme (24.2) under the hypotheses(H):

(h_1)   There exists $x_0 \in D$ such that $F'(x_0)^{-1} \in L(E_2, E_1)$ and

$$\|F'(x_0)^{-1}\| \leq \xi.$$

(h_2) $\|F'(x_0)^{-1}F(x_0)\| \leq \eta$
(h_3) $\|F'(x) - F'(x_0)\| \leq M_0\|x - x_0\|$ for each $x \in D$

Set
$$D_0 = D \cap U(x_0, \frac{1}{\xi M_0})$$

(h_4) $\|F''(x)\| \leq M$
(h_5) $\|F''(y) - F''(x)\| \leq K\|x - y\|$ for each $x, y \in D_0$.

We assume that
$$M_0 \leq M \tag{24.4}$$

If $M < M_0$, then simply replace M by M_0 in the results that follow. Condition (h_3) is used to define D_0 needed in (h_4) and (h_5) that is $M = M(D_0)$ and $K = K(D_0)$. In the literature [1, 2] for this type of studies instead of (h_3)–(h_4) the following conditions are used:

(h_3)′ $\|F''(x)\| \leq M_1$ for each $x \in D$
(h_4)′ $\|F''(y) - F''(x)\| \leq K_1\|y - x\|$ for each $x, y \in D$.

We have that

$$M_0 \leq M_1 \tag{24.5}$$

$$M \leq M_1 \tag{24.6}$$

and

$$K \leq K_1, \tag{24.7}$$

since

$$D_0 \subseteq D. \tag{24.8}$$

It follows that M, K can replace M_1, K_1 in the analysis that follows. Hence, the proofs are omitted.

As in earlier works [1, 2] we use the cubic polynomial defined by

$$q(t) = \frac{K}{6}t^3 + \frac{M}{2}t^2 - \frac{1}{\beta}t + \frac{\eta}{\beta} \tag{24.9}$$

together with the cubic polynomial considered in [1, 2] given by

$$q_1(t) = \frac{K_1}{6}t^3 + \frac{M_1}{2}t^2 - \frac{1}{\beta}t + \frac{\eta}{\beta} \tag{24.10}$$

Notice that

$$q(t) \leq q_1(t), \quad \text{for each } t. \tag{24.11}$$

Define parameters δ and γ by

$$\delta = \frac{4K + M^2\beta - M\beta\sqrt{M^2 + 2K\beta}}{3\beta K(M + \sqrt{M^2 + 2K\beta})} \tag{24.12}$$

and

$$\gamma = \frac{4K_1 + M_1^2\beta - M_1\beta\sqrt{M_1^2 + 2K_1\beta}}{3\beta K_1(M_1 + \sqrt{M_1^2 + 2K_1\beta})}. \tag{24.13}$$

Then, in view of (24.1), (24.7), (24.12) and (24.13) we have

$$\gamma \leq \delta, \tag{24.14}$$

since it can easily be seen that these parameters are decreasing with respect to K and M.

The sufficient convergence criterion for scheme (24.3) is given by

$$\eta \leq \gamma \tag{24.15}$$

but ours is

$$\eta \leq \delta. \tag{24.16}$$

It follows from (24.14)-(24.16) that

$$\eta \leq \gamma \Rightarrow \eta \leq \delta \tag{24.17}$$

but not necessarily vice versa unless, if $\gamma = \delta$. Hence, in view of (24.11) and (24.17), we have extended the applicability of scheme (24.3) with advantages as already stated in the introduction of this article(see also the numerical example).

Under the hypothesis (24.16) the cubic polynomial $q(t)$ has two roots t_* and $t_{**}(t_* \leq t_{**})$. We can choose a and b such that $0 < a < t_*$ and $b > 2/(M\beta + \sqrt{M^2\beta^2 + 2K\beta})$.

Moreover, we suppose

$(HG1)$ $\|L_F(x)^2 G_F(x)\| \leq L_q(t)^2 G_p(t)$, for $\|x - x_0\| \leq t - t_0 \leq t_* - t_0$,
$(HG2)$ $1 + L_q(t) + L_q(t)^2 G_q(t) \geq$,
$(HG3)$ $h'(t) > 0$, in $[a, t_*[$, where

$$h(t) = t - \frac{q(t)}{q'(t)} - (1 + L_q(t) + L_q(t)^2 G_q(t)) \frac{q(t - (q(t)/q'(t)))}{q'(t)}.$$

Notice that G_q satisfies the last three hypotheses. We need an auxillary result on majorizing sequence for scheme (24.3).

Lemma 24.1. l The sequence

$$s_n = t_n - \frac{q(t_n)}{q'(t_n)},$$
$$t_{n+1} = s_n - (1 + L_q(t_n) + L_q(t_n)^2 G_q(t_n)) \frac{q(s_n)}{q'(t_n)}, \qquad (24.18)$$

starting from t_0 monotonically converges to t_*, the simple solution of $q(t) = 0$ in [a,b].

Next we present the main semilocal convergence analysis for scheme (24.3) under the preceding hypotheses and notation.

Theorem 24.1. Let us assume $x_0 \in D$ and $t_0 \in [a, t_*]$ verifying the hypotheses (h_1)–(h_4) and $(HG1)$–$(HG3)$ with η satisfying (24.16),

$$U(x_0, t_* - t_0) \subset D. \qquad (24.19)$$

Then, the sequence (24.2) is well defined and converges to x^* which is the only solution of $F(x) = 0$ in $\overline{U(x_0, t_* - t_0)}$. Moreover, the error bounds hold

$$\|x^* - x_n\| \leq t_* - t_n, \qquad (24.20)$$

for each $n = 0, 1, 2, \ldots$ where sequence $\{t_n\}$ is given in (24.18).

Remark 24.1. Suppose that

$$\beta M_0 \eta < 1. \qquad (24.21)$$

Then, we can define

$$D_0^1 := D \cap U(y_0, \frac{1}{\beta M_0} - \|F'(x_0)^{-1}F(x_0)\|), \qquad (24.22)$$

where $y_0 = x_0 - F'(x_0)^{-1}F(x_0)$. Suppose further that (h_4) and (h_5) hold on D_0^1 with corresponding constants $\overline{M}, \overline{K}$, respectively. Then, since

$$D_0^1 \subseteq D_0, \tag{24.23}$$

we have that

$$\overline{M} \leq M \tag{24.24}$$

$$\overline{K} \leq K. \tag{24.25}$$

Moreover, set

$$\overline{q(t)} = \frac{\overline{K}}{6}t^3 + \frac{\overline{M}}{2}t^2 - \frac{1}{\beta}t + \frac{\eta}{\beta} \tag{24.26}$$

and (24.16) can be replaced by the even weaker

$$\eta \leq \overline{\delta}, \tag{24.27}$$

provided that (24.21) holds. It is worth noticing that (24.21) is the sufficient convergence criterion for the modified Newton's method

$$z_{n+1} = z_n - F'(z_0)^{-1}F(z_n), \tag{24.28}$$

where $z_0 = x_0$.

24.2 APPLICATIONS

We present a numerical example to test the new and earlier convergence criterion for scheme (24.3)

Example 24.1. Let $E_1 = E_2 = \mathbb{R}, D = \bar{U}(x_0, 1 - p), x_0 = 1$ and $p \in [0, \frac{1}{2}]$. Define function F on D by

$$F(x) = x^3 - p \tag{24.29}$$

Using the hypotheses (HG) and (24.29), we get that $\beta = 1/3$, $M_0 = 3(3 - p)$, $M = 6(1 + \frac{1}{\beta M_0})$, $\eta = \frac{1}{3}(1 - p)$, $M_1 = 6(2 - p)$, and $K = K_1 = 6$. Notice that

$$M_0 < M < M_1. \tag{24.30}$$

Then, estimate (24.15) is satisfied for each $p \in I_0 = [0.399, \frac{1}{2}]$
Moreover, (24.16) is satisfied for each $p \in I = [0.300481, \frac{1}{2}]$
Hence, the applicability of scheme (24.3) is expanded, since $I \subseteq I_0$.

24.3 EXERCISES

1. Solve the initial value problem

$$\dot{x}(t) = 1 + \cos(x(t)), x(0) = 0,$$

using the method (24.3).

2. The predator-prey population models describe the interaction of a prey population X and a predator population Y. Assume that their interaction is modeled by the system of ordinary differential equations

$$\dot{x} = x - \frac{1}{4}x^2 - \frac{1}{10}xy + 1$$
$$\dot{y} = -\frac{1}{4}y + \frac{1}{7}xy + \frac{2}{3}.$$

(Assume $x(0) = y(0) = 0$). Solve the system.

3. Solve the initial value problem

$$\dot{x} = (x(t) + t, x(0) = 1)$$

by the iteration method (24.3). Perform 3 steps.

4. Solve the Fredholm-type integral equation

$$x(t) = \int_0^1 \frac{t \cdot s}{10} ds + 1$$

by the iteration method (24.3). Perform 3 steps.

5. Solve the Volterra-type integral equation

$$x(t) = \int_0^t \frac{t \cdot s}{10} ds + 1$$

by the iteration method (24.3). Perform 3 steps.

6. Solve equation
$$x = \frac{\sin x}{2} + 1$$

by using algorithm (24.3).

7. Repeat the previous problem for equation

$$x = \frac{\sin x + \cos 2x}{10}.$$

REFERENCES

1. Amat, S., Bermudez, C., Busquier, S., Legaz, M.J., and Plaza, S. 2012. On a family of high-order iterative method under Kantorovich conditions and some Applications. *Abstract and Applied Analysis*, Article Id 782170, doi:10.1155/2012/782170.
2. Amat, S., and Busquier, S. 2007. Third order iterative methods under Kantorovich conditions. *Journal of Mathematical Analysis and Applications*, 336(1):243–261.
3. Amat, S., Busquier, S., and Gutirrez, J.M. 2006. An adaptive version of a fourth-order iterative method for quadratic equations. *Journal of Computational and Applied Mathematics*, 191(2):259–268.
4. Argyros, I.K. 2007. *Computational theory of iterative methods*.eds. Chui, C.K., Wuytack, L. Series: Studies in Computational Mathematics, 15. New York: Elsevier.
5. Argyros, I.K. 1998. Improving the order and rates of convergence for the super-Halley method in Banach spaces. *The Korean Journal of Computational and Applied Mathematics*, 5(2):465–474.
6. Argyros, I.K. 1993. The convergence of a Halley-Chebysheff-type method under Newton-Kantorovich hypotheses. *Applied Mathematics Letters*, 6(5):71–74.
7. Argyros, I.K., Magreñán, A.A. 2017. *Iterative Methods and their Dynamics with Applications*, New York: CRC Press.
8. Argyros, I.K., George, S., and Magrenan, A.A. 2015. Local convergence for multi-point-parametric Chebyshev-Halley-type methods of high convergence order. *Journal of Computational and Applied Mathematics*, 282:215–224.
9. Argyros, I.K., George, S., and Thapa, N. 2018. *Mathematical Modeling for the Solution of Equations and Systems of Equations with Applications*, Volume I, N.Y: Nova Science Publishers.
10. Argyros, I.K., George, S., and Thapa, N. 2018. *Mathematical Modeling for the Solution of Equations and Systems of Equations with Applications*, Volume II, N.Y: Nova Science Publishers.
11. Ezquerro, J.A., Sanchez, M.G., Grau, A., Hernandez, M.A., Noguera, M., and Romero, N. 2011. On iterative methods with accelerated convergence for solving systems of non-linear equations. *Journal of Optimization Theory and Applications*, 151(1):163–174.
12. Ezquerro, J.A., and Hernandez, M.A. 2005. New Kantorovich-type conditions for Halley's method. *Applied Numerical Analysis and Computational Mathematics*, 2(1):70–77.
13. Hernandez, M.A. and Salanova, M.A. 2000. Modification of the Kantorovich assumptions for semilocal convergence of the Chebyshev method. *Journal of Computational and Applied Mathematics*, 126(1-2):131–143.
14. Magréñan, A.A. 2014. A new tool to study real dynamics: The convergence plane. *Appl. Math. Comput.* 248:215–225.
15. Magréñan, A.A. 2015. Improved convergence analysis for Newton-like methods. *Numerical Algorithms*, 1–23.
16. Romero, N. 2006. Familias parametricas de procesos iterativos de alto orden de convergencia [Ph.D. thesis]. http://dialnet.unirioja.es/.
17. Traub, J.F. 1982. *Iterative methods for the solution of equations*. AMS Chelsea Publishing.

25 Semi-Local Convergence of a Derivative Free Method

In this chapter, we present the semi-local convergence analysis of a two step derivative free method for solving Banach space valued equations. The convergence criteria are based only on the first derivative and our idea of recurrent functions.

Let B_1, B_2 stand for Banach space, $\Omega \subset B_1$ be open and $U(x, \rho)$ denote a closed ball of center $x \in B_1$ and of radius $\rho > 0$. We denote the closure of $U(x, \rho)$ by $\bar{U}(x, \rho)$.

We are dealing with the problem of approximating a solution x_* of equation

$$F(x) = 0. \tag{25.1}$$

Solving equation (25.1) is very important, because many problems reduce to it by Mathematical modeling [1–8]. The solution methods are usually of iterative nature, since solutions in closed form are rarely obtained. In this article, we develop a derivative free method to generate a sequence approximating x_* under certain conditions. The method is defined for all $n \geq 0$ as

$$\begin{aligned} y_n &= x_n - B_n^{-1} F(x_n) \\ x_{n+1} &= x_n - A_n^{-1} F(x_n), \end{aligned} \tag{25.2}$$

where $A_n = [\frac{x_n + y_n}{2}, x_n; F](n \geq 0)$, $B_n = [\frac{x_{n-1} + y_{n-1}}{2}, x_n; F](n \geq 1)$ and $x_0, y_0 \in \Omega$ are initial points. Here, $[x, y; F] : \Omega \times \Omega \longrightarrow L(X, Y)$ denotes a divided difference of order one for operator F at the point $x, y \in D$ (see [2, 6, 8]). Method (25.2) is a usefull alternative to third order methods such as the method of tangent hyperbolas (Halley) or the method of tangent parabolas (Euler-Chebysheff) [1–8]. However, these methods are very expensive, since they require the evaluation of the second Fréchet-derivative at each step. Discretized versions of these methods such as Ulm's method use divided differences of order one [1–8].

The rest of the chapter is organized as follows: In Section 25.1, we present the semi-local convergence analysis of method (25.2), whereas in the concluding Section 25.2, we present the applications.

DOI: 10.1201/9781003128915-25

25.1   SEMI-LOCAL CONVERGENCE

It is based on the majorant sequence defined for $n = 1, 2, \ldots$ and some $\eta \geq 0$ and $s \geq 0$ as follows

$$
\begin{aligned}
t_0 &= 0, t_1 = \eta \geq 0, s_0 = s \geq 0, \\
s_n &= t_n + \frac{L(2(t_n - t_{n-1}) + (s_{n-1} - t_{n-1}))(t_n - t_{n-1})}{2[1 - \frac{L_0}{2}(t_{n-1} + s_{n-1} + 2t_n + s)]} \\
t_{n+1} &= t_n + \frac{L(2(t_n - t_{n-1}) + (s_{n-1} - t_{n-1}))(t_n - t_{n-1})}{2[1 - \frac{L_0}{2}(3t_n + s_n + s)]}.
\end{aligned}
\tag{25.3}
$$

Define the scalar cubic polynomial p as

$$
p(t) = L_0 t^3 + 3L_0 t^2 + 3Lt - 3L \text{ for some } L_0 > 0 \text{ and } L > 0. \tag{25.4}
$$

By this definition $p(0) = -3L$ and $p(1) = 4L_0$. It follows from the intermediate value theorem and the Descartes rule of signs that polynomial p has a unique root $\gamma \in (0, 1)$. Moreover, define α_0 and α_1 as

$$
\alpha_0 = \frac{L(2(t_1 - t_0) + s_0 - t_0)}{2[1 - \frac{L_0}{2}(3t_1 + s_1 + s)]}, \tag{25.5}
$$

$$
\alpha_1 = \frac{L(2(t_1 - t_0) + s_0 - t_0)}{2[1 - \frac{L_0}{2}(t_0 + s_0 + 2t_1 + s)]}, \tag{25.6}
$$

and set

$$
\gamma_0 = \max\{\alpha_0, \alpha_1\}. \tag{25.7}
$$

Next, we present a convergence result for majorizing sequence $\{t_n\}$.

Lemma 25.1. Suppose that there exists γ such that $L_0 s < 2$ and

$$
0 < \gamma_0 \leq \gamma \leq 1 - \frac{L_0 \eta}{1 - \frac{L_0}{2}s}. \tag{25.8}
$$

Then, sequence $\{t_n\}$ given by (25.3) is nondecreasing, bounded from above by $t_{**} = \frac{\eta}{1-\gamma}$ and converges to its unique least upper bounds t_* which satisfies $\eta \leq t_* \leq t_{**}$.

Proof. We shall show using induction that

$$
0 < \frac{L(2(t_{k+1} - t_k) + (s_k - t_k))}{2[1 - \frac{L_0}{2}(3t_{k+1} + s_{k+1} + s)]} \leq \gamma \tag{25.9}
$$

and

$$
0 < \frac{L(2(t_{k+1} - t_k) + (s_k - t_k))}{2[1 - \frac{L_0}{2}(t_k + s_k + 2t_{k+1} + s)]} \leq \gamma. \tag{25.10}
$$

Estimates (25.9) and (25.10) hold for $k = 0$ by (25.3), (25.5)–(25.8). Suppose that (25.9) and (25.10) hold for $j = 1, 2, \ldots, k-1$. Then, by (25.3), (25.9) and (25.10)

$$0 < t_{k+1} - t_k \leq \gamma(t_k - t_{k-1}) \leq \gamma^k \eta \implies t_{k+1} \leq \frac{1 - \gamma^{k+1}}{1 - \gamma} \eta, \qquad (25.11)$$

$$0 < s_k - t_k \leq \gamma(t_k - t_{k-1}) \leq \gamma^k \eta \implies s_k \leq \frac{1 - \gamma^k}{1 - \gamma} \eta + \gamma^k \eta = \frac{1 - \gamma^{k+1}}{1 - \gamma} \eta. \quad (25.12)$$

By (25.9) and (25.10) we must only complete the induction for (25.9). Evidently this will be true by (25.3), (25.11) and (25.12), if

$$\frac{L}{2}(2\gamma^k \eta + \gamma^k \eta) + \frac{\gamma L_0}{2}(3\frac{1 - \gamma^{k+1}}{1 - \gamma}\eta + \frac{1 - \gamma^{k+2}}{1 - \gamma}\eta + s) - \gamma \leq 0. \qquad (25.13)$$

Estimate (25.13) suggests to introduce functions φ_k on $[0, 1)$ as

$$\varphi_k(t) = \frac{3L}{2}t^{k-1}\eta + \frac{L_0}{2}[3(1 + t + \ldots + t^k) + (1 + t + \ldots + t^{k+1})]\eta. \qquad (25.14)$$

We seek a relationship between two consecutive functions φ_k. We can write in turn

$$
\begin{aligned}
\varphi_{k+1}(t) &= \frac{3L}{2}t^k\eta + \frac{L_0}{2}(3(1 + t + \ldots + t^{k+1}) \\
&\quad + (1 + t + \ldots + t^{k+2}))\eta + \frac{L_0}{2}s - 1 \\
&\quad - \frac{3}{2}Lt^{k-1}\eta - \frac{L_0}{2}(3(1 + t + \ldots + t^k) \\
&\quad + (1 + t + \ldots + t^{k+1}))\eta - \frac{L_0}{2}s + 1 + \varphi_k(t) \\
&= \varphi_k(t) + p(t)\frac{t^{k-1}\eta}{2}, \qquad (25.15)
\end{aligned}
$$

where $p(t)$ is given by (25.4).

In particular we get

$$\varphi_{k+1}(\gamma) = \varphi_k(\gamma) \qquad (25.16)$$

by the definition of γ. Therefore, (25.13) holds, if

$$\varphi_\infty(\gamma) \leq 0, \qquad (25.17)$$

where

$$\varphi_\infty(\gamma) = \lim_{k \to \infty} \varphi_k(\gamma). \qquad (25.18)$$

But by (25.13)

$$\varphi_\infty(\gamma) = \frac{2L_0\eta}{1 - \gamma} + \frac{L_0}{2}s - 1. \qquad (25.19)$$

Hence, (25.17) holds, if

$$\frac{2L_0\eta}{1-\gamma} + \frac{L_0}{2}s - 1 \leq 0 \tag{25.20}$$

or

$$\gamma \leq 1 - \frac{L_0\eta}{1 - \frac{L_0}{2}s} \tag{25.21}$$

which is true by (25.8). Then, sequence $\{t_n\}$ is nondecreasing and in view of (25.11) bounded from above by t_{**}. Hence, as such it converges to its unique least upper bound satisfying $\eta \leq t_* \leq t_{**}$.

$\square$

Next, we present the semi-local convergence for method (25.2).

Theorem 25.1. Assume:

(i) $F : \Omega \subset B_1 \longrightarrow B_2$ is a continuous operator with a standard divided differ-ence of order one such that $[.,.] : \Omega \times \Omega \longrightarrow L(B_1, B_2)$ and $x_0, y_0 \in \Omega$ are such that $A_0 = [\frac{x_0+y_0}{2}, x_0; F]$ is invertible. Let $\|x_1 - x_0\| \leq \eta$ and $\|y_0 - x_0\| \leq s$.

(ii) Hypotheses of Lemma 25.1 hold.

(iii) $\|A_0^{-1}([x, y; F] - A_0)\| \leq L_0(\|x - \frac{x_0+y_0}{2}\| + \|y - x_0\|)$ for all $x, y \in \Omega$ and some $L_0 > 0$. Set $\rho = \frac{1}{4}(\frac{2}{L_0} - s)$ and $\Omega_0 = \Omega \cap U(x_0, \rho)$.

(iv) $\|A_0^{-1}([x, y; F] - [z, y; F])\| \leq L\|x - z\|$ for all $x, y, z \in \Omega_0$ and some $L > 0$.

(v) $\bar{U}(x_0, t_*) \subseteq \Omega$.

Then, there exists a limit point $x_* \in \bar{U}(x_0, t_*)$ of the sequence $\{x_n\}$ such that $F(x_*) = 0$.

Proof. We use mathematical induction to show the estimates

$$\|x_{n+1} - x_n\| \leq t_{n+1} - t_n \tag{25.22}$$

and

$$\|y_n - x_n\| \leq s_n - t_n. \tag{25.23}$$

But these estimates are true by the initial conditions and (25.3) for $n = 0$. Suppose initial conditions and (25.3) hold for $n = 0$. Suppose that they are true for all $k = 0, 1, 2, \ldots n - 1$. Then, we have by (iii)

$$
\begin{aligned}
\|A_0^{-1}(B_k - A_0)\| &\leq L_0(\|\frac{x_{k-1} + y_{k-1}}{2} - \frac{x_0 + y_0}{2}\| + \|x_k - x_0\|) \\
&\leq \frac{L_0}{2}(\|x_{k-1} - x_0\| + \|y_{k-1} - y_0\| + 2\|x_k - x_0\|) \\
&\leq \frac{L_0}{2}(2\|x_{k-1} - x_0\| + \|y_{k-1} - x_{k-1}\| + \|y_0 - x_0\| + 2\|x_k - x_0\|) \\
&\leq \frac{L_0}{2}(2(t_{k-1} - t_0) + s_{k-1} - t_{k-1} + 2(t_k - t_0) + s) \\
&= \frac{L_0}{2}(t_{k-1} + s_{k-1} + 2t_k + s) < 1, \tag{25.24}
\end{aligned}
$$

which together with a Lemma by Banach on invertible operators give B_{k-1} is invertible and

$$\|B_k^{-1}A_0\| \leq \frac{1}{1 - \frac{L_0}{2}(t_{k-1} + s_{k-1} + 2t_k + s)}.$$ (25.25)

Similarly,

$$
\begin{aligned}
\|A_0^{-1}(A_k - A_0)\| &\leq L_0(\|\frac{x_k + y_k}{2} - \frac{x_0 + y_0}{2}\| + \|x_k - x_0\|) \\
&\leq \frac{L_0}{2}(\|x_k + y_k - (x_0 + y_0)\| + 2\|x_k - x_0\|) \\
&\leq \frac{L_0}{2}(\|x_k - x_0\| + \|y_k - y_0\| + 2\|x_k - x_0\|) \\
&\leq \frac{L_0}{2}(3\|x_k - x_0\| + \|y_k - y_0\|) \\
&\leq \frac{L_0}{2}(3\|x_k - x_0\| + \|y_k - x_k\| + \|x_k - x_0\| + \|x_0 - y_0\|) \\
&\leq \frac{L_0}{2}(4\|x_k - x_0\| + \|y_k - x_k\| + \|x_0 - y_0\|) \\
&\leq \frac{2L_0}{2}(4t_k + s_k - t_k + s) \\
&= \frac{L_0}{2}(3t_k + s_k + s) < 1,
\end{aligned}
$$

so

$$\|A_k^{-1}A_0\| \leq \frac{1}{1 - \frac{L_0}{2}(3t_k + s_k + s)}.$$ (25.26)

By method (25.2), we get the identity

$$F(x_k) = F(x_k) - F(x_{k-1}) - [\frac{x_{k-1} + y_{k-1}}{2}, x_{k-1}; F](x_k - x_{k-1})$$ (25.27)

so by (iii) and (25.27

$$
\begin{aligned}
\|A_0^{-1}F(x_k)\| &\leq L(\|x_k - \frac{x_{k-1} + y_{k-1}}{2}\| + \|x_k - x_{k-1}\|) \\
&\leq \frac{L}{2}\|2x_k - (x_{k-1} + y_{k-1})\|\|x_k - x_{k-1}\| \\
&\leq \frac{L}{2}(\|x_k - x_{k-1}\| + \|x_k - y_{k-1}\|)\|x_k - x_{k-1}\| \\
&\leq \frac{L}{2}(\|x_k - x_{k-1}\| + \|x_k - x_{k-1}\| + \|y_{k-1} - x_{k-1}\|)\|x_k - x_{k-1}\| \\
&\leq \frac{L}{2}(2(t_k - t_{k-1}) + (s_{k-1} - t_{k-1})(t_k - t_{k-1}).
\end{aligned}
$$ (25.28)

Then, by (25.3), (25.25), (25.26) and (25.28) we obtain

$$
\begin{aligned}
\|y_k - x_k\| &= \|[B_k^{-1}A_0][A_0^{-1}F(x_k)]\| \\
&\leq \|B_k^{-1}A_0\|\|A_0^{-1}F(x_k)\| \\
&\leq \frac{L(2(t_k - t_{k-1}) + (s_{k-1} - t_{k-1}))(t_k - t_{k-1})}{2[1 - \frac{L_0}{2}(t_{k-1} + s_{k-1} + 2t_k + s)]} = s_k - t_k
\end{aligned}
$$

(25.29)

and

$$
\begin{aligned}
\|x_{k+1} - x_k\| &= \|[A_k^{-1}A_0][A_0^{-1}F(x_k)]\| \\
&\leq \|A_k^{-1}A_0\|\|A_0^{-1}F(x_k)\| \\
&\leq \frac{L(2(t_k - t_{k-1}) + (s_{k-1} - t_{k-1}))(t_k - t_{k-1})}{2[1 - \frac{L_0}{2}(3t_k + s_k + s)]} = t_{k+1} - t_k
\end{aligned}
$$

(25.30)

completing the induction for (25.22) and (25.23). We also have

$$
\begin{aligned}
\|x_{k+1} - x_0\| &\leq \|x_{k+1} - x_k\| + \ldots + \|x_1 - x_0\| \\
&\leq t_{k+1} - t_k + \ldots + t_1 - t_0 = t_{k+1} < t_*
\end{aligned}
$$

and

$$
\begin{aligned}
\|y_k - x_0\| &\leq \|y_k - x_k\| + \|x_k - x_0\| \\
&\leq s_k - t_k + t_k - t_0 = s_k < t_*,
\end{aligned}
$$

so $y_k, x_{k+1} \in U(x_0, t_*)$. Moreover, sequence $\{t_k\}$ is fundamental by Lemma 25.1. Hence, sequence $\{x_k\}$ is fundamental too and as such it converges to some $x_* \in \bar{U}(x_0, t_*)$. By sending $k \longrightarrow \infty$ in (25.28) and using the continuity of F, we conclude $F(x_*) = 0$.

□

Concerning the uniqueness of the solution x_*, we have :

Proposition 25.1. Under the hypotheses of Theorem 25.1 further assume that

$$
L_0(3t_*^1 + t_*) < 2
$$

(25.31)

for some $t_*^1 \geq t_*$. Then, x_* is the only solution of equation $F(x) = 0$ in the set $\Omega_1 = \Omega \cap U(x_0, t_*^1)$.

Proof. Let $x_*^1 \in \Omega_1$ with $F(x_*^1) = 0$. Set $T = [x_*, x_*^1; F]$. Using (iii) and (25.31), we get

$$
\begin{aligned}
\|A_0^{-1}(T - A_0)\| &\leq L_0(\|x_* - \frac{x_0 - y_0}{2}\| + \|x_*^1 - x_0\|) \\
&\leq L_0(\frac{\|x_* - x_0\| + \|x_*^1 - x_0\|}{2} + \|x_*^1 - x_0\|) \\
&\leq L_0(\frac{t_* + t_*^1}{2} + t_*^1) < 1, \qquad (25.32)
\end{aligned}
$$

so $x_* = x_*^1$ is deduced, since T is invertible and

$$
T(x_* - x_*^1) = F(x_*) - F(x_*^1) = 0 - 0 = 0. \qquad (25.33)
$$

$\square$

Remark 25.1. We can compute the computational order of convergence (COC) defined by

$$
a = \ln\left(\frac{\|x_{n+1} - x^*\|}{\|x_n - x^*\|}\right) / \ln\left(\frac{\|x_n - x^*\|}{\|x_{n-1} - x^*\|}\right)
$$

or the approximate computational order of convergence

$$
b = \ln\left(\frac{\|x_{n+1} - x_n\|}{\|x_n - x_{n-1}\|}\right) / \ln\left(\frac{\|x_n - x_{n-1}\|}{\|x_{n-1} - x_{n-2}\|}\right).
$$

This way we obtain in practice the order of convergence in a way that avoids the existence of the high Fréchet derivatives for operator F and Taylor series used in other studies.

25.2 APPLICATIONS

Example 25.1. Let $B_1 = B_2 = \mathbb{R}^3$, $\Omega = U(0,1)$. Define F on Ω by

$$
F(x) = F(u_1, u_2, u_3) = (e^{u_1} - 1, \frac{e-1}{2}u_2^2 + u_2, u_3)^T. \qquad (25.34)
$$

For the points $u = (u_1, u_2, u_3)^T$, the Fréchet derivative is given by

$$
F'(u) = \begin{pmatrix} e^{u_1} & 0 & 0 \\ 0 & (e-1)u_2 + 1 & 0 \\ 0 & 0 & 1 \end{pmatrix}.
$$

Using the norm of the maximum of the rows for $x_0 = (10^{-3}, 10^{-3}, 10^{-3})^T$, $y = (10^{-4}, 10^{-4}, 10^{-4})^T$, we get $L_0 = 0.7(e-1)$, $L = e^\rho$, where $\rho = 0.4118$. Then, we have $s = 0.0156, \eta = 0.0015, \gamma_0 = 0.0035 < \gamma = 0.6245 < 1 - \frac{L_0\eta}{1 - \frac{L_0}{2}s} = 0.9985, t_{**} = 0.0015$. We have verified all the conditions of Theorem 25.1. Hence, we conclude that $\lim_{n \to \infty} x_n = x_* = (0,0,0)^T$.

25.3 EXERCISES

1. Solve the initial value problem

$$\dot{x}(t) = 1 + \cos(x(t)), x(0) = 0,$$

   using the method (25.2).

2. The predator-prey population models describe the interaction of a prey population X and a predator population Y. Assume that their interaction is modeled by the system of ordinary differential equations

$$\dot{x} = x - \frac{1}{4}x^2 - \frac{1}{10}xy + 1$$
$$\dot{y} = -\frac{1}{4}y + \frac{1}{7}xy + \frac{2}{3}.$$

   (Assume $x(0) = y(0) = 0$). Solve the system.

3. Solve the initial value problem

$$\dot{x} = (x(t) + t, x(0) = 1)$$

   by the iteration method (25.2). Perform 3 steps.

4. Solve the Fredholm-type integral equation

$$x(t) = \sum_{0}^{1} \frac{t \cdot s}{10} ds + 1$$

   by the iteration method (25.2). Perform 3 steps.

5. Solve the Volterra-type integral equation

$$x(t) = \sum_{0}^{t} \frac{t \cdot s}{10} ds + 1$$

   by the iteration method (25.2). Perform 3 steps.

6. Solve equation

$$x = \frac{\sin x}{2} + 1$$

   by using algorithm (25.2).

7. Repeat the previous problem for equation

$$x = \frac{\sin x + \cos 2x}{10}.$$

REFERENCES

1. Appell, J., De Pascale, E., Evkhuta, N.A., and Zabrejko, P.P. 1995. On the two step Newton method for the solution of nonlinear equations. *Math. Nachr.*, 172:5–14.
2. Argyros, I.K. 2007. *Computational Theory of Iterative Methods, Series: Studies in Computational Mathematics*, New York: Elsevier Publ. Company.
3. Argyros, I.K., and George, S. 2020. *Mathematical modeling for the solution of equations and systems of equations with applications*, Volume-IV, N.Y: Nova Science Publisher.
4. Ezquerro, J.A., and Hernandez, M. A. 2009. An improvement of the region of accessibility of Chebyshev's method from Newton's method. *Math. Comput.*, 78. 267:1613–1627.
5. Ezquerro, J.A., Hernandez, M.A., and Magrenan, A.A. 2018. Stating points for Newton's method under a center Lipschitz condition for the second derivative. *J. Comput. Appl. Math.*, 330, 721–731.
6. Kantorovich, L.V., and Akilov, G.P., 1982. *Functional Analysis.* Oxford: Pergamon.
7. Magreñán, A.A., and Argyros, I.K. 2014. Two-step Newton methods. *Journal of Complexity*, 30(4):533–553.
8. Potra, F.A., and Ptak, V., 1984. Nondiscrete induction and iterative processes. *Research notes in Mathematics*, 103. Boston: Pitman.

26 A new Higher-Order Iterative Scheme for the Solutions of Non-linear Systems

Many real life problems can be reduced to scalar and vectorial nonlinear equations by using mathematical modeling. We introduce a new iterative family of sixth-order for system of nonlinear equations. In addition, we present their converges analyses. Moreover, we present the computable radii for the guaranteed convergence of them for Banach space valued operators and error bounds based on the Lipschitz constants. Moreover, we shown the applicability of them on some real life problems e.g. kinematic syntheses, Bratu's, Fisher's, boundary value and Hammerstein integral problems. We wind up finally on the ground of achieved numerical experiments that they execute better than other competing schemes.

Establishment of higher-order efficient iterative schemes for finding the solutions

$$F(U) = 0, \tag{26.1}$$

(where $F : \mathbb{D} \subset \mathbb{R}^m \to \mathbb{R}^m$ is a differentiable mapping with open domain $\mathbb{D}$) is one of the foremost task in the area of numerical analysis and computation methods because of its wide application in real life situations. We can easily find several real life problems that were phrased into nonlinear system (26.1) along with the same fundamental properties. For example, transport theory, combustion, reactor, kinematics syntheses, steering, chemical equilibrium, neurophysiology and economics modeling problems were solved by formulated to $F(U) = 0$ and details can be seen in the research articles [7, 12, 16, 18, 26].

Analytical methods for these problems are rare. Therefore, authors developed iterative schemes that are based on iteration procedures. These iterative methods depends upon several things like starting initial guess/es, considered problem, body structure of the proposed method, efficiency, etc. (for the details, please go through [4, 5, 19, 20, 25]). Some authors [1, 6, 8, 9, 23, 29] given special concern to the development of higher-order multi-point iterative methods. The faster convergence toward the required root, better efficiency, less CPU time and fast accuracy are one of the main reasons behind the importance of multi-point methods.

The inspiration behind this work is to suggest a new 6th-order iterative technique based on weight function approach along with lower computational cost for large nonlinear systems. The beauty of using this approach is that it gives us the flexibility to produce new as well as some special cases of the earlier methods. A good variety

DOI: 10.1201/9781003128915-26

of applied science problems is considered in order to investigate the authenticity of our presented methods. Finally, using numerical experiments we show the superiority of our schemes when compared to others for computational cost, residual error and CPU-time. Moreover, they also shown the stable computational order of convergence and minimum asymptotic error constants in the contrast of exiting iterative methods.

26.1   MULTI-DIMENSIONAL CASE

Consider the following new scheme

$$V_\zeta = U_\zeta - \frac{2}{3}F'(U_\zeta)^{-1}F(U_\zeta),$$

$$W_\zeta = U_\zeta - P\Big(T(U_\zeta)\Big)\Big(F'(U_\zeta) + bF'(V_\zeta)\Big)^{-1}F(U_\zeta), \qquad (26.2)$$

$$X_{\zeta+1} = W_\zeta + 2Q(U_\zeta)^{-1}F(W_\zeta),$$

where $T : \mathbb{D} \subseteq \mathbb{R}^m \longrightarrow \mathbb{R}^m$ be sufficiently differentiable in $\mathbb{D}$ with $T(U_\zeta) = F'(U_\zeta)^{-1}F'(V_\zeta)$, where $Q(U_\zeta) = F'(U_\zeta) - 3F'(V_\zeta)$. We demonstrate the sixth order convergence in Theorem 26.1 by adopting the same procedure suggested in [8].

Let $F : \mathbb{D} \subseteq \mathbb{R}^m \longrightarrow \mathbb{R}^m$ be sufficiently differentiable in $\mathbb{D}$. The kth derivative of F at $u \in \mathbb{R}^m$, $k \geq 1$, is the k-linear function $F^{(k)}(u) : \mathbb{R}^m \times \cdots \times \mathbb{R}^m \longrightarrow \mathbb{R}^m$ with $F^{(k)}(u)(v_1, \ldots, v_k) \in \mathbb{R}^m$, we have

1. $F^{(k)}(u)(v_1, \ldots, v_{k-1}, \cdot) \in \mathscr{L}(\mathbb{R}^m)$
2. $F^{(k)}(u)(v_{\sigma(1)}, \ldots, v_{\sigma(k)}) = F^{(k)}(u)(v_1, \ldots, v_k)$, for each permutation σ of $\{1, 2, \ldots, k\}$,

that further yield

(a) $F^{(k)}(u)(v_1, \ldots, v_k) = F^{(k)}(u)v_1 \ldots v_k$
(b) $F^{(k)}(u)v^{k-1}F^{(p)}v^p = F^{(k)}(u)F^{(p)}(u)v^{k+p-1}$

This $\xi + h \in \mathbb{R}^m$ containing in the neighborhood of required root ξ of $F(x) = 0$, we have

$$F(\xi + h) = F'(\xi)\left[h + \sum_{k=2}^{p-1} C_k h^k\right] + O(h^p), \qquad (26.3)$$

where $C_k = (1/k!)[F'(\xi)]^{-1}F^{(k)}(\xi)$, $k \geq 2$, provided $F'(\xi)$ is invertible. We recognize that $C_k h^k \in \mathbb{R}^m$ since $F^{(k)}(\xi) \in \mathscr{L}(\mathbb{R}^m \times \cdots \times \mathbb{R}^m, \mathbb{R}^m)$ and $[F'(\xi)]^{-1} \in \mathscr{L}(\mathbb{R}^m)$. We can also write

$$F'(\xi + h) = F'(\bar{x})\left[I + \sum_{k=2}^{p-2} kC_k h^{k-1}\right] + O(h^{p-1}), \qquad (26.4)$$

I for being the identity and $kC_k h^{k-1} \in \mathscr{L}(\mathbb{R}^m)$.

By (26.4), we get

$$[F'(\xi + h)]^{-1} = [I + U_2 h + U_3 h^2 + U_4 h^4 + \cdots][F'(\xi)]^{-1} + O(h^p), \qquad (26.5)$$

with

$$U_2 = -2C_2,$$
$$U_3 = 4C_2^2 - 3C_3,$$
$$U_4 = -8C_2^3 + 6C_2 C_3 - 4C_4 + 6C_3 C_2,$$
$$\vdots$$

The $e_\zeta = U_\zeta - \xi$ denotes as the error in the nth step. Then,

$$e_{\zeta+1} = M e_\zeta^p + O(e_\zeta^{p+1})$$

where M is a p-linear function $M \in \mathscr{L}(\mathbb{R}^m \times \cdots \times \mathbb{R}^m, \mathbb{R}^m)$, is known the *error equation* and p is the *convergence order*. Observe that e_ζ^p is $(e_\zeta, e_\zeta, \cdots, e_\zeta)$.

Theorem 26.1. Suppose $F : \mathbb{D} \subseteq \mathbb{R}^m \to \mathbb{R}^m$ is sufficiently differentiable mapping with open domain $\mathbb{D}$ that consists the required zero ξ. Further, we assume $F'(x)$ is invertible and continuous around ξ. Moreover, we consider the starting guess X_0 is close enough to ξ for sure convergence. Then, scheme (26.2) attain maximum sixth-order convergence provided

$$P(I) = (1+b)I, \ P'(I) = \frac{b-3}{4}I, \ P''(I) = \frac{3(b+3)}{4}I, \qquad (26.6)$$

where I as identity matrix.

Proof. We can write $F(U_\zeta)$ and $F'(U_\zeta)$ as follow:

$$F(U_\zeta) = F'(\xi)\left[e_\zeta + C_2 e_\zeta^2 + C_3 e_\zeta^3 + C_4 e_\zeta^4 + C_5 e_\zeta^5 + C_6 e_\zeta^6\right] + O(e_\zeta^7) \qquad (26.7)$$

and

$$F'(U_\zeta) = F'(\xi)\left[I + 2C_2 e_\zeta + 3C_3 e_\zeta^2 + 4C_4 e_\zeta^3 + 5C_5 e_\zeta^4 + 6C_6 e_\zeta^5\right] + O(e_\zeta^6), \qquad (26.8)$$

where I is the identity matrix of size $m \times m$ and $C_m = \frac{1}{m!} F'(\xi)^{-1} F^{(m)}(\xi)$, $m = 2, 3, 4, 5, 6$.
By expressions (26.7) and (26.8), we get

$$F'(U_\zeta)^{-1} = \left[I - 2C_2 e_\zeta + (4C_2^2 - 3C_3)e_\zeta^2\right] F'(\xi)^{-1} \qquad (26.9)$$

and

$$F'(U_\zeta)^{-1} F(U_\zeta) = e_\zeta - C_2 e_\zeta^2 + (2C_2^2 - 2C_3)e_\zeta^3 + O(e_\zeta^4). \qquad (26.10)$$

Using expression (26.10) in (26.2), we have

$$V_\zeta - \xi = \frac{1}{3}e_\zeta + \frac{2}{3}C_2 e_\zeta^2 - \frac{2}{3}(2C_2^2 - 2C_3)e_\zeta^3 + O(e_\zeta^4), \qquad (26.11)$$

which further produce

$$F'(V_\zeta) = F'(\xi) \left[I + \frac{2}{3}C_2 e_\zeta + \frac{1}{3}(4C_2^2 + C_3)e_\zeta^2 \right] + O(e_\zeta^3) \tag{26.12}$$

and

$$F'(U_\zeta) + bF'(V_\zeta)$$
$$= F'(\xi) \left[(1+b)I + \frac{2(b+3)C_2}{3} e_\zeta + \frac{1}{3}(4bC_2^2 + (b+9)C_3)e_\zeta^2 \right] + O(e_\zeta^3). \tag{26.13}$$

We can easily obtain from the expressions (26.10) and (26.13), which is given as follow

$$T(U_\zeta) = F'(U_\zeta)^{-1}F'(V_\zeta) = I - \frac{4C_2 e_\zeta}{3} + \left(4C_2^2 - \frac{8C_3}{3} \right) e_\zeta^2 + O(e_\zeta^3). \tag{26.14}$$

We deduce from expression (26.14) that $T(U_\zeta) - I = O(e_\zeta)$. Moreover, we can write

$$P\left(T(U_\zeta) \right) = P(I) + P'(I)T(U_\zeta) + \frac{1}{2!}P''(I)T(U_\zeta)^2 + O\left(T(U_\zeta)^3 \right), \tag{26.15}$$

so

$$P(T(U_\zeta)) = (1+b)I - \frac{(b-3)}{3}C_2 e_\zeta + \frac{(5b-3)c_2^2 - 2(b-3)c_3}{3} e_\zeta^2 + O(e_\zeta^3). \tag{26.16}$$

By adopting the expressions (26.10) and (26.16), we have

$$P(T(U_\zeta))\left(F'(U_\zeta) + bF'(V_\zeta) \right)^{-1} F(U_\zeta)$$
$$= \left((1+b)I - \frac{(b-3)}{3}C_2 e_\zeta + \frac{(5b-3)c_2^2 - 2(b-3)c_3}{3} e_\zeta^2 + O(e_\zeta^3) \right)$$
$$\times \left(\frac{e_\zeta}{b+1} + \frac{(b-3)C_2}{3(b+1)^2} e_\zeta^2 - \frac{2(7b^2 + 6b - 9)C_2^2 - 6(b^2 - 2b - 3)C_3}{9(b+1)^3} e_\zeta^3 + O(e_\zeta^4) \right),$$
$$= e_\zeta + O(e_\zeta^4). \tag{26.17}$$

Then, using expression (26.17) in (26.2), we yield

$$W_\zeta - \xi = \alpha_1 e_\zeta^4 + \alpha_2 e_\zeta^5 + \alpha_3 e_\zeta^6 + O(e_\zeta^7), \tag{26.18}$$

where α_i, $i = 1, 2, 3$ depend on some b and C_i, $2 \le j \le 6$.
Moreover, we have

$$F(W_\zeta) = F'(\xi) \left[\alpha_1 e_\zeta^4 + \alpha_2 e_\zeta^5 + \alpha_3 e_\zeta^6 \right] + O(e_\zeta^7). \tag{26.19}$$

After some simple algebraic calculations, we have

$$2Q(U_\zeta)^{-1}F(W_\zeta) = 2\left(F'(U_\zeta) + F(V_\zeta) \right)^{-1} F(W_\zeta)$$
$$= -\alpha_1 e_\zeta^4 - \alpha_2 e_\zeta^5 + \left(-\alpha_3 + 2\alpha_1 c_2^2 - \alpha_1 c_3 \right) e_\zeta^6 + O(e_\zeta^7). \tag{26.20}$$

Finally, we have

$$X_{\zeta+1} - \xi = W_\zeta - \xi + 2Q(U_\zeta)^{-1}F(W_\zeta) = \alpha_1 \left(2c_2^2 - c_3\right)e_\zeta^6 + O(e_\zeta^7), \qquad (26.21)$$

where α_1 is a function of only b, C_2, C_3, C_4.   □

26.1.1 SPECIALIZATIONS

Some of the fruitful cases are mentioned below:
(1) We assume

$$P(U) = \frac{3}{2}U^2 - \frac{1}{2}U + 4I$$

for $b = 1$ which generate the following new sixth-order Jarratt type scheme

$$\begin{cases} V_\zeta = U_\zeta - \dfrac{2}{3}F'(U_\zeta)^{-1}F(U_\zeta), \\[2mm] W_\zeta = U_\zeta - \left(\dfrac{3}{2}\left(F'(U_\zeta)^{-1}F'(V_\zeta)\right)^2 - \dfrac{1}{2}\left(F'(U_\zeta)^{-1}F'(V_\zeta)\right) + 4I\right) \\[2mm] \qquad\quad \times \left(F'(U_\zeta) + F'(V_\zeta)\right)^{-1}F(U_\zeta), \\[2mm] U_{\zeta+1} = W_\zeta + 2Q(U_\zeta)^{-1}F(W_\zeta). \end{cases} \qquad (26.22)$$

(2) Consider the following weight function for $b = 0$.

$$P(U) = \frac{3U}{8} + \frac{9}{8}U^{-1} - \frac{1}{2}I.$$

leading to

$$\begin{cases} V_\zeta = U_\zeta - \dfrac{2}{3}F'(U_\zeta)^{-1}F(U_\zeta), \\[2mm] W_\zeta = U_\zeta - \left(\dfrac{3}{8}\left(F'(U_\zeta)^{-1}F'(V_\zeta)\right) + \dfrac{9}{8}\left(F'(U_\zeta)^{-1}F'(V_\zeta)\right)^{-1} - \dfrac{1}{2}I\right) \\[2mm] \qquad\quad \times F'(U_\zeta)^{-1}F(U_\zeta), \\[2mm] U_{\zeta+1} = W_\zeta + 2Q(U_\zeta)^{-1}F(W_\zeta). \end{cases} \qquad (26.23)$$

(3) Now, we assume another weight function (for $b = 0.5$)

$$P(U) = \left(44I - 84U\right)^{-1}(41I - 101U),$$

that yields

$$\begin{cases} V_\zeta = U_\zeta - \dfrac{2}{3}F'(U_\zeta)^{-1}F(U_\zeta), \\[2mm] W_\zeta = U_\zeta - \left(44I - 84F'(U_\zeta)^{-1}F'(V_\zeta)\right)^{-1}\left(41I - 101F'(U_\zeta)^{-1}F'(V_\zeta)\right) \\[2mm] \qquad\quad \times \left(F'(U_\zeta) + 0.5F'(V_\zeta)\right)^{-1}F(U_\zeta), \\[2mm] U_{\zeta+1} = W_\zeta + 2Q(U_\zeta)^{-1}F(W_\zeta), \end{cases}$$

$$(26.24)$$

is another new sixth-order scheme.

In like manner, we can obtain many familiar and advance 6th-order Jarratt type scheme by adopting different weight functions.

26.2 LOCAL CONVERGENCE ANALYSIS

It is well known that iterative methods defined on the real line or on the m−dimensional Euclidean space constitute the motivation for extending these methods to more abstract spaces such as Hilbert or Banach or other spaces. The local convergence analysis of method (26.2) after defining it for Banach space operators for all $\zeta = 0,1,2,3,\ldots$ as

$$V_\zeta = U_\zeta - \frac{2}{3} F'(U_\zeta)^{-1} F(U_\zeta),$$

$$W_\zeta = U_\zeta - P\big(T(U_\zeta)\big)\big(F'(U_\zeta) + bF'(V_\zeta)\big)^{-1} F(U_\zeta), \qquad (26.25)$$

$$U_{\zeta+1} = W_\zeta + 2Q(U_\zeta)^{-1} F(W_\zeta),$$

where $\mathbb{E}_1$, $\mathbb{E}_2$ are Banach spaces, $\Omega \subseteq \mathbb{E}_1$ is nonempty, convex and open subset of $\mathbb{E}_1$, $T(U) = F'(U)^{-1}F'(V)$, $Q(U) = F'(U) - 3F'(V)$, $b \in \mathbb{R} - \{-1\}$ and $T : \mathbb{E}_1 \to \mathbb{E}_2$ is such that $P(I) = (1+b)I$. Then, under certain hypotheses given later method (26.25) converges to a solution U_* of equation

$$F(U) = 0, \qquad (26.26)$$

where $F : \Omega \to \mathbb{E}_2$ is a continuously differentiable operator in the sense of Fréchet. For acceptable convergence analysis, we first to define some parameters and scalar functions. Let $\psi_0 : \mathbb{T} \to \mathbb{T}$ be a continuous and increasing function with $\psi_0(0) = 0$, where $\mathbb{T} = [0, +\infty)$. The equation

$$\psi_0(t) = 1 \qquad (26.27)$$

has minimum one positive zero. Denote by ρ_0 the smallest such solution and set $\mathbb{T}_0 = [0, \rho)$. Let $\psi : \mathbb{T}_0 \to \mathbb{T}$, $\psi_1 : \mathbb{T}_0 \to \mathbb{T}$ be continuous and increasing maps with $\psi(0) = 0$. Consider maps φ_1 and $\bar{\varphi}_1$ on $\mathbb{T}_0$ as

$$\varphi_1(t) = \frac{\int_0^1 \psi\big((1-\tau)t\big)d\tau + \frac{1}{3}\int_0^1 \psi_1(\tau t)d\tau}{1 - \psi_0(t)}$$

and

$$\bar{\varphi}_1(t) = \varphi_1(t) - 1.$$

Suppose that

$$\frac{\psi_1(t)}{3} < 1. \qquad (26.28)$$

Then by the definition of function $\bar{\varphi}_1$ and (26.26), we have $\bar{\varphi}_1(0) = \frac{\psi_1(t)}{3} - 1 < 0$ and $\bar{\varphi}_1(t) \to +\infty$, as $t \to \rho_0^-$. Then, by the mean value theorem there exists at least one solution of $\bar{\varphi}_1(t) = 0$ in the interval $(0, \rho_0)$. Denoted by R_1 the smallest such solution.

Suppose

$$\lambda(t) = 1 \qquad (26.29)$$

where

$$\lambda(t) = \frac{1}{|1+b|}\left(\psi_0(t) + |b|\psi_0(\varphi_1(t)t)\right)$$

has minimum one positive zero. Dented by ρ_λ the smallest such solution and set $\mathbb{T}_1 = [0, \rho_1)$, $\rho_1 = \min\{\rho_0, \rho_\lambda\}$.

Further, we consider functions φ_2 and $\bar{\varphi}_2$ on $\mathbb{T}_1 = [0, \rho)$ by

$$\varphi_2(t) = \frac{\int_0^1 \psi((1-\tau)t)d\tau}{1 - \psi_0(t)} + \frac{\left(\psi_0(t) + \psi_0(\varphi_1(t)t)\right)A(t) + \psi_1(\varphi_1(t)t)B(t)\int_0^1 \psi_1(\tau t)d\tau}{(1 - \lambda(t))(1 - \psi_0(t))}$$

and

$$\bar{\varphi}_2(t) = \varphi_2(t) - 1,$$

where $A : \mathbb{T}_1 \to \mathbb{T}$ and $B : \mathbb{T}_1 \to \mathbb{T}$ are continuous and increasing functions. We also get $\bar{\varphi}_2(0) = -1$ and $\bar{\varphi}_2(t) \to +\infty$, as $t \to \rho_0^-$. Recall by R_2 the smallest solution of equation $\bar{\varphi}_2(t) = 0$ in $(0, \rho_1)$. Suppose that the equations

$$g(t) = 1, \quad \psi_0(\varphi_2(t)t) = 1 \qquad (26.30)$$

have at least one positive solution, where $g(t) = \frac{1}{2}\left(\psi_0(t) + 3\psi_0(\varphi_1(t)t)\right)$. Denote by ρ_g, ρ_h the smallest such solutions and set $\mathbb{T}_2 = [0, \rho)$, $\rho = \min\{\rho_0, \rho_\lambda, \rho_g, \rho_h\}$. Define functions ψ_3 and $\bar{\psi}_3$ on the interval $\mathbb{T}_3 = [0, \rho)$ by

$$\varphi_3(t) = \left(\frac{\int_0^1 \psi((1-\tau)\Lambda)d\tau}{1 - \psi_0(\varphi_2(t))} + \frac{\left(\psi_0(t) + 3\psi_0(\varphi_1(t)t) + 2\psi_0(\Lambda)\right)\int_0^1 \psi_1(\tau\Lambda)d\tau}{1 - \psi_0(\Lambda)}\right)\varphi_2(t)$$

and

$$\bar{\varphi}_3(t) = \varphi_3(t) - 1.$$

where,

$$\Lambda = \varphi_2(t)t$$

We also get $\bar{\varphi}_3(0) = -1$ and $\bar{\varphi}_3(t) \to +\infty$, as $t \to \rho-$. Denote by R_3 the smallest solution of equation $\bar{\varphi}_3(t) = 0$ in $(0, \rho)$. Define a radius of convergence R by

$$R = \min\{R_i\}, \ i = 1, 2, 3. \qquad (26.31)$$

It follows from each $t \in [0, R)$

$$0 \le \psi_0(t) < 1, \qquad (26.32)$$

$$0 \le \psi_0(\varphi_1(t)t) < 1, \qquad (26.33)$$

$$0 \le \psi_0(\varphi_2(t)t) < 1, \qquad (26.34)$$

$$0 \le \varphi_1(t) < 1, \qquad (26.35)$$

$$0 \le \lambda(t) < 1 \qquad (26.36)$$

$$0 \le \varphi_2(t) < 1, \qquad (26.37)$$

$$0 \le g(t) < 1 \qquad (26.38)$$

and

$$0 \le \varphi_3(t) < 1 \qquad (26.39)$$

Define $K(U,a) = \{V \in \mathbb{E}_1 \text{ such that } \|U - V\| < a\}$. Let also $\bar{K}(U,a)$ stands for the closure of $K(U,a)$. By $\mathscr{LB}(\mathbb{E}_1, \mathbb{E}_2)$ denote the space of bounded linear operators from $\mathbb{E}_1$ in to $\mathbb{E}_2$.

The following conditions (H) are the base for the study of local convergence analysis:

(h_1) $F : \Omega \to \mathbb{E}_2$ with $U_* \in \Omega$ such that $F(U_*) = 0$ and $F'(U_*)^{-1}$ is invertible.

(h_2) There exists function $\psi_0 : \mathbb{T} \to \mathbb{T}$ continuous, increasing with $\psi_0(0) = 0$ such that for each $x \in \Omega$

$$\left\| F'(U_*)^{-1} \left(F'(U) - F'(U_*) \right) \right\| \le \psi_0(\|U - U_*\|).$$

Set $\Omega_0 = \Omega \cap S(U_*, \rho_0)$, where ρ_0 is given in (26.25).

(h_3) There exist functions $\psi : \mathbb{T}_0 \to \mathbb{T}$, $\psi_1 : \mathbb{T}_0 \to \mathbb{T}$, $A : \mathbb{T}_0 \to \mathbb{T}$ and $B : \mathbb{T}_0 \to \mathbb{T}$ continuous and increasing with $\psi(0) = 0$ such that for each $X, V \in \Omega_0$

$$\left\| F'(U_*)^{-1} \left(F'(V) - F'(U) \right) \right\| \le \psi(\|V - U\|),$$

$$\left\| F'(U_*)^{-1} F'(U) \right\| \le \psi_1(\|V - U\|),$$

$$\|I - K(T(U))\| \le A(\|V - U\|),$$

$$\|K(I) - K(T(U))\| \le B(\|V - U\|),$$

and

$$K(I) = (1 + b)I.$$

(h_4) $\bar{K}(U_*, R) \subset \Omega$, $\rho_0, \rho_\lambda, \rho_g$ exist, given by (26.25), (26.29), (26.30), respectively and R is defined in (26.31).

(h_5) There exists $R_* \ge R$ such that

$$\int_0^1 \psi_0(\tau R_*) d\tau < 1.$$

Set $\Omega_1 = \Omega \cap \bar{K}(U_*, R_*)$.

Next, we provide the local convergence analysis of method (26.25) using the hypotheses (H) and the previously developed notations.

Theorem 26.2. Assume the hypotheses (H) hold and $U_0 \in K(U_*,R) - \{U_*\}$. Then, the sequence $\{U_\zeta\} \subset K(U_*,R)$, $\lim\limits_{\zeta \to \infty} U_\zeta = U_*$ and

$$\|V_\zeta - U_*\| \le \varphi_1(\|U_\zeta - U_*\|)\|U_\zeta - U_*\| \le \|U_\zeta - U_*\| < R, \qquad (26.40)$$

$$\|W_\zeta - U_*\| \le \varphi_2(\|U_\zeta - U_*\|)\|U_\zeta - U_*\| \le \|U_\zeta - U_*\| \qquad (26.41)$$

and

$$\|X_{\zeta+1} - U_*\| \le \varphi_3(\|U_\zeta - U_*\|)\|U_\zeta - U_*\| \le \|U_\zeta - U_*\|. \qquad (26.42)$$

Moreover, U_* is the only solution of $F(x) = 0$ in the set Ω_1 given in (h_5).

Proof. Estimates (26.40)–(26.42) are shown utilizing the hypotheses (H) and mathematical induction. By (h_1), (h_2), (26.31) and (26.32), we have turn that

$$\|F'(U_*)^{-1}(F'(U) - F'(U_*))\| \le \psi_0(\|U - U_*\|) < \psi_0(R) < 1, \qquad (26.43)$$

for each $U \in K(U_*,R) - \{U_*\}$, so $F'(U)^{-1} \in \mathscr{LB}(\mathbb{E}_2,\mathbb{E}_1)$ and

$$\|F'(U)^{-1}F'(U_*)\| \le \frac{1}{1 - \psi_0(\|U - U_*\|)}, \qquad (26.44)$$

by the Banach perturbation lemma on invertible operators [4, 5]. Then, V_0 is well defined by method (26.25) for $\zeta = 0$. By convexity, we have that $U_* + \tau(U - U_*) \in K(U_*,R)$ for each $U \in K(U_*,R)$, by adopting (h_1), we get

$$F(U) = F(U) - F(U_*) = \int_0^1 F'(U_* + \tau(U - U_*))d\tau(U - U_*). \qquad (26.45)$$

From the hypotheses (h_1) and (h_3), we get

$$\|F'(U_*)^{-1}F(U)\| \le \int_0^1 \psi_1(\tau\|U - U_*\|)d\tau\|U - U_*\|. \qquad (26.46)$$

Using the first substep of method (26.25) for $\zeta = 0$, (26.31), (26.35), (h_3), (26.44) (for $U = U_0$), and (26.46) (for $U = U_0$), we have in turn that

$$\|V_0 - U_*\| = \left\| (U_0 - U_*F'(U_0)^{-1}F(U_0)) + \frac{1}{3}F'(U_0)^{-1}F(U_0) \right\|$$

$$\le \|F'(U_0)^{-1}F'(U_*)\| \left\| \int_0^1 F'(U_*)^{-1}(F'(U_* + \tau(U_0 - U_*)) - F'(U_0))d\tau \right\|$$

$$+ \frac{1}{3}\|F'(U_0)^{-1}F'(U_0)\|\|F'(U_0)^{-1}F(U_0)\|$$

$$\le \frac{\left[\int_0^1 \psi((1 - \tau)\|U_0 - U_*\|)d\tau + \frac{1}{3}\int_0^1 \psi_1(\tau\|U_0 - U_*\|)d\tau \right]}{1 - \psi_0(\|U_0 - U_*\|)}$$

$$= \varphi_1(\|U_0 - U_*\|)\|U_0 - U_*\| \le \|U_0 - U_*\| < R,$$

$$(26.47)$$

so (26.40) holds for $\zeta = 0$ and $V_0 \in K(U_*, R)$. We must show that $\left(F'(U_0) + bF'(V_0) \right)^{-1} \in \mathscr{LB}(\mathbb{E}_2, \mathbb{E}_1)$.

By (26.31), (26.36), (h_2) and (26.47), we have

$$
\begin{aligned}
&\left\| \left((1+b)F'(U_*) \right)^{-1} \left[(F'(U_0) - F'(U_*)) + b(F'(V_0) + F'(U_*)) \right] \right\| \\
&\leq \frac{1}{|1+b|} \left[\left\| F'(U_*)^{-1} \left(F'(U_0) - F'(U_*) \right) \right\| + |b| \left\| F'(U_*)^{-1} \left(F'(V_0) - F'(U_*) \right) \right\| \right] \\
&\leq \frac{1}{|1+b|} \left[\psi_0(\|U_0 - U_*\|) + |b| \psi_0(\|V_0 - U_*\|) \right] \\
&\leq \lambda(\|U_0 - U_*\|) < \lambda(R) < 1,
\end{aligned}
$$

(26.48)

so

$$
\left\| \left(F'(U_0) + bF'(V_0) \right)^{-1} F'(U_*) \right\| \leq \frac{1}{|1+b|(1 - \lambda(\|U_0 - U_*\|))}.
$$

(26.49)

Then, since W_0 is well defined by (26.24) and the second substep of method (26.25), we can write

$$
\begin{aligned}
W_0 - U_* =\ & W_0 - U_* - F'(U_0)^{-1} F(U_0) \\
& + \left[F'(U_0)^{-1} - P(T(U_0)) \left(F'(U_0) + bF'(V_0) \right)^{-1} \right] F(U_0)
\end{aligned}
$$

(26.50)

We need an estimate on the expression inside the bracket in (26.50):

$$
\begin{aligned}
&F'(U_0)^{-1} - P(T(U_0)) \left(F'(U_0) + bF'(V_0) \right)^{-1} \\
&= F'(U_0)^{-1} \left[I - F'(U_0) P(T(U_0)) \left(F'(U_0) + bF'(V_0) \right)^{-1} \right] \\
&= F'(U_0)^{-1} \left[F'(U_0) + bF'(V_0) - F'(U_0) P(T(U_0)) \right] \left(F'(U_0) + bF'(V_0) \right)^{-1} \\
&= F'(U_0)^{-1} \left[\left(F'(U_0) - F'(U_*) \right) + \left(F'(U_*) - F'(V_0) \right) ((I - P(T(U_0))) \right. \\
&\quad \left. + F'(V_0) \left(P(I) - P(T(U)) \right) \right] \left(F'(U_0) + bF'(V_0) \right)^{-1},
\end{aligned}
$$

(26.51)

so by (h_3), (26.44) and (26.49), we have in turn that (26.51) in norm is bounded above by

$$
\|F'(U_0)^{-1} F'(U_*)\| \left[\left(\|F'(U_*)^{-1}(F'(U_0) - F'(U_*))\| + \|F'(U_*)^{-1}(F'(V_0) - F'(U_*))\| \right) \right.
$$
$$
\left. \times \|I - P(T(U_0))\| \|F'(U_*)^{-1} F'(V_0)\| \|P(I) - P(T(U_0))\| \right] \frac{1}{\lambda(\|U_0 - U_*\|)},
$$

(26.52)

and (26.31), (26.37), (26.44) (for $U = U_0$), (26.46) (for $U = U_0$), (26.47), (26.49) and (26.52), we yield

$$
\begin{aligned}
&\|W_0 - U_*\| \\
&= \|U_0 - U_* - F'(U_0)^{-1}F(U_0)\| + \|\left[F'(U_0)^{-1} - P(T(U_0))\left(F'(U_0) + bF'(V_0)\right)^{-1}\right] \\
&\quad \times F'(U_*)\|\|F'(U_*)^{-1}F(U_0)\| \\
&\leq \left[\frac{\int_0^1 \psi((1-\tau)\|U_0 - U_*\|)d\tau}{1 - \psi_0(\|U_0 - U_*\|)} + \frac{1}{(1 - \psi_0(\|U_0 - U_*\|))(1 - \lambda(\|U_0 - U_*\|))} \right. \\
&\quad \times \left[\left(\psi_0(\|U_0 - U_*\| + \psi_0(\varphi_1(\|U_0 - U_*\|)\|U_0 - U_*\|))\right)A(\|U_0 - U_*\|) \right. \\
&\quad \left.\left. + \psi_1(\varphi_1(\|U_0 - U_*\|)\|U_0 - U_*\|)B(\|U_0 - U_*\|)\int_0^1 \psi_1(\tau\|U_0 - U_*\|)d\tau\right]\right]\|U_0 - U_*\| \\
&= \varphi_2(\|U_0 - U_*\|)\|U_0 - U_*\| \leq \|U_0 - U_*\| < R,
\end{aligned}
$$

(26.53)

so (26.41) holds for $\zeta = 0$ and $W_0 \in K(U_*, R)$. Next, we must show that $Q(U_0)^{-1} \in \mathscr{LB}(\mathbb{E}_2, \mathbb{E}_1)$. Using (26.31), (26.38), (h_2) and (26.47), we get that

$$
\begin{aligned}
&\|(-2F'(U_*)^{-1})\left(F'(U_0) - F'(U_*) - 3(F'(V_0) - F'(U_*))\right)\| \\
&= \frac{1}{2}\left[\|F'(U_*)^{-1}(F'(U_0) - F'(U_*))\| + 3\|F'(U_*)^{-1}(F'(V_0) - F'(U_*))\|\right] \\
&= \frac{1}{2}\left[\psi_0(\|U_0 - U_*\|) + 3\psi_0(\|V_0 - U_*\|)\right] \\
&\leq g(\|U_0 - U_*\|) < 1,
\end{aligned}
$$

(26.54)

So, $Q(U_0)^{-1} \in \mathscr{LB}(\mathbb{E}_2, \mathbb{E}_1)$ and

$$
\|Q(U_0)^{-1}F'(U_*)\| \leq \frac{1}{2(1 - g(\|U_0 - U_*\|))} \tag{26.55}
$$

Hence, U_1 is well defined by the last substep of method (26.25). By adopting (26.31), (26.39), (26.53) and (26.55), we have

$$
\begin{aligned}
U_1 - U_* &= W_0 - U_* - F'(W_0)^{-1}F(W_0) + (F'(W_0)^{-1} + 2Q(U_0)^{-1})F(W_0) \\
&= W_0 - U_* - F'(W_0)^{-1}F(W_0) + F'(W_0)^{-1}\left[(F'(U_0) - F'(U_*))\right. \\
&\quad \left. - 3(F'(V_0) - F'(U_*)) + 2(F'(W_0) - F'(U_0))\right]Q(U_0)^{-1}F(W_0)
\end{aligned}
$$

(26.56)

from which, we get that

$$\|U_1 - U_*\|$$
$$\leq \|W_0 - U_* - F'(W_0)^{-1} F(W_0)\| + \|F'(W_0)^{-1} F'(U_*)\|$$
$$\times \Big[\|(F'(U_*)^{-1}(F'(U_0) - F'(U_*)))\|$$
$$+ 3\|(F'(U_*)^{-1}(F'(V_0) - F'(U_*))\| + 2\|(F'(U_*)^{-1}(F'(W_0) - F'(U_*)))\| \Big]$$
$$\times \|Q(U_0)^{-1} F'(U_*)\| \|F'(U_*)^{-1} F(W_0)\|$$
$$\leq \frac{\int_0^1 \psi((1-\tau)\|U_0 - U_*\|)d\tau}{1 - \psi_0(\|U_0 - U_*\|)}$$
$$+ \frac{\Big(\psi_0(\|U_0 - U_*\|) + 3\psi_0(\|V_0 - U_*\|) + 2\psi_0(\|W_0 - U_*\|) \Big) \int_0^1 \psi_1(\tau\|U_0 - U_*\|)d\tau}{(1 - \psi_0(\|W_0 - U_*\|))(1 - g(\|U_0 - U_*\|))}$$
$$\leq \varphi_3(\|U_0 - U_*\|)\|U_0 - U_*\| \leq \|U_0 - U_*\| < R,$$

$$(26.57)$$

so (26.42) holds for $\zeta = 0$ and $U_1 \in K(U_*, R)$.

Then, replace U_0, V_0, W_0, U_1 by U_j, V_j, W_j, U_{j+1} in the preceding estimations to finish the induction for (26.40) – (26.42). In view of the estimate

$$\|U_{j+1} - U_*\| \leq r\|U_j - U_*\| < R, \quad r = \psi_3(\|U_0 - U_*\|) \in [0, 1), \qquad (26.58)$$

we conclude that $\lim_{j \to \infty} U_j = U_*$ and $U_{j+1} \in K(U_*, R)$. Let us consider that $V_* \in \Omega$ be such that $F(V_*) = 0$. Using $K1 = \int_0^1 F'(V_* + \tau(U_* - V_*))d\tau$, (h_2) and (h_5), we have

$$\|F'(U_*)^{-1}(K - F'(U_*))\| \leq \| \int_0^1 \psi_0(\tau\|U_* - V_*\|)d\tau$$
$$\leq \int_0^1 \psi_0(\tau R)d\tau < 1,$$

$$(26.59)$$

so $K1^{-1} \in \mathscr{LB}(\mathbb{E}_1, \mathbb{E}_2)$. Therefore, by the identity

$$0 = F(U_*) - F(V_*) = K(U_* - V_*), \qquad (26.60)$$

we deduce that $U_* = V_*$ $\qquad\qquad\qquad\qquad\qquad\qquad\qquad\qquad\qquad\quad\Box$

Application 3.2: Let us see how functions A and B can be chosen when P is given above (26.22). We get in turn

$$I - P(T(U_\zeta))$$
$$= I - \frac{3}{2} T(U_\zeta)^2 + \frac{1}{2} T(U_\zeta) - 4I$$
$$= \frac{1}{2} \Big(T(U_\zeta) - I \Big) - \frac{3}{2} \Big(T(U_\zeta)^2 - I \Big) - I$$
$$= \frac{1}{2} \Big(F'(U_\zeta)^{-1} F'(V_\zeta) - I \Big) - \frac{3}{2} \Big((F'(U_\zeta)^{-1} F'(V_\zeta))^2 - I \Big) - I$$
$$= \frac{1}{2} F'(U_\zeta)^{-1} \Big(F'(V_\zeta) - F'(U_\zeta) \Big)$$
$$+ \frac{3}{2} \Big[\Big(F'(U_\zeta)^{-1}(F'(V_\zeta) - F'(U_\zeta)) \Big)^2 + 2F'(U_\zeta)^{-1} F'(V_\zeta) \Big] - I$$

$$(26.61)$$

so

$$\|I - P(T(U_\zeta))\|$$
$$= \frac{1}{2} \frac{\psi_0(\|V_\zeta - U_*\|) + \psi_0(\|U_\zeta - U_*\|)}{1 - \psi_0(\|U_\zeta - U_*\|)}$$
$$+ \frac{3}{2} \left[\left(\frac{\psi_0(\|V_\zeta - U_*\|) + \psi_0(\|U_\zeta - U_*\|)}{1 - \psi_0(\|U_\zeta - U_*\|)} \right)^2 + \frac{2\psi_0(\|V_\zeta - U_*\|)}{1 - \psi_0(\|U_\zeta - U_*\|)} \right] + 1$$

Hence, function A can be defined by

$$A(t) = \frac{1}{2} \frac{\psi_0(\varphi_1(t)t) + \psi_0(t)}{1 - \psi_0(t)} + \left[\left(\frac{\psi_0(\varphi_1(t)t) + \psi_0(t)}{1 - \psi_0(t)} \right)^2 + \frac{2\psi_1(\varphi_1(t)t)}{1 - \psi_0(t)} \right] + 1$$

$$(26.62)$$

Similarly

$$P(I) - P(T(U_\zeta)) = \frac{3}{2}I^2 - \frac{1}{2}I + 4I - \frac{3}{2}T(U_\zeta) + \frac{1}{2}T(U_\zeta) - 4I$$
$$= \frac{3}{2} \left(I^2 - (F'(U_\zeta)^{-1}F'(V_\zeta))^2 \right) - \frac{1}{2} \left(I - F'(U_\zeta)^{-1}F'(V_\zeta) \right),$$

$$(26.63)$$

so, we can define B by

$$B(t) = \frac{3}{2} \left[\left(\frac{\psi_0(\varphi(t)t) + \psi_0(t)}{1 - \psi_0(t)} \right)^2 + \frac{\psi_1(\varphi_1(t)t)}{1 - \psi_0(t)} \right]$$
$$+ \frac{1}{2} \frac{\psi_0(\varphi_1(t)t) + \psi_0(t)}{1 - \psi_0(t)},$$

$$(26.64)$$

$$= A(t) - 1.$$

Remark 26.1. The results in this section are obtained using hypotheses only on the first derivative in contrast to the results in Theorem 26.1 where hypotheses up to seventh derivative of F are used to show the convergence order six. Hence, we have extended the usage of method (26.25) in Banach space valued operators. Notice also that there are even simple functions defined on the real line, where the hypotheses of Theorem 26.1 do not hold. Hence, the method 26.2 may or may not converge. As a motivational and academic example (see example 6 in the next section). Then, notice that the third derivative of F does not exist. Using the approach of Theorem (26.2), we bypass the computation of higher order than one derivatives. But, we can still obtain the convergence order (COC) or the approximate computational order (ACOC) that do not use higher than one derivatives.

26.3 APPLICATIONS

Here, we demonstrate the suitability of our iterative methods on real life complications. In addition, we also want to validate our theoretical results which were

presented in earlier sections. Therefore, we consider four real life issues, namely,
Bratu's one-dimension, Fisher's, kinematic synthesis and Hammerstein integration
problems), fifth one is the standard academic problem and sixth one is a motivational
problem. The corresponding starting initial approximation and zeros are depicted in
examples (26.1)–(26.6).

Next, we consider our schemes namely, (26.22), (26.23) and (26.24) recalled as
($NM1$), ($NM2$) and ($NM3$), respectively to investigate the computational conduct of
them with existing techniques. We contrast them with sixth-order schemes given by
Hueso et al. [13] and Lotfi et al. [17], out of them we consider the expressions nam-
ley, (14–15) $\left(\text{for } t_1 = -\frac{9}{4} \text{ and } s_2 = \frac{9}{8}\right)$ and (5), known as (HU) and (LO), respectively. In addition, we also compare them with an Ostrowski type method proposed
by Grau-Sánchez et al. [11], among them we choose the iterative scheme (7), denoted
by (GR). Further, we contrast them with sixth-order iterative schemes presented by
Sharma and Arora [22] (expression-18) and Abbasbandy et al. [2] (expression-8),
called by (SA) and (AB), respectively. Furthermore, we contrast them with solution
techniques of order six designed by Soleymani et al. [24] (method-5) and Wang and
Li [28] (method-6), known as (SO) and (WL), respectively.

In the Tables 26.1, 26.2, 26.4, 26.6 and 26.7, we report our findings iteration
indexes (n), $(\|F(x_\zeta)\|)$, $\|U_{\zeta+1} - U_\zeta\|$, $\rho = \frac{\log\left[\|U_{\zeta+1}-X_\zeta\|/\|U_\zeta-U_{\zeta-1}\|\right]}{\log\left[\|U_\zeta-U_{\zeta-1}\|/\|U_{\zeta-1}-U_{\zeta-2}\|\right]}$, $\frac{\|U_{\zeta+1}-U_\zeta\|}{\|U_\zeta-U_{\zeta-1}\|^6}$
and η by using Mathematica (Version 9) with multiple precision arithmetic and min-
imum 300 digits of mantissa that minimizing the rounding-off errors. Further, the
variable η is the last obtained value of $\frac{\|U_{\zeta+1}-U_\zeta\|}{\|U_\zeta-U_{\zeta-1}\|^6}$. The α $(\pm\beta)$ indicates $\alpha \times 10^{(\pm\beta)}$.

Example 26.1. Bratu Problem
We can find the huge applicability of Bratu Problem [3] in the areas of thermal re-
action, the Chandrasekhar model of the expansion of the universe, chemical reactor
theory, radiative heat transfer, fuel ignition model of thermal combustion and nan-
otechnology [10, 14, 15, 27]. The mathematical formulation of this problem is given
below:

$$y'' + Ce^y = 0, \ y(0) = y(1) = 0. \tag{26.65}$$

By adopting following central difference

$$y''_\sigma = \frac{y_{\sigma-1} - 2y_\sigma + y_{\sigma+1}}{h^2}, \ \sigma = 1, 2, \ldots, 51,$$

that yields the following nonlinear system of size 50×50 from BVP (26.65) with
step size $h = 1/50$

$$h^2 C \exp(x_\sigma) + (x_{\sigma-1} + x_{\sigma+1} - 2x_\sigma) = 0, \ \sigma = 1, 2, \ldots, 51. \tag{26.66}$$

We consider $C = 3$ and initial value $(sin(\pi h), sin(2\pi h), \ldots, sin(50\pi h))^T$ for this
problem and computational out comings are depicted in table 26.1.

Methods	ζ	$\|F(U_\zeta)\|$	$\|U_{\zeta+1}-U_\zeta\|$	ρ^*	$\dfrac{\|U_{\zeta+1}-U_\zeta\|}{\|U_\zeta-U_{\zeta-1}\|^6}$	η
4*HU	1	2.2($-$4)	1.2($-$2)			
	2	2.1($-$11)	1.1($-$8)		1.083162484($-$4)	3.174505914($+$1)
	3	1.3($-$46)	7.0($-$44)	5.3632	3.174505914($+$1)	
4*LO	1	2.2($-$4)	1.2($-$1)			
	2	5.7($-$13)	3.1($-$10)		1.87422921($-$4)	5.428184301($+$1
	3	9.7($-$58)	2.8($-$56)	5.3634	5.428184301($+$1)	
4*GR	1	1.3($-$5)	7.1($-$3)			
	2	9.0($-$21)	4.8($-$18)		3.793760743($-$5)	3.818203150($-$5)
	3	8.9($-$112)	4.8($-$109)	5.9998	3.818203150($-$5)	
4*SA	1	1.1($-$3)	5.5($-$1)			
	2	7.4($-$9)	4.0($-$6)		1.486642520($-$4)	3.377015100($-$4)
	3	2.6($-$39)	1.4($-$36)	5.9306	3.377015100($-$4)	
4*AB	1	1.9($-$3)	9.1($-$1)			
	2	1.4($-$7)	7.6($-$5)		1.306885506($-$4)	5.224036587($-$4)
	3	1.8($-$31)	9.9($-$29)	5.8525	5.224036587($-$4)	
4*SO	1	3.0($-$6)	1.6($-$3)			
	2	7.3($-$25)	3.9($-$22)		2.402458013($-$5)	2.407289926($-$5)
	3	1.5($-$136)	8.2($-$134)	6.0000	2.407289926($-$5)	
4*WL	1	8.2($-$4)	4.2($-$1)			
	2	1.5($-$9)	8.1($-$7)		1.427111276($-$4)	2.712173991($-$4)
	3	1.4($-$43)	7.4($-$41)	5.9512	2.712173991($-$4)	
4*NM1	1	6.7($-$5)	3.5($-$2)			
	2	3.6($-$16)	1.9($-$13)		9.787819663($-$5)	1.020710352($-$4)
	3	9.7($-$84)	5.2($-$81)	5.9984	1.020710352($-$4)	
4*NM2	1	3.7($-$5)	2.0($-$2)			
	2	8.2($-$18)	4.4($-$15)		6.899443783($-$5)	7.050512396($-$5)
	3	9.3($-$94)	5.0($-$91)	5.9993	7.050512396($-$5)	
4*NM3	1	1.2($-$6)	6.5($-$4)			
	2	1.3($-$27)	7.5($-$25)		9.459851406($-$6)	9.479844812($-$6)
	3	2.5($-$153)	1.3($-$150)	6.0000	9.479844812($-$6)	

Table 26.1: Conduct of different techniques on Bratu's problem 26.1.

Example 26.2. Here, we choose another well known Fisher's equation [21]

$$\theta_t = D\theta_{xx} + \theta(1-\theta) = 0,$$

with homogeneous Neumann's boundary conditions

$$\theta(x,0) = 1.5 + 0.5\cos(\pi x), 0 \le x \le 1, \tag{26.67}$$

$$\theta_x(0,t) = 0, \forall t \ge 0,$$

$$\theta_x(1,t) = 0, \forall t \ge 0,$$

where D is the diffusion parameter. We adopt finite difference discretization technique, in order to convert the above differential equation (26.67) in to a system of nonlinear equations. So, we choose $w_{i,j} = \theta(x_i,t_j)$ as the required solution at the grid points of the mesh. In addition, x and t are the numbers of steps in the direction of M and N, respectively. Moreover, h and k, respectively are corresponding step size of M and N. By adopting central, backward and forward difference, we have

$$\theta_{xx}(x_i, t_j) = (w_{i+1,j} - 2w_{i,j} + w_{i-1,j})/h^2,$$
$$\theta_t(x_i, t_j) = (w_{i,j} - w_{i,j-1})/k,$$
and
$$\theta_x(x_i, t_j) = (w_{i+1,j} - w_{i,j})/(h), \quad t \in [0,1],$$

that leading to

$$\frac{w_{1,j} - w_{i,j-1}}{k} - w_{i,j}\left(1 - w_{i,j}\right) - D\frac{w_{i+1,j} - 2w_{i,j} + w_{i-1,j}}{h^2}, \tag{26.68}$$

where

$$h = \frac{1}{M}, k = \frac{1}{N}, i = 1, 2, 3, \ldots, \text{ and } M, j = 1, 2, 3, \ldots, N,.$$

For particular values of $M = 9$, $N = 9$, $h = \frac{1}{9}, k = \frac{1}{9}$ and $D = 1$ that leads us to a nonlinear system of size 81×81, with the starting point $x_0 = (i/81)^T, i = 1, 2, \ldots, 8$ convergence towards the following solution

$$u(x_i, t_j) = \begin{pmatrix} 1.6017\ldots, 1.4277\ldots, 1.3328\ldots, 1.2740\ldots, 1.2331\ldots, 1.2022\ldots, 1.1772\ldots, 1.1563\ldots, 1.1385\ldots \\ 1.5726\ldots, 1.4159\ldots, 1.3277\ldots, 1.2717\ldots, 1.2322\ldots, 1.2017\ldots, 1.1770\ldots, 1.1563\ldots, 1.1384\ldots \\ 1.5203\ldots, 1.3940\ldots, 1.3182\ldots, 1.2676\ldots, 1.2303\ldots, 1.2009\ldots, 1.1767\ldots, 1.1561\ldots, 1.1384\ldots \\ 1.4521\ldots, 1.3648\ldots, 1.3055\ldots, 1.2619\ldots, 1.2278\ldots, 1.1998\ldots, 1.1762\ldots, 1.1559\ldots, 1.1383\ldots \\ 1.3771\ldots, 1.3321\ldots, 1.2911\ldots, 1.2556\ldots, 1.2250\ldots, 1.1985\ldots, 1.1756\ldots, 1.1556\ldots, 1.1381\ldots \\ 1.3045\ldots, 1.2998\ldots, 1.2768\ldots, 1.2492\ldots, 1.2221\ldots, 1.1973\ldots, 1.1750\ldots, 1.1554\ldots, 1.1380\ldots \\ 1.2429\ldots, 1.2719\ldots, 1.2642\ldots, 1.2436\ldots, 1.2196\ldots, 1.1961\ldots, 1.1745\ldots, 1.1551\ldots, 1.1379\ldots \\ 1.1990\ldots, 1.2514\ldots, 1.2550\ldots, 1.2395\ldots, 1.2178\ldots, 1.1953\ldots, 1.1742\ldots, 1.1550\ldots, 1.1379\ldots \\ 1.1768\ldots, 1.2406\ldots, 1.2501\ldots, 1.2373\ldots, 1.2168\ldots, 1.1949\ldots, 1.1740\ldots, 1.1549\ldots, 1.1378\ldots \end{pmatrix}.$$

We depicted the numerical out coming in table 26.2.

Methods	ζ	$\|F(U_\zeta)\|$	$\|U_{\zeta+1}-U_\zeta\|$	ρ^*	$\dfrac{\|U_{\zeta+1}-U_\zeta\|}{\|U_\zeta-U_{\zeta-1}\|^6}$	η
4*HU	1	4.0	8.9(−1)			
	2	1.7(−5)	1.8(−61)		3.767384807(−6)	1.887278364
	3	8.4(−34)	7.6(−35)	4.9968	1.887278364	
4*LO	1	6.8	1.3			
	2	2.6(−5)	2.7(−6)		6.210769679(−7)	6.028784826(−3)
	3	2.7(−35)	2.6(−36)	5.2967	6.028784826(−3)	
4*GR	1	4.8(−1)	1.1(−1)			
	2	1.5(−12)	1.6(−13)		1.337154575(−13)	1.444113141(−13)
	3	1.5(−83)	1.4(−84)	5.9996	1.444113141(−13)	
4*SA	1	2.0(+1)	3.5			
	2	4.7(−3)	4.9(−4)		2.612933396(−7)	1.099716041(−4)
	3	1.5(−25)	1.5(−26)	5.8382	1.099716041(−4)	
4*AB	1	1.4(+1)	2.9			
	2	2.4(−3)	3.1(−4)		5.378390973(−7)	1.597885339(−6)
	3	1.5(−26)	1.4(−27)	5.8809	1.597885339(−6)	
4*SO	1	2.7(−1)	6.8(−2)			
	2	4.5(−14)	4.8(−15)		1.923091087(−12)	2.166790648(−12)
	3	4.8(−93)	4.7(−94)	5.9994	2.166790648(−12)	
4*WL	1	2.0(+1)	3.5			
	2	4.7(−3)	4.9(−4)		1.666475363(−16)	1.692929278(−16)
	3	1.5(−25)	1.5(−26)	5.9999	1.692929278(−16)	
4*NM1	1	1.4	4.1(−1)			
	2	1.3(−8)	1.8(−9)		2.468042987(−7)	2.892790896(−7)
	3	1.1(−58)	1.1(−59)	5.9917	2.892790896(−7)	
4*NM2	1	1.0	2.4(−1)			
	2	2.1(−10)	2.1(−11)		6.595078065(+5)	3.781460299(+4)
	3	1.9(−70)	1.8(−71)	6.2266	3.781460299(+4)	
4*NM3	1	3.6(−1)	8.7(−2)			
	2	3.6(−13)	3.9(−14)		2.391680319(+5)	7.549898932(+3)
	3	4.1(−87)	4.0(−88)	6.1921	7.549898932(+3)	

Table 26.2: Conduct of different techniques on Fisher's equation (26.2).

Example 26.3. Here, we choose a remarkable kinematic synthesis problem that related to steering as mentioned in [7, 26], which is defined as follows

$$[E_i (u_2 \sin(\psi_i) - u_3) - F_i (u_2 \sin(\phi_i) - u_3)]^2$$
$$+ [F_i (u_2 \cos(\phi_i) + 1) - F_i (u_2 \cos(\psi_i) - 1)]^2$$
$$- [u_1 (u_2 \sin(\psi_i) - u_3)(u_2 \cos(\phi_i) + 1) - u_1 (u_2 \cos(\psi_i) - u_3)(u_2 \sin(\phi_i) - u_3)]^2 = 0,$$
$$(26.69)$$

where

$$E_i = -u_3 u_2 (\sin(\phi_i) - \sin(\phi_0)) - u_1 (u_2 \sin(\phi_i) - u_3) + u_2 (\cos(\phi_i) - \cos(\phi_0)),$$
$$F_i = -u_3 u_2 \sin(\psi_i) + (-u_2) \cos(\psi_i) + (u_3 - u_1) u_2 \sin(\psi_0) + u_2 \cos(\psi_0) + u_1 u_3,$$

and

$$i = 1, 2, 3,$$

and The values of ψ_i and ϕ_i (in radians) are depicted in Table 26.3 and behavior of methods in Table 26.4. We choose the starting approximation $u_0 = (0.7, 0.7, 0.7)$ that converges to

$$\xi = (0.9051567\ldots, 0.6977417\ldots, 0.6508335\ldots)^T.$$

i	ψ_i	ϕ_i
0	1.3954170041747090114	1.7461756494150842271
1	1.7444828545735749268	2.0364691127919609051
2	2.0656234369405315689	2.2390977868265978920
3	2.4600678478912500533	2.4600678409809344550

Table 26.3: The parameters ψ_i and ϕ_i (in radians) used in example 26.3.

Methods	ζ	$\|F(U_\zeta)\|$	$\|U_{\zeta+1} - U_\zeta\|$	ρ^*	$\frac{\|U_{\zeta+1} - U_\zeta\|}{\|U_\zeta - U_{\zeta-1}\|^6}$	η
4*HU	1	3.0(−4)	7.0(−3)			
	2	8.7(−9)	9.3(−7)		7.812154542(+6)	8.147793628(+9)
	3	1.4(−28)	5.2(−27)	5.2217	8.147793628(+9)	
4*LO	1	1.2(−4)	2.7(−3)			
	2	4.8(−13)	3.0(−11)		7.926557348(+4)	2.247398482(+11)
	3	6.9(−53)	1.6(−52)	5.1887	2.247398482(+11)	
4*GR	1	1.2(−3)	3.7(−2)			
	2	6.3(−6)	6.1(−4)		2.198663751(+5)	2.496661657(+8)
	3	2.1(−13)	1.3(−11)	4.2910	2.496661657(+8)	
4*SA	1	7.5(−4)	4.9(−3)			
	2	3.3(−9)	6.5(−7)		4.841793597(+7)	1.033201098(+6)
	3	1.7(−33)	7.7(−32)	6.4311	1.033201098(+6)	
4*AB	1	1.1(−3)	5.3(−3)			
	2	9.1(−9)	1.3(−6)		5.600775513(+7)	2.359206614(+6)
	3	2.0(−31)	1.0(−29)	6.3799	2.359206614(+6)	
4*SO	1	4.3(−5)	5.1(−3)			
	2	3.8(−11)	2.1(−9)		1.188878144(+5)	1.015723644(+4)
	3	2.0(−50)	1.0(−48)	6.1675	1.015723644(+4)	
4*WL	1	5.8(−4)	3.9(−3)			
	2	1.0(−9)	2.0(−7)		5.475067615(+7)	8.521077783(+5)
	3	1.5(−36)	5.8(−35)	6.4215	8.521077783(+5)	
4*NM1	1	1.8(−4)	5.6(−3)			
	2	8.7(−10)	6.0(−8)		1.569574095(+6)	7.322404948(+4)
	3	1.2(−40)	4.1(−39)	6.2643	7.322404948(+4)	
4*NM2	1	1.3(−4)	5.6(−3)			
	2	3.8(−10)	2.5(−8)		7.167106213(+5)	4.034357383(+4)
	3	2.9(−43)	1.0(−41)	6.2301	4.034357383(+4)	
4*NM3	1	9.9(−5)	4.9(−3)			
	2	8.2(−11)	5.2(−9)		2.391680319(+5)	7.549898932(+3)
	3	1.2(−47)	4.9(−46)	6.1987	7.549898932(+3)	

Table 26.4: Conduct of different techniques on kinematic synthesis problem 26.3.

Example 26.4. We choose here a distinguished problem of applied science that is popular as Hammerstein integral equation (see [19, pp. 19–20], is given as follows:

$$x(s) = 1 + \frac{1}{5} \int_0^1 F(s,t)x(t)^3 dt \qquad (26.70)$$

where $x \in C[0,1]$, $s,t \in [0,1]$ and the kernel F is

$$F(s,t) = \begin{cases} (1-s)t, t \le s, \\ s(1-t), s \le t. \end{cases}$$

To convert the expression (26.70) into a finite-dimensional problem. We adopt the following Gauss Legendre quadrature formula:

$$\int_0^1 f(t)dt \simeq \sum_{j=1}^8 w_j f(t_j),$$

where t_j and w_j are the abscissas and the weights respectively. We recall $x(t_i)$ by $x_i (i = 1, 2, ..., 8)$, we have

$$5x_i - 5 - \sum_{j=1}^8 a_{ij} x_j^3 = 0, \text{ where} i = 1, 2, ..., 8$$

and

$$a_{ij} = \begin{cases} w_j t_j (1 - t_i), j \le i, \\ w_j t_i (1 - t_j), i < j. \end{cases}$$

The parameters t_j and w_j are mentioned depicted in Table 26.5.

j	t_j	w_j
1	0.0198550717512318841582195 7...	0.0506142681451881295762656 7...
2	0.101666761293186630204223 03...	0.111190517226687235272178 00...
3	0.237233795041835507091130 47...	0.156853322938943643668981 10...
4	0.408282678752175097530261 93...	0.181341891689180991482575 22...
5	0.591717321247824902469738 07...	0.181341891689180991482575 22...
6	0.762766204958164492908869 52...	0.156853322938943643668981 10...
7	0.898333238706813369795776 96...	0.111190517226687235272178 00...
8	0.980144928248768115841780 43...	0.050614268145188129576265 67...

Table 26.5: (Abscissas and weights for $t = 8$)

The desired root is $\xi^* = (1.002096..., 1.009900..., 1.019727..., 1.026436..., 1.026436..., 1.019727..., 1.009900..., 1.002096...)^T$. We depicted the numerical out coming in Table 26.6 on the ground of the starting approximation $U_0 = \left(\frac{1}{2}, \frac{1}{2}, \frac{1}{2}, \frac{1}{2}, \frac{1}{2}, \frac{1}{2}, \frac{1}{2}, \frac{1}{2}\right)^T$.

Methods	ζ	$\|F(U_\zeta)\|$	$\|U_{\zeta+1}-U_\zeta\|$	ρ^*	$\frac{\|U_{\zeta+1}-U_\zeta\|}{\|U_\zeta-U_{\zeta-1}\|^6}$	η
4*HU	1	3.0(−5)	6.5(−6)			
	2	6.9(−31)	1.5(−31)		1.994799598	9.220736175(+25)
	3	4.4(−159)	9.4(−160)	4.9991	9.220736175(+25)	
4*LO	1	8.6(−6)	1.8(−6)			
	2	8.4(−37)	1.7(−37)		4.445532921(−3)	1.232404905(+31)
	3	1.4(−189)	3.0(−190)	4.9924	1.232404905(+31)	
4*GR	1	6.7(−6)	1.4(−6)			
	2	1.0(−40)	2.1(−41)		2.526393028(−6)	2.741190361(−6)
	3	1.2(−249)	2.6(−250)	5.9990	2.741190361(−6)	
4*SA	1	9.4(−6)	2.0(−6)			
	2	1.5(−39)	3.2(−40)		4.964844066(−6)	5.324312398(−6)
	3	2.7(−242)	5.7(−243)	5.9991	5.324312398(−6)	
4*AB	1	5.5(−7)	1.2(−7)			
	2	8.3(−47)	1.8(−47)		6.557370741(−6)	7.0685748(−6)
	3	1.1(−285)	2.3(−286)	5.9992	7.0685748(−6)	
4*SO	1	5.4(−6)	1.2(−6)			
	2	1.9(−41)	4.1(−42)		1.622600124(−6)	1.775041548(−6)
	3	3.7(−254)	8.0(−255)	5.9999	1.775041548(−6)	
4*WL	1	1.6(−6)	3.3(−7)			
	2	9.0(−45)	1.9(−45)		1.377597868(−6)	1.431074607(−6)
	3	3.3(−274)	7.1(−275)	5.9996	1.431074607(−6)	
4*NM1	1	6.9(−6)	1.5(−6)			
	2	1.4(−40)	2.9(−41)		2.737455233(−6)	2.968285627(−6)
	3	9.1(−249)	2.0(−249)	5.9990	2.968285627(−6)	
4*NM2	1	6.4(−6)	1.4(−6)			
	2	7.1(−41)	1.5(−41)		2.315331324(−6)	2.514101750(−6)
	3	1.5(−250)	3.1(−251)	5.9990	2.514101750(−6)	
4*NM3	1	5.8(−6)	1.2(−6)			
	2	3.3(−41)	7.1(−42)		2.391680319(+5)	7.549898932(+3)
	3	1.3(−252)	2.7(−253)	6.1921	7.549898932(+3)	

Table 26.6: Conduct of different techniques on Hammerstein integral problem 26.4.

Example 26.5. Finally, we choose

$$F(U) = \begin{cases} u_j^2 u_{j+1} - 1 = 0, & 1 \le j \le \zeta, \\ u_\zeta^2 \zeta_1 - 1 = 0. \end{cases} \tag{26.71}$$

We pick $\zeta = 110$ in order to deduce a huge system of 110×110. In addition, we select the starting guess $U_0 = (1.25,\ 1.25,\ 1.25,\ \cdots,\ (110\ times))^T$ that converges to $\xi = (1,\ 1,\ 1,\ \cdots,\ (110\ times))^T$ and results are depicted in Table 26.7.

Methods	ζ	$\|F(U_\zeta)\|$	$\|U_{\zeta+1}-U_\zeta\|$	ρ^*	$\dfrac{\|U_{\zeta+1}-U_\zeta\|}{\|U_\zeta-U_{\zeta-1}\|^6}$	η
4*HU	1	2.2(−1)	7.4(−3)			
	2	1.5(−14)	5.0(−15)		2.953163177(−2)	4.440762811(+6)
	3	2.0(−75)	6.6(−76)	4.9998	4.440762811(+6)	
4*LO	1	2.5(−1)	8.4(−3)			
	2	1.7(−16)	5.7(−17)		1.658701079(−4)	1.670091725(−4)
	3	1.7(−101)	5.7(−102)	5.9998	1.670091725(−4)	
4*GR	1	9.1(−3)	3.0(−3)			
	2	8.0(−20)	2.7(−20)		3.496581094(−5)	3.502158271(−5)
	3	3.7(−122)	1.2(−122)	6.0000	3.502158271(−5)	
4*SA	1	2.9(−2)	9.6(−3)			
	2	4.7(−16)	1.6(−16)		2.066540292(−4)	2.083784171(−4)
	3	9.6(−99)	3.2(−99)	5.9997	2.083784171(−4)	
4*AB	1	4.0(−2)	1.3(−2)			
	2	6.1(−15)	2.0(−15)		3.387336833(−4)	3.430169462(−4)
	3	7.0(−92)	2.3(−92)	5.9996	3.430169462(−4)	
4*SO	1	2.0(−3)	6.6(−4)			
	2	8.7(−25)	2.9(−25)		3.501864006(−6)	3.502158271(−6)
	3	6.5(−153)	2.2(−153)	6.0000	3.502158271(−6)	
4*WL	1	3.1(−2)	1.0(−2)			
	2	8.7(−16)	2.9(−16)		2.342171924(−4)	.363956833(−4)
	3	4.2(−97)	1.4(−97)	5.9937	2.363956833(−4)	
4*NM1	1	1.1(−2)	3.8(−3)			
	2	4.2(−19)	1.4(−19)		4.950410601(−5)	4.961390884(−5)
	3	1.1(−117)	3.6(−118)	5.9999	4.961390884(−5)	
4*NM2	1	6.8(−3)	2.3(−3)			
	2	9.3(−21)	3.1(−21)		2.235218683(−5)	2.237490006(−5)
	3	5.9(−128)	2.0(−128)	6.0000	2.237490006(−5)	
4*NM3	1	2.5(−3)	8.3(−4)			
	2	7.2(−24)	2.4(−24)		7.590662313(−6)	7.588009587(−6)
	3	4.3(−147)	1.4(−147)	6.0000	7.588009587(−6)	

Table 26.7: Conduct of different techniques on Example 26.5.

Example 26.6. For a counter example, we pick a function F on $\mathbb{E}_1 = \mathbb{E}_2 = \mathbb{R}$, $\Omega = [-\frac{1}{\pi}, \frac{2}{\pi}]$ by

$$F(x) = \begin{cases} x^3 \log(\pi^2 x^2) + x^5 \sin\left(\dfrac{1}{x}\right), & x \neq 0 \\ 0, & x = 0 \end{cases},$$

that leads to

$$F'(x) = 2x^2 - x^3 \cos\left(\frac{1}{x}\right) + 3x^2 \log(\pi^2 x^2) + 5x^4 \sin\left(\frac{1}{x}\right),$$

$$F''(x) = -8x^2 \cos\left(\frac{1}{x}\right) + 2x(5 + 3\log(\pi^2 x^2)) + x(20x^2 - 1)\sin\left(\frac{1}{x}\right)$$

and

$$F'''(x) = \frac{1}{x}\left[(1 - 36x^2)\cos\left(\frac{1}{x}\right) + x\left(22 + 6\log(\pi^2 x^2) + (60x^2 - 9)\sin\left(\frac{1}{x}\right)\right)\right].$$

Surely, we can say that $F'''(x)$ is not bounded on Ω in the neighborhood of point $x = 0$. This means the study prior to Section 5 is not applicable. Especially, hypotheses on the seventh order derivative of F or even higher are considered to demonstrate the convergence of proposed scheme in section 3. Because of this section, we now demand the hypotheses on first order.

Further, we have

$$H = \frac{80 + 16\pi + (\pi + 12\log 2)\pi^2}{2\pi + 1}, \ b = 1, \psi_0(t) = \psi(t) = Ht, \ \psi_1(t) = 1 + Ht$$

and functions A and B as given in Application 3.2. The desired solution of (26.6) is $x^* = \frac{1}{\pi}$. The distinct radii, U_0, COC (ρ) and *CPU time* are stated in Table 26.8.

2*Schemes	Distinct parameters that appease the Theorem 26.1						
	R_1	R_2	R_3	R	U_0	ρ	CPU Time
NM1	0.011971	0.0016362	0.0032737	0.0016362	0.3198	6	0.156250
NM2	0.011971	0.00096269	0.00064786	0.00064785	0.3177	6	0.12500
NM3	0.011971	0.00041256	0.0011889	0.00041256	0.3179	6	0.156250

Table 26.8: Different radii of convergence

IM/Ex	Ex. 26.1	Ex. 26.2	Ex. 26.3	Ex. 26.4	Ex. 26.5	TT	AT
HU	39.4198	17.40530	1.29392	28.05384	51.16305	109.28202	21.856405
LO	21.3660	8.03168	0.54339	14.07596	26.15242	70.16949	14.033898
GR	28.0788	5.59796	0.57942	14.08696	22.25168	70.59480	11.765800
SA	27.8556	11.14688	2.48477	82.84556	53.57176	177.90459	35.580918
AB	42.7671	14.15601	0.90264	27.91375	53.92799	139.66752	27.933505
SO	27.7756	10.18220	0.94667	27.78366	33.67275	100.41370	20.082739
WL	22.3137	7.79853	0.53439	13.98790	27.28121	72.91573	14.583146
NM1	18.5871	7.95962	0.20717	0.10908	28.15583	55.01879	11.003757
NM2	13.2373	5.76506	0.19013	0.10708	14.22802	33.52763	6.705525
NM3	13.0102	5.75208	0.18615	0.10507	8.03466	27.08812	5.417624

(TT: stands for total time for all examples to the corresponding iterative method. AT: means average time taken by corresponding iterative method.)

Table 26.9: Consumption of CPU time by distinct schemes

26.4 EXERCISES

1. Solve the initial value problem

$$\dot{x}(t) = 1 + \cos(x(t)), x(0) = 0,$$

using the method (26.2).

2. The predator-prey population models describe the interaction of a prey population X and a predator population Y. Assume that their interaction is modeled by the system of ordinary differential equations

$$\dot{x} = x - \frac{1}{4}x^2 - \frac{1}{10}xy + 1$$
$$\dot{y} = -\frac{1}{4}y + \frac{1}{7}xy + \frac{2}{3}.$$

(Assume $x(0) = y(0) = 0$). Solve the system.

3. Solve the initial value problem

$$\dot{x} = (x(t) + t, x(0) = 1)$$

by the iteration method (26.2). Perform 3 steps.

4. Solve the Fredholm-type integral equation

$$x(t) = \sum_{0}^{1} \frac{t \cdot s}{10} ds + 1$$

by the iteration method (26.2). Perform 3 steps.

5. Solve the Volterra-type integral equation

$$x(t) = \sum_{0}^{t} \frac{t \cdot s}{10} ds + 1$$

by the iteration method (26.2). Perform 3 steps.

6. Solve equation

$$x = \frac{\sin x}{2} + 1$$

by using algorithm (26.2).

7. Repeat the previous problem for equation

$$x = \frac{\sin x + \cos 2x}{10}.$$

REFERENCES

1. Abad, M.F., Cordero, A., and Torregrosa, J.R. 2014. A family of seventh-order schemes for solving nonlinear systems. *Bull. Math. Soc. Sci. Math. Roum.*, 57:133–145.
2. Abbasbandy, S., Bakhtiari, P., Cordero, A., Torregrosa, J.R., and Lotfi, T. New efficient methods for solving nonlinear systems of equations with arbitrary even order. *Appl. Math. Comput.*, 287–288:94–103.
3. Alaidarous, E.S., Ullah, M.Z., Ahmad, F., and Al-Fhaid, A.S. 2013. An Efficient Higher-Order Quasilinearization Method for Solving Nonlinear BVPs. *J. Appl. Math.* Article ID 259371. 11 pages, http://dx.doi.org/10.1155/2013/259371.
4. Argyros, I.K. 2008. *Convergence and Application of Newton-type Iterations*, Springer.
5. Argyros, I.K., and Hilout, S. 2013. *Numerical methods in Nonlinear Analysis*, New Jersey: World Scientific Publ. Comp.
6. Artidiello, S., Cordero,A., Torregrosa, J.R., and Vassileva, M.P. 2015. Multidimensional generalization of iterative methods for solving nonlinear problems by means of weight-function procedure. *Appl. Math. Comput.*, 268:1064–1071.
7. Awawdeh, F. 2010. On new iterative method for solving systems of nonlinear equations. *Numer. Algor.*, 54:395–409.
8. Cordero, A., Hueso, J.L., Martínez, E., and Torregrosa, J.R. 2010. A modified Newton-Jarratt's composition. *Numer. Algor.*, 55:87–99.
9. Cordero, A., Maimó, J.G., Torregrosa,J.R., and Vassileva, M.P. Solving nonlinear problems by Ostrowski-Chun type parametric families. *J. Math. Chem.*, 52: 430–449.
10. Gelfand, I.M. 1963. Some problems in the theory of quasi-linear equations. *Trans. Amer. Math. Soc. Ser.*, 2:295–381.
11. Grau-Sánchez, M., Grau, Á., and Noguera, M. 2011. Ostrowski type methods for solving systems of nonlinear equations. *Appl. Math. Comput.*, 218:2377–2385.
12. Grosan, C., and Abraham, A. 2008. A new approach for solving nonlinear equations systems. *IEEE Trans. Syst. Man Cybernet Part A: Syst. Humans*, 38:698–714.
13. Hueso, J.L., Martínez, E., and Teruel, C. 2015. Convergence, efficiency and dynamics of new fourth and sixth order families of iterative methods for nonlinear system. *J. Comput. Appl. Math.*, 275:412–420.
14. Jacobsen, J., and Schmitt, K. 2002. The Liouville Bratu Gelfand problem for radial operators. *J. Diff. Equat.*, 184:283–298.
15. Jalilian, R. 2010. Non-polynomial spline method for solving Bratu's problem. *Comput. Phys. Comm.*, 181:1868–1872.
16. Lin, Y., Bao, L., and Jia, X. 2010. Convergence analysis of a variant of the Newton method for solving nonlinear equations. *Comput. Math. Appl.*, 59:2121–2127.
17. Lotfi, T., Bakhtiari, P., Cordero, A., Mahdiani, K., and Torregrosa, J.R. 2015. Some new efficient multipoint iterative methods for solving nonlinear systems of equations. *Int. J. Comput. Math.*, 92(9):1921–1934.
18. Moré, J.J. 1990. A collection of nonlinear model problems, in:E.L. Allgower, K. Georg(Eds.), Computational Solution of Nonlinear Systems of Equations Lectures in Applied Mathematics, *Amer. Math. Soc.*, 26:723–762.
19. Ortega, J.M., and Rheinboldt, W.C. 1970. *iterative solution of nonlinear equations in several variables.*, New-York: Academic Press.
20. Petković, M.S., Neta, B., Petković, L.D., and Džunić, J. 2012. *multi-point methods for solving nonlinear equations*, Academic Press.
21. Sauer, T. 2012. *Numerical analysis, 2nd edn.* USA: Pearson.

22. Sharma, J.R., and Arora, H. 2014. Efficient Jarratt-like methods for solving systems of nonlinear equations. *Calcolo*, 51:193–210.
23. Sharma, J.R., Gua, R.K., and Sharma, R. 2013. An efficient fourth order weighted-Newton method for systems of nonlinear equations. *Numer. Algor.*, 2:307–323.
24. Soleymani, F., Lotfi, T., and Bakhtiari, P. 2014. A multi-step class of iterative methods for nonlinear systems. *Optim. Lett.*, 8:1001–1015.
25. Traub, J.F. 1964. *Iterative methods for the solution of equations.* Prentice- Hall Series in Automatic Computation, Englewood Cliffs, N.J.
26. Tsoulos, I.G., and Stavrakoudis, A. 2010. On locating all roots of systems of nonlinear equations inside bounded domain using global optimization methods., *Nonlinear Anal. Real World Appl.*, 11:2465–2471.
27. Wan, Y.Q., Guo, Q., and Pan, N. 2004. Thermo-electro-hydrodynamic model for electro-spinning process. *Int. J. Nonlinear Sci. Numer. Simul.*, 5:5–8.
28. Wang, X., and Li, Y. 2017. An efficient sixth-order Newton-type method for solving non-linear systems. *Algor.*, 10(2):45, doi:10.3390/a10020045.
29. Wang, X., and Zhang, T. 2013. A family of Steffensen type methods with seventh-order convergence. *Numer. Algor.*, 62:429–444.

Glossary

iff	if and only if
w.r.t	with respect to
$\neq$	non-equality
$\emptyset$	empty set
$\in, \notin$	belong to and does not belong to
$\Rightarrow$	implication
$\Leftrightarrow$	if and only if
max	maximum
min	minimum
sup	supremum (least upper bound)
inf	infimum (greatest lower bound)
for all n	for all $n \in N$
$\mathbb{R}^n$	Real n-dimensional space
$\mathbb{C}^n$	Complex n-dimensional space
$X \times Y, X \times X = X^2$	Cartesian product space of X and Y
$e^1, \ldots, e^n$	The coordinate vector of $\mathbb{R}^n$
$x = (x_1, \ldots, x_n)^T$	Column vector with component x_i
x^T	The transpose of x
$\{x_n\}^{n \geq 0}$	Sequence of point from X
$\|\cdot\|$	Norm on X
$\|\cdot\|_p$	L_p norm
$\|\cdot\|$	Absolute value symbol
$/\cdot/$	Norm symbol of a generalized Banach space X
$U(x_0, R)$	Open ball $\{z \in X \,\|\, \|x_0 - z\| < R\}$
$\overline{U}(x_0, R)$	Closed ball $\{z \in X \,\|\, \|x_0 - z\| \leq R\}$
$U(R) = U(x_0, R)$	Ball centered at the zero
	element on X and of radius R
$U, \overline{U}$	Open, closed balls, respectively
	no particular reference to X, x_0 or R
I	Identity matrix operator
L	Linear operator
L^{-1}	Inverse
$M = \{m_{ij}\}$	Matrix $1 \leq i, J \leq n$
M^{-1}	Inverse of M
det M or $|M|$	Determinant of M
$\sum$	Summation symbol
$\prod$	Product of factors symbol
$\int$	Integration symbol
$\in$	Element inclusion
$\subset, \subseteq$	Strict and non-strict set inclusion
$\forall$	For all
$\Rightarrow$	implication
$\cup, \cap$	Union, intersection
$A - B$	Difference between set A and B
$F : D \subseteq X \to Y$	An operator with domain D included in X,
	and values in Y
$F'(x), F''(x)$	First, second Fréchet-derivatives of F evaluated at x

Index

For Product Safety Concerns and Information please contact our EU
representative GPSR@taylorandfrancis.com
Taylor & Francis Verlag GmbH, Kaufingerstraße 24, 80331 München, Germany

www.ingramcontent.com/pod-product-compliance
Lightning Source LLC
Chambersburg PA
CBHW060426220326
41598CB00021BA/2301